AF755028

PHARMACOLOGY – RESEARCH, SAFETY TESTING AND REGULATION

GLUTATHIONE

BIOSYNTHESIS, FUNCTIONS AND BIOLOGICAL IMPLICATIONS

PHARMACOLOGY – RESEARCH, SAFETY TESTING AND REGULATION

Additional books and e-books in this series can be found on Nova's website under the Series tab.

PHARMACOLOGY – RESEARCH, SAFETY TESTING AND REGULATION

GLUTATHIONE

BIOSYNTHESIS, FUNCTIONS AND BIOLOGICAL IMPLICATIONS

PÁL PERJÉSI
EDITOR

nova
Medicine & Health
New York

Copyright © 2019 by Nova Science Publishers, Inc.
All rights reserved. No part of this book may be reproduced, stored in a retrieval system or transmitted in any form or by any means: electronic, electrostatic, magnetic, tape, mechanical photocopying, recording or otherwise without the written permission of the Publisher.

We have partnered with Copyright Clearance Center to make it easy for you to obtain permissions to reuse content from this publication. Simply navigate to this publication's page on Nova's website and locate the "Get Permission" button below the title description. This button is linked directly to the title's permission page on copyright.com. Alternatively, you can visit copyright.com and search by title, ISBN, or ISSN.

For further questions about using the service on copyright.com, please contact:
Copyright Clearance Center
Phone: +1-(978) 750-8400 Fax: +1-(978) 750-4470 E-mail: info@copyright.com.

NOTICE TO THE READER

The Publisher has taken reasonable care in the preparation of this book, but makes no expressed or implied warranty of any kind and assumes no responsibility for any errors or omissions. No liability is assumed for incidental or consequential damages in connection with or arising out of information contained in this book. The Publisher shall not be liable for any special, consequential, or exemplary damages resulting, in whole or in part, from the readers' use of, or reliance upon, this material. Any parts of this book based on government reports are so indicated and copyright is claimed for those parts to the extent applicable to compilations of such works.

Independent verification should be sought for any data, advice or recommendations contained in this book. In addition, no responsibility is assumed by the publisher for any injury and/or damage to persons or property arising from any methods, products, instructions, ideas or otherwise contained in this publication.

This publication is designed to provide accurate and authoritative information with regard to the subject matter covered herein. It is sold with the clear understanding that the Publisher is not engaged in rendering legal or any other professional services. If legal or any other expert assistance is required, the services of a competent person should be sought. FROM A DECLARATION OF PARTICIPANTS JOINTLY ADOPTED BY A COMMITTEE OF THE AMERICAN BAR ASSOCIATION AND A COMMITTEE OF PUBLISHERS.

Additional color graphics may be available in the e-book version of this book.

Library of Congress Cataloging-in-Publication Data

ISBN: 978-1-53614-740-7
Library of Congress Control Number: 2018965096

Published by Nova Science Publishers, Inc. † New York

CONTENTS

PREFACE

This year we celebrate the 131st anniversary of the discovery of glutathione (L-γ-glutamy-L-cysteinyl-glycine). The major intracellular thiol compound was first described in the literature in 1888 by J. De-Rey Pailhade [1]. He has isolated the compound from various sources, such as yeast, asparagus, fish, egg, and livestock, that was able to release hydrogen sulfide, bleached several dyes, such as methylene blue, and reacted with various compounds. He named the substance *phylothion*; the Greek expression for sulfur-loving. F.G. Hopkins, who was awarded the Nobel Prize in Physiology or Medicine in 1929 for "his discovery of the growth-stimulating vitamins", isolated the compound from liver, skeletal muscle and yeast in 1921 and described it as "autooxidizable constituent of the cell" [2].

The structure of glutathione was controversial for several years. In 1927, G. Hunter and B.A. Eagles described a product, isolated using the same procedure employed by Hopkins, which had significantly less sulfur per mass than Hopkins had reported and was possibly a tripeptide [3]. In 1929, after developing a new procedure for preparing crystalline glutathione, Hopkins recognized that "Hunter and Eagles were right in doubting that the substance is a simple dipeptide of glutamic acid and cysteine...." [4]. He then showed that glutathione is indeed a tripeptide of glutamic acid, glycine, and cysteine.

Over the past 131 years there have appeared a huge number of publications, reviews and books on the chemistry, biochemistry and pharmacology of glutathione. The enormous field of this topic is clearly shown by the following facts: during the last 10 years (2008–2018), around 57000 scientific papers and more than 2500 review articles were listed in the PubMed database on glutathione. Its wide use as ingredient of dietary supplements and applications by the cosmetic industry has increased the demand of new biotechnological processes, state of the art analytical methods and better understanding of the fate and effects of the exogenously applied substance. Recent randomized controlled trial of oral glutathione supplementation indicated that *per os* glutathione supplementation was effective at increasing body compartment stores of GSH [5]. The increased demand for pharmacopoeial grade glutathione signals importance of new, economical biotechnology-based technologies for production and pharmacopoeial qualification of glutathione preparations. In parallel, several research projects are focusing on investigation of endogenous glutathione metabolism, cellular glutathione redox homeostasis and role of glutathione in cell cycle control, immunological defense and pathological abnormalities. Accordingly, results of the new biochemical studies can provide the molecular basis of application of glutathione as a nutritional supplement. From this point, pharmaceutical qualification of the synthetic or fermentation products is essential not only for the pharmaceutical but for the food and the cosmetic industry as well. Besides using qualified glutathione products as a pharmakon, it can be successfully applied in a variety of other industries where the reversible redox and antioxidant properties of the compound can be utilized.

The objective of this book is to provide a comprehensive summary on the biochemistry of the most important endogenous (antioxidant, cytoprotective) functions, biotechnological production, pharmaceutical qualification and some industrial (cryoprotective, electrochemical sensor) application of glutathione. Separate chapters are devoted to summarizing the chemistry and biochemistry of glutathione dependent biotransformation of xenobiotics, which could be the basis of toxication or detoxication of the foreign substrates (such as drugs and their metabolites). Several chapters are dealing with GSH-related enzymes whose malfunctions are associated with different pathological abnormalities (e.g., cardiac dysfunctions, neurological disorders, kidney injuries and different types of cancer) The information presented represents a valuable source for advanced students, scholars, teachers and scientist (academic and industrial) interested or already engaged in research that involves glutathione.

As editor of this book, I am highly thankful to the experienced and well-versed contributors, who cordially accepted my invitation to write their chapters. I would also like to extend my thanks to Ms. Nadya Columbus and the editorial staff of Nova Science Publishers, Hauppauge NY, who initiated this book project. I do believe that the information covered in this book will make a sound contribution to the research of this fascinating compound.

References

[1] Lillig, C. H.; Berndt, C. Preface. Cellular functions of glutathione. *Biochim. Biophys. Acta*. 2013, 1830, 3137-3138.

[2] Hopkins, F. G. On an autoxidizable constituent of the cell. *Biochem. J.* 1921, 15, 286-305.

[3] Hunter, G.; Eagles, B. A. Glutathione. A critical study. *J. Biol. Chem.* 1927, 72, 147-166.

[4] Hopkins, F. G. On glutathione, a reinvestigation. *J. Biol. Chem.* 1929, 84, 269-320.

[5] Richie, J. P. Jr.; Nichenametla, S.; Neidig, W.; Calcagnotto, A.; Haley, J. S.; Schell, T. D.; Muscat, J. E. Randomized controlled trial of oral glutathione supplementation on body stores of glutathione. *Eur. J. Nutr.* 2015, 54, 251-263.

SESSION I. SYNTHESIS -BIOSYNTHESIS

In: Glutathione
Editor: Pál Perjési
ISBN: 978-1-53614-740-7
© 2019 Nova Science Publishers, Inc.

Chapter 1

PHYSICO-CHEMICAL CHARACTERIZATION OF GLUTATHIONE (GSH)

Pál Perjési[1,*], Caridad Noda Perez[1,2], Hamilton B. Napolitano[3] and Hawsar O. Mohammed[1,4]

[1]Institute of Pharmaceutical Chemistry, University of Pécs, Pécs, Hungary
[2]Instituto de Química, Universidade Federal de Goiás, Goiânia, Goiás, Brazil
[3]Ciências Exatas e Tecnológicas, Universidade Estadual de Goiás, Anápolis, Goiás, Brazil
[4]College of Veterinary Medicine, University of Sulaimani, Sulaymaniyah, Kurdistan Region, Iraq

ABSTRACT

Glutathione (GSH) is the most abundant intracellular thiol as well as a major cellular antioxidant. Besides its redox activity it is a powerful nucleophilic agent. Both physico-chemical properties of the compound are related to the thiol (SH) function of the cysteinyl moiety of the tripeptide. Because of its central role in maintaining the cellular homeostatis and protection against diverse (toxic) electrophilic species, it is also a component of several nutritional products. As such, knowledge of physico-chemical characteristics of the compound is essential for both pharmacopoeal characterization and molecular-level understanding of its main biological actions . The chapter summarizes the spectroscopic (IR, MS, X-ray, UV) and the relevant physico-chemical characteristics (acid-base and redox properties) that are applied in its pharmacopoeial characterization and biochemical/toxicological studies related to the compound.

* Corresponding Author E-mail: pal.perjesi@gytk.pte.hu.

Keywords: glutathione, pharmacopoeia, mass spectrometry, X-ray, IR spectroscopy, NMR spectroscopy, acid-base properties, redox properties

1. INTRODUCTION

Glutathione (GSH) is the most abundant intracellular thiol as well as a major cellular antioxidant. Besides its redox activity, it is a powerful nucleophilic agent. Both physico-chemical properties of the compound are related to the thiol (SH) function of the cysteinyl moiety of the tripeptide. Because of its central role in maintaining the cellular homeostasis and protection against diverse (toxic) electrophilic species, it is also a component of several nutritional products. As such, knowledge of physico-chemical characteristics of the compound is essential for both pharmacopoeial characterization and molecular-level understanding its main biological actions. The chapter summarizes the spectroscopic (IR, MS, X-ray, UV) and relevant physico-chemical characteristics (acid-base and redox properties) that are applied in its pharmacopoeial characterization and biochemical/ toxicological studies related to the compound. Each section briefly introduces the theory of the relevant analytical technique or physico-chemical property.

2. CHEMICAL CHARACTERIZATION

2.1. Structure

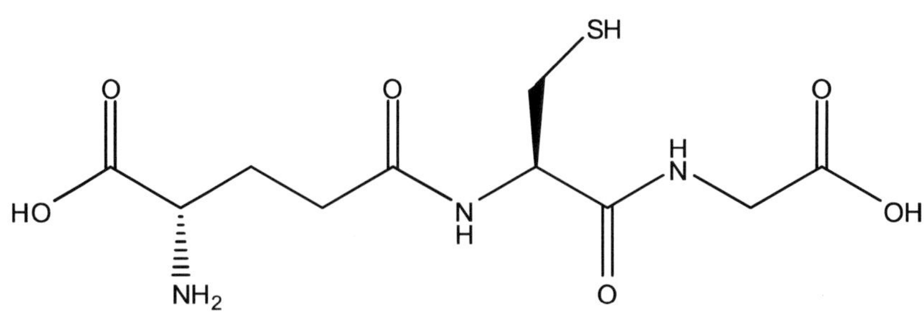

Figure 1. Structure of reduced glutathione (GSH).

Empirical formula: $C_{10}H_{17}N_3O_6S$
Molecular weight: 307.32
CAS number: 70-18-8
Melting point: 195 °C (with decomposition) [1]
Nomenclature

(2S)-2-amino-5-[[(2R)-1-(carboxymethylamino)-1-oxo-3-sulfanylpropan-2-yl]amino]-5-oxopentanoic acid,
(2S)-2-amino-4-{[(1R)-1-[(carboxymethyl)carbamoyl]-2 sulfanylethyl]carbamoyl}butanoic acid,
γ-L-Glutamyl-L-cysteinylglycine.
Nonproprietary names
Glutathione, L-Glutathione, Reduced glutathione, Glutathione-SH, GSH
Proprietary names

Agifutol®; Atomolan®; Beamthion®; Copren®; Deltathione®; Glutamed®; Glutasan®; Glutathin®; Glutathiol®; Glutathion®; Gluthion®; Glutanil®; Glutoxil®; Ipatox®; Sethion®; Maglut®; Neuthion®; Novatox®; Ridutox®; Rition®; Scavenger®; Tad®; Tathiclon®; Tathion®; Tationil®; Thioxene®; Tition®; Tripride®

3. Pharmacopoeial Characterization

3.1. European Pharmacopoeia Edition 9.0

The European Pharmacopoeia Edition 9.0 (Ph. Eur. 9.0) [2] lists *Glutathione* and describes its *Definition*, *Content*, *Characters*, *Identification*, *Tests*, *Assay Storage*, and *Impurities* as follows.

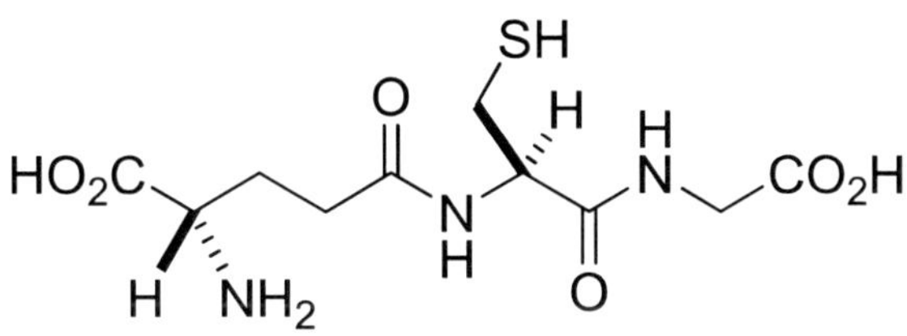

Definition

$C_{10}H_{17}N_3O_6S$. M_r 307.3. L-γ-Glutamyl-L-cysteinylglycine. Fermentation product. CAS: 70-18-8.

Content

98.0 per cent to 101.0 per cent (dried substance).

Characters

Appearance

White or almost white, crystalline powder or colourless crystals.

Solubility

Freely soluble in water, very slightly soluble in ethanol (96 per cent) and in methylene chloride.

Identification

A

Specific optical rotation.

B

Infrared absorption spectrophotometry.

Comparison

Glutathione CRS.

Tests (Selections):

Solution S

Dissolve 5.0 g in *distilled water R* and dilute to 50 ml with the same solvent.

Appearance of solution

Solution S is clear and colourless.

Specific optical rotation.

15.5 to -17.5 (dried substance). Dissolve 1.0 g in *water R* and dilute to 25.0 ml with the same solvent.

Related substances

Capillary electrophoresis.

Prepare the solutions immediately before use.

Internal standard solution (a). Dissolve 0.100 g of *phenylalanine R* in the electrolyte solution and dilute to 50.0 ml with the same solution.

Internal standard solution (b). Dilute 10.0 ml of internal standard solution (a) to 100.0 ml with the electrolyte solution.

Test solution (a). Dissolve 0.200 g of the substance to be examined in the electrolyte solution and dilute to 10.0 ml with the same solution.

Test solution (b). Dissolve 0.200 g of the substance to be examined in internal standard solution (b) and dilute to 10.0 ml with the same solution.

Reference solution (a). Dissolve 20.0 mg of the substance to be examined in internal standard solution (a) and dilute to 10.0 ml with the same solution.

Reference solution (b). Dilute 5.0 ml of reference solution (a) to 50.0 ml with the electrolyte solution.

Reference solution (c). Dissolve 0.200 g of the substance to be examined in 5 ml of the electrolyte solution. Add 1.0 ml of internal standard solution (a), 0.5 ml of a 2 mg/ml solution of *L-cysteine R* (impurity B) in the electrolyte solution, 0.5 ml of a 2 mg/ml solution of *oxidised L-glutathione R* (impurity C) in the electrolyte solution and 0.5 ml of a 2 mg/ml solution of *L-γ-glutamyl-L-cysteine R* (impurity D) in the electrolyte solution. Dilute to 10.0 ml with the electrolyte solution.

Capillary.

-material: uncoated fused silica;

-size: length to the detector cell = 0.5 m; total length = 0.6 m; Ø = 75 μm.

Temperature: 25°C.

Electrolyte solution. Dissolve 1.50 g of anhydrous sodium dihydrogen phosphate R in 230 ml of water R and adjust to pH 1.80 with phosphoric acid R. Dilute to 250.0 ml with water R. Check the pH and, if necessary, adjust with phosphoric acid R or dilute sodium hydroxide solution R.

Detection: spectrophotometer at 200 nm.

Preconditioning of a new capillary: rinse the new capillary before the first injection with 0.1 M hydrochloric acid at 138 kPa for 20 min and with water R at 138 kPa for 10 min; for complete equilibration, condition the capillary with the electrolyte solution at 350 kPa for 40 min, and subsequently at a voltage of 20 kV for 60 min.

Preconditioning of the capillary: rinse the capillary with the electrolyte solution at 138 kPa for 40 min.

Between-run rinsing: rinse the capillary with water R at 138 kPa for 1 min, with 0.1 M sodium hydroxide at 138 kPa for 2 min, with water R at 138 kPa for 1 min, with 0.1 M hydrochloric acid at 138 kPa for 3 min and with the electrolyte solution at 138 kPa for 10 min.

Injection: test solutions (a) and (b), reference solutions (b) and (c) and the electrolyte solution (blank): under pressure (3.45 kPa) for 5 s.

Migration: apply a voltage of 20 kV.

Run time: 45 min.

Relative migration with reference to the internal standard (about 14 min): impurity A = about 0.77; impurity B = about 1.04; impurity E = about 1.2; impurity C = about 1.26; impurity D = about 1.3.

System suitability:

-resolution: minimum 1.5 between the peaks due to the internal standard and impurity B in the chromatogram obtained with reference solution (c); if necessary, increase the pH with dilute sodium hydroxide solution R;

-peak-to-valley ratio: minimum 2.5, where H_p=height above the baseline of the peak due to impurity D and H_v= height above the baseline of the lowest point of the curve separating this peak from the peak due to glutathione in the chromatogram obtained with reference solution (c); if necessary, lower the pH with phosphoric acid R;

-check that in the electropherogram obtained with test solution (a) there is no peak with the same migration time as the internal standard (in such case correct the area of the phenylalanine peak).

Limits: test solution (b):

-*corrected areas*: divide all the peak areas by the corresponding migration times;

-*correction factors*: for the calculation of content, multiply the ratio of time-corrected peak areas of impurity and the internal standard by the corresponding correction factor: impurity B = 3.0; impurity D = 1.4;

-*impurity C*: not more than 1.5 times the ratio of the area of the peak due to glutathione to the area of the peak due to the internal standard in the electropherogram obtained with reference solution (b) (1.5 per cent);

-*impurity D*: not more than the ratio of the area of the peak due to glutathione to the area of the peak due to the internal standard in the electropherogram obtained with reference solution (b) (1.0 per cent);

-*impurities A, B, E*: for each impurity, not more than 0.5 times the ratio of the area of the peak due to glutathione to the area of the peak due to the internal standard in the electropherogram obtained with reference solution (b) (0.5 per cent);

-*any other impurity*: for each impurity, not more than 0.2 times the ratio of the area of the peak due to glutathione to the area of the peak due to the internal standard in the electropherogram obtained with reference solution (b) (0.2 per cent);

-*total*: not more than 2.5 times the ratio of the area of the peak due to glutathione to the area of the peak due to the internal standard in the electropherogram obtained with reference solution (b) (2.5 per cent);

-*disregard limit*: 0.05 times the ratio of the area of the peak due to glutathione to the area of the peak due to the internal standard in the electropherogram obtained with reference solution (b) (0.05 per cent)

Assay

In a ground-glass-stoppered flask, dissolve 0.500 g of the substance to be examined and 2 g of *potassium iodide R* in 50 ml of *water R*. Cool the solution in iced water and add 10 ml of *hydrochloric acid R1* and 20.0 ml of *0.05 M iodine*. Stopper the flask and allow to stand in the dark for 15 min. Titrate with *0.1 M sodium thiosulphate* using 1 ml of *starch solution R*, added towards the end of the titration, as indicator. Carry out a blank titration. 1 ml of *0.05 M iodine* is equivalent to 30.73 mg of $C_{10}H_{17}N_3O_6S$.

Storage

Protected from light.

Impurities

Specified impurities: A, B, C, D, E.

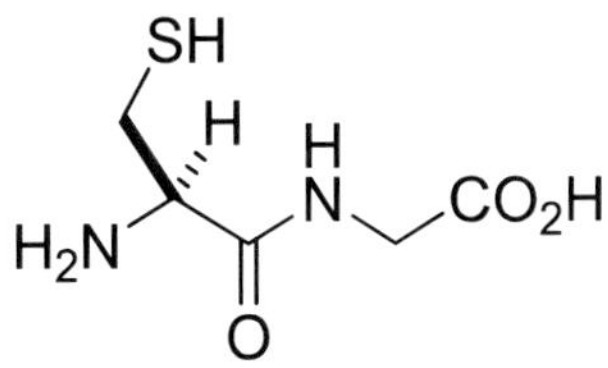

Impurity A. L-cysteinyl-glycine,

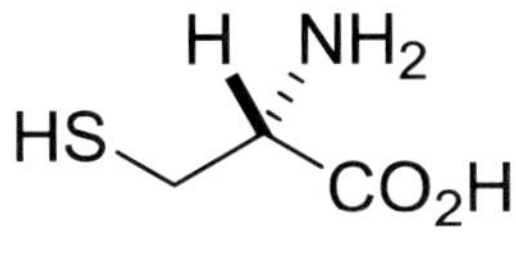

Impurity B. Cysteine,

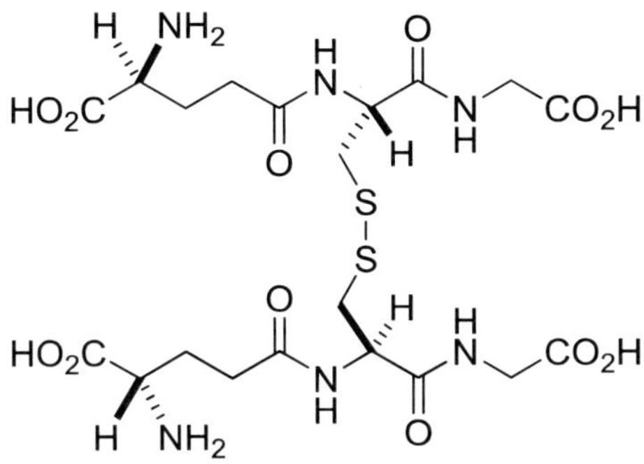

Impurity C. Oxidized glutathione (GSSG),

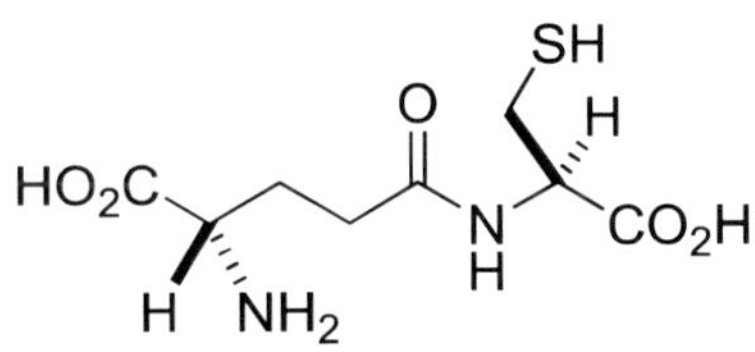

Impurity D. L-g-glutamyl-L-cysteine,

Impurity E. Impurity with unidentified structure.

3.2. The United States Pharmacopoeia – Dietary Supplement Compendium

The United States Pharmacopoeia (USP) Dietary Supplement Compendium (2015) monograph [3] includes the *name* of the ingredient or preparation; the *definition*;

packaging, *storage*, and *labeling* requirements; and the *specification*. The specification consists of a series of *tests*, *procedures* for the tests, and *acceptance criteria*.

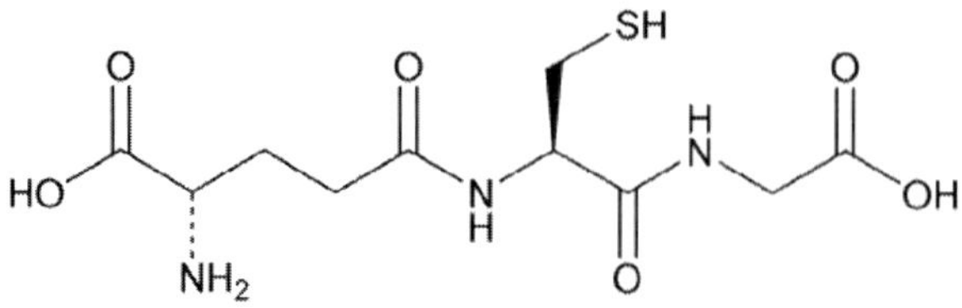

$C_{10}H_{17}N_3O_6S$, 307.32
Pentanoic acid, 2-amino-5-[(R)-1-(carboxymethylamino)-3-mercapto-1-oxopropan-2-ylamino]-5-oxo, (S); N-(N-l-γ-Glutamyl-l-cysteinyl)glycine [70-18-8].

Definition

Glutathione contains NLT 98.0% and NMT 101.0% of $C_{10}H_{17}N_3O_6S$, as glutathione, calculated on the dried basis.

Identification

A

Infrared Absorption

B

Optical Rotation, Specific Rotation

Sample solution

40 mg/mL in water. Acceptance criteria: -15.5° to -17.5°, at 20°.

ASSAY

Procedure

Sample

500 mg of glutathione previously dried

Titrimetric system

Mode: Direct titration

Titrant: 0.1 N iodine VS

Endpoint detection: Visual

Blank: 50 mL of metaphosphoric acid (1 in 50)

Analysis: Dissolve the Sample in 50 mL of metaphosphoric acid (1 in 50) and titrate with the Titrant.

Calculate the percentage of glutathione ($C_{10}H_{17}N_3O_6S$) in the portion of Glutathione taken:

Result = [(V-B) × N × F × 100]/W

where

V = titrant volume of the Sample (mL)

B = titrant volume of the Blank (mL)

N = titrant normality (mEq/mL)

F = equivalency factor, 307.32 mg/mEq

W = weight of the Sample (mg)

Acceptance criteria: 98.0%–101.0% on the dried basis

Impurities

Inorganic Impurities

Ammonium

Standard solution: 10 μg of ammonium from a diluted ammonium chloride solution

Sample solution: 50 mg of Glutathione

Analysis: Transfer the Sample solution and the Standard solution to separate 25-mL jars fitted with caps, and dissolve in 1 mL of water. Add 0.30 g of magnesium oxide. Close immediately after placing a piece of silver manganese paper 5-mm square, wetted with a few drops of water, under the caps. Swirl, avoiding projections of liquid, and allow to stand at 40° for 30 min.

Acceptance criteria: If the silver manganese paper shows a gray color, it is not more intense than the standard (NMT 200 ppm).

Arsenic

NMT 2 ppm

Chloride and Sulfate, Chloride: Dissolve 0.7 g with water to make 15 mL. The solution shows no more chloride than corresponds to 0.20 mL of 0.020 N hydrochloric acid (NMT 200 ppm).

Chloride and Sulfate

Sulfate: Dissolve 0.8 g in water to make 15 mL. The solution shows no more sulfate than corresponds to 0.25 mL of 0.020 N sulfuric acid (NMT 300 ppm).

Heavy Metals

Method I: NMT 10 ppm

Iron

NMT 10 ppm

Residue on Ignition

NMT 0.1%

Organic Impurities

Procedure

Mobile phase: 6.8 g/L of potassium dihydrogen phosphate with 2.02 g/L of sodium 1-heptane sulfonate. Adjust with phosphoric acid to a pH of 3.0. Mix 970 mL of this solution with 30 mL of methanol.

System suitability solution: 0.1 mg/mL of USP l-Phenylalanine RS, 0.5 mg/mL of USP Glutathione RS, and 0.5 mg/mL of USP Ascorbic acid RS in Mobile phase

Standard solution: 0.01 mg/mL of USP Glutathione RS in Mobile phase. [Note-This solution has a concentration equivalent to 2.0% of that of the Sample solution.]

Sample solution: 50 mg of glutathione in 100 mL of Mobile phase. [Note—Allow the solution to stand for 5 min before use.]

Chromatographic system

Mode: LC

Detector: UV 210 nm

Column: 4.6-mm × 15-cm; 5-μm packing L1

Column temperature: 30°

Flow rate: Adjust so that the retention time of glutathione is about 5 min.

Injection size: 10 μL

System suitability

Sample: System suitability solution

Suitability requirements

Resolution: NLT 5.0 between the ascorbic acid and glutathione peaks; and NLT 5.0 between the glutathione and l-phenylalanine peaks

Relative standard deviation: NMT 1.5% for replicate injections

Analysis

Samples: Standard solution and Sample solution

Calculate the percentage of any impurity in the portion of Glutathione taken:

Result = (rU/rS) × (CS/CU) × 100

where,

r_U = peak response of any peak from the Sample solution other than glutathione,

r_S = peak response of the glutathione peak from the Standard solution,

C_S = concentration of USP Glutathione RS in the Standard solution (mg/mL),

C_U = concentration of Glutathione in the Sample solution (mg/mL)

Acceptance criteria

Individual impurity: NMT 1.5% for the impurity with the relative retention time of about 4.

Total impurities: NMT 2.0%.

Specific Tests

Clarity and Color of Solution

Sample solution

0.1 g/mL in water

Analysis

Using identical tubes of colorless, transparent, neutral glass with a flat base and an internal diameter of 15–25 mm, compare the liquid to be examined with water, the depth of the layer being 40 mm. Compare the colors in diffused daylight, viewing vertically against a white background.

Acceptance criteria

The solution is clear and colorless.

Loss on Drying

NMT 0.5%, 105° for 3 h

Additional Requirements

Packaging and Storage

Preserve in tight containers.

4. Gas Phase Characterization

4.1. Mass Spectrometry

Mass spectrometry is a versatile, generally used analytical technique in the field of bioanalysis. It can be applied for determination of molecular weight, molecular formula (HRMS), structure (from fragmentation fingerprint), isotopic incorporation/distribution, and protein sequence (MS-MS), just to mention its most important applications [4-7].

The most important parts of the mass spectrometers are the sample *inlet*, the *ion source*, the *mass analyzer*, the *detector* and the *data analyzer* unit. The sample introduction could be direct (through a gate) or indirect (from a separation unit; e.g., GC, HPLC, SFC, CE). In the ion source the analyte particles are ionized (converted to positive or negative ions) by direct (with high speed electrons) or indirect (with neutral or charged particles; with charge transfer; or with laser photons) methods. The charged particles are separated in the mass analyzer unit. The choice of mass analyzer should be based upon the application, cost, and performance desired. There is no ideal mass analyzer that is good for all applications [8].

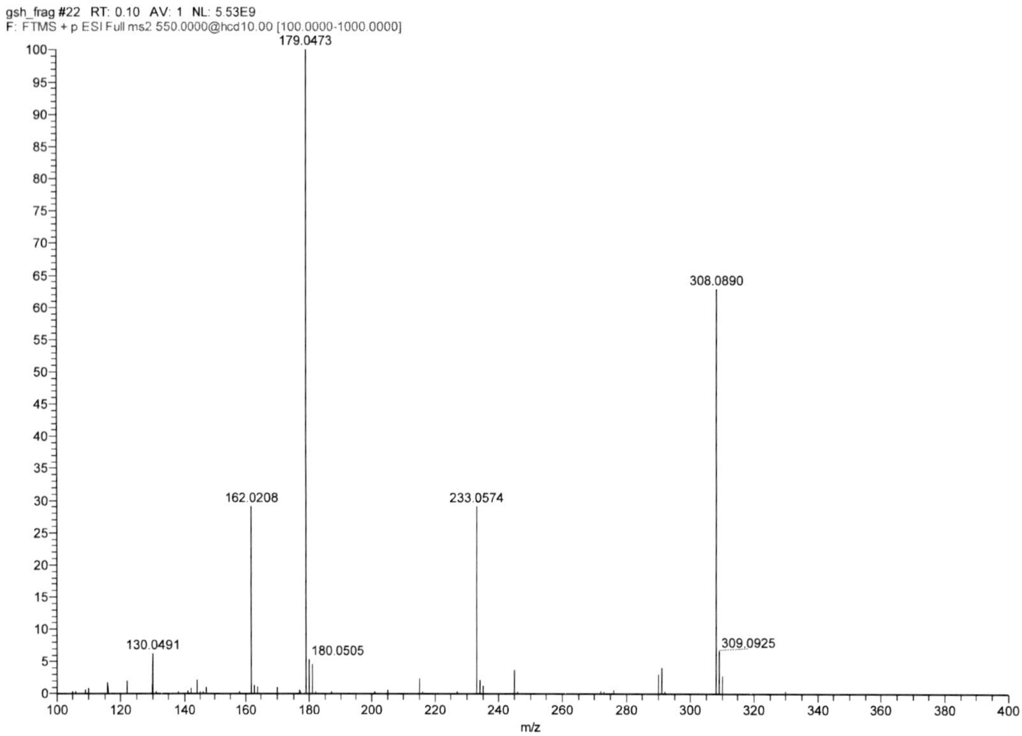

Figure 2. Positive ion mode HESI fragmentation pattern of reduced glutathione (GSH).

The commonly used mass analyzer could be a magnetic (B) or electrostatic (E) sector analyzer; a quadrupole mass filter (Q) or a time-of-flight (TOF) analyzer; an ion-trap (Penning, Paul, Kingdon traps, a quadrupole ion trap (Q) or orbitrap (OT)); or an ion-cyclotron resonance (ICR) analyzer. The mass analyzers separate the ions according to their mass-to-charge (*m/z*) ratio. In the detector, the signal intensity is directly proportional to the number of the ions. The most commonly used detectors are the photographic plate detectors, the Faraday Cups, and the (photo) electron multipliers [9].

The mass spectrum is presented in terms of ion abundance vs. *m/z* ratio (mass). If a molecule loses only one electron in the ionization process, a molecular ion is observed that gives its molecular weight – this is designated as M^+ in the spectrum. In most cases, when a molecule loses a valence electron, bonds are broken, or the ion formed quickly fragments to lower energy ions. The masses of charged ions are recorded as fragment ions by the spectrometer (see Figure 2 and Figure 3).

Tandem mass spectrometry (MS/MS or MS^2) is a technique to break down selected fragment ions (precursor ions) into fragments (product ions). The fragments then reveal aspects of the chemical structure of the precursor ion [10].

Figure 2 illustrates the positive ion mode HESI fragmentation pattern of reduced glutathione (GSH) $[M+H]^+$ = 308.1. The mass spectrometer was a Q Exactive™ Focus hybrid quadrupole-Orbitrap™ instrument (Thermo Scientific, Bremen, Germany). In the spectrum, three main intense peaks observed at *m/z* 233.1, 179.0 and 162.0 from the fragmentation of the $[M+H]^+$ species. The product ion at 233.1 arises by the loss of a neutral glycine residue (-75.0 Da), while the m/z 179.0 results from the loss of a glutamyl residue (-129.1 Da). The m/z 162.0 fragment ion is formed as a result of the loss of a neutral glutamate residue [11].

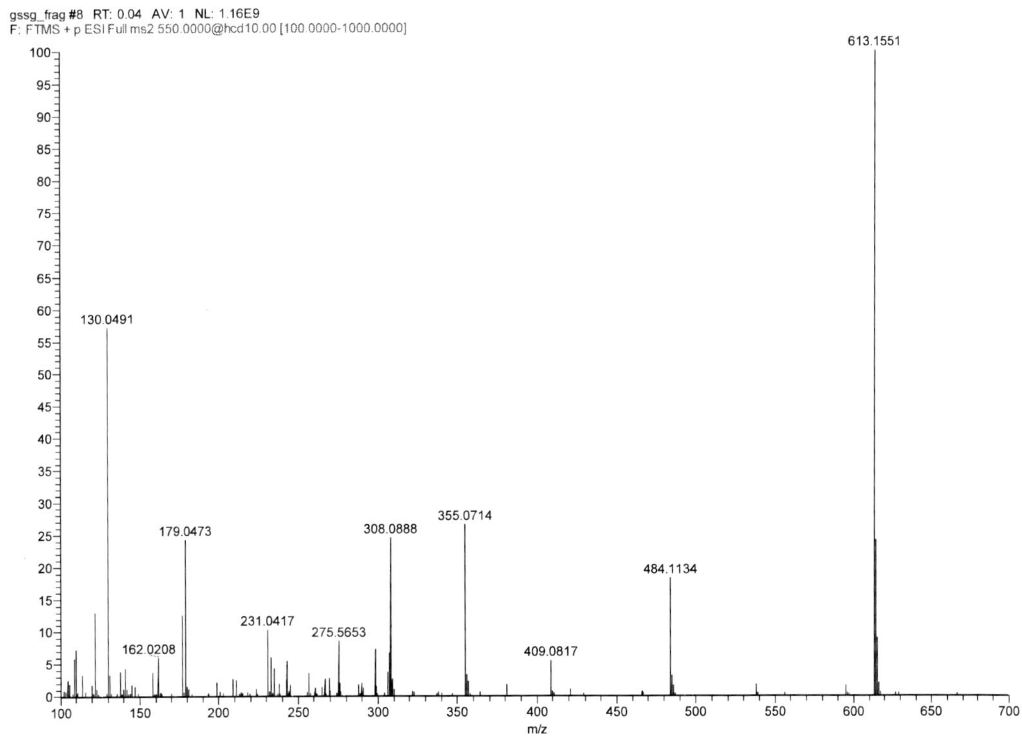

Figure 3. Positive ion mode HESI fragmentation pattern of oxidized glutathione (GSSG).

Figure 3 illustrates the positive ion mode HESI fragmentation pattern of oxidized glutathione (GSSG) $[M+H]^+$ = 613.2. In the spectrum, three main intense peaks observed at m/z 484.1, 355.1 and 308.1 from the fragmentation of the $[M+H]^+$ species. The product ion at 484.1 arises by the loss of a glutamyl residue (-129.1 Da), while the *m/z* 355.1 results from the loss of two glutamyl residues leaving the dipeptide Cys-Gly fragments covalently linked *via* a disulfide bond. This result is in agreement with previous HPLC-MS investigation of GSSG. The third one, m/z 308.1, corresponds to protonated reduced glutathione (GSH) $[M+H]^+ = 308.1$. Formation of this latter fragment (and the fragments of this ion) could not be observed under the previously used ionization conditions [12].

5. Solid State Characterization

5.1. Infrared Spectroscopy

Infrared spectroscopy is the analysis of infrared light interacting with a molecule. The main use of this technique is in organic and inorganic chemistry. It is used to determine functional groups in molecules. The infrared absorption spectra of organic small molecules are as distinctive and unique as a fingerprint [13].

Light with wavelengths from 800 nm to about 1 millimeter (1000 μm) is called infrared light. This wide band of the electromagnetic spectrum is divided into three regions; the near-, mid- and far-infrared. The higher-energy *near-IR*, approximately 12500–4000 cm^{-1} (0.8–2.5 μm wavelength), can excite overtone or harmonic vibrations. The *mid-infrared*, approximately 4000–400 cm^{-1} (2.5–25 μm), may be used to study the fundamental vibrations and associated rotational-vibrational structure. This is the region, which is most commonly used in the organic analytical practice. The *far-infrared*, approximately 400–10 cm^{-1} (25–1000 μm), has low energy and may be used for rotational spectroscopy [14].

Infrared spectroscopy exploits the fact that molecules absorb frequencies that are characteristic of their structure. These absorptions occur at resonant frequencies, i.e., the frequency of the absorbed radiation matches the vibrational frequency. The atoms in a CH_2X_2 group - commonly found in organic molecules - where X can represent any other atom, can vibrate in nine (3N-6) different ways. Six of these vibrations involve only the CH_2 portion: symmetric and antisymmetric stretching, scissoring, rocking, wagging and twisting. The spectra generally show the transmittance of the sample over the investigated wavenumber (cm^{-1}) range (see Figure 4 and Figure 5).

Since the observed set of absorption bands are, in essence, a display of some of the resonant vibrational frequencies of the molecule, the infrared spectrum is directly related to molecular parameters such as bond lengths and angles, bond groups ("functional groups") and molecular symmetry. Thus, an infrared spectrum not only provides information on the identity of "unknown" compounds, but also yields definite data on its molecular structure [15].

Infrared spectra may be obtained from samples in all phases (liquid, solid and gaseous). Liquids are usually examined as a thin film sandwiched between two polished NaCl plates. If solvents are used to dissolve solids, care must be taken to avoid obscuring important spectral regions by solvent absorption. Solids may either be incorporated in a thin KBr disk, prepared under high pressure, or mixed with a little non-volatile liquid and ground to a paste (or mull) that is smeared between salt plates.

The analytical infrared absorption spectra can be divided into four distinct regions [16]:

1) 4000-2800 cm^{-1}. The region of the OH, NH and CH stretching frequencies. (The OH and NH bands are more intense than the CH bands.)
2) 2800-2000 cm^{-1}. The region of rare functional groups (e.g., triple bonds (C≡N, C≡C), cumulated double bonds).
3) 2500-1500 cm^{-1}. The region of the double bonds (e.g., C=O, C=N, C=C and aromatic skeletal stretching).
4) 1500-400 cm^{-1}. The so called "*fingerprint region.*" This region usually contains a rather complicated series of absorptions. These are mainly due to all manner of bending vibrations within the molecule. It is said so because it can be used to identify unknown or two different organic compounds by comparing the troughs in this region.

Infrared (IR) spectroscopy is a universally used instrumental method for identification of pharmacopoeial substances. The method is based on comparison of the IR spectra of the sample to be identified with the corresponding *Chemical Reference Standard* (CRS) (Ph. Eur.)/*Reference Standard* (USP RS) (USP). The IR spectrum of reduced glutathione (GSH) recorded in KBr is shown in Figure 4. The observed absorption frequencies are listed in Table 1. The spectra were recorded with an Impact 400 (Nicolet) FT-IR spectrophotometer using a KBr pellet as the background reference spectrum.

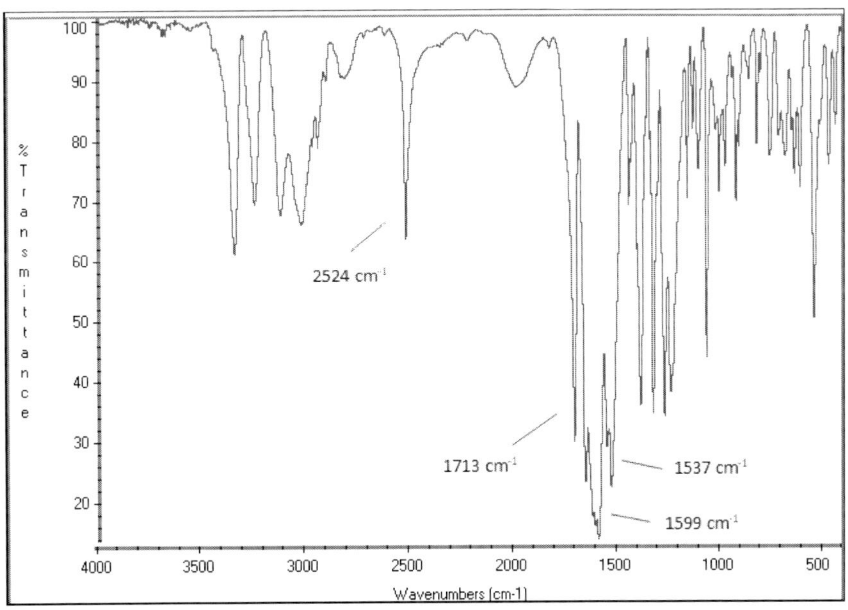

Figure 4. IR spectrum of reduced glutathione (GSH) recorded in KBr.

As it is shown, there are well separated sharp resonance peaks in the functional group region of the spectrum. Above the 3000 cm^{-1} region, several associated νOH, νNH, νNH_2 (3346, 3249, and 3126) and νCH (3024 cm^{-1}) bands can be seen. The band at 2524 cm^{-1} can be associated with the cysteine thiol (SH) group. The 1713 cm^{-1} band corresponds to the νC=O band of the terminal carboxyl groups. The 1661 cm^{-1} band can be assigned as the amid I band. The bands at 1599 and 1537 cm^{-1} could be associated with the overlapping amid II (νC=O (amide)) and δNH_3^+ resonances. The fingerprint region (≤ 1500 cm^{-1}) shows several unique resonances, which form an intricate pattern that can be used as a fingerprint to determine the compound.

Table 1. Characteristic IR bands of reduced (GSH) and oxidized (GSSG) glutathione

IR bands	GSH (cm^{-1})	GSSG (cm^{-1})
ν(OH, NH)	3346, 3249, 3126	3400-2700
ν(CH)	3024	3400-2700
ν(SH)	2524	None
ν(C=O, carboxyl)	1713	1736
ν((C=O, amide I)	1661	1655
ν(CO, amide II), $\delta(NH_3^+)$	1599 and 1537	1524

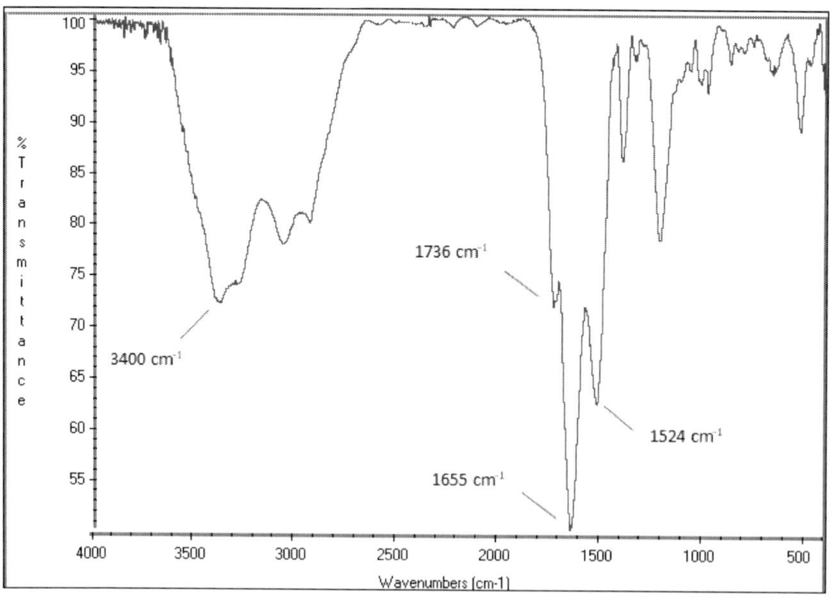

Figure 5. IR spectrum of oxidized glutathione (GSSG) recorded in KBr.

The IR spectrum of the oxidized form of glutathione shows broad, not well separated bands. In the 3400-2700 cm^{-1} region a broad, overlapping νOH, νNH, νNH_2 and νCH absorption can be seen. According to the modified chemical structure, no SH band appears in the spectrum. The 1736 cm^{-1} band corresponds to the νC=O band of the terminal carboxyl groups. The 1655 cm^{-1} band can be assigned as the amid I band. The broad band at 1524 cm^{-1} could be associated with the overlapping amid II (νC=O (amide)) and δNH_3^+ bands. The fingerprint region (≤ 1500 cm^{-1}) shows much less characteristic separated bands than the spectrum of GSH.

5.2. Crystallography

5.2.1. X-Ray Diffraction

Crystallography is an experimental methodology based on X-ray diffraction by crystalline materials and studies atomic arrangement on solid state samples [17]. X-ray crystallography is essentially dependent on analysis of the diffraction patterns of one crystal by an X-ray beam. The intensities and directions of the diffracted waves on diffraction pattern are correlated to position and atomic species into the crystal. The atomic species and positions are correlated by the maximum values of the electronic density distribution on the crystal, which can be obtained through Fourier transform on diffracted X-ray function [18-20]. Because of atomic positions, information such as (1) bond distances, (2) bond angles, (3) molecular packing, (4) supramolecular arrangement and (5) charge distribution can easily be obtained. This way, once the molecular structure of a compound is obtained, its physico-chemical properties can be described [21].

Since X-rays have shorter wavelengths than those of the IR and UV radiations, they have higher energy and more penetrative power [19]. Each X-ray wave on diffraction pattern interacts with one another either constructively or destructively. The observed intensity originated from the constructive interaction of the diffracted waves and energy (**1**), understood as a beam of photons.

$$E = h\nu = \frac{hc}{\lambda} \tag{1}$$

where h is the Planck's constant, c is the speed of light, ν is the frequency and λ is the wavelength of the radiation.

As a consequence of monochromatic X radiation striking a crystalline material, the electrons are forced to oscillate due to the electric field that makes up the incident radiation, making it a new source of scattering in all directions. The diffraction arises from the constructive interference of these waves scattered by the various discrete atoms. William L. Bragg described the constructive conditions of the X-ray diffraction, when a wavelength comparable to atomic bond distances is scattered in a pattern dependent on atomic pattern distribution in the crystalline sample. Diffraction process in the crystal can be regarded as if the X-rays were "reflected" by sets of parallel planes in the crystal [18-20]. Equation (**2**) represents Bragg´s law in a simplified way.

$$2 \text{ x } d \text{ x } \sin\theta = n \text{ x } \lambda \quad (2)$$

where d is the interplanar distance, θ is the incident angle, λ is the wavelength of the incident beam and n is the order of the diffraction.

5.2.2. Crystallographic Molecular Model

Once the X-ray radiation is diffracted by a single crystal and Bragg's law indexes every single direction on diffraction pattern, the mathematical model used to describe this phenomenon and obtain the molecular model (structure) is based on Fourier methods. A crystallographic classical example is the calculation of an electron density map from waves diffracted by all components of the unit cell. In turn, Fourier analysis is the decomposition of an arbitrary function into its component cosine and sine harmonics [21-24]. On crystallography, the electron density functions correspond to the positions and X-ray scattering powers of the atoms, and then perform a Fourier transform on the electron density to obtain the structure factors. The measured diffraction pattern of a crystal is described by Fourier analysis and electron density distribution (structure) is described by the Fourier synthesis. The electron density function ($\rho(\mathbf{r})$) is related by equation (**3**)

$$\rho(\boldsymbol{r}) = \frac{1}{V}\sum_h \quad |F(\boldsymbol{h})|\, exp[-2\pi i(\boldsymbol{h}.\boldsymbol{r}) + i\varphi(\boldsymbol{h})] \quad (3)$$

where V is the volume of the unit cell, $F(\mathbf{h})$ is the structure factor with magnitude $|F(\mathbf{h})|$ proportional to the reflection intensity measured and **h** indexes of each direction. The quantity **h.r** corresponds to the scalar product between the scattering vector in the reciprocal space **h** and the position vector in the direct space **r**.

The classical phase problem is evident in equation (**3**) since the full knowledge of electron density $\rho(\mathbf{r})$ depends on the knowledge of the phases $\varphi(\mathbf{h})$ that are lost in the measurement process. In general, the solution of the phase problem in crystallography of small molecules is obtained by the Patterson or direct methods. Once the draft of the crystallographic molecular model is first obtained [25-27], the following step is to refine this model throughout least squares procedures where the (1) accurate atomic position, (2) atomic species, (3) atomic displacement behavior, (4) solvent content, (5) chemical consistency and other important information are all assigned. In general, the programs used are SHELXL [28], SIR [29] and OLEX2 [30].

The first glutathione molecular structure, described by W. B. Wright [31], is shown in Figure 6.

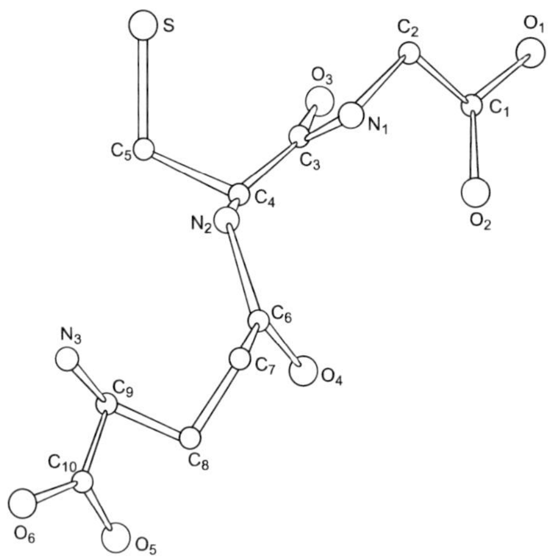

Figure 6. X-ray structure of reduced glutathione (GSH). Based on reference [31].

The X-ray analysis showed a tripeptide with a gamma peptide linkage between the carboxyl group of the glutamate side chain and the amine group of cysteine. Wright also described a molecular conformation adopting a helical configuration with almost planar structure between glycine carboxyl and amino-carboxyl groups. The glutathione crystallized in the orthorhombic space group $P2_12_12_1$ showing the carboxyl group of cysteine attached by a normal peptide linkage to glycine, and the sulphur is forming a zigzag chain [31, 32]. The molecules form a three-dimensional network of hydrogen bonds on molecular packing.

6. In Solution Characterization

6.1. UV-Vis Spectroscopy

UV-Vis spectroscopy is a spectroscopic method to characterize electronic structure of organic molecules. Electronic transition energies correspond to the energies of the analytical UV (200-400 nm) and the visible (400-700 nm) range photons. Absorption of visible and ultraviolet (UV) radiation is associated with excitation of electrons, in both atoms and molecules, from lower to higher energy levels. Since the energy levels of matter are quantized, only light with the precise amount of energy can cause transitions from one level to another will be absorbed.

The absorption of UV or visible radiation corresponds to the excitation of outer electrons. There are three types of electronic transitions to be considered: (a) transitions involving p, s and n electrons, (b) transitions involving charge-transfer electrons, and (c) transitions involving d and f electrons [33, 34]. In this chapter only possible transitions involving p, s, and n electrons are discussed.

The energy associated with electromagnetic radiation is defined by the following equation:

$$E = h \times \nu \tag{4}$$

where E is the energy (J), h is the Planck's constant (6.62×10^{-34} Js), and ν is the frequency (1/s).

There are a number of transitions possible involving the bonding (s and p) and nonbonding (n) electrons. The transitions generally occur from the highest occupied molecular orbital (HOMO) to the lowest unoccupied molecular orbital (LUMO) in a molecule. The larger the gap between the energy levels, the greater the energy required to promote the electron to the higher energy level; resulting in light of higher frequency, and therefore shorter wavelength, being absorbed.

Of the possible electronic transitions (s→s*, s→p*, p→s*, n→s*, p→p* and n→p*), only the two lowest energy ones (p→p* and n→p*) can be achieved by the energies available in the 200-700 nm region (Figure 7). Thus, only those molecules are expected to absorb in the UV-Vis region, which contains p-electrons and may also have atoms with nonbonding electrons [33, 34].

These transitions should result in very narrow absorbance bands at wavelengths highly characteristic of the difference in energy levels of the absorbing species. This is true, however, only for atoms. For molecules, vibrational and rotational energy levels are superimposed on the electronic energy levels. Because many transitions with different energies can occur, the bands are broadened (see Figures 8 and 9). The broadening is even greater in solutions owing to solvent-solute interactions [33]. Accordingly, UV-visible spectra generally show only a few broad absorbance bands. Compared with techniques such as infrared spectroscopy, which produces many narrow bands, UV-visible spectroscopy provides a limited amount of qualitative information. The method is mainly used for quantitative measurements [34, 35].

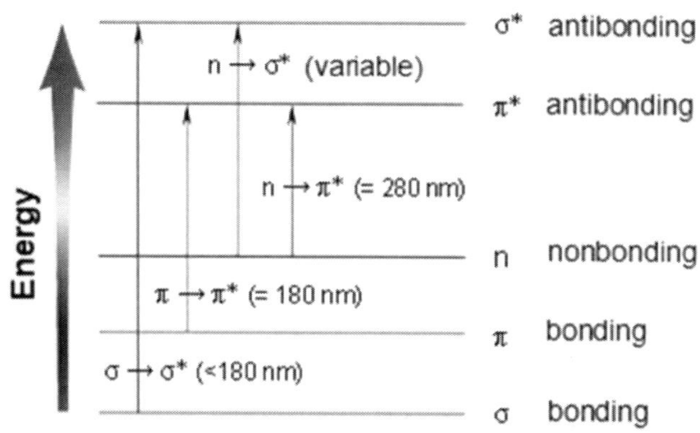

Figure 7. Possible UV-Vis electronic transitions.

The UV spectra of reduced (GSH) and oxidized (GSSG) glutathione recorded in aqueous solution are shown in Figure 8 and Figure 9. The spectra were recorded in a Jasco V-750 Spectrophotometer, using water as a solvent.

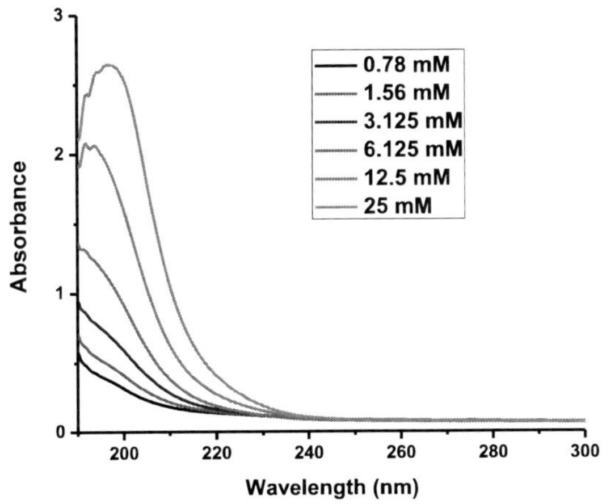

Figure 8. UV spectrum of reduced glutathione (GSH) recorded in aqueous solution.

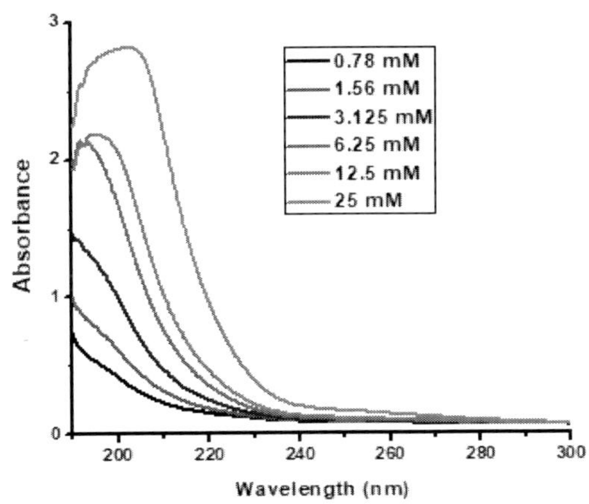

Figure 9. UV spectrum of oxidized glutathione (GSSG) recorded in aqueous solution.

Glutathione exhibits an UV absorption maximum at 197 nm. The absorption band corresponds the n→p* electronic transitions in the carboxyl and the amide bonds. GSSG has got very similar UV spectrum with an absorption maximum of 202 nm.

6.2. NMR Spectroscopy

NMR spectroscopy is the most powerful non-destructive analytical technique available for organic structure determination in solution. The method is based on the fact that some atomic nuclei possess a nuclear spin not equal to zero. A nucleus with an odd atomic number or an odd mass number has a nuclear spin. The overall spin of the nucleus is determined by the spin quantum number S. If the numbers of both the protons and neutrons in a given nuclide are even then $S = 0$, i.e., there is no overall spin. Then, just as electrons pair up in nondegenerate atomic orbitals, so do even numbers of protons or even numbers of neutrons giving zero overall spin. When the number of protons or neutrons or both of them are uneven, however, the nucleus has an overall spin. This is because protons and neutrons will have lower energy when their spins are parallel, not anti-parallel [36, 37]. The most commonly used nuclei are summarized in Table 2.

Nuclear spins are oriented randomly in the absence of an external magnetic field but have a specific orientation in the presence of an external field, $\boldsymbol{B}$. Some nuclear spins are aligned parallel to the external field, which is the more likely, lower energy orientation; some nuclear spins are aligned antiparallel to the external field, which is the less likely, higher energy orientation.

Table 2. Characteristics of the most common NMR active nuclei

Nuclei	Unpaired protons	Unpaired neutrons	Net spin	γ (MHz/T)
^{1}H	1	0	1/2	42.58
^{2}H	1	1	1	6.54
^{13}C	0	1	1/2	10.71
^{14}N	1	1	1	3.08
^{19}F	1	0	1/2	40.08
^{23}Na	1	2	3/2	11.27
^{31}P	1	0	1/2	17.25

When nuclei that are aligned parallel with an external magnetic field are irradiated with the proper *frequency* (ν) *of electromagnetic radiation* (radiofrequency range), the energy is absorbed and the nuclei "spin-flips" to the higher-energy antiparallel alignment. The frequency (ν) depends on the *gyromagnetic ratio* (γ) of the particle (**5**).

Nuclei that undergo "spin-flips" in response to the applied radiation are said to be in resonance with the applied radiation - *nuclear magnetic resonance*. The frequency necessary for the resonance depends on the strength of the external field ($\boldsymbol{B}$) and the identity (γ) of the nuclei

$$\nu = \gamma \boldsymbol{B} \tag{5}$$

where ν is the frequency of electromagnetic radiation (Hz), γ is the gyromagnetic ratio (MHz/T), and $\boldsymbol{B}$ is the magnetic field strength (T).

If the oriented nuclei are exposed to radiofrequency pulses (generated by the spectrometer console) one will be able to gain information about the magnetic and chemical environment of the studied nuclei. Nuclei are surrounded by electrons that shield them from the external field. Circulating electrons create an induced magnetic field that opposes the external magnetic field. Accordingly, each nucleus with different molecular environment has a different effective magnetic field and resonates with photons of different energy (frequency).

According to the above equation, the energy difference ($\Delta E=h\nu$) between nuclear spin states depends on the strength of the applied magnetic field ($\boldsymbol{B}$). Consequently, the higher the magnetic field strength, the higher the resolution of two nuclei with similar ΔE value. This relationship could make it difficult to compare NMR spectra taken on spectrometers operating at different field strengths. The term *chemical shift* was developed to avoid this problem.

The chemical shift of a nucleus is the difference between the resonance frequency of the nucleus and a standard, relative to the standard. This quantity is reported in ppm and given the symbol delta (δ).

$$\delta = (\nu - \nu_{REF}) \times 10^6 / \nu_{REF}$$

where δ is the chemical shift (ppm), ν is the resonance frequency of the nucleus (Hz), and ν_{REF} is the resonance frequency of the reference (Hz).

In ^1H NMR spectroscopy, this standard is often tetramethylsilane, $Si(CH_3)_4$, abbreviated as TMS. The chemical shift is a very precise metric of the chemical environments around a nucleus.

Nuclei experiencing the same chemical environment or chemical shift are called *equivalent*. Those nuclei experiencing different environments or having different chemical shifts are *nonequivalent*. Nuclei that are close to one another exert an influence on each other's effective magnetic field. This effect shows up in the NMR spectrum when the nuclei are nonequivalent. If the distance between the non-equivalent nuclei is less than or equal to three bond lengths, this effect is observable. This effect is called spin-spin coupling or *J coupling*.

The difference in ΔE of nonequivalent nuclei originated from the spin-spin coupling results in *splitting* of the NMR signal. If a ^1H signal is split by N equivalent protons, it is split into $N+1$ peaks. (The N+1 rule). The value (given in Hz) of the vicinal ^1H-^1H splitting (^3J-coupling) is related to the dihedral torsion angle of the two protons (Karplus equation) [39].

It is unlikely that a ^{13}C would be adjacent to another ^{13}C, so splitting by carbon is negligible. Carbon-13 nuclei will magnetically couple, however, with attached protons and adjacent protons. These complex splitting patterns are difficult to interpret. Therefore, in a "normal" (*broadband, proton decoupled*) ^{13}C *spectra*, these couplings are removed by applying a continuous second radio frequency signal of a broad frequency range that excites all the ^{1}H nuclei and cancels out the coupling patterns due to the interaction of the ^{1}H with the ^{13}C. This means that each carbon atom is seen as a single line. In *off-resonance decoupling* all the carbon-hydrogen couplings are removed from a signal except those between the nuclei that are directly bonded to one another. Thus, a methine (CH) appears as a doublet, a methylene (CH_2) appears as a triplet, and a methyl (CH_3) shows a quartet [40].

The one-dimensional ^{1}H NMR spectra of reduced (GHS) and oxidized (GSSG) glutathione are shown in Figure 10 and Figure 11, respectively. (The spectra were recorded with a Bruker Avance III 500 (500.15 MHz for ^{1}H) spectrometer. Chemical shifts were referenced to the residual solvent signal. Measurements were run at a probe temperature of 298 K in D_2O solutions). The spectroscopic data are summarized in Table 3.

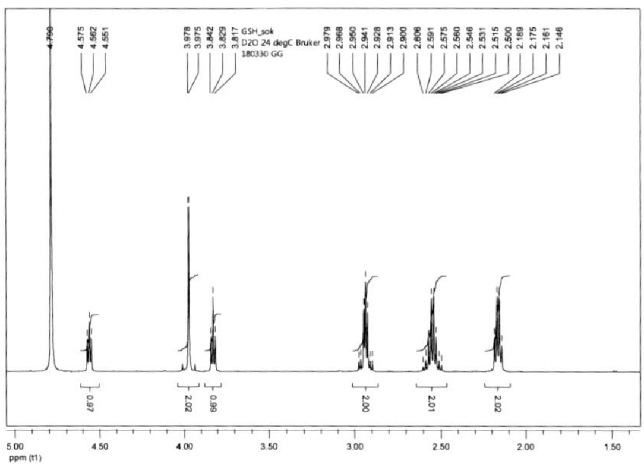

Figure 10. ^{1}H-NMR (500 MHz) spectrum of reduced glutathione (GSH) recorded in D_2O.

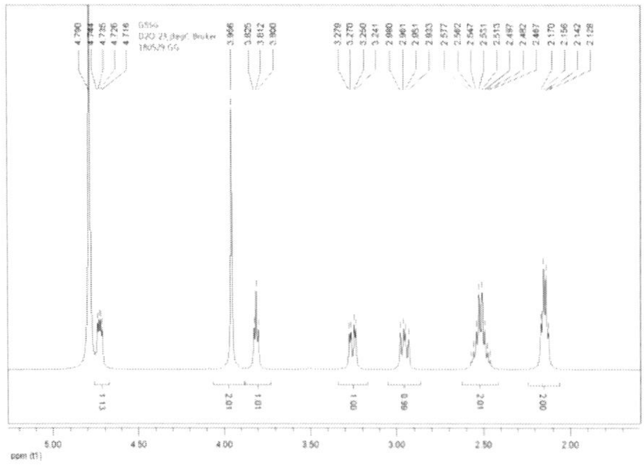

Figure 11. ^{1}H-NMR (500 MHz) spectrum of oxidized glutathione (GSSG) recorded in D_2O.

Table 3. ^{1}H-NMR (500 MHz) data of reduced (GSH) and oxidized (GSSG) glutathione recorded in D_2O

	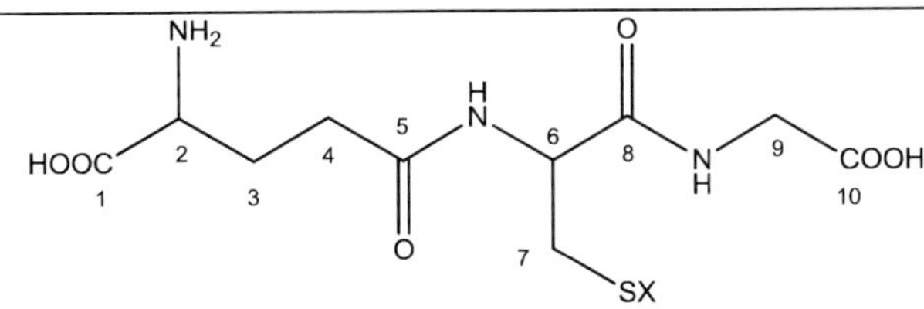	
Proton (Number of protons)	**GSH**	**GSSG**
H-3 (2H)	2.17 ppm	2.15 ppm
H-4 (2H)	2.56 ppm	2.52 ppm
H-7a (1H)	2.94 ppm	2.96 ppm (9.5, 14.2 Hz))
H-7b (1H)	2.94 ppm	3.26 ppm (4.5, 14.2 Hz)
H-2 (1H)	3.83 ppm	3.81 ppm
H-9 (2H)	3.98 ppm	3.96 ppm
H-6 (1H)	4.56 ppm	4.74 ppm

The methyl proton of the glutamate moiety (H-2) gives a doublet-of-doublets at 3.83 ppm, and the methylene protons - H-3 and H-4 - give two separate multiplets at 2.17 and 2.56 ppm, respectively. The methylene (H-7) and the methyne (H-6) protons of the cysteinyl moiety forms an AA'X spin system with three doublet-of-doublets at 2.94 (H-7) and 4.56 (H-6) ppm. The methylene protons of the glycine moiety (H-9) appear a singlet at 3.98 ppm. The respective protons in the oxidized glutathione (GSSG) show similar chemical shifts and multiplicity as those of the respective protons in the reduced form (GSH).

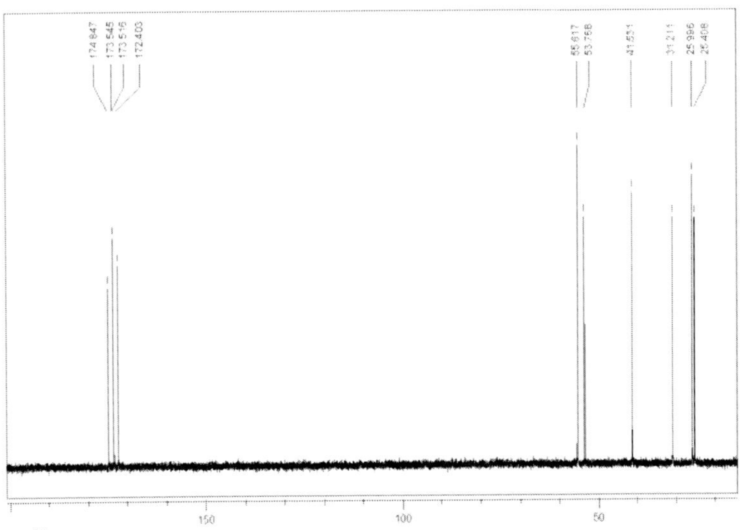

Figure 12. ^{13}C-NMR (125 MHz) spectrum of reduced glutathione (GSH) recorded in D_2O.

The one-dimensional ^{13}C NMR spectra with broadband proton decoupling of reduced (GHS) and oxidized (GSSG) glutathione are shown in Figure 12 and Figure 13, respectively. (The spectra were recorded with a Bruker Avance III 500 (125.04 MHz for ^{13}C) spectrometer. Measurements were run at a probe temperature of 298 K in D_2O solutions.) The spectroscopic data are summarized in Table 4.

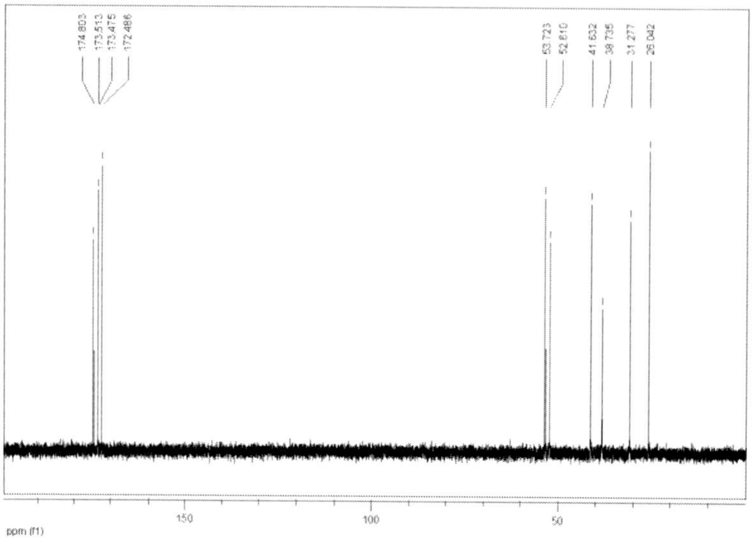

Figure 13. ^{13}C-NMR (125 MHz) spectrum of oxidized glutathione (GSSG) recorded in D_2O.

Table 4. ^{13}C-NMR (125 MHz) data of reduced (GSH) and oxidized (GSSG) glutathione recorded in D_2O

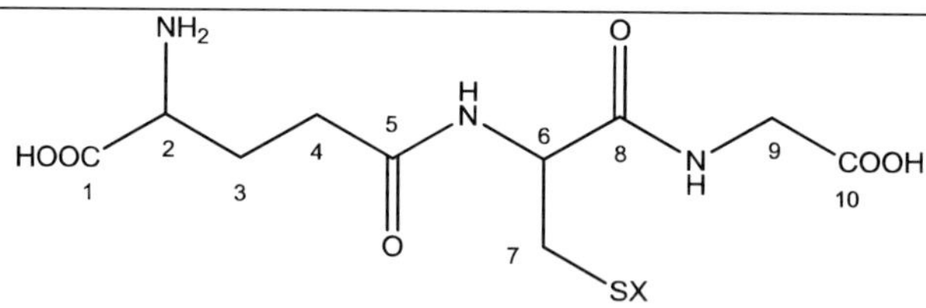

Carbon	GSH	GSSG
C-1	173.52 ppm	173.48 ppm
C-2	53.77 ppm	53.72 ppm
C-3	26.00 ppm	26.04 ppm
C-4	31.21 ppm	31.28 ppm
C-5	173.55 ppm	173.51 ppm
C-6	55.62 ppm	52.61 ppm
C-7	25.41 ppm	38.74 ppm
C-8	172.40 ppm	172.49 ppm
C-9	41.53 ppm	41.63 ppm
C-10	174.85 ppm	174.80 ppm

7. ACID-BASE PROPERTIES

7.1. Acid-Base Theories

Definition of acids and bases has been altered several times. One of the first classification of acids and bases was made by Arrhenius (1888). In Arrhenius's definitions, an *acid* is any substance that, when dissolved in water, increases concentration of the hydrogen (hydronium) ions, while a *base* is any substance that, when dissolved in water, increases concentration of the hydroxide ions [41].

J. N. Brønsted and T. M. Lowry extended the acid-base theory of Arrhenius and emphasized the role of hydrogen ion (proton) in the acid-base reactions. According to the Brønsted and Lowry theory, an *acid* is a species that donates a proton to another species in a proton-transfer reaction. They defined a *base* as a species that accepts a proton in a proton-transfer reaction [41].

According to their theory, acids are converted into a base by releasing a proton and a base is transformed to an acid by accepting a proton. (See examples below.) An acid donates a proton only in the presence of a proton acceptor base. Accordingly, conversion of an acid to the respective base and conversion of a base to its respective acid takes place in the same reaction.

G. N. Lewis realized that the concept of acids and bases could be generalized to include reactions (e.g., reaction of acidic and basic oxides) that do not fit either the Arrhenius or the Brønsted-Lowry concepts [42]. According to this theory, a Lewis acid is a species (atom, molecule, or ion) that can form a covalent bond by accepting a pair of electrons from another species, while a Lewis base is a species (atom, molecule, or ion) that can form a covalent bond by donating an electron pair to another species.

7.2. The Importance of Acid-Basic Properties in Drugs Development

In any protonation process, one positive charge is transferred to the base. Therefore, the proton transfer affects the physico-chemical properties such as charge, solubility, polarizability, and dipole of the base (B or A^-) [43]. Most of the present-day drug molecules are acids or bases. When they are dissolved in water, they can undergo proton transfer reactions of different degree.

$$B + H_2O \leftrightarrow BH^+ + OH^-$$

and

$HA + H_2O \leftrightarrow H_3O^+ + A^-$

where B is a monoprotic base, HA is a monoprotic acid

The reactions are reversible, and the composition of the equilibrium can be characterized by the pK_a value of the equilibrium (the ionization constant of the molecule). While analyzing acid-base properties of 907 drug molecules, it was found that 64% of these compounds contained an ionizable group. Within this group of ionizable compounds, 34% contained a single basic group while 20% contained a single acidic functional group [44]. Focusing on oral drugs showed that 78.6% of compounds contained an ionizable group, while 11.9% were neutral, 4.3% always ionized and the remainder (5.2%) was made up of salts, miscellaneous compounds (e.g., mixtures) and high molecular weight substance [45].

Efforts to determine the physicochemical properties that relate to long-term compound viability have been conducted in concert with gathering biology data on attributes such as cell toxicity, efflux liability, metabolic stability or inhibition, cell permeability, bioavailability, CNS permeability, protein binding, brain tissue binding, promiscuity, clearance and volumes of distribution. Among them, the charge state of the molecules turned to be one of the fundamental physico-chemical properties. Accordingly, the acid/base properties of a molecule are among the most fundamental ones for drug actions [45, 46].

7.3. Glutathione - Acid-Basic Properties

Dissociation constants of weak acids are commonly evaluated by analyzing their titration curves. The titration curves of monoprotic acids can be analyzed conveniently on the basis of Henderson-Hasselbalch equation using different techniques e.g., potentiometry, UV/VIS spectroscopy, HPLC, NMR spectroscopy, conductometry, calorimetry, capillary zone electrophoresis, and computational methods [47]. On the other hand, the pK_a values of polyprotic acids are evaluated with great difficulty using software methods involving complex algebraic calculations and approximate formulae [48].

The acid-base chemistry of small organic molecules is usually characterized in terms of macroscopic acid dissociation/protonation constants obtained from pH-titration data. If the molecule contains two or more functional groups of comparable acidity, the pH ranges for titration of the two groups overlap and the titration can occur by two independent pathways. In such cases, the macroscopic constants are composites of the microscopic constants that describe the acid-base chemistry at the molecular level. One of the widely used methods for determining microscopic acid dissociation constants of such molecules is NMR spectroscopy [49]. Considering the number of ionizable functions of a molecule, in the case of two (e.g., morphine; phenolic OH-tertiary N) there are two macroscopic protonation constants and four microspecies with different protonation state. In the case of three ionizable groups (e.g., thyroxine; phenolic OH-amino-carboxyl), the number of macroscopic protonation constants is 3 and the number of different microspecies is 8. These figures in the case of a compound with four ionizable groups (e.g., glutathione) are 4 and 16, respectively [50].

The acid-base properties of glutathione (GSH) have long been in the focus of scientific interest. It has three acidic (thiol, glycinyl carboxyl, glutamyl carboxyl) and one basic (amino) functional groups. Accordingly, in aqueous solution glutathione can exist in five different macroscopic protonation state as follows.

$$L^{3-} \underset{}{\overset{K_1}{\rightleftharpoons}} HL^{2-} \overset{K_2}{\rightleftharpoons} H_2L^{-} \overset{K_3}{\rightleftharpoons} H_3L \overset{K_4}{\rightleftharpoons} H_4L^{+}$$

where L^{3-} is the fully deprotonated glutathione, H_4L^{+} is the fully protonated glutathione.

Since the HL^{2}- and the H_3L forms have 4 protonation isomers (microspecies) each, and the H_2L^{-} form has 6 microspecies, the molecule has got altogether 16 different protonation states (microspecies) [50].

The micro- and sub-micro protonation constants characterize the acid-base properties at the submolecular level [51]. These constants allow quantification of the proton binding capacity of submolecular basic units, when the protonation states of all other sites are defined in the molecule [52]. Group constants are special micro-constants, when the rest of the groups in the molecule are far enough apart and their protonation does not affect the basicity of the group [53]. The rotational state of the flexible parts of the molecules is defined by the sub-micro constants, when protonation occurs [54]. The correct characterization of basicity of the sites of protonation of multidentate ligands can be done using the micro and sub-micro constants. In addition, this group of constants is used to measure the concentration of different forms of protonation, of which the principal form is not always the reactive form in chemical and biological processes. [55-59]. The macroscopic protonation constants (K_1-K_4) determined by ^{1}H NMR-pH titrations were as follows: logK_1 9.65; logK_2 8.78; logK_3 3.52; logK_4 2.22 [50].

The obtained values were found to be very similar to those determined in earlier works of Pirie and Pinhey [60] (9.62, 8.66, 3.53, 2.12), Li et al. [61] (9.65, 8.75, 3.59), and Martin and Edsall [62] (9.62, 8.74). The results demonstrated that the first and the second protonation constants were predominated by the overlapping protonation of the amino and the thiolate site, the amino being typically more favored. The carboxylate groups also protonated in an overlapping fashion, the glycinyl carboxylate being the more basic. It is worth mentioning, that it is the protonated amino group that makes the inherently more basic glutamyl carboxylate more acidic [50].

It is important to note that the physico-chemical properties (e.g., complex formation, nucleophilic reactivity, redox properties) and biological functions of glutathione could be significantly different in different protonation states (i.e., in solutions with different pH values) [63-66]. Furthermore, ionic strength and the nature of the ionic media also affect the acid-base characteristics of glutathione [69].

8. Redox Properties

8.1. Redox Potentials

Redox reactions involve an electron transfer from one particle onto another. *Oxidation* means a half-reaction in which a substance (atomic, ionic or molecular) releases electron(s). In the *reduction* half-reaction electron(s) is/are accepted. Thus, oxidation and reduction are antiparallel and simultaneously occurring electron-transfer processes. In organic and biochemical reactions oxidation is frequently accompanied by gaining oxygen or loosing hydrogen atoms; while reduction is manifested as losing oxygen or gaining hydrogen atoms.

Half-cell potentials of reversible redox systems can be calculated according to the *Nernst* equation, of which form referring to 25°C as follows:

$$E = E^o - \frac{RT}{nF} ln \frac{[red]}{[ox]} \quad (6)$$

where E is the actual redox potential (V), E^o is the standard redox potential (V), R is the gas constant (8.314 J mol^{-1} K^{-1}), T is the temperature (K), n is the number of transferred electrons (per particle), F is the Faraday constant (96 485 C), and

[red] = concentration (activity) of the reduced form of the redox system (mol/dm^3).
[ox] = concentration (activity) of the oxidized form of the redox system (mol/dm^3).

The direction of flow of electrons in a spontaneous redox reaction can be determined by comparison of the electrode and/or redox potentials of the reversible redox systems (ox_1/red_1 and red_2/ox_2) of the reactions.

$$Ox_1 + Red_2 = Red_1 + Ox_2 \quad (7)$$

When $E(red_1/ox_1) > E(red_2/ox_2)$ the direction of a spontaneous redox reaction corresponds to the above generalized chemical reaction. In other words, in spontaneous redox reaction the redox system with the more positive potential oxidizes the more negative one. The redox system with the more negative potential reduces the more positive one.

Redox potentials of reversible redox systems can only be determined by measurements of potential differences. At first, the redox potentials were determined against the potential of the *standard hydrogen electrode* (SHE) - with 1 M hydronium ion (H_3O^+) concentration (activity) and 1 atmosphere of hydrogen (H_2) gas pressure on it - of which potential was arbitrary set to be zero ($E^o = 0.00$ V). The standard redox potential of the hydrogen is zero at pH 0 (1 M H_3O^+), however, for biologic systems, the biologic (apparent) half-cell redox potential (E'^o) is normally expressed at pH 7 [70].

8.2. The Glutathione (GSH) and Glutathione Disulfide (GSSG) Reversible Redox System - Chemistry

Although the importance of the GSH/GSSG redox system in biological redox processes has been long recognized, there is some uncertainty as to its half-cell potential. This is because direct measurement of the GSH/GSSG redox system (and another thiol/disulfide redox systems) by standard electrochemical methods is not possible due to formation of stable metal-thiolate complexes at the electrode surfaces. Thus, the redox potentials of GSH and other thiols can only be determined indirectly by measurement of equilibrium constants for their reaction with redox systems of known redox potentials. Willis et al. reported the E'^o of the GSSG/GSH half-cell potential to be -0.263 V (pH 7, 298 K, 0.1 M phosphate buffer) determined by ^{1}H NMR method [71].

According to the second law of thermodynamics, the criterion for spontaneous change and equilibrium in a closed system with a single chemical reaction at constant temperature and pressure is given by the reaction Gibbs energy ($\Delta_r G$). Gibbs energy is a state function, meaning that the change in Gibbs energy, when a system changes state, is independent of the path from the initial to the final state. This means that the standard reaction Gibbs energy ($\Delta_r G^o$) of a given reaction is the same whether it takes place in one or more steps. It follows that

$$\Delta_r G^o = \Sigma \, \Delta_f G^o \tag{8}$$

where $\Delta_r G^o$ is the standard reaction Gibbs energy, $\Delta_f G^o$ is the standard Gibbs energy of formation.

The standard reaction Gibbs energy of a reaction is related to its equilibrium constant K by

$$\Delta_r G^o = - RT \ln K \tag{9}$$

where $\Delta_r G^o$ is the standard reaction Gibbs energy, R is the gas constant, T is the temperature (K), K is the equilibrium constant.

This relation, which follows from the fact that at equilibrium $\Delta_r G=0$, enables derivation of $\Delta_r G^o$ from experimentally measured K. The *standard reaction Gibbs energy* of a reaction could theoretically be derived, by experimentally measuring the concentration of all reactants at equilibrium and plugging the above equation. It frequently occurs, however, that the reactants have ionizable functional groups. In such cases, if the equilibrium is set at a pH near their pK_a values, several ionic species of the reactants may be present. Different species of the same reactant have different standard Gibbs energies of formation [70].

Accordingly, using the different species of reactant concentrations in equation (9), it gives neither the standard reaction Gibbs energy nor the equilibrium constant K. Instead, it gives what is known as *standard transformed reaction Gibbs energy* ($\Delta_r G'^o$) and *apparent equilibrium constant* (K'). Accordingly, in a closed system with a single biochemical reaction at constant temperature and pressure, and specified pH, the criterion for spontaneous change and equilibrium is given by the transformed reaction Gibbs energy

$$\Delta_r G' \leq 0 \quad (10)$$

To date, the most comprehensive tables of estimated $\Delta_r G'^o$ values for biochemical compounds were compiled by Alberty [72, 73].

It is the importance of knowledge of standard transformed Gibbs energies of formation of reactants that they can be used to calculate standard apparent reduction potentials (E'^o) at specified pH and ionic strength. Conversely, standard apparent reduction potentials at specified pH can be used to calculate standard transformed Gibbs energies of formation of one or two reactants in the half reaction [74]. The calculated standard apparent (biochemical) reduction potentials (E'^o) of some biochemically important reversible redox systems (pH 7, T=298.15 K, ionic strength = 0.25 M) are listed in Table 5.

Table 5. Calculated standard apparent reduction potentials (pH 7, 298.15 K, 0.25 M ionic strength) of some biochemically important reversible redox systems [75]

System	**E'_o (V)**
$O_2(aq) + 4 H^+ + 4 e^- = 2 H_2O$	+0.849
$O_2(aq) + 2 H^+ + 2 e^- = H_2O_2$	+0.357
Cytochrome c(Fe^{3+}) + e^- = Cytochrome c(Fe^{2+})	+0.212
Glutathione(ox) + 2 H^+ + 2 e^- = Glutathione(red)	-0.288
Thioredoxin(ox) + 2 H^++ 2e^- = Thioredoxin (red)	-0.289
NAD^+(ox) + H^+ + 2 e^- = NADH(red)	-0.316
$NADP^+$(ox) + H^+ + 2 e^- = NADPH(red)	-0.317
Cystine + 2 H^+ + 2 e^- = 2 Cysteine	-0.363
$2 H^+ + 2 e^- = H_2(g)$	-0.423

The pH dependent half-cell reduction potential of the GSH/GSSG reversible redox system draws attention to the reduction potentials of the variously protonated microspecies existing under different acid-base conditions. By means of determination of the species-specific protonation constants, and the conditional equilibrium constants of the pH-dependent redox equilibria, the standard redox potentials of six biogenic thiols were determined by ^1H NMR method [76]. It was found, that that even minor changes in the sites of protonation could lead to differences in redox characteristics. Furthermore, not only the protonation of the thiolate itself, its protonation fraction, and the concomitant redox behavior found to be sensitive to minor pH changes, but the protonation status of the adjacent basic moieties could also influence the redox behavior by various inductive effects on the thiolate moiety. The highest thiolate oxidizability of GSH (−383 mV) belonged to the microspecies in which all other basic sites in the molecule were in non-protonated form. The easiest reducibility of GSSG (−325 mV) was born by the microspecies of complete amino and carboxylate protonation. The correlation between the thiolate basicity and the standard redox potentials verified the previous claims that thiolate basicity and oxidizability are proportional parameters [76].

8.3. The Glutathione (GSH) and Glutathione Disulfide (GSSG) Reversible Redox System – Biological Implications

The glutathione (GSH) and glutathione disulfide (GSSG) reversible redox system

$$GSSG + 2\,e^- + 2\,H^+ \leftrightarrow 2\,GSH \qquad (11)$$

is crucially participating and having a vital role in maintaining the environment of the intracellular redox system, antioxidant defense system, and cellular signaling processes [77-82].

The intracellular concentration of GSH is quite high and ranges from approx. 0.5 to 15 mM. The concentration of GSSG is usually several orders of magnitude lower [80, 83]. The reduced form (GSH) is a key contributor to the cellular antioxidant defense system and to maintaining the intracellular redox milieu for the preservation of the thiol-disulfide redox states of proteins. In general, disruption in homeostasis of the cellular redox system will influence the normal physiology of cells such as differentiation, proliferation, cell signaling and programed cell death [82, 84]

There are several redox couples in a cell that work together to maintain the redox environment, being the GSSG/2GSH couple the most abundant one in a cell. Within mammalian cells, glutathione exists mainly (98%) in the thiol-reduced form (GSH), whereas glutathione disulfide (GSSG) is usually less than 1% of GSH [85]. Its reduced form is largely maintained by the NADPH-dependent glutathione reductase [86]. The average cellular concentration of glutathione and NADPH are about 5 mM, and 0.1 mM, and the half-cell reduction potential (E'_0) of GSSG/2GSH and NADP/NADPH, (pH 7.0), are reported to be -288 mV and -317 mV, respectively (Table 5).

The GSH/GSSG ratio is a critical mechanism for cell survival; in fact, it is known that it varies in association with proliferation, differentiation, and apoptosis [87]. As a result, the GSH/GSSG ratio is considering as a most common biomarker for the detection of cellular oxidative stress [84, 88, 89].

Shafer and Buettner pointed to the important feature of the cellular redox homeostasis [85]. They defined the term *cellular redox environment*, which is the summation of the products of the reduction potential and reducing capacity of the linked redox couples present in a biological fluid, organelle, cell or tissue. The *redox state* (E_{hc}) of a redox couple is defined by the half-cell reduction potential and the reducing capacity of that couple. In contrast to many other physiological redox systems, the redox state of the GSSG/GSH system not only depends on the GSH/GSSG ratio, the temperature and the pH, but also on the actual concentration of glutathione. This is in contrast with the $NAD(P)^+/NAD(P)H$ couples, where only the ratio of the interconvertible oxidized and reduced forms of these redox couples is enough.

Both GSH and GSSG concentrations depend on the subcellular compartment, the cell type and the organism. Accordingly, the redox potential of the GSH/GSSG system varies from tissue to tissue, from organism to organism. This relays on the proportion of GSH and GSSG, and the total concentration of glutathione, which (their actual *in vivo* concentration and ratio) is quite difficult to estimate [80, 90]. For example, taking the local pH and GSH/GSSG ratios into consideration, cytosolic $E_{pH7.0}$ = -289 mV (or even lower), mitochondrial matrix $E_{pH7.0}$ = -296 mV (or even lower), and human plasma $E_{pH7.4}$ = -118 mV half-cell reduction potentials have been estimated [86]. Furthermore, correlation has been found between the cell cycle, the condition of the cell (stressed, apoptotic, etc.), and the GSH/GSSG ratio. For instance, in cell proliferation (E_{hc} = ~-240 mV), in cell differentiation (E_{hc} = ~-200 mV), and in apoptosis (E_{hc}= ~-170 mV), which can be applicable for a better understanding of cellular oxidative stress [82, 91].

Furthermore, glutathione reversibly reacts with sulfhydryl group (SH) of thiol-containing proteins modifying them into an active or inactive form, in a process, which is known as S-glutathionylation. The aim of this alteration is for the protection of the sensitive thiol-containing proteins from irreversible oxidation. In the end, the process of S-glutathiolation (also called S-glutathionylation) affects the activity of proteins and the cellular signaling process, in both ways; regulation or deregulation [79].

In the case of an oxidative challenge, interaction of GSSG with reactive cysteine residues of proteins to form mixed disulfides can occur [92]. By means of such thiol-disulfide exchange, the oxidized form of glutathione (GSSG) can react with thiol-containing proteins to form glutathione-protein disulfides (Pr-SSG) (12) or protein disulfides (Pr-SS-Pr) (13) [90]. The GSH/GSSG redox regulation of proteins happens under circumstances of enzymatic actions and metabolic systems to control the exchange reaction and to sustain the proportion of GSH/GSSG [80].

$$2\ GSSG + Pr\text{-}SH \leftrightarrow GSH + Pr\text{-}SSG \qquad (12)$$

$$GSSG + 2\ Pr\text{-}SH \leftrightarrow 2\ GSH + Pr\text{-}SS\text{-}Pr \qquad (13)$$

Oxygen, as a natural electron acceptor, is a reactive molecule altering redox states of cellular compartments. The cellular antioxidant defense system maintains reduced redox environment *via* modifying ROS to less harmless molecules [93]. Glutathione (GSH), being cosubstrate of GSH peroxidase (EC 1.11.1.9), takes part in elimination of hydrogen peroxide (H_2O_2) (14). The formed GSSG can be reduced back to GSH by the NADPH-dependent GSH reductase (EC 1.8.1.7), maintaining the physiological redox environment of the cell (15). This GSH redox cycle is the major mechanism of the maintenance of the cellular GSH homeostasis during oxidant challenge that occurs in the case of peroxide exposure [91].

$$H_2O_2 + 2\ GSH \rightarrow 2\ H_2O + GSSG \qquad (14)$$

$$GSSG + NADPH + H^+ \rightarrow 2\ GSH + NADP^+ \qquad (15)$$

8.4. Oxidation Reactions of GSH and Other Cellular Thiols

Both redox reactions and nucleophilic substitution or addition reactions of GSH are associated with the thiol chemistry of its cysteinyl moiety. They do not necessarily require enzyme catalysis to proceed and can occur in aqueous media under a wide range of conditions.

The cysteine side chain, with its high nucleophilic capacity, appears to be the principal target of reactive oxygen and nitrogen species (ROS/RNS) in cells. In the thiolate form, it can react with H_2O_2 to generate a sulfenic acid (RSOH), and this oxidized form can be considered as a central species among thiol modifications. Reaction of GSH with H_2O_2 results in formation of the respective sulfenic acid (RSOH) [94].

The heme-containing enzymes myeloperoxidase (MPO) and lactoperoxidase (LPO) use H_2O_2 to oxidize halides (Cl^-, Br^- and I^-) and thiocyanate (SCN^-) into their respective hypohalous acids (HOXs). These species are more reactive than H_2O_2 and can chemically modify cysteine and methionine residues of peptides and proteins. Thiols can react with HOCl and HOBr *via* formation of a very unstable sulfenyl halide (16, 17), which readily rearranges to sulfenic acid (RSOH) [95]. Otherwise, HOSCN generates a sulfonyl-thiocyanate (18), which can react with another thiol to yield a disulfide and SCN^- (19) [96]

$$R\text{-}S^- + HOCl = R\text{-}S\text{-}Cl + OH^- \quad (16)$$

$$R\text{-}S^- + HOBr\text{-} = R\text{-}S\text{-}Br + OH^- \quad (17)$$

$$R\text{-}S^- + HOSCN = R\text{-}S\text{-}SCN + OH^- \quad (18)$$

$$R_1\text{-}S^- + R\text{-}S\text{-}SCN = R_1\text{-}S\text{-}S\text{-}R + SCN^- \quad (19)$$

Nitrogen monoxide (NO), one of the most important RNS, is formed in the nitric oxidesynthase(NOS)-catalyzed oxidation of guanidine moiety of L-arginine. [97]. NO such as carbon monoxide (CO) and dihydrogen sulfide (H_2S) is a small endogenous signaling molecule implicated in several biological processes. NO is a free radical with a relatively long half-life [98], which can freely diffuse across the membranes. The primary targets of NO include the soluble guanylyl cyclase, leading to production of cGMP and thus affecting its downstream targets. NO may also activate or inhibit proteins serving as a signal transducer by promoting S-nitrosylation, tyrosine nitration and binding to metal centers in proteins. The nitrosylation species are generated *via* further oxidation of NO. The main nitrosylation species is N_2O_3, of which formation can be explained by the following series of reactions (20-22).

$$2\ NO + O_2 = 2\ NO_2 \quad (20)$$

$$NO_2 + NO = N_2O_3 \quad (21)$$

$$R\text{-}S^- + N_2O_3 = RS\text{-}NO + NO_2^- \quad (22)$$

Reaction of NO with superoxide radical ($O_2{\cdot}^-$) converts the relatively unreactive NO to peroxynitrous acid (ONOOH). The pK_a value of ONOOH is 6.8 [99], thus it is in equilibrium with its deprotonated form under physiological pH. Both forms are strong oxidant and can be involved in both one- or two-electron oxidations. It can be involved in oxidation of thiols to disulfides (RSSR) and sulfenic acid (RSOH) and S-nitrosylation of thiols, involving GSH [100].

Sulfenic acids (RSOH) are directly formed *in vivo* through oxidation of thiols with hydrogen peroxide or peroxynitrous acid, or alkyl peroxides. Several enzymes form -SOH intermediates during their catalytic cycles [94]. Furthermore, RSOH can be formed by hydrolysis of sulfonyl halides [95], sulfonyl thiocyanates [101], thiosulfinates [102] and S-nitrosothiols [103]. Sulfenic acids can react with thiols to form disulfides (23) and the redox reaction represents an important biological process. The reaction allows recycling of sulfenic acids since the formed disulfides can be reduced to thiols through the action of cellular reducing agents, such as glutathione (GSH).

$$R\text{-}SOH + R_1\text{-}SH = R\text{-}S\text{-}S\text{-}R_1 + H_2O \quad (23)$$

Sulfenic acids can be further oxidized to sulfinic acids ($R\text{-}SO_2H$) in the presence of excess of ROS or RNS. Sulfenic acids are relatively stable species with pK_a values around 2. Thus, they can be found in deprotonated forms under physiological pH [104]. In deprotonated form, the sulfinate group (($R\text{-}SO_2^-$) can behave as a soft nucleophile [105] and can react with several electrophilic species such as alkyl halides [106] and α,β-unsaturated aldehydes [107]. Sulfinic acids can be further oxidized to sulfonic acids ($R\text{-}SO_3H$) but cannot be reduced by thiols. Sulfonic acid represents the highest S-oxidized form of cysteine and no biological pathway is known to reduce it [101].

ACKNOWLEDGMENTS

The present contribution is dedicated to the 650th anniversary of the foundation of the University of Pécs, Hungary. The authors express their special thanks to Professor Tamás Lóránd and Associate Professor Gergely Gulyás (University of Pécs, Institute of Biochemistry and Medical Chemistry) for recording and assistance in interpretation of the IR and the NMR spectra, respectively. The authors express their special thanks to Zsuzsanna Moravecz for her skilled assistance in preparation of the manuscript.

REFERENCES

[1] Alanazi, A. M.; Mostafa, G. A. E.; Al-Badr, A. A. Glutathione. In *Profiles of drug substances, excipients, and related methodologies*. Vol. 40.; Academic Press: Burlington, MA, 2015.; pp 43-158.

[2] *European Pharmacopeia*, 9th ed.; EDQM: Strasbourg, France, 2016.

[3] United States Pharmacopoeia, (USP 38, NF 33) *Dietary Supplement Compendium*; The United States Pharmacopoeial Convention: Rockville, MD, 2015.

[4] Downard, K. *Mass spectrometry: A foundation course*, 1st ed.; The Royal Society of Chemistry: Cambridge, UK 2004; pp 22-66.

[5] Skoog, D. A.; Holler, F. J.; Crouch, S. R. *Principles of instrumental analysis*, 7th ed.; Cengage Learning: Boston, MA, 2018; pp. 501-536.

[6] Herbert, C. G.; Johnstone, R. A. W. *Mass spectrometry basics*.; CRC Press: Boca Raton, FL, 2003; pp. 275-286.

[7] De Hoffmann, E.; Stroobant, V. *Mass spectrometry: Principles and applications*, 3rd ed.; Wiley and Sons: Chichester, UK, 2007; pp 85-174.

[8] Brunnée, C. The ideal mass analyzer: Fact or fiction? *Int. J. Mass Spectr. Ion Proc.* 1987, 76, 125–237.

[9] Clark, W. Mass spectrometry in the clinical laboratory: Determining the need and avoiding pitfalls. Chapter 1. pp 1-15. In *Mass spectrometry for the clinical laboratory.*; Nair, H.; Clark. W., Eds.; Academic Press: London, 2017.

[10] Madeira, P. J. A.; Florencio, M. H. Applications of tandem mass spectrometry: From structural analysis to fundamental studies. In *Tandem mass spectrometry: Applications and principles.;* Jeevan P., Ed.; InTech: Rijeka, Croatia, 2012.; Chapter 1. pp 3-32.

[11] Feng, S.; Zheng, X.; Wang, D.; Gong, Y.; Wang, Q.; Deng, H. Systematic analysis of reactivities and fragmentation of glutathione and its isomer GluCysGly. *J. Phys. Chem. A.* 2014, 118, 8222–8228.

[12] Takur, S. S.; Balaram, P. Fragmentation of peptide disulfides under conditions of negative ion mass spectrometry: Studies of oxidized glutathione and contryphan. *J. Am. Mass. Spectrom.* 2008, 19, 358-366.

[13] Skoog, D. A.; Holler, F. J.; Crouch, S. R. *Principles of instrumental analysis*, 7th ed.; Cengage Learning: Boston, MA, 2018; pp. 389-436.

[14] Theophanides T. Introduction to infrared spectroscopy. In *Infrared spectroscopy–Materials science, engineering and technology.;* Theophile T., Ed.; InTech: Rijeka, Croatia, 2012.; Chapter 1. pp 1-11.

[15] Günzler, H.; Gremlich, H.-U. *Infrared spectroscopy: An introduction.* Wiley-VCH: Weinheim, Germany, 2002.

[16] Roeges, N. P. G. *A Guide to the complete interpretation of infrared spectra of organic structures.* John Wiley and Sons: New York, NY, 1994.

[17] Clegg, W., Ed. *Crystal structure analysis: Principle and practices,* 2nd ed. IUCr-Oxford University Press: New York, NY, 2009.

[18] Giacovazzo, C., Ed. *Fundamentals of crystallography,* 3rd ed. IUCr-Oxford University Press: New York, NY, 2011.

[19] Glusker, J. P.; Lewis, M.; Rossi, M. *Crystal structure analysis for chemists and biologists.* Wiley-VCH: New York, NY, 1994.

[20] Hammond, C. *The basics of crystallography and diffraction*, 4th ed. IUCr-Oxford University Press: New York, NY, 2015.

[21] Bruno, I. J.; Cole, J. C.; Kessler, M.; Luo, J.; Motherwell, W. D. S.; Purkis, L. H.; Smith, B. R.; Taylor, R.; Cooper, R. I.; Harris, S. E.; Orpen, A. G. Retrieval of crystallographically-derived molecular geometry information. *J. Chem. Inf. Comput. Sci.* 2004, 44, 2133–2144.

[22] Müller, P.; Irmer, R. H.; Spek, A. L.; Schneider, T. R.; Sawaya, M. R. *Crystal structure refinement: A crystallographer's guide to SHELXL.* IUCr-Oxford University Press: New York, NY, 2006.

[23] Napolitano, H. B.; Camargo, A. J.; Mascarenhas, Y. P.; Vencato, I.; Lariucci, C. Análise da difração dos Raios X. *Revista Processos Químicos* [Analysis of X-ray diffraction. *Chemical Processes Journal*]. Goiânia. 2007, 1, 35-45.

[24] Schwarzenbach, D. *Crystallography.* John Wiley and Sons: Chichester, UK, 1996.

[25] Shmueli, U. *Theories and techniques of crystal structure determination.* IUCr-Oxford University Press: New York, NY, 2007.

[26] Spek, A. L. Structure validation in chemical crystallography. *Acta Cryst.* (2009). D65, 148-155.

[27] Stout, G. H.; Jensen, L. H.; *X-Ray structure determination - A practical guide*, 2nd ed. Wiley-Interscience, New York, NY, 1989.

[28] Sheldrick, G. M. *The SHELX-97 Tutorial.* 1997.

[29] Burla, M. C.; Caliandro, R.; Carrozzini, B.; Cascarano, G. L.; Cuocci, C.; Giacovazzo, C.; Mallamo, M.; Mazzone, A.; Polidori, G. Crystal structure determination and refinement via SIR2014. *J. Appl. Crystallogr.* 2015, 48, 306–309.

[30] Dolomanov, O. V.; Bourhis, L. J.; Gildea, R. J.; Howard, J. A. K.; Puschmann, H. OLEX2: A complete structure solution, refinement and analysis program. *J. Appl. Crystallogr.* 2009, 42, 339–341.

[31] Wright, W. B. The crystal structure of glutathione. *Acta Cryst.* 1958, 11, 632-642.

[32] Moggach, S. A.; Lennie, A. R.; Morrison, C. A.; Richardson, P.; Stefanowicz, F. A.; Warren, J. E. Pressure induced phase transitions in the tripeptide glutathione to 5.24 GPa: The crystal structure of glutathione-II at 2.94 GPa and glutathione-III at 3.70 GPa. *Cryst. Eng. Commun.* 9, 2010.

[33] Owen, T. *Fundamentals of UV-visible spectroscopy. A primer.* Hewlett-Packard Co., Germany, 1996.

[34] Clark, B.J.; Frost, T.; Russell, M.A.; Eds. *UV Spectroscopy. Techniques, instrumentation and data handling.;* Chapman and Hall: London, UK, 1993.

[35] Skoog, D. A.; Holler, F. J.; Crouch, S. R. *Principles of instrumental analysis*, 7th ed.; Cengage Learning: Boston, MA, 2018; pp. 303-360.

[36] High Resolution NMR. *Theory and Chemical Applications.* 3rd Edition. D. Becker, Editor. Academic Press, San Diego, CA, 2000.

[37] Günther, H. *NMR spectroscopy: basic principles, concepts and applications in chemistry*, 3rd ed.; Wiley-VCH: Weinheim, Germany, 2013.

[38] Hoffman, R. E.; Standardization of chemical shifts of TMS and solvent signals in NMR solvents. *Magn. Reson. Chem.* 2006, 44, 606-616.

[39] Karplus, M. Vicinal proton coupling in nuclear magnetic resonance. *J. Am. Chem. Soc.* 1963, 85, 2870–2871.

[40] Sohár, P. *Nuclear magnetic resonance spectroscopy I-III*. CRC Press: Boca Raton, FL, 1983.

[41] Ebbing, D. D.; Gammon, S. D. *General chemistry*. Houghton Mifflin: Boston, MA, 2009.; pp 624-651.

[42] Atkins, P. W.; Overton, T. L.; Rourke, J.; Weller, M. T.; Armstrong, F. A. Shriver and Atkins' *Inorganic Chemistry,* 5th ed.; Oxford University Press: Oxford, UK, 2010.; pp 111-146.

[43] Rajak, S. K., Islam N.; Ghosh, D. C. Modeling of the chemico-physical process of protonation of molecules entailing some quantum chemical descriptors. *J. Quant. Inform. Sci.* 2011, 1, 87–95.

[44] Manallack, D. T. The acid-base profile of a contemporary set of drugs: Implications for drug discovery. *SAR QSAR Environ. Res.* 2009, 7-8, 611-655.

[45] Manallack, D. T.; Prankerd, R. J.; Yuriev, E.; Oprea, T. I.; Chalmers, D. K. The significance of acid/base properties in drug discovery. *Chem. Soc. Rev.* 2013, 42, 485–496.

[46] Charifson, P. S.; Walters, W. P. Acidic and basic drugs in medicinal chemistry: A perspective. *J. Med. Chem.,* 2014, 57, 9701–9717.

[47] Pathare, B.; Tambe, V.; Patil, V. A review on various analytical methods used in determination of dissociation constant. *Intl. J. Pharm. Pharm. Sci.* 2014, 6, 26-34.

[48] Khan, A. S. A. A simple method for analysis of titration curves of polyprotic acids. *Nucleus.* 2014, 51, 448-454.

[49] Sayer, T. L.; Rabenstein, D. L. Nuclear magnetic resonance studies of the acid-base chemistry of amino acids and peptides. 111. Determination of the microscopic and macroscopic acid dissociation constants of α,ω-diaminocarboxylic acids. *Can. J. Chem.* 1976, 54, 3392-3400.

[50] Mirzahosseini, A.; Somlyay, M.; Noszál, B. The comprehensive acid–base characterization of glutathione. *Chem. Phys. Lett.* 2015, 622: 50–56.

[51] Mazák, K.; Noszál, B. Advances in microspeciation of drugs and biomolecules: Species-specific concentrations, acid-base properties and related parameters. *J. Pharm. Biomed. Anal.* 2016, 130, 390-403.

[52] Bjerrum, N. Dissociation constants of polybasic acids and their application to the calculation of molecular dimensions. *Z. Phys. Chem.* 1923, 106, 219–242.

[53] Noszál, B. Group constant: A measure of submolecular basicity. *J. Phys. Chem.* 1986, 90, 4104-4110.

[54] Fujiwara, S.; Ishizuka, H.; Fudano, S. NMR study of amino acids and their derivatives. Dissociation constant of each rotational isomer of amino acids. *Chem. Lett.* 1974, 1281-1284.

[55] Noszál, B.; Scheller-Krattiger, V.; Martin, R. B. A unified view of carbon bound hydrogen exchange of H(2) in imidazoles and H(8) in purine nucleosides and their metal ion complexes. *J. Am. Chem. Soc.* 1982. 104, 1078–1081.

[56] Noszál, B.; Rabenstein, D. L. Nitrogen-protonation microequilibria and C(2)-deprotonation microkinetics of histidine, histamine, and related compounds. *J. Phys. Chem.* 1991, 95, 4761–4765.

[57] Szakács, Z.; Noszál, B. Determination of dissociation constants of folic acid, methotrexate, and other photolabile pteridines by pressure-assisted capillary electrophoresis. *Electrophoresis.* 2006. 27, 3399–3409.

[58] Orgován, G.; Tihanyi, K.; Noszál. B. NMR analysis, protonation equilibria and decomposition kinetics of tolperisone- *J. Pharm. Biomed. Anal.* 2009, 50, 718–723.

[59] Tóth, G.; Baska, F.; Schretner, A.; Rácz, Á.; Noszál. B. Site-specific basicities regulate molecular recognition in receptor binding: In silico docking of thyroid hormones. *Eur. Biophys. J.* 2013, 42, 721–730.

[60] Pirie, N. W.; Pinhey, K. G. The titration curve of glutathione. *J. Biol. Chem.* 1929, 84, 321-333.

[61] Li, N. C.; Gawron, O.; Bascuas, G. Stability of zinc complexes with glutathione and oxidized glutathione. *J. Am. Chem. Soc.* 1954, 76, 225-229.

[62] Martin, R. B.; Edsall, J. T. Glutathione: ionization in basic solutions and molecular rearrangement in strongly acid solution. *Bull. Soc. Chim. Biol.* 1958, 40, 1763-1771.

[63] Noszál, B.; Szakács, Z. Microscopic protonation equilibria of oxidized glutathione. *J. Phys. Chem. B.* 2003, 107, 5074–5080.

[64] Dorcák, V.; Kręzel, A. Correlation of acid–base chemistry of phytochelatin PC2 with its coordination properties towards the toxic metal ion Cd(II). *Dalton Trans.* 2003, 11, 2253–2259.

[65] Wang, X.; Li, K.; Yang, X. D.; Wang, L. L.; Shen, R. F. Complexation of Al(III) with reduced glutathione in acidic aqueous solutions. *J. Inorg. Biochem.* 2009, 103, 657–66.

[66] Mah, V.; Jalilehvand, F. Mercury(II) complex formation with glutathione in alkaline aqueous solution. *J. Biol. Inorg. Chem.* 2008, 13, 541–553.

[67] Madej, E.; Wardman, P. The oxidizing power of the glutathione thiyl radical as measured by its electrode potential at physiological pH. *Arch. Biochem. Biophys.* 2007, 462, 94–102.

[68] Gough, J. D.; Lees, W. J. Effects of redox buffer properties on the folding of a disulfide-containing protein: Dependence upon pH, thiol pK_a, and thiol concentration. *J. Biotechnol.* 2005, 115, 279–290.

[69] Cigala, R. M.; Crea, F.; De Stefano, C.; Lando, G.; Milea, D.; Sammartano, S. Modeling the acid–base properties of glutathione in different ionic media, with particular reference to natural waters and biological fluids. *Amino Acids.* 2012, 43, 629–648.

[70] Atkins, P.; De Paula, J. Atkins' *Physical Chemistry*, 9th ed.; Oxford University Press: Oxford, UK, 2010.; pp 156-245.

[71] Willis, K. K.; Weaver, K. H.; Rabenstein, D. L. Oxidation/reduction potential of glutathione. *J. Org. Chem.* 1993, 58, 4144-4146.

[72] Alberty, R. A. *Thermodynamics of biochemical reactions*; Wiley-Interscience: Hoboken, NJ, 2003.

[73] Alberty, R. A. *Biochemical thermodynamics: Applications of Mathematica.;* Wiley-Interscience: Hoboken, NJ, 2006.

[74] Alberty, R. A. Calculation of standard transformed formation properties of biochemical reactants and standard apparent reduction potentials of half reactions. *Arch. Biochem. Biophys.* 1998, 358, 25-39.

[75] Alberty, R. A. Standard apparent reduction potentials of biochemical half reactions and thermodynamic data on the species involved. *Biophys. Chem.* 2004, 111, 115-122.

[76] Mirzahosseini, A.; Noszál, B. Species-specific standard redox potential of thiol-disulfide systems: A key parameter to develop agents against oxidative stress. *Sci. Rep.* 2016, 6, 37596.

[77] Meng, D.; Zhang, P.; Zhang, L, Wang, H.; Ho, C.-T.; Li, S.; Shahidi, F.; Zhao, H. Detection of cellular redox reactions and antioxidant activity assays. *J. Funct. Foods.* 2017, 37, 467–479.

[78] Fraternale, A.; Paoletti, M. F.; Casabianca, A.; Nencioni, L.; Garaci, E.; Palamara, A. T.; Magnani, M. GSH and analogs in antiviral therapy. *Mol. Asp. Med.* 2009, 30, 99–110.

[79] Lu, S. C. Glutathione synthesis. *Biochim. Biophys. Acta.* 2013, 1830, 3143–3153.

[80] Kojer, K.; Bien, M.; Gangel, H.; Morgan, B.; Dick, T. P, Riemer, J. Glutathione redox potential in the mitochondrial intermembrane space is linked to the cytosol and impacts the Mia40 redox state. *EMBO J.* 2012, 31, 3169–82.

[81] Dalton, T. P.; Chen, Y.; Schneider S. N.; Nebert, D. W.; Shertzer, H. G. Genetically altered mice to evaluate glutathione homeostasis in health and disease. *Free Rad. Biol. Med.* 2004, 37, 1511–1526.

[82] Jones, D. P. Redox potential of GSH/GSSG couple: Assay and biological significance; In *Methods in Enzymology,* Vol. 348; Sies H.; Lester Packer, L. Eds.; Academic Press: San Diego CA, 2002; pp 93-112.

[83] Kalinina, E. V.; Chernov, N. N.; Novichkova, M. D. Role of glutathione, glutathione transferase, and glutaredoxin in regulation of redox-dependent processes. *Biochemistry* (Moscow) 2014, 79, 1562–1583.

[84] Giustarini, D.; Colombo, G.; Garavaglia, M. L.; Astori, E.; Portinaro, N. M.; Reggiani F.; Badalamenti, S.; Aloisi, A. M.; Santucci, A.; Rossi, R.; Milzani, A.; Dalle-Donne, I. Assessment of glutathione/glutathione disulphide ratio and S-glutathionylated proteins in human blood, solid tissues, and cultured cells. *Free Rad. Biol. Med.* 2017, 112, 360–375.

[85] Schafer, F. Q.; Garry R. Buettner, G. R. Redox Environment of the cell as viewed through the redox state of the glutathione disulfide/glutathione couple. *Free Rad. Biol. Med.* 2001, 30, 1191–1212.

[86] Deponte M. Glutathione catalysis and the reaction mechanisms of glutathione-dependent enzymes. *Biochim. Biophys. Acta.* 2013, 1830, 3217–3266.

[87] Moran, L. K.; Guttereidge, J. M. C.; Quinlan, G. J. Thiols in cellular redox signalling and control. *Curr. Med. Chem.* 2001, 8, 763 – 772.

[88] Owen, J. B.; Butterfield, D. A. Measurement of oxidized/reduced glutathione ratio. *Methods Mol. Biol.* 2010, 648, 269–277.

[89] Colombo, G.; Dalle-Donne, I.; Giustarini D.; Gagliano, N.; Portinaro, N.; Colombo, R.; Rossi, R.; Milzani, A. cellular redox potential and hemoglobin S-glutathionylation in human and rat erythrocytes: A comparative study. *Blood Cells Mol. Dis.* 2010, 44, 133–39.

[90] López-Mirabal, H. R.; Winther, J. R. Redox characteristics of the eukaryotic cytosol. *Biochim. Biophys. Acta – Mol. Cell Res.* 2008, 783, 629–640.

[91] Aw, T. Y. Cellular redox: A modulator of intestinal epithelial cell proliferation. *Physiology* 2003, 18, 201–204.

[92] Dalle-Donne, I.; Rossi, R., Giustarini, D., Colombo, R.; Milzani, A. S-glutathionylation in protein redox regulation. *Free Rad. Biol. Med.* 2007, 43, 883-898.

[93] Aon, M. A., Cortassa, S.; Maack, C.; O'Rourke, B. Sequential opening of mitochondrial ion channels as a function of glutathione redox thiol status. *J. Biol. Chem.* 2007, 282, 21889–21900.

[94] Lim, J. C.; You, Z.; Kim, G.; Levine, R. L. Methionine sulfoxide reductase A is a stereospecific methionine oxidase. *Proc. Natl. Acad. Sci. USA*, 2011, 108, 10472-10477.

[95] Nagy, P., Ashby, M. T. Reactive sulfur species: Kinetics and mechanisms of the oxidation of cysteine by hypohalous acid to give cysteine sulfenic acid. *J. Am. Chem. Soc.* 2007, 129, 14082-14091.

[96] Ashby, M. T.; Aneetha, H. Reactive sulfur species: Aqueous chemistry of sulfenyl thiocyanates. *J. Am. Chem. Soc.* 2004, 126, 10216-10217.

[97] Alderdon, W. K.; Cooper, C. E.; Knowles, R. G. Nitric oxide synthases: structure, function and inhibition. *Biochem. J.* 2001, 357, 593-615.

[98] Pacher, P.; Beckman, J. S.; Liaudet, L. Nitric oxide and peroxynitrite in health and disease. *Physiol. Rev.* 2007, 87, 315-424.

[99] Kissner, R.; Nauser, T.; Bugnon, P.; Lye, P. G.; Koppenol, W. H. Formation and properties of peroxynitrite as studied by laser flash photolysis, high-pressure stopped-flow technique, and pulse radiolysis. *Chem. Res. Toxicol.* 1997, 10, 1285-1292.

[100] Van der Vliet, A.; Hristova, M.; Cross, C. E.; Eiserich, J. P.; Goldkorn, T. Peroxynitrrite induces covalent dimerization of epidermal growth factor receptors in A431 epidermoid carcinoma cells. *J. Biol. Chem.* 1998, 273, 31860-31866.

[101] Lo Conte, M.; Carroll, K. S. The chemistry of thiol oxidation and detection. In *Oxidative stress and redox regulation*; Jakob, U.; Reichmann, D., Eds.; Springer: Dordrecht, 2013; pp 1-42.

[102] Nagy, P.; Ashby, M. T. Reactive sulfur species: Kinetics and mechanisms of the oxidation of cysteine by hypohalous acid to give cysteine sulfenic acid. *J. Am. Chem. Soc.* 2007, 129, 14082-14091.

[103] Percival, M. D.; Ouellet, M.; Campagnolo, C.; Claveau, D.; Li, C. Inhibition of cathepsin K by nitric oxide donors: evidence for the formation of mixed disulfides and a sulfenic acid. *Biochemistry.* 1999, 38, 13574-13583.

[104] Burkhard, R. K.; Sellers, D. E.; DeCou, F.; Lambert, J. L. The pKa's of aromatic sulfinic acids. *J. Org. Chem.* 1959, 24, 767-769.

[105] Reddie, K. G.; Carroll, K. S. Expanding the functional diversity of proteins: through cysteine oxidation. *Curr. Opin. Chem. Biol.* 2008, 12, 746-754.

[106] Cremlyn, R. J. *An Introduction to Organosulfur Chemistry.* John Wiley and Sons: Chichester, UK, 1996.; pp 41-62.

[107] Ogata, Y.; Sawaki, Y.; Isono, M. Kinetics of the addition of substituted benzenesulfinic acids to *p*-benzoquinone. *Tetrahedron.* 1970, 26, 731-736.

BIOGRAPHICAL SKETCHES

Pál Perjési

Affiliation

Institute of Pharmaceutical Chemistry, University of Pécs, Pécs, Hungary

Education

1979: Pharmacy, MS, Albert Szent-Györgyi University, Szeged, Hungary;

1983: Pharmacy, University Doctor (DrUniv), Albert Szent-Györgyi University, Szeged, Hungary;

1994: Chemistry: Candidate of Sciences (CSc), Hungarian Academy of Sciences, Budapest, Hungary;

1996: Chemistry: PhD, Albert Szent-Györgyi University, Szeged, Hungary;

2000: Chemistry: Habilitation, University of Pécs, Medical School, Pécs, Hungary

Research and Professional Experience

1979-1981: Institute of Pharmaceutical Chemistry, Albert Szent-Györgyi University, Szeged, Hungary;

1981-2001: Institute of Medical Chemistry, University of Pécs, Medical School, Pécs, Hungary;

2002- Institute of Pharmaceutical Chemistry, University of Pécs, Faculty of Medicine, Pécs, Hungary;

2016- Institute of Pharmaceutical Chemistry, University of Pécs, Faculty of Pharmacy, Pécs, Hungary

Professional Appointments

2002: Head of department, Institute of Pharmaceutical Chemistry, University of Pécs, Pécs, Hungary;

2016: Dean, University of Pécs, Faculty of Pharmacy, Pécs, Hungary;

2018: Editor-in-Chief, Journal of Pharmaceutical and Biopharmaceutical Research, SyncSci Publishing, Singapore

Honors

1990-1991: Fogarty, International Research Fellowship, National Institutes of Health; Department of Medicinal Chemistry, University of Washington, Seattle, WA, USA;

2001: Postdoctoral Fellowship, Center for Drug Discovery, University of Florida, Gainesville, FL, USA;

2003: Postdoctoral Fellowship, Department of Medicinal Chemistry, University of Florida, Gainesville, FL, USA;

2012: Postdoctoral Fellowship, German Academic Exchange Service (DAAD); Kekule Institute, University of Bonn, Bonn, Germany

Publications from the Last 3 Years (selection):

1. Almasi, A.; Pinto, E. D. L. N.; Kovacs, N. P.; Fischer, T.; Markovics, Z.; Fischer, E.; Perjési, P. Changes in hepatic metabolic enzyme activities and biliary excretion of 4-nitrophenol in streptozotocin induced diabetic rats. *Braz. J. Pharm. Sci.* 2018, 54, e17347.
2. Bernardes, A.; D'Oliveira, G. D. C.; Silezin, A.; Kuzma, M.; Molnar, S.; Noda-Perez, C.; Perjési, P. Reagent-induced asymmetric induction in addition reaction of reduced glutathione onto bis-Mannich chalcones. *Archiv Pharm.* 2018, 351, e1700386.
3. Kuzma, M.; Kovács, N.; Sziva, L.; Maasz, G.; Avar, P.; Perjési, P. Oxidation of hydroxy- and dihydroxybenzoic acids under the Udenfriend's conditions. An HPLC study. *Open Med. Chem. J.* 2018, 12, 13-22.
4. Rozmer, Z.; Marton, E.; Perjési, P. (*E*)-2-Benzylidenecyclanones: part XIV. Study on interaction of some (*E*)-2-benzylidenebenzosuberone derivatives with calf thymus

DNA by TLC and UV–Vis methods, a DNA cleavage study. *Med. Chem. Res*. 2017, 26, 2172-2179.
5. Rozmer Zs, Perjési P. Naturally occurring chalcones and their biological activities. *Phytochem. Rev*. 2016, 15, 87-120.
(For detailed information on list of publications, please address to https:/m2.mtmt.hu/gui2/?type=authors&mode=browse&sel=10008303)

Caridad Noda Perez

Present Position

Professor

Affiliation

Chemical Institute, Federal University of Goiás, Goiânia, Goiás, Brazil

Education

Graduated in Chemical Engineering in 1983, Master's (MSc) degree (1997) and PhD (2001) in Chemical Engineering.

Research and Professional Experience

Synthesis of chalcones and their derivatives. Synthesis of Mannich bases derived from chalcones. Physic-chemical characterization of synthesized compounds. Evaluation of antitumor activity of chalcones and their derivatives. Study of genotoxic and antigenotoxic activities of chalcones and derivatives. Study of basic solid catalysts for the synthesis of chalcones and their derivatives.

Short Biography

Graduated in Chemical Engineering from Instituto Superior Politécnico José Antonio Echevarria (1983), in Havana, Cuba. Master's (MSc) degree (1997) and PhD (2001) in Chemical Engineering at COPPE by Federal University of Rio de Janeiro. Currently is professor of the Graduate Programs in Chemical Engineering and Chemistry and lecturer in the Chemical Engineering course at the Federal University of Goias. Has experience in kinetics and heterogeneous catalysis. Research in the area of basic solid catalysts for the condensation reactions of aldehydes and ketones. The main area of research is synthesis, characterization and antitumor evaluation of chalcones and their derivatives. Within this line of research she works on Michael addition reactions of glutathione and chalcones and their derivatives.

Publications from the Last 3 Years

1 Michelini, L. J.; Castro, M. R. C.; Custodio, J. M. F.; Naves, L. F. N.; Vaz, W. F.; Lobón, G. S.; Martins, F. T.; Perez, C. N.; Napolitano, H. B. A Novel Potential Anticancer Chalcone: Synthesis, Crystal Structure and Cytotoxic Assay. *J. Mol. Struct.* 2018, 1168, 309–315.

2 Lemes, S. R.; eSilva, C. R.; Véras, J. H.; Chen-Chen, L.; Lima, R. S.; Perez, C. N.; Sousa, M. A. M. de; Reis, P. R. de M.; Junior, N. J. da S. Presence of antigenotoxic and anticytotoxic effects of the chalcone 1*E*,4*E*-1-(4-chlorophenyl)-5-(2,6,6-trimethylcyclohexen-1-yl)penta-1,4-dien-3-one using in vitro and in vivo assays. *Drug Chem. Toxicol.* 2018. https://doi.org/10.1080/01480545.2018.1497046.

3 Bernardes, A.; d'Oliveira, G. D. C.; Silezin, A.; Kuzma, M.; Molnár, S.; Pérez, C. N.; Perjési, P. Reagent-induced asymmetric induction in addition reaction of reduced glutathione onto bis-Mannich chalcones. *Arch. Pharm.* 2018, 351, 1700386.

4 Custodio, J. M. F.; Michelini, L. J.; Castro, M. R. C. de; Vaz, W. F.; Neves, B. J.; Cravo, P. V. L.; Barreto, F. S.; Filho, M. O. M.; Perez, C. N.; Napolitano, H. B. Structural insights into a novel anticancer sulfonamide chalcone. *New J. Chem.* 2018, 42, 3426–3434.

5 d'Oliveira, G.; Moura, A.; de Moraes, M.; Perez, C.; Lião, L. Synthesis, characterization and evaluation of in vitro antitumor activities of novel chalcone-quinolinone hybrid compounds. *J. Brazil. Chem. Soc.* 2018, 29, 2308-2325.

6 da Silva Lima, D. C.; do Vale, C. R.; Véras, J. H.; Bernardes, A.; Pérez, C. N.; Chen-Chen, L. Absence of genotoxic effects of the chalcone (*E*)-1-(2-hydroxyphenyl)-3-(4-methylphenyl)-prop-2-en-1-one) and its potential chemoprevention against DNA damage using in vitro and in vivo assays. *PloS one*. 2017, 12, e01712247.

7. Carneiro, C. C.; Véras, J. H.; Góes, B. R. L.; Pérez, C. N.; Chen-Chen, L.; Carneiro, C. C.; Véras, J. H.; Góes, B. R. L.; Pérez, C. N.; Chen-Chen, L. Mutagenicity and antimutagenicity of *Salacia Crassifolia* (Mart. Ex. Schult.) *G. Don.* evaluated by Ames test. *Brazil. J. Biol.* 2018, 78, 345–350.

8. Damasceno, C. T.; Jordao, G. R.; Noda-Perez, C. Simulação de uma unidade piloto de destilação de água para uso em laboratórios químicos empregando energia solar. *Rev. Proc. Quim.*, 2017, 22, 11-24.

9. de Castro, M. R.; Naves, R. F.; Bernardes, A.; da Silva, C. C.; Perez, C. N.; Moura, A. F.; Moraes, M. O.; Martins, F. T. Tandem chalcone-sulfonamide hybridization, cyclization and further Claisen–Schmidt condensation: Tuning molecular diversity through reaction time and order and catalyst. *Arab. J. Chem.* 2017. https://doi.org/10.1016/j.arabjc.2017.11.005

10. Winter, C.; Caetano, J. N.; Araújo, A. B. C.; Chaves, A. R.; Ostroski, I. C.; Vaz, B. G.; Pérez, C. N.; Alonso, C. G. Activated carbons for chalcone production: Claisen-Schmidt condensation reaction. *Chem. Engin. J.* 2016, 303, 604–610.

11. Lima, R. S.; Perez, C. N.; da Silva, C. C.; Santana, M. J.; Queiroz Júnior, L. H. K.; Barreto, S.; de Moraes, M. O.; Martins, F. T. Structure and cytotoxic activity of terpenoid-like chalcones. *Arab. J. Chem.* 2016. https://doi.org/10.1016/ j.arabjc.2016.02.013.

12. Bernardes, A.; Pérez, C.; Mayer, M.; da Silva, C.; Martins, F.; Perjési, P. Study of reactions of two Mannich bases derived of 4'-hydroxychalcones with glutathione by RP-TLC, RP-HPLC and RP-HPLC-ESI-MS analysis. *J. Brazil. Chem. Soc.* 2017, 28, 1048-1062.

Hamilton Barbosa Napolitano

Present Position

Associate Professor, State University of Goias, Goiana, Goias, Brazil

Affiliation

State University of Goias, Goiania, Goias, Brazil

Education

Ph.D. in Applied Physics (Crystallography) in 2004 at São Paulo University, Sao Paulo, Brazil

Research and Professional Experience

Crystallography of small organic molecules such as chalcones and their derivatives. Solid state Studies on supramolecular arrangements and intermolecular interactions.

Short Biography

Graduated in Biomolecular Physics at São Carlos Physics Institute of the University of São Paulo. Postdoctoral training at Hauptman-Woodward Medical Research Institute of the State University of New York at Buffalo. Served for two consecutive terms (2009-2015) as a Member of the Superior Council of the Foundation for Research Support of the State of Goias and serve as a member of the Mathematical and Theoretical Crystallography (MaThCryst) committee of the International Union of Crystallography (2017-2020). Has experience on molecular crystals engineering of drugs and biological inhibitors and works in several technological development projects in partnership with the industrial production sector.

Professional Appointments

Principal Research Associate of the Structural and Theoretical Chemistry Research Group of the State University of Goias

Publications from the Last 3 Years

1. Custodio, J. M. F.; Santos, F. G.; Vaz, W. F.; Cunha, C. E. P.; Silveira, R. G.; Anjos, M. M.; Campos, C. E. M.; Oliviera, G. R.; Martins, F. T.; Da Silva, C. C.; Valverde, C.; Baseia, B.; Napolitano, H. B. Molecular structure of hybrid imino-chalcone in the

solid state: X-ray diffraction, spectroscopy study and third-order nonlinear optical properties. *J. Mol. Struct.* 2018, 1157, 210-221.

2. Custodio, J.; Michelini, L.; Castro, M.; Fonseca Vaz, Wneves, B. Jr.; Cravo, Pbarreto, F.; De Moraes, M. O.; Perez, C.; Napolitano, H. B. Structural Insights on a Novel Anticancer Sulfonamide Chalcone. *New J. Chem.* 2018, 42, 3426-3434
3. Michelini, L. J.; Castro M. R. C.; Custodio, J. M. F.; Nunes, L. F.; Vaz, W. F.; Lobón, G. S.; Martins, F. T.; Perez, C. N.; Napilotano, H. B. A novel potential anticancer chalcone: Synthesis, crystal structure and cytotoxic assay. *J. Mol. Struct.* 2018, doi: 10.1016/j.molstruc.2018.05.010.
4. Custodio, J. M. F.; Vaz, W. F.; De Castro, M. R. C.;Bernardes, A.; Naves, R. F.; Moura, A. F.; De M. Filho, M. O.; Da Silva, C. C.; Martins, F. T.; Perez, C. N.; Napolitano, H. B. Solvent-driven structural adaptation in a novel anticancer sulfonamide chalcone. *J. Mol. Struct.* 2018, 1175, 389-397.
5. Oliveira, S. S.; Santin, L. G.; Almeida, L. R.; Malaspina, L. A.; Lariucci, C.; Silva, J. F.; Fernandes, W. B.; Aquino, G. L. B.; Gargano, R.; Camargo, A. J.; Napolitano, H. B. Synthesis, characterization, and computational study of the supramolecular arrangement of a novel cinnamic acid derivative. *J. Mol. Modeling.* 2017, 23, 1-12.
6. Custodio, J. M. F.; Vaz, W. F.; De Andrade, F. M; Camargo, A. J.; Oliveira, G. R.; Napolitano, H. B. Substitution effect on a hydroxylated chalcone: Conformational, topological and theoretical studies. *J. Mol. Struct.* 2017, 1136, 69-79.
7. Carvalho, P. S.; Custodio, J. M. F.; Vaz, W. F.; Cirilo, C. C.;Cidade, A. F.; Aquino, G. L. B.; Campos, D. M. B.; Cravo, P.; Coelho, C. J.; Oliveira, S. S.; Camargo, A. J.; Napolitano, H. B. Conformation analysis of a novel fluorinated chalcone. *J. Mol. Modeling.* 2017, 23, 97.
8. Custodio, J.; Faria, E.; Sallum, L.; Duarte, V.; Vaz, W.;De, Aquino, G.; Carvalho Jr., P.; Napolitano, H. The influence of methoxy and ethoxy groups on supramolecular arrangement of two methoxy-chalcones. *J. Brazil. Chem. Soc.* 2017, 28, 2180-2191.
9. Rodrigues, R. F. N.; Almeida, L. R.; Santos, F. G. D.; Carvalho, P. S.; Souza, W. C. De.; Moreira, K. S.; Aquino, G. L. B.; Valverde, C.; Napolitano, H. B.; Baseia, B. Solid state characterization and theoretical study of non-linear optical properties of a fluoro-N-acylhydrazide derivative. *PLoS One.* 2017, 12, e0175859.
10. Castro, A. N.; Osório, F. A. P.; Ternavisk, R. R.; Napolitano, H. B.; Valverde, C.; Baseia, B. Theoretical investigations of nonlinear optical properties of two crystalline acetamides structures including polarization effects of their environment. *Chem. Phys. Lett.* 2017, 681, 110-123.
11. Custodio, J.; Moreira, C. C.; Valverde, C.; De Aquino, G.; Baseia, B.; Napolitano, H. Hirshfeld surfaces and nonlinear optics on two conformers of a heterocyclic chalcone. *J. Brazil. Chem. Soc.* 2018, 29, 258-268.
12. Valverde, C.; Fonseca Vaz, W.; Custódio, J.; Duarte, V.; Carvalho Jr., P. S.; Figueredo, A.; Aquino, G. L. B.; Baseia, B.; Napolitano, H. Solid state structure and

environmental polarization effect of a novel asymmetric azine. *New J. Chem.* 2017, 41, 11361-11371.
13. Almeida, L. R.; Carvalho Jr. P. S.; Napolitano, H. B.; Oliveira, S. S.; Camargo, A. J.; Figueredo, A.; Aquino, G. L. B.; Carvalho-Silva, V. H. The contribution of directional dihydrogen interactions in the supramolecular assembly of single crystals: Quantum chemical and structural investigation of $C_{17}H_{17}N_3O_2$ azine. *Cryst. Growth Des.* 2017, 17, 5145.
14. Vaz, W. F.; Custodio, J. M. F.; Rodrigues, N. M. N.; Santin, L. G.; Oliveira, S. S.; Gargano, R.; Osório, F. A. P.; Aquino, G. L. B.; Camargo, A. J.; Oliveira, M. S.; Napolitano, H. B. A novel dihydrocoumarin under experimental and theoretical characterization. *J. Mol. Modeling.* 2017, 23, 315.
15. Almeida, L.; Anjos, M.; Ribeiro, G.; Valverde, C.; Machado, D.; Oliveira, G.; Napolitano, H.; Oliveira, H. C. B. Synthesis, structural characterization and computational study of a novel amino chalcone: a potential nonlinear optical material. *New J. Chem.* 2017, 41, 1744-1754.
16. Castro, A. N.; Almeida, L. R.; Anjos, M. M.; Oliveira, G. R.; Napolitano, H B.; Valverde, C.; Baseia, B. Theoretical study on the third-order nonlinear optical properties and structural characterization of 3-acetyl-6-bromocoumarin. *Chem. Phys. Lett.* 2016, 653, 122-130.
17. Carvalho, P. S.; Almeida, L. R.; Araújo Neto, J. H.; Medina, A. C. Q. D.; Menezes, A. C. S.; Sousa, J. E. F.; Oliveira, S. S.; Camargo, A. J.; Napolitano, H. B. Structural and theoretical investigation of anhydrous 3,4,5-triacetoxybenzoic acid. *Plos One.* 2016, 11, e0158029.
18. Vaz, W. F.; Custodio, J. M. F.; Silveira, R. G.; Castro, A. N.; Campos, C. E. M.; Anjos, M. M.; Oliveira, G. R.; Valverde, C.; Baseia, B.; Napolitano, H. B. Synthesis, characterization, and third-order nonlinear optical properties of a new neolignane analog. *RSC Adv.* 2016, 6, 79215-79227.
19. Ribeiro, G. C.; Almeida, L. R.;Napolitano, H. B.; Valverde, C.; Baseia, B. Polarization effects on the third-order nonlinear optical properties of two polymorphs of enamine derivative. *Theor. Chem. Acc.* 2016, 135, 244.
20. Cravo, P.; Napolitano, H.; Culleton, R. How genomics is contributing to the fight against artemisinin-resistant malaria parasites. *Acta Tropica.* 2015, 148, 1-7.

Hawsar Othman Mohammed

Present Position

PhD Student, University of Pécs, Institute of Pharmaceutical Chemistry, University of Pécs, Pécs, Hungary.

Affiliation

Department of Pharmaceutical Chemistry, Doctoral School of the University of Pécs, Pécs, Hungary

Department of Basic Sciences, College of Veterinary Medicine, University of Sulaimani, Iraq

Education

B.V.M&S, MSc - College of Veterinary Medicine/ University of Sulaimani, Iraq

Research and Professional Experience

Plant and drug toxicity, Lipid peroxidation, Diabetes

Short Biography

Graduated in 2009 with a bachelor degree (B.V.M&S) from the College of Veterinary Medicine/ University of Sulaimani, and had been appointed as a demonstrator at the same place for three years. In 2012, I got a scholarship to study abroad and had a chance to study in the United Kingdom. Hence, in 2014, I graduated from the University of Birmingham with a master degree in Toxicology. From 2015, I have been appointed as assistant lecture at the College of Veterinary Medicine/ University of Sulaimani until 2018. Currently, I am a PhD student at the Department of Pharmaceutical Chemistry of the Doctoral School of the University of Pecs, Pécs, Hungary.

Publications from the Last 3 Years

1. Phytochemical analysis and augmentation of pulmonary adenocarcinoma in BALB/c mice model treated with flaxseed oil (Accepted conference paper 2018).
2. Acute toxicity Study of Acorns (Quercus nut) in Sprague Dawely Rat Model (In progress).
3. The resistant of sheep liver flukes to Rafoxanide, Trichlabendazole and Closantel in Sharazour district (In progress).

In: Glutathione
Editor: Pál Perjési

ISBN: 978-1-53614-740-7
© 2019 Nova Science Publishers, Inc.

Chapter 2

Glutathione and Related Compounds Produced by Glutathione-Synthesizing Pathway

Junichi Fujii**, Sho Kobayashi and Takujiro Homma
Department of Biochemistry and Molecular Biology,
Graduate School of Medical Science, Yamagata University,
Yamagata, Japan

Abstract

Glutathione (GSH) plays pleiotropic functions, as an antioxidant by donating electrons to glutathione peroxidases and in the detoxification of toxic compounds by forming a GSH conjugate. Two ATP-requiring enzymes, γ-glutamylcysteine synthetase (γGCS) and glutathione synthetase (GSS) are involved in the synthesis of GSH. γGCS catalyzes the first, slow step to produce γ-glutamylcysteine from glutamate (Glu) and cysteine (Cys), and GSS produces GSH by adding a glycine (Gly) unit. Among the three constituent amino acids, cells maintain relatively low Cys levels, which, together with feedback inhibition of γGCS, limits GSH production. GSH synthesis is activated in situations where GSH is being consumed or the production of γGCS is induced, such as under oxidative stress. As a cellular nutrient, Cys, the reduced form, is taken up *via* neutral amino acid transporters, while the oxidized form of Cys, referred to as cystine, is taken up *via* systems x_c^- or $b^{0,+}$. Nutritional deprivation induces autophagy in which cellular constituents are degraded, leading to the production of Cys as well as other amino acids. Cys is also synthesized in conjunction with methionine metabolism by the transsulfuration pathway in which cystathionine β-synthase and cystathionine γ-lyase are participants. The transsulfuration pathway catalyzes the production of cystathionine from homocysteine, followed by the cleavage of cystathionine to Cys and 2-butyric acid with the concomitant release of

* Corresponding Author E-mail: jfujii@med.id.yamagata-u.ac.jp.

ammonia. The resulting 2-butyric acid is then converted to 2-aminobutyric acid by the action of aspartate transaminase and is then utilized to produce ophthalmic acid by the GSH-synthesizing machinery. Ophthalmic acid is produced due to lower substrate specificity of γGCS, although the physiological significance of ophthalmic acid remains unclear. Numerous other γ–glutamyl peptides in addition to ophthalmic acid are synthesized not only by γGCS but also by γ–glutamyltransferase and are regarded as markers for pathological conditions in the liver and other organs. Herein, we provide an overview of the regulation of glutathione synthesis by Cys-supply systems and discuss on the pathway for production of γ–glutamyl peptides.

Keywords: cysteine, cystine, γ-glutamylcysteine synthetase, glutathione synthetase, γ–glutamyltransferase

INTRODUCTION

Cysteine (Cys) is unique, in that it possesses a sulfhydryl group that is involved in reduction-oxidation (redox) reactions in a variety of cellular events. The dimeric form of Cys linked by disulfide bound, referred to as cystine, is formed in the extracellular milieu. While neutral amino acid transporters are involved in the uptake of extracellular Cys, a specific system x_c^- or a Na^+-independent system $b^{0,+}$ is required for the uptake of cystine, which is eventually reduced to Cys within the cell [1]. Cys is also formed *via* the transsulfuration pathway in conjunction with methionine (Met) metabolism [2].

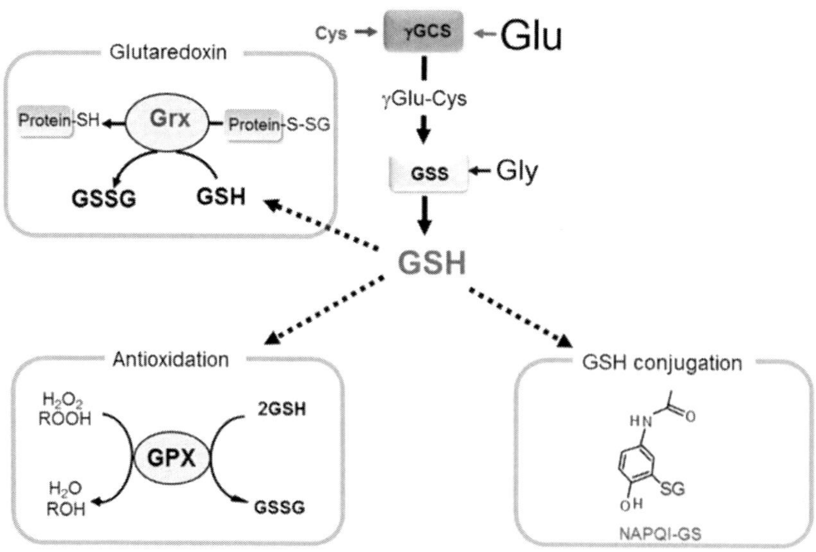

Figure 1. Synthesis and representative functions of GSH. GSH-synthesizing pathway together with three major functions of GSH in the mammalian body are depicted. γGCS, γ-glutamylcysteine synthetase; GSS, glutathione synthetase; Grx, glutaredoxin; GPX, glutathione peroxidase; NAPQI, *N*-acetyl-*p*-benzoquinone imine.

Glutathione (GSH), a tripeptide composed of Cys, Glu, and Gly, is maintained in millimolar levels in most cells and plays pleiotropic functions in antioxidant defense by donating electrons to glutathione peroxidases (GPX), in detoxification of cytotoxic compounds by forming GSH conjugates, and in disulfide exchange by glutaredoxin [3-5] (Figure 1). S-nitrosoglutathione (GSNO) is formed by the interaction of GSH with nitric oxide (NO) and constitutes a major nitrosothiol compound in plasma as well as in cells [6]. While GSNO has vasodilation, cardioprotection, and signal transduction properties *via* transnitrosation, it is also regarded as an effective glutathionylating agent [7, 8]. Oxidized glutathione, commonly designated as GSSG, is either recycled by means of an NADPH-dependent reduction catalyzed by glutathione reductase (GSR) within cells or is exported by the multidrug resistance protein (MRP) [9]. Because the recycling system alone is unable to meet the demand for GSH, GSH levels are largely maintained by its continuous synthesis from constituent amino acids [10]. In animals, GSH in the circulation is rapidly consumed by γ-glutamyltransferase (GGT) that is localized on the luminal surface of the renal brush-border membrane [11].

A decrease in the Cys levels is frequently observed in the plasma of older people or certain types of disease states, which implies that GSH redox system supported by Cys is not well-maintained in elderly people and patients [12]. In this review chapter, we overview the regulatory mechanism for GSH synthesis from the viewpoint of Cys-supply systems, recycling, metabolic use and to its roles in the regulation of ferroptosis. We also discuss on the pathway for production of γ–glutamyl peptides, which are applicable as surrogate markers [13, 14].

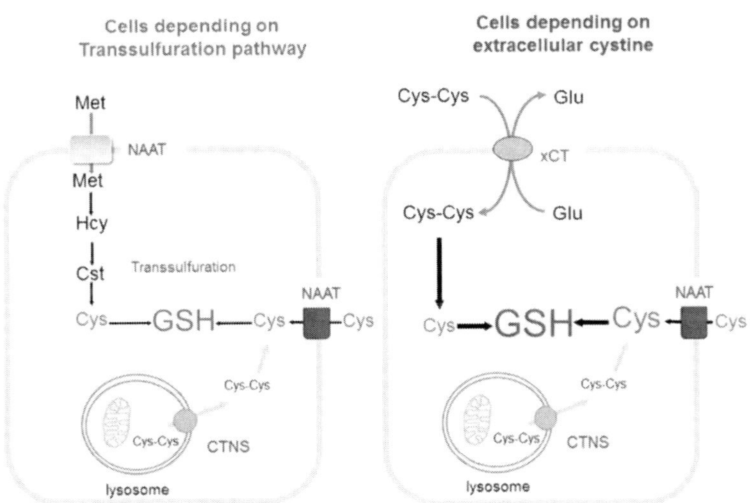

Figure 2. Two types of cells depicting three main pathways for providing Cys for GSH synthesis. Mammalian cells can be classified into two types regarding Cys securement; one type of cells depends on the transsulfuration pathway and the other depends on the availability of extracellular cystine, notably *via* xCT. Cystine that is produced by proteolytic degradation in lysosomes is transported out *via* cystinosine (CTNS). NAAT, a neutral amino acid transporter; Hcy, homocysteine; Cst, cystathionine.

CYS SUPPLY SYSTEMS THAT SUPPORT GSH SYNTHESIS

Cys is a component of GSH and other essential biological molecules such as coenzyme A [15]. Hydrogen sulfide and polysulfides produced from the sulfur in Cys or its precursor also play roles in our body [16-18]. Some intermediary compounds and end products, such as cysteamine [19] and taurine [20], are also produced from Cys and have biological functions. However, the intracellular concentration of Cys is maintained at relatively low levels, compared to other amino acids, partly due to its cytotoxic properties. Several cellular systems function coordinately to supply Cys in the case of need (Figure 2). There are essentially three sources for Cys *in vivo*; free amino acids from extracellular sources, metabolic synthesis from methionine (Met), and amino acid recruitment *via* degradation of proteins [21].

Cys and Cystine Uptake Systems in Plasma Membranes

Cys in its neutral state can be taken up *via* neutral amino acid transporters (NAAT), such as ASCT-1 and -2 and LAT-2, which are located in the plasma membrane [22, 23]. NAATs are responsible for the uptake of not only Cys but also for other neutral amino acids; so that they are expressed constitutively in most cells. The dimeric, oxidized form of the Cys, referred to as cystine, is about six-times more abundant than Cys in extracellular fluids [24]. Cystine is taken up by cells *via* a Na^+-independent system $b^{0,+}$ localized in the kidney or the specific system x_c^- [1, 25]. The dibasic amino acid transporter system $b^{0,+}$ consisting of SLC3A1 and/or SLC7A9, transports cystine, lysine, arginine, and ornithine in renal proximal tubules and the small intestine. A defect in the system $b^{0,+}$ causes cystinuria due to accumulation of cystine and dibasic amino acids in the urine, a condition that frequently leads to the formation of renal stones [26]. System x_c^- consists of two protein components; xCT/Slc7a11 and 4F2hc/SLC3A2. xCT is a principle component of system x_c^- and transports cystine into cells using Glu as the counter amino acid [27]. xCT has also been reported to transport cystathionine [28]. The constitutive expression of xCT is observed in limited organs; namely, the thymus, spleen, and the cerebral meninges, while most other organs do not express xCT [1]. However, xCT is induced in many organs, such as the kidney [29], lung [30], and the liver [31, 32], and serves as a backup system for supplying Cys to the redox system in an emergency, typically under conditions of oxidative stress and GSH depletion.

A phenotype of the xCT-null ($xCT^{-/-}$) mouse is normal under conventional breeding conditions except for a redox imbalance in the blood plasma [33]. Because xCT is expressed at high levels in the brain [34], notably in a subpopulation of astrocytes according to a recent observation [35], the potential roles of xCT in the central nervous system have been extensively studied. The released Glu as the counter amino acid of cystine *via* xCT exerts antithetical effects on surrounding neuronal cells. While Glu is the major excitatory neurotransmitter in the brain, excessive levels of Glu in extracellular space may cause oxidative Glu toxicity in neuronal cells, which is primarily mediated by the N-methyl-D-aspartate (NMDA) receptor [36]. However, $xCT^{-/-}$ mice show neither changes in GSH content, oxidative stress, nor neuronal dysfunctions; but the levels of extracellular Glu in the hippocampus are decreased, which influences spatial working memory and causes susceptibility to limbic seizures [37]. Excessive Glu activates the NMDA receptor, which then causes an intense transient influx of Ca^{2+}, resulting in the release of cytochrome c, depletion of ATP, the production of reactive oxygen species (ROS), and consequently cell death [38]. Meanwhile, in dopaminergic neurons, the loss of xCT decreases the level of extracellular Glu, leading to neuroprotection [39]. Despite the absence of any abnormality in motor, sensorimotor function, or basic visual acuity, anxiolytic and antidepressive-like effects have been reported in $xCT^{-/-}$ mice [40]. Precise electrophysiological analyses of $xCT^{-/-}$ mice, however, show an increase in the number of α-amino-3-hydroxy-5-methyl-4-isoxazolepropionic acid (AMPA) receptors, one of the Glu receptor subtypes, in neurons in the CA3-CA1 area in the hippocampus [41]. Both activated and resting microglial cells express xCT and release Glu *via* xCT [42]. Exposure to lipopolysaccharide-activated microglial cells causes oligodendrocyte injury in mixed glial cultures, which is prevented by the blocking of AMPA receptors or xCT [42, 43]. xCT knockout decreases the production of pro-inflammatory/neurotoxic factors, including TNFα and nitric oxide but increases anti-inflammatory/neuroprotective markers such as Ym1/Chil3 in microglia [43]. It thus appears that xCT mediates neurotoxicity moderately by elevating extracellular Glu levels in the brain. An xCT deficiency may cause abnormal sexual behavior in male mice, although they are apparently fertile, *via* these influences on the central nervous system [44]. In the case of tumor cells, however, the stable expression of xCT appears to promote tumor growth by elevating the redox potential in *in vivo* situations [45].

Unlike *in vivo* conditions, the function of xCT appears to be required in supplying Cys to cells in culture systems [33]. Cys in the culture medium is spontaneously oxidized to cystine by atmospheric oxygen under conventional culture conditions and is taken up by cells solely *via* xCT. Hepatocytes produce a large fraction of the GSH and provide it, together with multiple nutritional components, to peripheral tissues *via* the blood, but they do not express xCT *in vivo* [28, 33]. The redox status of hepatocytes undergoes tremendous environmental change after isolation and, as a result, can meet the demand for Cys under primary culture conditions. Comparative studies using mouse primary hepatocytes and mouse hepatoma-derived line Hepa 1-6 cells indicate that $xCT^{-/-}$ hepatocytes can survive normally in Cys-free media for at least a few days because Cys is mainly supplied by the transsulfuration pathway. Hepa 1-6 cells, however, largely depend on xCT for their survival and hence undergo ferroptosis in Cys-free media [46].

Cys Formation *via* the Transsulfuration Pathway in Conjunction with Met Metabolism

The amino acids Met and Cys both contain sulfur atoms among the twenty amino acids encoded by genetic codons. While Met is an essential amino acid, Cys is synthesized by the transsulfuration pathway in conjunction with Met metabolism. During the metabolic formation of Met, other important metabolic intermediates are also produced. Among them, S-adenosylmethionine functions as a methyl group donor required for nucleic acid synthesis and protein methylation in the epigenetic process [47]. Cystathionine β-synthase (CBS) and cystathionine γ-lyase (CSE) both participate in the transsulfuration pathway [2]. CBS catalyzes the production of cystathionine from homocysteine and serine. CSE then catalyzes the cleavage of cystathionine to Cys and 2-oxobutyric acid (2OB) with the concomitant release of ammonia [48]. The transsulfuration pathway plays as a prerequisite in providing Cys in many organs, most notably the liver [2]. Moreover, both CBS and CSE are also known to produce hydrogen sulfide (H_2S), a gaseous transmitter molecule, and reactive persulfides and polysulfides [17, 18].

Proteins as a Source for Cys

Protein synthesis and degradation are continuously ongoing processes within cells and maintain cellular homeostasis. Aged or damaged proteins are typically poly-ubiquitinated and are then degraded by proteasomes [49]. The resulting amino acids are then recycled for syntheses of proteins and other compounds, including GSH. This recycling system occurs constitutively and is enhanced in response to a variety of stress conditions. However, while this mechanism contributes to the removal of damaged or aged proteins, amino acid supply does not appear to be the main role of the system. Proteins that are incorporated from extracellular fluid *via* endocytosis are largely degraded in lysosomes by means of proteases and can serve as a source for amino acids in metabolism as well as nutrients [50]. The lysosomal-mediated proteolytic degradation of intracellular constituents also provides amino acids including Cys for the synthesis of GSH.

Autophagy, which typically occurs under conditions associated with an insufficient supply of essential amino acids and nutritional starvation, involves inhibition of the mammalian target of rapamycin (mTOR) [21]. Free Cys that is released together with other amino acids from proteins within lysosomes/autophagosomes is oxidized to cystine due to the oxidative microenvironment in the organelle. The resulting cystine is transported to the cytosol *via* cystinosin (CTNS), which is mainly localized on the lysosomal membrane and functions to transport both cystine and protons (H^+) out of the lysosome [51]. A CTNS deficiency results in cystinosis, a rare hereditary disorder caused by accumulation of cystine crystals within lysosomes [52]. Cystine crystals, but not a shortage of Cys, appears to cause the symptoms of cystinosis. Cysteamine forms a mixed disulfide with Cys and then exported out of the lysosome *via* a lysine transporter. Hence cysteamine is utilized in the therapeutic treatment of cystinosis [52].

PHYSIOLOGIC RELEVANCE OF THE GSH SYNTHESIZING MACHINERY

GSH is prerequisite for cellular redox homeostasis, and GSH-deficiency can cause a variety of pathological conditions and, in severe cases, eventually cell death. Thus, understanding how GSH synthesis is regulated at the level of metabolism as well as at the stage of gene expression is an important issue. N-acetyl-L-cysteine (NAC) commonly exerts beneficial functions in cellular antioxidant defense, mainly as a precursor for Cys and, subsequently, GSH [14, 53].

Machinery for and the Regulation of GSH Synthesis

GSH is synthesized by two ATP-requiring enzymes, γ-glutamylcysteine synthetase (γGCS) and glutathione synthetase (GSS). While γGCS catalyzes the first, slow step in the production of γ-glutamylcysteine from Glu and Cys, GSS produces GSH by adding a Gly unit to resulting γ-glutamylcysteine [3, 10]. γ-GCS is a heterodimeric enzyme composed of a catalytic subunit encoded by the GCLC gene and a modifier subunit encoded by the GCLM gene and is inhibited by the specific inhibitor buthionine sulfoximine [54]. γGCS expression is induced under oxidative conditions *via* Nrf2, which is one of the most important transcriptional regulatory factors that can be activated by oxidative stress and electrophilic compounds [55]. Meantime γGCS suffers from GSH-responsive feedback inhibition [56], which may play a role in suppressing the futile reaction that is mediated by production of 5-oxoproline, also known as pyroglutamic acid [57]. These mechanisms coordinately involve maintaining appropriate cellular GSH levels.

Phenotypes of γ-GCS and GSS Deficiency

Genetic defects in GSH-metabolizing enzymes in human are rare, but pathological phenotypes caused by their deficiency are well documented [58]. While GSS deficiencies are more frequent than γ-GCS and GGT deficiencies, γ-GCS and GSS deficiencies are commonly associated with decreased GSH levels and hemolytic anemia [59, 60]. Because mouse models of GSH deficiencies in which the γ-GCS, GSS, and GGT as well as GSR genes are modified, have been well overviewed [61, 62], here we describe their phenotypic properties only briefly.

GCLC-null mutant mice are lethal at their embryonic stage at day 7.5 due to accelerated cell death [63, 64]. A 20% reduction in GSH levels and a compensatory increase (about 30%) in ascorbate levels are observed in $GCLC^{+/-}$ mice. Supplementing culture media with GSH (2.5 mM) or NAC (1.25-5 mM) rescue the fatal nature of fetal embryonic fibroblasts (MEF) isolated from $GCLC^{-/-}$ mouse embryos. Mice that lack GCLC in a hepatocyte-specific manner exhibit steatosis and decreased ATP levels associated with mitochondrial injury and elevated lipid peroxidation [65].

Contrary to a GCLC deficiency, no apparent phenotypes are observed in $GCLM^{-/-}$ mice but the levels of GSH in the liver, lung, pancreas, erythrocytes, and plasma are 9-16% of wild-type mice [66]. However, the sensitivity of MEF isolated from $GCLM^{-/-}$ mice to oxidants, such as hydrogen peroxide and arsenic, are increased [67]. $GCLM^{-/-}$ mice are also sensitive to acetaminophen-induced liver damage, which is rescued by administration of NAC [68], and are sensitive to treatment with domoic acid, the causative agent for amnesic shellfish poisoning by mussels [69, 70]. Agents such as curcumin, quercetin, and *tert*-butylhydroquinone, which increase cellular GSH levels, induce GCLM much more than GCLC but fail to increase γ-GCS activity and GSH levels in $GCLM^{-/-}$ mice [71]. The transgenic mouse with an elevated expression of GCLM has increased resistance to acetaminophen-induced liver damage [72] and hydrogen peroxide-induced single-strand DNA breakage in cells [73]. $GCLM^{-/-}$ fibroblasts show diminished cellular proliferation and increased senescence accompanied by an increase in intracellular ROS levels [74]. This result suggests that GCLM plays a role in the upregulation of γ-GCS activity in response to these stimuli. GSH levels are increased by administration of NAC, leading to the prevention of premature senescence, suggesting that an increased GSH can protect against cellular aging.

Patients with a GSS deficiency show 5-oxoprolineuria, hemolytic anemia, and neurological dysfunction [75, 76]. Fibroblasts that were isolated from 9 patients with a GSS deficiency contained decreased levels of GSH and increased levels of cysteine and γ-glutamylcysteine [77].

A Variety of Byproducts Is Produced by the γGCS Reaction

γGCS combines Glu and Cys to generate γ-glutamylcysteine by a two-step reaction [78]. Glu is first activated by ATP-dependent phosphorylation, resulting in formation of the high energy intermediate γ-glutamylphosphate, which is then combined with cysteine to from γ-glutamyl cysteine. However, if Cys or other appropriate amino acids are not available, 5-oxoproline is produced by autocyclization of γ-glutamylphosphate. 5-Oxoprolinase cleaves the 5-oxoproline structure to regenerate Glu *via* an ATP consuming reaction [79]. Hence, when Cys or other acceptable amino acids are not available, the ATP-dependent formation of γ-glutamylphosphate, its autocyclization to 5-oxoproline-reformation of Glu is repeated with consumption of two ATP molecules per reaction cycle. The resulting decrease in the ATP level then suppresses the 5-oxoprolinase reaction, leading to the accumulation of 5-oxoproline.

Meanwhile, γ-glutamyl cyclotransferase catalyzes hydrolysis of γ-glutamyl amino acids to 5-oxoproline and the respective amino acid. The enzyme 5-oxoprolinase then converts 5-oxoproline to glutamic acid at the expense of one molecule of ATP, which causes the futile cycle for the cells as described above. Thus, the cyclic reaction of GGT coupled with γ-glutamyl cyclotransferase also produces 5-oxoproline, continuously consuming ATP, a process that can lead to the development of metabolic acidosis [57].

Products Produced by the GSH-Synthesizing Machinery as Potential Biomarkers for a Cys-GSH Insufficiency

It would be advantageous for maintaining health if it were possible to accurately maintain redox homeostasis supported by the Cys-GSH potential. However, the reliable quantification of compounds that contain a sulfhydryl group is difficult, especially in biological samples that have been stored for a while, because of their high susceptibility to being oxidized. The presence of surrogate markers that are present in stable amounts and reflect the Cys-GSH potential would be advantageous in this regard. Technical development regarding the unequivocal detection of minor metabolites would contribute to unraveling the roles of γGCS and GSS in producing other metabolic compounds that can be useful as surrogate markers [80].

Ophthalmic acid (OPT) is produced by the GSH-synthesizing machinery under conditions involving the marked consumption of GSH but an insufficient supply of Cys (Figure 3), as typically seen in cases of an overdose of acetaminophen [81-83]. 2-Oxobutyric acid (2OB) together with Cys and ammonia are produced by the catalytic cleavage of cystathionine (CST) by CSE in the transsulfuration pathway, as outlined below. 2OB is also produced during the catabolism of threonine (Thr) in addition to the transsulfuration pathway. 2-Aminobutyrate (2AB) is then synthesized from 2OB by the action of a transaminase [84]. The GSH-synthesizing machinery then utilizes 2AB as a substitute for Cys under Cys-insufficient conditions, to produce OPT in the liver and kidney [85]. It is noteworthy that under fasting conditions, being insufficient Cys supply, OPT levels increased despite the fact that GSH levels were decreased [86]. This suggests that 2AB production is stimulated under fasting conditions *via* activated amino acid catabolism.

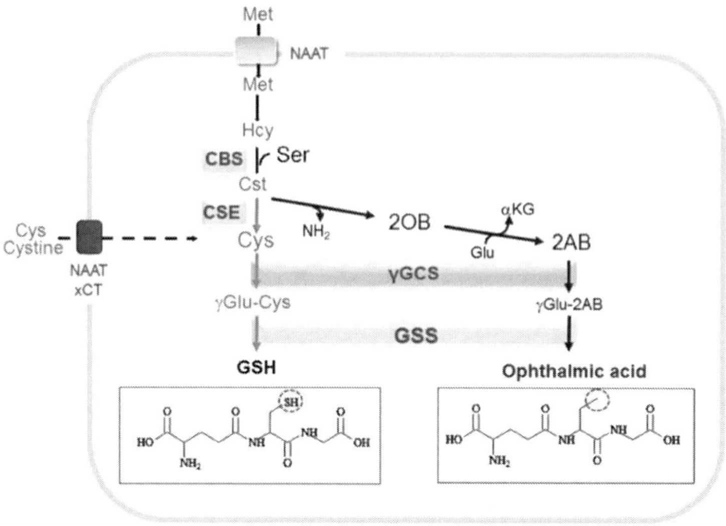

Figure 3. Alternative production of GSH and OPT by the GSH-synthesizing machinery.When 2AB is used by the GSH-synthesizing machinery under Cys insufficient conditions, OPT instead of GSH is produced. NAAT, a neutral amino acid transporter; Hcy, homocysteine; Cst, cystathionine.

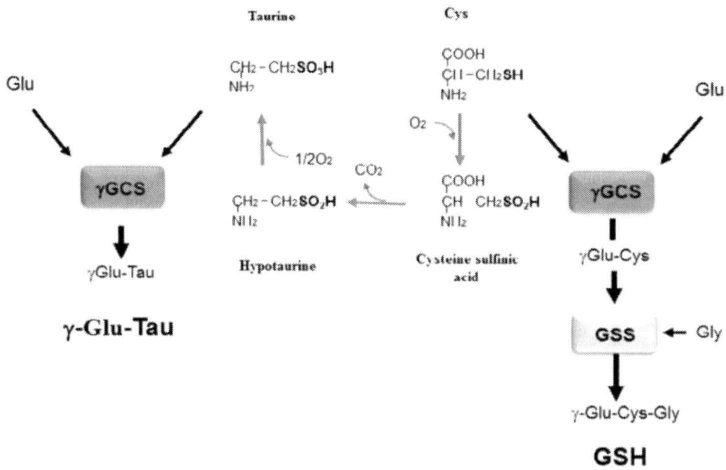

Figure 4. Hypothetical pathway for the production of γ-glutamyltaurine (γ-Glu-Tau) by γ GCS. Taurine (Tau) is synthesized from Cys by two enzymatic reactions and may become substrate for catalytic action of γGCS to produce γ-Glu-Tau.

In addition, other amino acids or related compounds can also be used as Cys alternatives by the GSH-synthesizing machinery and the resulting γ-glutamylated peptides can be regarded as surrogate markers of liver diseases [87]. Taurine (Tau) is produced by the enzymatic oxidation of the sulfhydryl group and decarboxylation of Cys and is the most abundant free amino acid in the animal body [20]. γ-Glutamyl taurine (γ-Glu-Tau), which appears to play notable roles in neuronal systems, is reportedly produced by a transglutamylation reaction catalyzed by GGT [88]. However, it may be necessary to reconsider this mechanism. The levels of γ-Glu-Tau as well as other γ-glutamyl peptides are elevated in patients with a liver injury [87] despite the low GGT activity in the liver under most disease conditions. These observations suggest that γ-Glu-Tau together with other γ-glutamyl peptides may be produced by a γGCS-catalyzed reaction like OPT in liver (Figure 4). Although Tau is commonly classified as an amino acid, it is not an amino acid in a precise sense, due to the absence of a carboxyl group [20]. The absence of a carboxyl group prevents addition of the third amino acid by GSS, which leads to the accumulation of γ-Glu-Tau.

METABOLISM AND FATE OF GSH *IN VIVO*

Under conditions of oxidative stress, GSH is oxidized to GSSG, which is then either recycled by glutathione reductase (GSR) using NADPH provided largely from the pentose phosphate pathway within cells or exported out of cells by the multidrug resistance protein (MRP) [9]. Consumption of GSH also occurs when glutathionylated proteins, such as glutathione S-conjugates, and S-nitrosoglutathione are formed [7, 8, 89, 90]. The radical form of GSH and S-nitrosoglutathione tend to participate in S-glutathionylation of proteins and other compounds. GSH in the circulation undergoes rapid degradation by GGT on the luminal surface of the renal brush-border membrane [11], and the resulting amino acids can be recycled.

Recycle of Oxidized GSH by GSR

GSSG is catalytically reduced by GSR, which is a member of the thioredoxin reductase family and a homodimeric flavoprotein that uses NADPH as an electron donor [91]. While GSH synthesis is the primary cellular action in antioxidant defense, GSR is induced in an Nrf2-dependent manner and plays a crucial role in cell survival under conditions of oxidative stress [92]. 1,3-Bis(2-chloroethyl)-1-nitorsourea (carmustine) is an anticancer agent with relatively high toxicity and a low selectivity [93]. Because carmustine is an inhibitor for GSR, a decrease in GSH levels due to the inhibition of GSR would exert tumorigenic effects but also would cause considerable side effects in normal cells.

The $Gr1^{a1Neu}$ mouse is a mutant strain that is generated by treating male mice with a mutagen, isopropyl methanesulfonate, and exhibits less than a 10% GSR activity in the liver [94]. Nucleotides corresponding to exons 2 through 5 of GSR are deleted in the mouse, which causes a frame shift in exon 6 with introduction of a premature stop codon in exon 7. This deletion mutation in the $Gr1^{a1Neu}$ mouse results in a dysfunctional GSR protein [95]. The $Gr1^{a1Neu}$ mice show renal proximal tubule injury when treated with a redox cycling toxicant, diquat [96], while they are less susceptible to acute lung injury from continuous exposure to 95% oxygen [97]. There is another pyridine nucleotide disulfide oxidoreductase, referred to as thioredoxin/glutathione reductase, which reductively recycles both the oxidized thioredoxin and GSH [98]. This enzyme may explain the remaining 10% GSR activity of Gr1^{a1Neu} mice, although its expression is reportedly limited to the testes [99, 100].

Transporters for GSSG and Conjugated Compounds

A subset of proteins belonging to the ATP-biding cassette transporter (ABC transporter) protein superfamily, which contains 49 members in humans [101], is involved in the export of a variety of compounds, including GSSG, GSNO, and GSH conjugates (GSR). The molecular mechanisms and physiological functions of MRP1 (gene symbol *ABCC1*) among the MRP members have been extensively studied from the view point of the transport of GSH and GSH conjugates [102]. Some GSH-related compounds are transported *via* certain forms of MRP2 (*ABCC2*), MRP4 (*ABCC4*), MRP5 (*ABCC5*), MRP6 (*ABCC6*), MRP8 (*ABCC8*), and CFTR (*ABCC7*; cystic fibrosis transmembrane conductance regulator) and *via* some members of the solute carrier (SLC) superfamily of proteins, resulting in different physiological functions [102].

Extracellular Metabolism of GSH by GGT

Extracellular GSH is hydrolyzed to its γ-glutamyl moiety and cysteinylglycine by GGT. Transferring reaction of γ-glutamyl moieties from GSH to other compounds is also catalyzed by GGT, which leads to formation of a variety of γ-glutamyl compounds [3]. The human GGT family is systematically designated by the HUGO Gene Nomenclature Committee [103]. GGT1, which is the best-characterized form of the enzyme family, is anchored to the cellular plasma membranes. Less than 10 cases of a GGT deficiency in humans has been reported worldwide, and no mutations have been detected either in GGT1 or in other family members. These patients show a low GGT activity and commonly have glutathionuria, an increased plasma GSH levels, and the presence of γ-glutamylcysteine and cysteine in the urine [104]. Deficiencies in LTD_4, which is generated by a GGT-catalyzed reaction from the primary cysteinyl leukotriene LTC_4, have also been reported in some subjects [105]. Regarding experimental animals with genetic deficiency, dwg and dwg (Bayer) mutations in *GGT1* have been identified in a spontaneous mutant mouse [106]. Treatment with N-ethyl-N-nitrosourea has been reported to generate *GGT^{enu1}* mice that have a point mutation within the protein coding region of *GGT1* [107]. Two genetically modified mice with the targeted disruption of *GGT1* [108] or *GGT5* [109] also have been generated. To understand the functions of the GGT isoforms more deeply, please refer to the review articles [61, 62] that have overviewed phenotypic observations on these mice more precisely.

ROLES OF CYS-GSH AXIS IN CELL SURVIVAL

Antioxidant defense is of principal importance for the survival of cells and to protect individuals against aging. GSH plays central roles by either the direct interaction with ROS or by donating reducing equivalents to GPX. When ROS are produced beyond the antioxidant capacity comprised of GSH and other redox systems, cells suffer from oxidative stress and die in severe cases. The cell death is typically caused by Cys and/or GSH deficiency, dependent on iron and accompanied by elevated lipid peroxidation; a condition that is referred to as ferroptosis [110]. The absence of caspase activation clearly distinguishes it from apoptosis [111, 112].

Antioxidant Function of GSH *via* GSH-Dependent Peroxidases

While GSH plays pleiotropic roles, its antioxidant function appears to be absolutely required in many cases [3]. Donation of electrons to other redox systems, such as GPX and Grx exerts an antioxidant function [4, 62]. Conventional GPX, comprised of 4 members, contains selenocysteine as the catalytic center in mammals [113]. The high catalytic activity of GPX can be attributed to the selenocysteine residue since replacing the selenocysteine moiety with cysteine results in a much lower GPX activity. Among the GPX family members, GPX4, which is the only member with lipid hydroperoxide-detoxifying activity, appears to be essential because deficiency of the gene causes embryonic death [114]. In addition to the selenocysteine-containing GPX, other enzymes such as glutathione S-transferase (GST), also exhibit GSH-dependent peroxidase activity, albeit with much lower efficiency [89]. Because GST catalyzes formation of GSH adducts with various electrophilic endogenous and exogenous (xenobiotic) compounds, detoxification appears to be the major function of the GST family [5, 115].

Ferroptosis Caused by Insufficient Cys Supply for GSH System

Pivotal roles of xCT in the Cys supply for the GSH redox system are typically observed in cancer cells that stably express enhanced levels of xCT and utilize it for tumorigenic growth [45]. The suppression of xCT functions or a decline in GSH levels has recently been linked to a novel form of non-apoptotic cell death (ferroptosis) [110]. While the inhibition of xCT by sulfasalazine or erastin induces ferroptosis *via* a decrease in intracellular GSH levels [110, 116], ferrostatin-1 suppresses lipid peroxidation and rescues the cell death [117]. While the phospholipid hydroperoxidase GPX4 effectively suppresses ferroptosis [118, 119], a loss of GPX4 activity results in ferroptotic death of a type of high mesenchymal therapy-resistant cell [120]. A specific type of interneuron in mice expressing mutant GPX4 that contains Cys instead of selenocysteine at the catalytic site are more sensitive to irreversible overoxidation, implying that selenocysteine has a critical role against oxidative stress [121]. Thus, an insufficient Cys supply appears to be responsible for the erastin-induced cell death. Because the cytotoxic effect of erastin is less severe in cells cultured in Met-free media [46] and cell growth stops under such conditions - which characteristically occurs in cancer cells [122] - the results indicate that erastin induces cell death more effectively in proliferating cells. Cancer cells that express a variant form of CD44 (CD44v) are reportedly resistant to EGF receptor-targeted therapy because the cells express xCT and hence possess a high redox potential [45]. Sulfasalazine, an inhibitor of xCT, depletes CD44v-expressing malignant tumors [123]. Ferroptosis is caused by suppression of the Cys-GSH axis under elevated lipid peroxidation, and depends on iron that triggers production of hydroxyl radicals *via* Fenton reactions. These collective findings suggest that drugs that induce ferroptosis in some diseases, notably cancer, might be of some benefit [111].

PERSPECTIVES

The availability of Cys determines cellular production of GSH that exerts pleiotropic functions in our body. Antioxidant defense *via* GSH directly determines the fate of cells, i.e., survival or death. Hence, a knowledge of the cellular Cys-GSH status is of importance for maintaining health against aging and diseases. Some γ−glutamyl compounds are produced by γGCS under conditions of Cys deficiency and can serve as surrogate markers for a Cys-GSH insufficiency.

REFERENCES

[1] Angeli, J. P. F.; Schneider, M.; Proneth, B.; Tyurina, Y. Y.; Tyurin,V. A.; Hammond, V. J.; Herbach, N.; Aichler, M.; Walch, A.; Eggenhofer, E.; Basavarajappa, D.; Rådmark, O.; Kobayashi, S.; Seibt, T.; Beck, H.; Neff, F.; Esposito, I.; Wanke, R.; Förster, H.; Yefremova, O.; Heinrichmeyer, M.; Bornkamm, G. W.; Geissler, E. K.; Thomas, S. B.; Stockwell, B. R.; O'Donnell, V. B.; Kagan, V. E.; Schick, J. A.;

Conrad, M. Inactivation of the ferroptosis regulator Gpx4 triggers acute renal failure in mice. *Nat. Cell Biol.* 2014, 16, 1180-1191.

[2] Awasthi, Y. C.; Chaudhary, P.; Vatsyayan, R.; Sharma, A.; Awasthi, S.; Sharma, R. Physiological and pharmacological significance of glutathione-conjugate transport. *J. Toxicol. Environ. Health B Crit. Rev.* 2009, 12, 540-551.

[3] Azad, R. K.; Shulaev, V. Metabolomics technology and bioinformatics for precision medicine. *Brief Bioinfo*rm. 2018, bbx170.

[4] Ballatori, N.; Krance, S. M.; Marchan, R.; Hammond, C. L. Plasma membrane glutathione transporters and their roles in cell physiology and pathophysiology. *Mol. Aspects Med.* 2009, 30, 13-28.

[5] Beger, R. D.; Bhattacharyya, S.; Yang, X.; Gill, P. S.; Schnackenberg, L. K.; Sun, J.; James, L. P. Translational biomarkers of acetaminophen-induced acute liver injury. *Arch. Toxicol.* 2015, 89, 1497-522.

[6] Bentea, E.; Demuyser, T.; Van Liefferinge, J.; Albertini, G.; Deneyer, L.; Nys, J.; Merckx, E.; Michotte, Y.; Sato, H.; Arckens, L.; Massie, A.; Smolders, I. Absence of system x_c^- in mice decreases anxiety and depressive-like behavior without affecting sensorimotor function or spatial vision. *Prog. Neuropsychopharmacol. Biol. Psychiatry.* 2015, 59, 49–58.

[7] Beutler, E.; Moroose, R.; Kramer, L.; Gelbart, T.; Forman, L. Gamma-glutamylcysteine synthetase deficiency and hemolytic anemia. *Blood.* 1990, 75, 271-273.

[8] Bittner, S.; Win, T.; Gupta, R. gamma-L-glutamyltaurine. *Amino Acids.* 2005, 28, 343-56.

[9] Botta, D.; Shi, S.; White, C. C.; Dabrowski, M. J.; Keener, C. L.; Srinouanprachanh, S. L.; Farin, F. M.; Ware, C. B.; Ladiges, W. C.; Pierce, R. H.; Fausto, N.; Kavanagh, T. J. Acetaminophen-induced liver injury is attenuated in male glutamate-cysteine ligase transgenic mice. *J. Biol. Chem.* 2006, 281, 28865-28875.

[10] Carroll, W. R.; Stacy, G. W.; Du Vigneaud, V. α-Ketobutyric acid as a product in the enzymatic cleavage of cystathionine, *J. Biol. Chem.* 1949, 180, 375-382.

[11] Chaleckis, R.; Murakami, I.; Takada, J.; Kondoh, H.; Yanagida, M. Individual variability in human blood metabolites identifies age-related differences. *Proc. Natl. Acad. Sci. USA.* 2016, 113, 4252-4259.

[12] Chen, Y.; Yang, Y.; Miller, M. L.; Shen, D.; Shertzer, H. G.; Stringer, K. F.; Wang, B.; Schneider, S. N.; Nebert, D. W.; Dalton, T. P. Hepatocyte-specific Gclc deletion leads to rapid onset of steatosis with mitochondrial injury and liver failure. *Hepatology.* 2007, 45, 1118-1128.

[13] Cole, S. P.; Deeley, R. G. Transport of glutathione and glutathione conjugates by MRP1. *Trends Pharmacol. Sci.* 2006, 27, 438-446.

[14] Conrad, M.; Sato, H. The oxidative stress-inducible cystine/glutamate antiporter, system x_c^- : cystine supplier and beyond. *Amino Acids.* 2012, 42, 231-246.

[15] Chen, Y.; Johansson, E.; Fan, Y.; Shertzer, H. G.; Vasiliou, V.; Nebert, D. W.; Dalton, T. P. Early onset senescence occurs when fibroblasts lack the glutamate-cysteine ligase modifier subunit. *Free Rad. Biol. Med.* 2009, 47, 410-418.

[16] Dahl, N.; Pigg, M.; Ristoff, E.; Gali, R.; Carlsson, B.; Mannervik, B.; Larsson, A.; Board, P. Missense mutations in the human glutathione synthetase gene result in severe metabolic acidosis, 5-oxoprolinuria, hemolytic anemia and neurological dysfunction. *Hum. Mol. Genet.* 1997, 6, 1147-1152.

[17] Dalle-Donne, I.; Rossi, R.; Colombo, G.; Giustarini, D.; Milzani, A. Protein S-glutathionylation: a regulatory device from bacteria to humans. *Trends Biochem. Sci.* 2009, 34, 85-96.

[18] Dalton, T. P.; Chen, Y.; Schneider, S. N.; Nebert, D. W.; Shertzer, H. G. Genetically altered mice to evaluate glutathione homeostasis in health and disease. *Free Rad. Biol. Med.* 2004, 37, 1511-1526.

[19] Dalton, T. P.; Dieter, M. Z.; Yang, Y.; Shertzer, H. G.; Nebert, D. W. Knockout of the mouse glutamate cysteine ligase catalytic subunit (Gclc) gene: embryonic lethal when homozygous, and proposed model for moderate glutathione deficiency when heterozygous. *Biochem. Biophys. Res. Commun.* 2000, 279, 324-329.

[20] De Bundel, D.; Schallier, A.; Loyens, E.; Fernando, R.; Miyashita, H.; Van Liefferinge, J.; Vermoesen, K.; Bannai, S.; Sato, H.; Michotte, Y.; Smolders, I.; Massie, A. Loss of system x(c)- does not induce oxidative stress but decreases extracellular glutamate in hippocampus and influences spatial working memory and limbic seizure susceptibility. *J. Neurosci.* 2011, 31, 5792–5803.

[21] De Luca, A.; Pierno, S.; Camerino, D. C. Taurine: the appeal of a safe amino acid for skeletal muscle disorders. *J. Transl. Med.* 2015, 13, 243.

[22] Dean, M.; Allikmets, R. Transporter Complete characterization of the human ABC gene family. *J. Bioenerg. Biomembr.* 2001, 33, 475-479.

[23] del Amo, E. M.; Urtti, A.; Yliperttula, M. Pharmacokinetic role of L-type amino acid transporters LAT1 and LAT2. *Eur. J. Pharm. Sci.* 2008, 35, 161–174.

[24] Dixon, S. J.; Lemberg, K. M.; Lamprecht, M. R.; Skouta, R.; Zaitsev, E. M.; Gleason, C. E.; Patel, D. N.; Bauer, A. J.; Cantley, A. M.; Yang, W. S.; Morrison, B. 3rd.; Stockwell, B. R. Ferroptosis: an iron-dependent form of nonapoptotic cell death. *Cell.* 2012, 149, 1060-1072.

[25] Domercq, M.; Sánchez-Gómez, M. V.; Sherwin, C.; Etxebarria, E.; Fern, R.; Matute, C. System x_c^- and glutamate transporter inhibition mediates microglial toxicity to oligodendrocytes. *J. Immunol.* 2007, 8, 6549–6556.

[26] Emmett, M. Acetaminophen toxicity and 5-oxoproline (pyroglutamic acid): a tale of two cycles, one an ATP-depleting futile cycle and the other a useful cycle. *Clin. J. Am. Soc. Nephrol.* 2014, 9, 191-200.

[27] Flohé, L. Selenium in mammalian spermiogenesis. *Biol. Chem.* 2007, 388, 987-995.

[28] Fontecave, M.; Atta, M.; Mulliez, E. S-adenosylmethionine: nothing goes to waste. *Trends Biochem. Sci*. 2004, 29, 243-249.

[29] Foster, M. W.; Hess, D. T, Stamler, J. S. Protein S-nitrosylation in health and disease: a current perspective. *Trends Mol. Med.* 2009, 15, 391-404.

[30] Franco, R.; Cidlowski, J. A. Apoptosis and glutathione: beyond an antioxidant. *Cell Death Differ.* 2009, 16, 1303-1314.

[31] Franco, R.; Schoneveld, O. J.; Pappa, A.; Panayiotidis, M. I. The central role of glutathione in the pathophysiology of human diseases. *Arch. Physiol. Biochem.* 2007, 113, 234-58.

[32] Fujii, J.; Ito, J. I.; Zhang, X.; Kurahashi, T. Unveiling the roles of the glutathione redox system *in vivo* by analyzing genetically modified mice. *J. Clin. Biochem. Nutr.* 2011, 49, 70–78.

[33] Fujii, J.; Iuchi, Y.; Matsuki, S.; Ishii T. Cooperative function of antioxidant and redox systems against oxidative stress in male reproductive tissues. *Asian J. Androl.* 2003, 5, 231-242.

[34] Gahl, W. A.; Thoene, J. G.; Schneider, J. A. Cystinosis. *N. Engl. J. Med.* 2002, 347, 111-121.

[35] Gallego-Villar, L.; Hannibal, L.; Häberle, J.; Thöny B.; Ben-Omran, T.; Nasrallah, G. K.; Dewik, A. N.; Kruger, W. D.; Blom, H. J. Cysteamine revisited: repair of arginine to cysteine mutations. *J. Inherit. Metab. Dis*. 2017, 40, 555-567.

[36] Gao, M.; Jiang, X. To eat or not to eat-the metabolic flavor of ferroptosis. *Curr. Opin. Cell Biol.* 2017, 51, 58-64.

[37] Gerashchenko, M. V.; Su, D.; Gladyshev, V. N. CUG start codon generates thioredoxin/glutathione reductase isoforms in mouse testes. *J. Biol. Chem.* 2010, 285, 4595-4602.

[38] Giordano, G.; White, C. C.; McConnachie, L. A.; Fernandez, C.; Kavanagh, T. J.; Costa, L. G. Neurotoxicity of domoic acid in cerebellar granule neurons in a genetic model of glutathione deficiency. *Mol. Pharmacol.* 2006, 70, 2116-2126.

[39] Giordano, G.; White, C. C.; Mohar, I.; Kavanagh, T. J.; Costa, L. G. Glutathione levels modulate domoic acid induced apoptosis in mouse cerebellar granule cells. *Toxicol. Sci.* 2007, 100, 433-444.

[40] Griffith, O. W, Biologic and pharmacologic regulation of mammalian glutathione synthesis, *Free Rad. Biol. Med.* 1999, 27, 922-935.

[41] Guo, H.; Lishko, V. K.; Herrera, H.; Groce, A.; Kubota, T.; Hoffman, R. M. Therapeutic tumor-specific cell cycle block induced by methionine starvation *in vivo*. *Cancer Res.* 1993, 53, 5676-5679.

[42] Hah, R.; Wendel, A.; Flohé, L. The fate of extracellular glutathione in the rat, *Biochim. Biophys. Acta*. 1978, 53, 324-337.

[43] Hamashima, S.; Homma, T.; Kobayashi, S.; Ishii, N.; Kurahashi, T.; Watanabe, R.; Kimura, N.; Sato, H.; Fujii, J. Decreased reproductive performance in xCT-knockout male mice. *Free Rad. Res.* 2017, 51, 851-860.

[44] Harding, C. O.; Williams, P.; Wagner, E.; Chang, D. S.; Wild, K.; Colwell, R. E.;Wolff, J. A. Mice with genetic gamma-glutamyl transpeptidase deficiency exhibit glutathionuria, severe growth failure, reduced life spans, and infertility. *J. Biol. Chem.* 1997, 272, 12560-12567.

[45] Hangauer, M. J.; Viswanathan, V. S.; Ryan, M. J.; Bole, D.; Eaton, J. K.; Matov, A.; Galeas, J.; Dhruv, H. D.; Berens, M. E.; Schreiber, S. L.; McCormick, F.; McManus, M. T. Drug-tolerant persister cancer cells are vulnerable to GPX4 inhibition. *Nature.* 2017, 551, 247-250.

[46] Harvey, C. J.; Thimmulappa, R. K.; Singh, A.; Blake, D. J.; Ling, G.; Wakabayashi, N.; Fujii, J.; Myers, A.; Biswal, S. Nrf2-regulated glutathione recycling independent of biosynthesis is critical for cell survival during oxidative stress. *Free Rad. Biol. Med.* 2009, 46, 443-453.

[47] Heisterkamp, N.; Groffen, J.; Warburton, D.; Sneddon, T. P. The human gamma-glutamyltransferase gene family. *Hum. Genet.* 2008, 123, 321-332.

[48] Hess, D. T.; Matsumoto, A.; Kim, S. O.; Marshall, H. E.; Stamler, J. S. Protein S-nitrosylation: purview and parameters. *Nat. Rev. Mol. Cell Biol.* 2005, 6, 150-166.

[49] Hochstrasser, M. Ubiquitin-dependent protein degradation. *Annu. Rev. Genet.* 1996, 30, 405-39.

[50] Homma, T.; Fujii J. Application of glutathione as anti-oxidative and anti-aging drugs. *Curr. Drug Metab.* 2015, 16, 560-571.

[51] Huang, C. S.; Chang, L. S.; Anderson, M. E.; Meister, A. Catalytic and regulatory properties of the heavy subunit of rat kidney gamma-glutamylcysteine synthetase. *J. Biol. Chem.* 1993, 268, 19675-19680.

[52] Ida, T.; Sawa, T.; Ihara, H.; Tsuchiya, Y.; Watanabe, Y.; Kumagai, Y.; Suematsu, M.; Motohashi, H; Fujii, S.; Matsunaga, T.; Yamamoto, M.; Ono, K.; Devarie-Baez, N. O.; Xian, M.; Fukuto, J. M.; Akaike, T. Reactive cysteine persulfides and S-polythiolation regulate oxidative stress and redox signaling. *Proc. Natl. Acad. Sci. USA.* 2014, 111, 7606-7611.

[53] Ingold, I.; Berndt, C.; Schmitt, S.; Doll, S.; Poschmann, G.; Buday, K.; Roveri, A.; Peng, X.; Porto Freitas, F.; Seibt, T.; Mehr, L.; Aichler, M.; Walch, A.; Lamp, D.; Jastroch, M.; Miyamoto, S.; Wurst, W.; Ursini, F.; Arnér, E. S. J.; Fradejas-Villar, N.; Schweizer, U.; Zischka, H.; Friedmann Angeli, J. P.; Conrad, M. Selenium utilization by GPX4 is required to prevent hydroperoxide-induced ferroptosis. *Cell.* 2017, 172, 409-422.

[54] Irino, Y.; Toh, R.; Nagao, M.; Mori, T.; Honjo, T.; Shinohara, M.; Tsuda, S.; Nakajima, H.; Satomi-Kobayashi, S.; Shinke, T.; Tanaka, H.; Ishida, T.; Miyata, O.;

Hirata, K. I. 2-Aminobutyric acid modulates glutathione homeostasis in the myocardium, *Sci. Rep.* 2016, 6, 36749.

[55] Ishimoto, T.; Nagano, O.; Yae, T.; Tamada, M.; Motohara, T.; Oshima, H.; Oshima, M.; Ikeda, T; Asaba, R.; Yagi, H.; Masuko, T.; Shimizu, T.; Ishikawa, T.; Kai, K.; Takahashi, E.; Imamura, Y.; Baba, Y.; Ohmura, M.; Suematsu, M.; Baba, H.; Saya, H. CD44 variant regulates redox status in cancer cells by stabilizing the xCT subunit of system x_C^- and thereby promotes tumor growth. *Cancer Cell.* 2011, 19, 387-400.

[56] Kabil, O.; Motl, N.; Banerjee, R. H_2S and its role in redox signaling. *Biochim. Biophys. Acta.* 2014, 1844, 1355-1366.

[57] Kamat, P. K.; Rai, S.; Swarnkar, S.; Shukla, R.; Ali, S.; Najmi, A. K.; Nath, C. Okadaic acid-induced Tau phosphorylation in rat brain: role of NMDA receptor. *Neuroscience.* 2013, 238, 97–113.

[58] Kanai, Y.; Clémençon, B.; Simonin, A.; Leuenberger, M.; Lochner, M.; Weisstanner, M.; Hediger, M. A. The SLC1 high-affinity glutamate and neutral amino acid transporter family. *Mol. Aspects Med.* 2013, 34, 108–120.

[59] Kanai, Y.; Endou, H. Heterodimeric amino acid transporters: molecular biology and pathological and pharmacological relevance. *Curr. Drug Metab.* 2001, 2, 339-354.

[60] Kang, E. S.; Lee, J.; Homma, T.; Kurahashi, T.; Kobayashi, S.; Nabeshima, A.; Yamada, S.; Seo, H. G.; Miyata, S.; Sato, H.; Fujii, J. xCT deficiency aggravates acetaminophen-induced hepatotoxicity under inhibition of the transsulfuration pathway. *Free Rad. Res.* 2017, 51, 80-90.

[61] Kann, S.; Estes, C.; Reichard, J. F.; Huang, M. Y.; Sartor, M. A.; Schwemberger, S.; Chen, Y.; Dalton, T. P.; Shertzer, H. G.; Xia, Y.; Puga, A. Butylhydroquinone protects cells genetically deficient in glutathione biosynthesis from arsenite-induced apoptosis without significantly changing their prooxidant status. *Toxicol. Sci.* 2005, 87, 365-384.

[62] Kaur, G.; Leslie, E. M.; Tillman, H.; Lee, W. M.; Swanlund, D. P.; Karvellas, C. J.; US Acute Liver Failure Study Group. Detection of ophthalmic acid in serum from acetaminophen-induced acute liver failure patients is more frequent in non-survivors. *PLoS One*. 2015, 10, e0139299.

[63] Klatt, P.; Lamas, S. Regulation of protein function by S-glutathiolation in response to oxidative and nitrosative stress. *Eur. J. Biochem.* 2000, 267, 4928-4944.

[64] Kobayashi, S.; Kuwata, K.; Sugimoto, T.; Igarashi, K.; Osaki, M.; Okada, F.; Fujii, J.; Bannai, S.; Sato, H. Enhanced expression of cystine/glutamate transporter in the lung caused by the oxidative-stress-inducing agent paraquat. *Free Rad. Biol. Med.* 2012, 53, 2197-2203.

[65] Kobayashi, S.; Sato, M.; Kasakoshi, T.; Tsutsui, T.; Sugimoto, M.; Osaki, M.; Okada, F.; Igarashi, K.; Hiratake, J.; Homma, T.; Conrad, M.; Fujii, J.; Soga, T.; Bannai, S.; Sato, H. Cystathionine is a novel substrate of cystine/glutamate transporter: implications for immune function. *J. Biol. Chem.* 2015, 290, 8778-8788.

[66] Kobayashi, S.; Lee, J.; Takao, T.; Fujii, J. Increased ophthalmic acid production is supported by amino acid catabolism under fasting conditions in mice. *Biochem. Biophys. Res. Commun.* 2017, 491, 649-655.

[67] Konrad, P. N.; Richards, F. II.; Valentine, W. N.; Paglia, D. E. γGlutamyl-cysteine synthetase deficiency. A cause of hereditary hemolytic anemia. *N. Engl. J. Med.* 1972, 286, 557-561.

[68] Kurahashi, T.; Lee, J.; Nabeshima, A.; Homma, T.; Kang, E. S.; Saito, Y.; Yamada, S.; Nakayama, T.; Yamada, K. I.; Miyata, S.; Fujii, J. Ascorbic acid prevents acetaminophen-induced hepatotoxicity in mice by ameliorating glutathione recovery and autophagy. *Arch. Biochem. Biophys.* 2016, 604, 36-46.

[69] Lavoie, S.; Chen, Y.; Dalton, T. P.; Gysin, R.; Cuénod, M.; Steullet, P.; Do, K. Q. Curcumin, quercetin, and tBHQ modulate glutathione levels in astrocytes and neurons: importance of the glutamate cysteine ligase modifier subunit. *J. Neurochem.* 2009, 108, 1410-1422.

[70] Lee, J.; Kang, E. S.; Kobayashi, S.; Homma, T.; Sato, H.; Seo, H. G.; Fujii, J. The viability of primary hepatocytes is maintained under a low cysteine-glutathione redox state with a marked elevation in ophthalmic acid production. *Exp. Cell Res.* 2017, 361, 178-191.

[71] Lieberman, M. W.; Wiseman, A. L.; Shi, Z. Z.; Carter, B. Z.; Barrios, R.; Ou, C. N.; Chévez-Barrios, P.; Wang, Y.; Habib, G. M.; Goodman, J. C.; Huang, S. L.; Lebovitz, R. M.; Matzuk, M. M. Growth retardation and cysteine deficiency in gamma-glutamyl transpeptidase-deficient mice. *Proc. Natl. Acad. Sci. USA.* 1996, 93, 7923-7926.

[72] Mani, S.; Cao, W.; Wu, L.; Wang, R. Hydrogen sulfide and the liver. *Nitric Oxide.* 2014, 41, 62-71.

[73] Massie, A.; Schallier, A.; Kim, S. W.; Fernando, R.; Kobayashi, S.; Beck, H.; De Bundel, D.; Vermoesen, K.; Bannai, S.; Smolders, I.; Conrad, M.; Plesnila, N.; Sato, H.; Michotte, Y. Dopaminergic neurons of system x_c^--deficient mice are highly protected against 6-hydroxydopamine-induced toxicity. *FASEB J.* 2011, 25, 1359–1369.

[74] Mayatepek, E.; Okun, J. G.; Meissner, T.; Assmann, B.; Hammond, J.; Zschocke, J.; Lehmann, W. D. Synthesis and metabolism of leukotrienes in gamma-glutamyl transpeptidase deficiency. *J. Lipid. Res.* 2004, 45, 900-904.

[75] McBean, G. J. The transsulfuration pathway: a source of cysteine for glutathione in astrocytes. *Amino Acids.* 2012, 42, 199-205.

[76] McConnachie, L. A.; Mohar, I.; Hudson, F. N.; Ware, C. B.; Ladiges, W. C.; Fernandez, C.; Chatterton-Kirchmeier, S.; White, C. C.; Pierce, R. H.; Kavanagh, T. J. Glutamate cysteine ligase modifier subunit deficiency and gender as determinants of acetaminophen-induced hepatotoxicity in mice. *Toxicol. Sci.* 2007, 99, 628-636.

[77] Meister, A. Glutathione biosynthesis and its inhibition. *Methods. Enzymol.* 1995, 252, 26-30.

[78] Mesci, P.; Zaïdi, S.; Lobsiger, C. S.; Millecamps, S.; Escartin, C.; Seilhean, D.; Sato, H.; Mallat, M.; Boillée, S. System x_C^- is a mediator of microglial function and its deletion slows symptoms in amyotrophic lateral sclerosis mice. *Brain.* 2015, 138, 53–68.

[79] Meyer, Y.; Buchanan, B. B.; Vignols, F.; Reichheld, J. P. Thioredoxins and glutaredoxins: unifying elements in redox biology. *Annu. Rev. Genet.* 2009, 43: 335-367.

[80] Mizushima, N.; Yoshimori, T.; Levine, B. Methods in mammalian autophagy research, *Cell.* 2010, 140, 313–326.

[81] Murphy, T. H.; Miyamoto, M.; Sastre, A.; Schnaar, R. L.; Coyle, J. T. Glutamate toxicity in a neuronal cell line involves inhibition of cystine transport leading to oxidative stress. *Neuron.* 1989, 2, 1547-1558.

[82] Njålsson, R. Glutathione synthetase deficiency. *Cell Mol. Life Sci.* 2005, 62, 1938-1945.

[83] Orlowski, M.; Meister, A. Partial reactions catalyzed by -glutamylcysteine synthetase and evidence for an activated glutamate intermediate. *J. Biol. Chem.* 1971, 246, 7095-7105.

[84] Orlowski, M.; Wilk, S. Synthesis of ophthalmic acid in liver and kidney *in vivo. Biochem. J.* 1978, 170, 415–419.

[85] Ottestad-Hansen, S.; Hu, Q. X.; Follin-Arbelet, V. V.; Bentea, E.; Sato, H.; Massie, A.; Zhou, Y.; Danbolt, N. C. The cystine-glutamate exchanger (xCT, Slc7a11) is expressed in significant concentrations in a subpopulation of astrocytes in the mouse brain. *Glia.* 2018, 66, 951-970.

[86] Pretsch, W. Glutathione reductase activity deficiency in homozygous Gr1a1Neu mice does not cause haemolytic anaemia. *Genet. Res.* 1999, 73, 1-5.

[87] Richman, P. G.; Meister, A.; Regulation of gamma-glutamyl-cysteine synthetase by nonallosteric feedback inhibition by glutathione. *J. Biol. Chem.* 1975, 250, 1422-1426.

[88] Ristoff, E.; Mayatepek, E.; Larsson, A. Long-term clinical outcome in patients with glutathione synthetase deficiency. *J. Pediatr.* 2001, 139, 79-84.

[89] Ristoff, E.; Hebert, C.; Njålsson, R.; Norgren, S.; Rooyackers, O.; Larsson, A. Glutathione synthetase deficiency: is gamma-glutamylcysteine accumulation a way to cope with oxidative stress in cells with insufficient levels of glutathione? *J. Inherit. Metab. Dis.* 2002, 25, 577-584.

[90] Rogers, L. K.; Tamura, T.; Rogers, B. J.; Welty, S. E.; Hansen, T. N.; Smith, C. V. Analyses of glutathione reductase hypomorphic mice indicate a genetic knockout. *Toxicol. Sci.* 2004, 82, 367-373.

[91] Rogers, L. K.; Bates, C. M.; Welty, S. E.; Smith, C. V. Diquat induces renal proximal tubule injury in glutathione reductase-deficient mice. *Toxicol. Appl. Pharmacol.* 2006, 217, 289-298.

[92] Ruivo, R.; Bellenchi, G. C.; Chen, X.; Zifarelli, G.; Sagné, C.; Debacker, C; Pusch, M. Supplisson S, Gasnier B. Mechanism of proton/substrate coupling in the heptahelical lysosomal transporter cystinosin. *Proc. Natl. Acad. Sci. USA*. 2012, 109, E210-E217.

[93] Rushworth, G. F.; Megson, I. L. Existing and potential therapeutic uses for N-acetylcysteine: the need for conversion to intracellular glutathione for antioxidant benefits. *Pharmacol. Ther.* 2014, 141, 150-159.

[94] Sato, H.; Tamba, M.; Okuno, S.; Sato, K.; Keino-Masu, K.; Masu, M.; Bannai, S. Distribution of cystine/glutamate exchange transporter, system x(c)-, in the mouse brain. *J. Neurosci.* 2002, 22, 8028-8033.

[95] Sato, H.; Tamba, M.; Ishii, T.; Bannai, S. Cloning and expression of a plasma membrane cystine/glutamate exchange transporter composed of two distinct proteins. *J. Biol. Chem.* 1999, 274, 11455-11458.

[96] Sato, H.; Tamba, M.; Kuriyama-Matsumura, K.; Okuno, S.; Bannai, S. Molecular cloning and expression of human xCT, the light chain of amino acid transport system x_c^-. *Antioxid. Redox Sign.* 2000, 2, 665-671.

[97] Sato, H.; Shiiya, A.; Kimata, M.; Maebara; K.; Tamba, M.; Sakakura, Y.; Makino, N.; Sugiyama, F.; Yagami, K.; Moriguchi, T.; Takahashi, S.; Bannai, S. Redox imbalance in cystine/glutamate transporter-deficient mice. *J. Biol. Chem.* 2005, 280, 37423-37429.

[98] Sato, M.; Kusumi, R.; Hamashima, S.; Kobayashi, S.; Sasaki, S.; Komiyama, Y.; Izumikawa, T.; Conrad, M.; Bannai, S.; Sato, H. The ferroptosis inducer erastin irreversibly inhibits system x_c- and synergizes with cisplatin to increase cisplatin's cytotoxicity in cancer cells. *Sci. Rep.* 2018, 8, 968.

[99] Sau, A.; Pellizzari Tregno, F.; Valentino, F.; Federici, G.; Caccuri, A. M. Glutathione transferases and development of new principles to overcome drug resistance. *Arch. Biochem. Biophys.* 2010, 500, 116-122.

[100] Shi, Z. Z.; Osei-Frimpong, J.; Kala, G.; Kala, S. V.; Barrios, R. J.; Habib, G. M.; Lukin, D. J.; Danney, C. M.; Matzuk, M. M.; Lieberman, M. W. Glutathione synthesis is essential for mouse development but not for cell growth in culture. *Proc. Natl. Acad. Sci. USA*. 2000, 97, 5101-5106.

[101] Shi, S.; Hudson, F. N.; Botta, D.; McGrath, M. B.; White, C. C.; Neff-LaFord, H. D.; Dabrowski, M. J.; Singh, N. P.; Kavanagh, T. J. Over expression of glutamate cysteine ligase increases cellular resistance to H_2O_2-induced DNA single-strand breaks. *Cytometry A*. 2007, 71, 686-692.

[102] Shi, Z. Z.; Han, B.; Habib, G. M.; Matzuk, M. M.; Lieberman, M. W. Disruption of gamma-glutamyl leukotrienase results in disruption of leukotriene D(4) synthesis *in vivo* and attenuation of the acute inflammatory response. *Mol. Cell Biol.* 2001, 21, 5389-5395.

[103] Shibasaki, T.; Iuchi, Y.; Okada, F.; Kuwata, K.; Yamanobe, T.; Bannai, S.; Tomita, Y.; Sato, H.; Fujii, J. Aggravation of ischemia-reperfusion-triggered acute renal failure in xCT-deficient mice. *Arch. Biochem. Biophys.* 2009, 490, 63-69.

[104] Skouta, R.; Dixon, S. J.; Wang, J.; Dunn, D. E.; Orman, M.; Shimada, K.; Rosenberg, P. A.; Lo, D. C.; Weinberg, J. M.; Linkermann, A.; Stockwell, B. R. Ferrostatins inhibit oxidative lipid damage and cell death in diverse disease models. *J. Am. Chem. Soc.* 2014, 136, 4551-4556.

[105] Soga, T.; Baran, R.; Suematsu, M.; Ueno, Y.; Ikeda, S.; Sakurakawa, T.; Kakazu, Y.; Ishikawa, T.; Robert, M.; Nishioka, T.; Tomita, M. Differential metabolomics reveals ophthalmic acid as an oxidative stress biomarker indicating hepatic glutathione consumption. *J. Biol. Chem.* 2006, 281, 16768–16776.

[106] Soga, T.; Sugimoto, M.; Honma, M.; Mori, M.; Igarashi, K.; Kashikura, K.; Ikeda, S.; Hirayama, A.; Yamamoto, T.; Yoshida, H.; Otsuka, M.; Tsuji, S.; Yatomi, Y.; Sakuragawa, T.; Watanabe, H.; Nihei, K.; Saito, T.; Kawata, S.; Suzuki, H.; Tomita, M.; Suematsu, M. Serum metabolomics reveals γ-glutamyl dipeptides as biomarkers for discrimination among different forms of liver disease. *J. Hepatol.* 2011, 55, 896-905.

[107] Stankiewicz-Kranc, A.; Miltyk, W.; Skrzydlewska, E. Comparison of influence of carmustine and new proline analog of nitrosourea on antioxidant system in breast carcinoma cells (MCF-7). *Drug Chem. Toxicol.* 2010, 33, 55-63.

[108] Stockwell, B. R.; Friedmann Angeli, J. P.; Bayir, H.; Bush, A. I.; Conrad, M.; Dixon, S. J.; Fulda, S.; Gascón, S.; Hatzios, S. K.; Kagan, V. E.; Noel, K.; Jiang, X.; Linkermann, A.; Murphy, M. E.; Overholtzer, M.; Oyagi, A.; Pagnussat, G. C.; Park, J.; Ran, Q.; Rosenfeld, C. S.; Salnikow, K.; Tang, D.; Torti, F. M.; Torti, S. V.; Toyokuni, S.; Woerpel, K. A.; Zhang, D. D. Ferroptosis: A Regulated Cell Death Nexus Linking Metabolism, Redox Biology, and Disease. *Cell.* 2017, 171, 273-285.

[109] Su, D.; Novoselov, S. V.; Sun, Q. A.; Moustafa, M. E.; Zhou, Y.; Oko, R.; Hatfield, D. L.; Gladyshev, V. N. Mammalian selenoprotein thioredoxin-glutathione reductase. Roles in disulfide bond formation and sperm maturation. *J. Biol. Chem.* 2005, 280: 26491-26498.

[110] Sun, Q. A.; Kirnarsky, L.; Sherman, S.; Gladyshev, V. N. Selenoprotein oxidoreductase with specificity for thioredoxin and glutathione systems. *Proc. Natl. Acad. Sci. USA.* 2001, 98, 3673-3678.

[111] Thomas, K.; Wong, K.; Withington, J.; Bultitude, M.; Doherty, A. Cystinuria - a urologist's perspective. *Nat. Rev. Urol.* 2014, 11, 270-277.

[112] Tipple, T. E.; Welty, S. E.; Rogers, L. K.; Hansen, T. N.; Choi, Y. E.; Kehrer, J. P.; Smith, C. V. Thioredoxin-related mechanisms in hyperoxic lung injury in mice. *Am. J. Respir. Cell Mol. Biol.* 2007, 37, 405-413.

[113] Tsuji, T.; Yamada, K.; Kunieda, T. Characterization of the dwg mutations: dwg and dwg(Bayer) are new mutant alleles of the Ggt1 gene. *Mamm. Genome.* 2009, 20, 711-719.

[114] Van Der Werf, P.; Griffith, O. W.; Meister, A. 5-Oxo-L-prolinase (L-pyroglutamate hydrolase). Purification and catalytic properties. *J. Biol. Chem.* 1975, 250, 6686-6692.

[115] Wild, A. C.; Mulcahy, R. T. Regulation of gamma-glutamylcysteine synthetase subunit gene expression: insights into transcriptional control of antioxidant defenses. *Free Rad. Res.* 2000, 32, 281-301.

[116] Williams, L. E.; Featherstone, D. E. Regulation of hippocampal synaptic strength by glial xCT. *J. Neurosci.* 2014, 34, 16093–16102.

[117] Wright, E. C.; Stern, J.; Ersser, R.; Patrick, A. D. Glutathionuria: gamma-glutamyl transpeptidase deficiency. *J. Inherit. Metab. Dis.* 1980, 2, 3-7.

[118] Xu, H.; Ren, D. Lysosomal physiology. *Annu. Rev. Physiol.* 2015, 77, 57-80.

[119] Yang, W. S.; SriRamaratnam, R.; Welsch, M. E.; Shimada, K.; Skouta, R.; Viswanathan, V. S.; Cheah, J. H.; Clemons, P. A.; Shamji, A. F.; Clish, C. B.; Brown, L. M.; Girotti, A. W.; Cornish, V. W.; Schreiber, S. L.; Stockwell, B. R. Regulation of ferroptotic cancer cell death by GPX4. *Cell.* 2014, 156, 317-331.

[120] Yang, Y.; Dieter, M. Z.; Chen, Y.; Shertzer, H. G.; Nebert, D. W.; Dalton, T. P. Initial characterization of the glutamate-cysteine ligase modifier subunit Gclm(-/-) knockout mouse. Novel model system for a severely compromised oxidative stress response. *J. Biol. Chem.* 2002, 277, 49446-49452.

[121] Yant, L. J.; Ran, Q.; Rao, L.; Van Remmen, H.; Shibatani, T.; Belter, J. G; Motta, L.; Richardson, A.; Prolla, T. A. The selenoprotein GPX4 is essential for mouse development and protects from radiation and oxidative damage insults. *Free Rad. Biol. Med.* 2003, 34, 496-502.

[122] Yin, J.; Ren, W.; Yang, G.; Duan, J.; Huang, X.; Fang, R.; Li, C.; Li, T.; Yin, Y.; Hou, Y.; Kim, S. W.; Wu, G. L-Cysteine metabolism and its nutritional implications. *Mol. Nutr. Food Res.* 2016, 60, 134-146.

[123] Yoshikawa, M.; Tsuchihashi, K.; Ishimoto, T.; Yae, T.; Motohara, T.; Sugihara, E.; Onishi, N.; Masuko, T.; Yoshizawa, K.; Kawashiri, S.; Mukai, M.; Asoda, S.; Kawana, H.; Nakagawa, T.; Saya, H.; Nagano, O. xCT inhibition depletes CD44v-expressing tumor cells that are resistant to EGFR-targeted therapy in head and neck squamous cell carcinoma. *Cancer Res.* 2013, 73, 1855-1866.

BIOGRAPHICAL SKETCH

Junichi Fujii

Present Position

Professor

Affiliation

Department of Biochemistry and Molecular Biology, Graduate School of Medical Science, Yamagata University, Yamataga, Japan

Education

1982	B. S., Shizuoka University, Shizuoka, Japan
1984	M. S., Shizuoka University, Shizuoka, Japan
1988	Ph. D., Osaka University Medical School, Osaka, Japan

Research and Professional Experience

Biochemistry and molecular biology in redox biology

Short Biography

I have started working on muscular biochemistry by focusing on sarcoplasmic reticulum calcium transporter/channel. Then I have changed the research field to redox biology. If more information is needed, please visit my home page: http://www.id.yamagata-u.ac.jp/BiochemII/fujii.html).

Publications from the Last 3 Years

1. Guo, X.; Noguchi, H.; Ishii, N.; Homma, T.; Hamada, T.; Hiraki, T.; Zhang, J.; Matsuo, K.; Yokoyama, S.; Ishibashi, H.; Fukushige, T.; Kanekura, T; Fujii, J.; Uramoto, H.; Tanimoto, A.; Yamada, S. The Association of peroxiredoxin 4 with the initiation and progression of hepatocellular carcinoma. *Antioxid. Redox Signal.* 2018 (in press)
2. Fujii, J.; Homma, T.; Kobayashi, S.; Seo, H. G. Mutual interaction between oxidative stress and endoplasmic reticulum stress in the pathogenesis of diseases specifically focusing on non-alcoholic fatty liver disease. *World J. Biol. Chem.* 2018, 18, 1-15.
3. Hirai, A.; Aung, N. Y.; Ohe, R.; Nishida, A.; Kato, T.; Meng, H.; Ishizawa, K.; Fujii, J.; Yamakawa, M. Expression of TRPM8 in human reactive lymphoid tissues and mature B-cell neoplasms. *Oncol. Lett.* 2018, 16, 5930-5938.
4. Lee, J.; Homma, T.; Kobayashi, S.; Ishii, N.; Fujii, J. Unveiling systemic organ disorders associated with impaired lipid catabolism in fasted SOD1-deficient mice. *Arch. Biochem. Biophys.* 2018, 654, 163-171.

5. Homma, T.; Kurahashi, T.; Lee, J.; Nabeshima, A.; Yamada, S.; Fujii, J. Double knockout of peroxiredoxin 4 (Prdx4) and superoxide dismutase 1 (Sod1) in mice results in severe liver failure. *Oxid. Med. Cell. Longev.* 2018, 2018, 2812904.
6. Warang, P.; Homma, T.; Pandya, R.; Sawant, A.; Shinde, N.; Pandey, D.; Fujii, J.; Madkaikar, M.; Mukherjee, M. B. Potential involvement of ubiquitin-proteasome system dysfunction associated with oxidative stress in the pathogenesis of sickle cell disease. *Br. J. Haematol.* 2018, 182, 559-566.
7. Kobayashi, S.; Hamashima, S.; Homma, T.; Sato, M.; Kusumi, R.; Bannai, S.; Fujii, J.; Sato, H. Cystine/glutamate transporter, system x_c^-, is involved in nitric oxide production in mouse peritoneal macrophages. *Nitric Oxide.* 2018, 78, 32-40.
8. Homma, T.; Shirato, T.; Akihara, R.; Kobayashi, S.; Lee, J.; Yamada, K. I.; Miyata, S.; Takahashi, M.; Fujii, J. Mice deficient in aldo-keto reductase 1a (Akr1a) are resistant to thioacetamide-induced liver injury. *Toxicol. Lett.* 2018, 294, 37-43.
9. Lee, J.; Kang, E. S.; Kobayashi, S.; Homma, T.; Sato, H.; Seo, H. G.; Fujii, J. The viability of primary hepatocytes is maintained under a low cysteine-glutathione redox state with a marked elevation in ophthalmic acid production. *Exp. Cell Res.* 2017, 361, 178-191.
10. Hamashima, S.; Homma, T.; Kobayashi, S.; Ishii, N.; Kurahashi, T.; Watanabe, R.; Kimura, N.; Sato, H.;_Fujii, J. Decreased reproductive performance in xCT-knockout male mice. *Free Radic. Res.* 2017, 51, 851-860.
11. Kobayashi, S.; Lee, J.; Takao, T.; Fujii, J. Increased ophthalmic acid production is supported by amino acid catabolism under fasting conditions in mice. *Biochem. Biophys. Res. Commun.* 2017, 491, 649-655.
12. Lee, J.; Homma, T.; Fujii J. Mice in the early stage of liver steatosis caused by a high fat diet are resistant to thioacetamide-induced hepatotoxicity and oxidative stress. *Toxicol. Lett.* 2017, 277, 92-103.
13. Tasaki, E.; Matsumoto, S.; Tada, H.; Kurahashi, T.; Zhang, X.; Fujii, J.; Utsumi, T.; Iuchi, Y. Protective role of testis-specific peroxiredoxin 4 against cellular oxidative stress. *J. Clin. Biochem. Nutr.* 2017, 60, 156-161.
14. Kang, E. S.; Lee, J.; Homma, T.; Kurahashi, T.; Kobayashi, S.; Nabeshima, A.; Yamada, S.; Seo, H. G.; Miyata, S.; Sato, H.; Fujii, J. xCT deficiency aggravates acetaminophen-induced hepatotoxicity under inhibition of the transsulfuration pathway. *Free Radic. Res.* 2017. 51, 80-90.
15. Homma, T.; Akihara, R.; Okano, S.; Shichiri, M.; Yoshida, Y.; Yamada, K. I.; Miyata, S.; Nakajima, O.; Fujii J. Heightened aggressive behavior in mice deficient in aldo-keto reductase 1a (Akr1a). *Behav. Brain Res.* 2017, 319, 219-224.
16. Ito, J.; Ishii, N.; Akihara, R.; Lee, J.; Kurahashi, T.; Homma, T.; Kawasaki, R.; Fujii, J. A high-fat diet temporarily renders Sod1-deficient mice resistant to an oxidative insult. *J. Nutr. Biochem.* 2017, 40, 44-52.

17. Shirato, T.; Homma, T.; Lee, J.; Kurahashi, T.; Fujii, J. Oxidative stress caused by a SOD1 deficiency ameliorates thioacetamide-triggered cell death via CYP2E1 inhibition but stimulates liver steatosis. *Arch. Toxicol.* 2017, 91, 1319-1333.
18. Homma, T.; Fujii, J. Heat stress promotes the down-regulation of IRE1α in cells: An atypical modulation of the UPR pathway. *Exp. Cell Res.* 2016, 349, 128-138.
19. Akihara, R.; Homma, T.; Lee, J.; Yamada, K.; Miyata, S.; Fujii, J. Ablation of aldehyde reductase aggravates carbon tetrachloride-induced acute hepatic injury involving oxidative stress and endoplasmic reticulum stress. *Biochem. Biophys. Res. Commun.* 2016, 478, 765-771.
20. Kurahashi, T.; Lee, J.; Nabeshima, A.; Homma, T.; Kang, E. S.; Saito, Y.; Yamada, S.; Nakayama, T.; Yamada, K.; Miyata, S.; Fujii, J. Ascorbic acid prevents acetaminophen-induced hepatotoxicity in mice by ameliorating glutathione recovery and autophagy. *Arch. Biochem. Biophys.* 2016, 604, 36-46.
21. Otsuki, N.; Homma, T.; Fujiwara, H.; Kaneko, K.; Hozumi, Y.; Shichiri, M.; Takashima, M.; Ito, J.; Konno, T.; Kurahashi, T.; Yoshida, Y.; Goto, K.; Fujii, S.; Fujii, J. Trichloroethylene exposure aggravates behavioral abnormalities in mice that are deficient in superoxide dismutase. *Regul. Toxicol. Pharmacol.* 2016, 79, 83-90.
22. Otsuki, N.; Konno, T.; Kurahashi, T.; Suzuki, S.; Lee, J.; Okada, F.; Iuchi, Y.; Homma, T.; Fujii, J. The SOD1 transgene expressed in erythroid cells alleviates fatal phenotype in congenic NZB/NZW-F1 mice. *Free Radic. Res.* 2016, 50, 793-800.
23. Nawata, A.; Noguchi, H.; Mazaki, Y.; Kurahashi, T.; Izumi, H.; Wang, K. Y.; Guo, X.; Uramoto, H.; Kohno, K.; Taniguchi, H.; Tanaka, Y.; Fujii, J.; Sasaguri, Y.; Tanimoto, A.; Nakayama, T.; Yamada, S. Overexpression of peroxiredoxin 4 affects intestinal function in a dietary mouse model of nonalcoholic fatty liver disease. *PLoS One*. 2016, 11, e0152549.
24. Nagahisa, H.; Okabe, K.; Iuchi, Y.; Fujii, J.; Miyata, H. Characteristics of skeletal muscle fibers of SOD1 knockout mice. *Oxid. Med. Cell. Longev.* 2016, 2016, 9345970.

In: Glutathione
Editor: Pál Perjési

ISBN: 978-1-53614-740-7
© 2019 Nova Science Publishers, Inc.

Chapter 3

GENETIC ENGINEERING OF MICROORGANISMS FOR HIGH GLUTATHIONE PRODUCTION

***Jyumpei Kobayashi*[1], *Daisuke Sasaki*[1], *Kiyotaka Y. Hara*[2,3] *and Akihiko Kondo*[1,4,5,*]**

[1]Graduate School of Science, Technology, and Innovation, Kobe University, Hyogo, Japan
[2]Division of Environmental Sciences, Graduate School of Nutritional and Environmental Sciences, University of Shizuoka, Shizuoka, Japan
[3]Depertment of Environmental and Life Sciences, School of Food and Nutritional Sciences, University of Shizuoka, Shizuoka, Japan
[4]Department of Chemical Science and Engineering, Graduate School of Engineering, Kobe University, Hyogo, Japan
[5]RIKEN Center for Sustainable Resource Science, Kawasaki, Japan

ABSTRACT

Glutathione, the most abundant thiol-containing tripeptide in all living organisms, is widely used in medical, food, and cosmetic industries due to its various physiological functions, including antioxidant, detoxifier and immune booster properties. Because some kinds of yeasts can produce high concentration of glutathione in their cells and are considered to be safe for use as food-processing microorganisms, glutathione has been industrially produced mainly by fermentation of yeasts. As such, fermentation using genetically modified yeast strains is a promising method for the industrial mass-production of glutathione. However, the genetic modification of yeasts for intracellular glutathione production is challenging due to the inhibition of the catalytic activity of glutathione producing enzymes when glutathione is produced. To overcome this difficulty, many

[*] Corresponding Author E-mail: akondo@kobe-u.ac.jp.

studies focusing on glutathione production using genetically modified yeasts have been published, each employing various unique and effective strategies. This chapter aims to provide an overview of these genetic engineering strategies to produce glutathione using different types of yeasts and bacteria.

Keywords: bifunctional glutathione synthetase, *Escherichia coli*, γ-glutamylcysteine, glutathione production, glutathione synthetase, *Saccharomyces cerevisiae*, yeast, *Pichia pastoris*

1. INTRODUCTION

Glutathione is the most abundant thiol-containing tripeptide in all living organisms [1] and plays a pivotal role in maintaining redox balance in living cells. Therefore, glutathione is widely used in the medical, food, and cosmetic industries [2, 3], and exerts various effects such as antioxidant, detoxifying and immune booster properties [4–9] when ingested or applied on the skin. As a result, the demand for glutathione has increased in recent years. At present, glutathione is industrially produced by fermentation using microorganisms, especially the budding yeast *Saccharomyces cerevisiae*, which contains a high concentration of glutathione and has been generally accepted as a safe microorganism for food production. In production by fermentation, genetic engineering is one of the most effective strategies for improving yields of glutathione. Nevertheless, many studies have focused on optical fermentation conditions to increase glutathione productivity using *S. cerevisiae* or its mutant strains obtained by random mutagenesis [3]. Therefore, our current understanding of genetic engineering for high glutathione production is not enough and genetic engineering still has a great potential in terms of improving glutathione production. In this chapter, we provide a review of current studies that use genetic engineering for high glutathione fermentation to promote the use of effective strategies in future studies.

2. GLUTATHIONE PRODUCTION IN *S. CEREVISIAE*

2.1. Glutathione Synthetases

In *S. cerevisiae*, glutathione is generally and mainly synthesized by two consecutive adenosine triphosphate (ATP) dependent reactions catalyzed by γ-glutamylcysteine synthetase (GCS, EC 6.3.2.2) encoded by *GSH1*, and glutathione synthetase (GS, EC 6.3.2.3) encoded by *GSH2* from three precursor amino acids (GCS and GS pathway in Figure 1). GCS catalyzes the reaction to form γ-glutamylcysteine (γ-GC) from L-glutamic acid (L-Glu) and L-cysteine (L-Cys). GS catalyzes the reaction to form reduced glutathione

(GSH) from γ-GC and glycine (Gly). Overexpression of the *GSH1* and *GSH2* genes is the simplest and most effective genetic engineering strategy for high glutathione production in *S. cerevisiae* cells. Indeed, this strategy increased the intracellular glutathione content in *S. cerevisiae* [10], therefore, several studies employed this strategy to enhance glutathione production (Table 1) [11, 12]. In addition, in the series of glutathione synthesis, GCS is a rate-limiting step and is inhibited by GSH [1]. Therefore, the *GSH1* gene is often in the focus of mutation studies to improve its specific activity and avoid inhibitory effects of GSH on it. For example, *S. cerevisiae* HBU2.558 strain harboring mutant *GSH1* genes and containing amino acid replacements of L51P, E62V, A332E, and S653G in its gene products, produced 1.72-folds higher GSH content compared to the wild type strain [13]. Furthermore, a GCS variant enzyme with a C266A amino acid replacement showed a 1.6-fold higher K_m value for L-Glu, 2.2-fold higher K_i value for GSH, and almost the same V_{max} compared to those of parental GCS [14]. Thus, if the mutant *GSH1* gene coding this variant enzyme is overexpressed in *S. cerevisiae*, its gene product may enable the host strain to accumulate a high concentration of GSH.

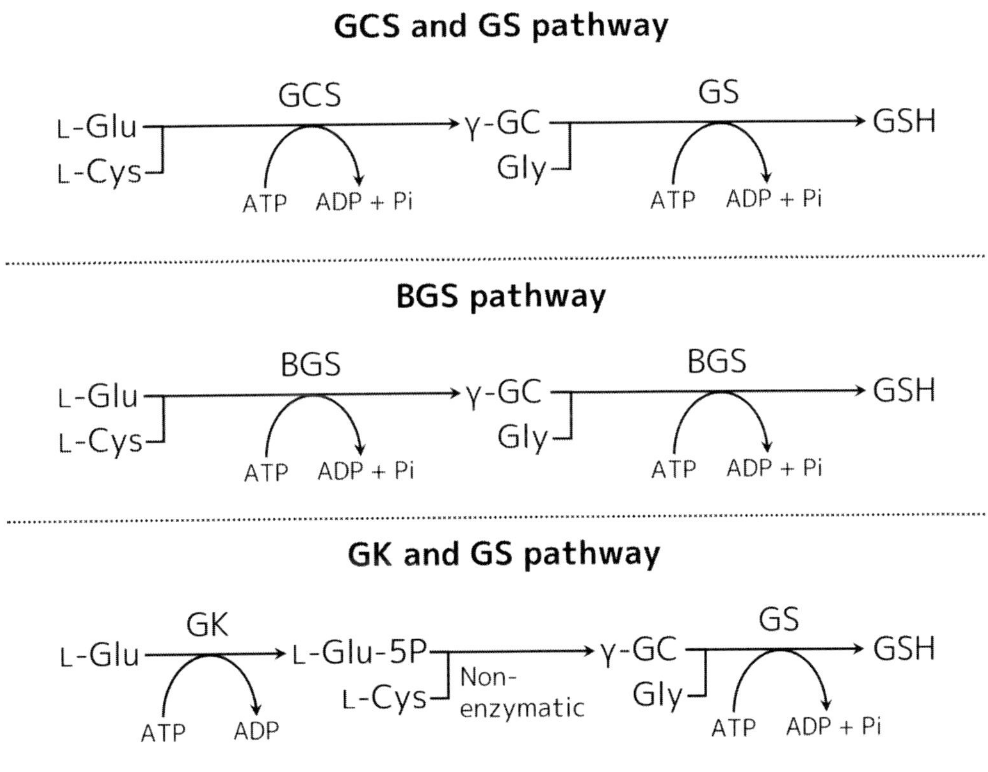

Figure 1. Three different glutathione synthetic pathways from three precursor amino acids. ADP: Adenosine diphosphate; ATP: Adenosine triphosphate; L-Cys: L-Cysteine; γ-GC: γ-Glutamylcysteine; GK: γ-Glutamyl kinase; L-Glu: L-Glutamic acid; L-Glu-5P: L-Glutamic acid-5-phosphate; Gly: Glycine; GSH: Reduced glutathione; GGCS: γ-Glutamylcysteine synthetase; GS: Glutathione synthetase; BGS: Bifunctional glutathione synthetase gene; Pi: Phosphate ion.

Table 1. Gene expression and/or disruption in *S. cerevisiae* strains for improving glutathione production

Host strain	Modified gene	Maximal yield (Relative yield to control strain)	Reference
YPH499	Expressing: *GSH1* (γ-Glutamylcysteine synthetase) from *S. cerevisiae* YPH499 and *GSH2* (Glutathione synthetase) from *S. cerevisiae* YPH499	40.6 mg/L (120%)	[10]
BY4717	Expressing: *GCSGS* (Bifunctional glutathione synthetase) from *Streptococcus thermophilus* ATCC19258	54.9 μmol/g dry weight (469%)	[17]
BY4717	Expressing: *GCSGS*[a] from *S. thermophilus* ATCC19258	321.8 mg/L (262%)[b]	[16]
GCI[c]	Expressing: *MET16* (3′-Phosphoadenylsulfate reductase) from *S. cerevisiae* YPH499	43.9 mg/L (108%)	[10]
GCI[c]	Expressing: *MET14* (Adenylyl-sulfate kinase) from *S. cerevisiae* YPH499	42.6 mg/L (105%)	[10]
GCI[c]	Disrupted: *GLR* (Glutathione reductase)	92.1 ± 11.6 mg/L (100%)	[39]
GCI[c]	Expressing: *ERV1*[d] (Thiol oxidase) from *S. cerevisiae* YPH499, Disrupted: *GLR*	99.8 ± 9.4 mg/L (108%)	[39]
GCI[c]	Expressing: *ERO1*[e] (Thiol oxidase) from *S. cerevisiae* YPH499, Disrupted: *GLR*	118.6 ± 2.0 mg/L (129%)	[39]
GCI[c]	Expressing: *ERV1* from *S. cerevisiae* YPH499 with S32A mutation, Disrupted: *GLR*	105.5 ± 5.9 mg/L (123%)	[40]
GCI[c]	Expressing: *ERV1* from *S. cerevisiae* YPH499 with S32T and N34A mutations, Disrupted: *GLR*	105.7 ± 5.2 mg/L (123%)	[40]
W303-1b	Expressing: *gshF* (Bifunctional glutathione synthetase) from *Actinobacillus pleuropneumoniae*	186.77 ± 3.19 mg/L (189%)	[18]
W303-1b	Expressing: *GSH1* from *S. cerevisiae* W303-1b and *GSH2* from *S. cerevisiae* W303-1b fused gene	145.72 ± 14.87 mg/L (148%)	[18]
W303-1b	Expressing: *PRO1* (γ-Glutamyl kinase) from *S. cerevisiae* W303-1b and *gshB* (Glutathione synthetase) from *Escherichia coli* fused gene	111.82 ± 11.34 mg/L (113%)	[18]
W303-1b	Expressing: *proB* (γ-Glutamyl kinase) from *E. coli* and *gshB* (Glutathione synthetase) from *E. coli* fused gene	104.24 ± 9.86 mg/L (106%)	[18]
W303-1b	Expressing: *gshF* from *A. pleuropneumoniae* and, *GSH1* from *S. cerevisiae* W303-1b and *GSH2* from *S. cerevisiae* W303-1b fused gene	206.70 ± 7.23 mg/L (209%)	[18]
W303-1b	Expressing: *gshF* from *A. pleuropneumoniae* and, *PRO1* from *S. cerevisiae* W303-1b and *gshB* from *E. coli* fused gene	190.57 ± 1.56 mg/L (193%)	[18]
W303-1b	Expressing: *gshF* from *A. pleuropneumoniae* and, *GSH1* from *S. cerevisiae* W303-1b and *GSH2* from *S. cerevisiae* W303-1b fused gene and *PRO1* from *S. cerevisiae* W303-1b and *gshB* from *E. coli* fused gene	216.50 ± 11.46 mg/L (219%)	[18]

Host strain	Modified gene	Maximal yield (Relative yield to control strain)	Reference
GCI[c]	Expressing: *GPX3* (Glutathione peroxidase) from *S. cerevisiae* YPH499	94.6 ± 18.7 mg/L	[38]
CEN.PK 113-7D	Expressing: *GSH1* from *S. cerevisiae* CEN.PK 113-7D	12.5 ± 2.0 μmol/g dry weight (142%)	[11]
BY4727	Expressing: *GSH1* from *S. cerevisiae* BY4727	41.06 ± 2.33 nmol/OD_{600} (139%)	[12]
BY4727	Expressing: *GSH1* from *S. cerevisiae* BY4727, Disrupted: *RTS1* (B-type regulatory subunit of protein phosphatase 2A)	56.45 ± 1.38 nmol/OD_{600} (191%)	[12]
BY4727	Expressing: *GSH1* from *S. cerevisiae* BY4727, Disrupted: *DEF1* (RNAPII degradation factor)	61.77 ± 3.00 nmol/OD_{600} (210%)	[12]
BY4727	Expressing: *GSH1* from *S. cerevisiae* BY4727, Disrupted: *UBP6* (Ubiquitin-specific protease)	57.90 ± 6.36 nmol/OD_{600} (196%)	[12]
BY4727	Expressing: *GSH1* from *S. cerevisiae* BY4727, Disrupted: *PEP12* (Target membrane receptor)	168.18 ± 55.85 nmol/OD_{600} (570%)	[12]
YPH499	Disrupted: *CIS2* (γ-Glutamyl transpeptidase) and *OPT1* (Proton-coupled oligopeptide transporter of the plasma membrane)	0.82 mg/L/OD_{600} (315%)	[48]
YPH499	Expressing: *GSH1* from *S. cerevisiae* YPH499, *GSH2* from *S. cerevisiae* YPH499 and *APD1* (ATP-dependent glutathione exporter) from *S. cerevisiae* YPH499	5.80 mg/L/OD_{600} (1710%) (Extracellular), 12.7 mg/L/OD_{600} (694%) (Intracellular)	[48]
YPH499	Expressing: *SBA*[a] (α-Amylase) from *Streptococcus bovis* and *GluRAG*[a] (Glucoamylase) from *Rhizopus oryzae*	89.7 ± 1.20 mg/L (1120%)[f]	[52]
YPH499	Expressing: *SBA*[a] (α-Amylase) from *S. bovis* and *GluRAG*[a] (Glucoamylase) from *R. oryzae*	32.9 mg/L (539%)[g]	[52]
GCI[c]	Expressing: *ManB*[a] (Mannanase) from *Bacillus licheniformis* and *Man2A*[a] (Mannosidase) from *Cellulomonas fimi*	14.0 mg/L (264%)[h]	[56]
YPH499	Expressing: *CPY* (Vacuolar carboxypeptidase Y) from *S. cerevisiae* YPH499	7.3 mg/L (456%)	[21]
YPH499	Expressing: *CPY* (Vacuolar carboxypeptidase Y) from *S. cerevisiae* YPH499	18.7 mg/L (334%)	[21]

[a]The gene product was displayed on the cell surface of the host. [b]Feather hydrolysate was used as an amino acid source. [c]*S. cerevisiae* GCI strain is the *GSH1* and *GSH2* over-expressing *S. cerevisiae* YPH499 strain. [d]The gene product is located on the mitochondrial inter-membrane space. [e]The gene product is located on endoplasmic reticulum. [f]Reagent-grade starch was used as the carbon source. [g]Cooked rice was used as the carbon source. [h]Locust bean gum was used as the carbon source. [i]Keratin was used as the amino acid source. [j]Soy protein was used as the amino acid source.

Furthermore, bifunctional glutathione synthetase (GshF), which simultaneously catalyzes γ-GC synthesis from L-Glu and L-Cys following GSH synthesis from γ-GC and

Gly, was recently found in several bacterium stains (BGS pathway in Figure 1) [15]. As mentioned above, the catalytic activity of GCS is competitively inhibited by GSH, and it is therefore difficult to accumulate GSH at high concentrations in cells, even if the *GSH1* gene is overexpressed. On the other hand, the inhibitory effect of GSH on BGS is much lower than that of on GCS [15]. Therefore, BGS is a prospective candidate for high glutathione production by genetic engineering in *S. cerevisiae*. Indeed, bacterial *GshF* genes coding BGS that were heterologously overexpressed in *S. cerevisiae* cells and *GshF* overexpressing *S. cerevisiae* strains successfully increased glutathione production, compared to the host strain [16, 17].

GSH can also be synthesized through L-glutamate-5-phosphate from L-Glu and ATP catalyzed by γ-glutamyl kinase (GK) (GK and GS pathway in Figure 1). GK is encoded by the *PRO1* gene in *S. cerevisiae* and the first enzyme in the L-proline synthetic pathway from L-Glu. The L-glutamate-5-phosphate (L-Glu-5P) is non-enzymatically converted to γ-GC by L-Cys [18] and the reaction to form GSH from γ-GC and glycin is catalyzed by GS, like in the GCS and GS pathway (Figure 1). Therefore, over-expression of the *PRO1* gene practically increased intracellular GSH in *S. cerevisiae* cells [18].

2.2. Sulfur Metabolism

Supply of several kinds of amino acids during glutathione fermentation increased the glutathione productivity in *S. cerevisiae* [19, 20]. Among them, addition of L-Cys was the most effective for increasing the intracellular glutathione content [19-21]. However, addition of L-Cys is not economically viable for industrial production because of its high cost. Thus, increasing the L-Cys supply by metabolic engineering would be a promising way to increase glutathione content of the cells without the increased cost of external addition of L-Cys. The sulfur atom in the cysteine molecule is obtained from sulfate through the sulfur assimilation pathway (Figure 2).

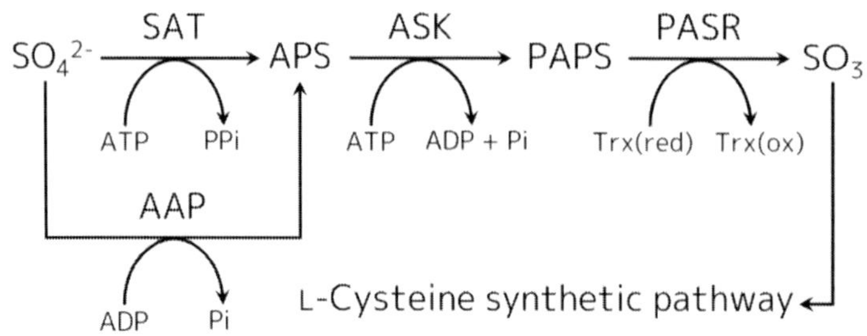

Figure 2. Sulfur ion metabolism leading to L-cysteine synthetic pathway. AAP: ATP adenyl transferase; ADP: Adenosine diphosphate; APS: Adenosine phosphosulfate; ASK: Adenylylsulfate kinase; ATP: Adenosine triphosphate; PAPS: Phosphoadenosine phosphosulfate; PASR: Phosphoadenylyl-sulfate reductase; Pi: Phosphate ion; PPi: Pyrophosphate ion; Trx(ox): Oxidized thioredoxin; Trx(red): Reduced thioredoxin; SAT: Sulfate adenylyltransferase.

The sulfur assimilation pathway, therefore, was previously modified by overexpressing *APA1*, *MET3*, *MET14*, and *MET16* genes coding ATP adenyl transferase (AAT, EC 2.7.7.53), sulfate adenylyltransferase (SAT, EC 2.7.7.4), adenylylsulfate kinase (ASK, EC 2.7.1.25), and phosphoadenylyl-sulfate reductase (PASR, EC 1.8.4.8), respectively [10]. In this study, overexpression of the *MET14* and *MET16* genes increased glutathione content in *S. cerevisiae* cells, indicating that adenylylsulfate kinase and phosphoadenyl-sulfate reductase are limiting steps in the sulfur assimilation pathway.

2.3. Redox Metabolism (Oxidized Glutathione)

In many cases, in living organisms glutathione exists mainly in reduced (GSH) form [22]. GSH is maintained by glutathione redox metabolism. Glutathione redox metabolism involves both oxidization of GSH to oxidized glutathione (GSSG) by thiol peroxidases (TPO, EC 1.11.1.15), thiol oxidase (TO, EC 1.8.3.2), glutathione S-transferase (GT, EC 2.5.1.18) or free radicals, and reduction of GSSG to GSH by glutathione reductase (GR, EC 1.8.1.7) [23]. Glutathione generally regulates activities of various proteins by regulation of disulfide bonds through the disulfide relay systems [24]. For example: (1) Erv1 is a flavin-linked thiol (sulfhydryl) oxidase located in the mitochondrial intermembrane space (IMS) and oxidizes Mia40, part of the disulfide relay system that promotes IMS retention of imported proteins [25, 26]; (2) Erv2 is a flavin-linked thiol oxidase localized in the lumen of the endoplasmic reticulum (ER) and involved in disulfide bond formation within the ER [27]; (3) Ero1 is a thiol oxidase required for oxidative protein folding in the ER and is essential for maintaining proper redox balance in the ER. Ero1 is activated by reduced Pdi1 through the direct reduction of Ero1 regulatory bonds [28-31]. Because glutathione functions not in its oxidized but its reduced form in living cells, several studies have been focused on GSH [32-34]. However, GSSG has some merit for production and applications of glutathione. In industrial mass production of glutathione, GSSG is preferable to GSH due to its higher stability. Moreover, GSSG shows the same antioxidant activity in the intestine after dietary intake [35] and promotes plant growth more potently than GSH [36, 37]. Therefore, in the present-day industrial practice, GSH is chemically oxidized to GSSG after its extraction from yeast. However, oxidization of GSH by chemical methods requires a large amount of acid, which increases the cost of production. Therefore, genetic engineering of high GSSG production is also important for biological glutathione production. Moreover, the catalytic activity of GCS is not inhibited by GSSG, thus GSSG production using genetic engineering has a potential for high total glutathione (GSH and GSSG) production, avoiding feedback inhibition by GSH.

Indeed, the GSSG and the total glutathione content is higher in strains lacking glutathione reductase and overexpressing the *GSH1* and *GSH2* genes [38]. Furthermore, in this report, overexpression of the *GPX3* gene, coding glutathione peroxidase, also increased GSSG and total glutathione content. In other examples, overexpressions of the *ERV1* and *ERO1* genes also improved GSSG and consecutive total glutathione production in *S. cerevisiae* cells [39]. GSSG content reached a maximum of 80% of the total glutathione content when the *ERV1* gene was overexpressed in an engineered platform strain lacking the *GLR1* gene coding glutathione reductase and overexpressing *GSH1* and *GSH2*. Furthermore, Kobayashi et al. enzymatically improved Erv1 for GSH oxidation by amino acid replacements, and consequently enhanced GSSG production in *S. cerevisiae* cells by overexpressing mutant *ERV1* genes [40].

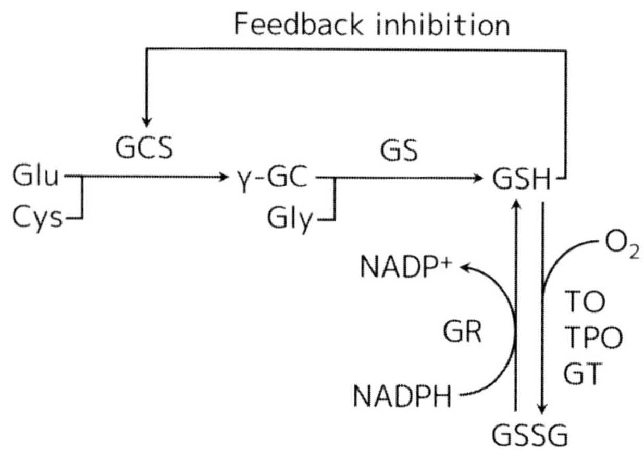

Figure 3. Glutathione redox metabolism. Cys: L-Cysteine; γ-GC: γ-Glutamylcysteine; GGCS: γ-Glutamylcysteine synthetase; Glu: L-Glutamate; Gly: Glycine; GR: Glutathione reductase; GS: Glutathione synthetase; GSH: Reduced glutathione; GSSG: Oxidized glutathione; GT: Glutathione *S*-transferase; $NADP^+$: Oxidized nicotinamide adenine dinucleotide phosphate; NADPH: Reduced nicotinamide adenine dinucleotide phosphate; TO: Thiol oxidase; TPO: Thiol peroxidase.

2.4. Glutathione Production Using Permeable Cells

In glutathione fermentation, improvement of glutathione productivity has spatial limitations and feedback inhibition because glutathione is basically produced within cells. In contrast to the fermentation methods, enzymatic glutathione production using permeated cells has the potential to produce a higher concentration of the product. This enzymatic glutathione biosynthesis is carried out by two consecutive ATP-dependent reactions catalyzed by GCS and GS from added amino acids, and ATP is regenerated by glycolysis in the permeated yeast cells. However, to date, enzymatic glutathione production has not been commercialized because of low glutathione productivity. One of the possible reasons for low glutathione productivity is degradation of glutathione by γ-glutamyl transferase (GTF, EC 2.3.2.2) [41, 42] and another is the low efficiency of the ATP regeneration [43]. Yoshida et al. demonstrated that overexpression of the *GSH1* and *GSH2* genes and deletion of the *CIS2* gene coding GTF increased enzymatic glutathione production in *S. cerevisiae* cells. Additional enhancement of glycolytic ATP regeneration increased productivity up to 2.6-fold compared with that of the parental strain [44]. These results clearly show that metabolic engineering is useful for glutathione production using yeast permeable cells as a whole-cell biocatalyst.

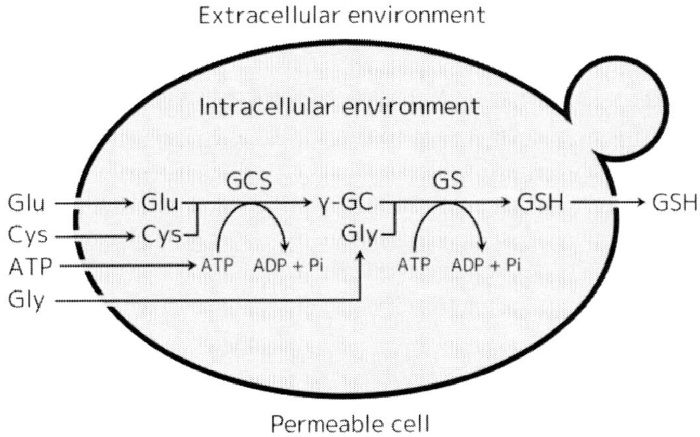

Figure 4. Glutathione synthesis from three precursor amino acids using permeable cell. ADP: Adenosine diphosphate; ATP: Adenosine triphosphate; Cys: L-Cysteine; γ-GC: γ-Glutamylcysteine; Glu: L-Glutamate; Gly: Glysine; GSH: Reduced glutathione; GGCS: γ-Glutamylcysteine synthetase; GS: Glutathione synthetase; Pi: Phosphate ion.

2.5. Glutathione Transport

As mentioned above, enzymatic glutathione production using permeated cells was developed to increase glutathione production without the problems associated with spatial limitations and feedback inhibition [44]. However, enzymatic glutathione production requires external addition of ATP and substrate amino acids, which in turn increases the overall cost of production. Thus, extracellular glutathione fermentation is an attractive approach to produce high amount of glutathione because it encompasses advantages of both the general (intracellular) fermentation and the enzymatic methods, since there is no need for substrate amino acids and no spatial limitations, respectively. Extracellular glutathione fermentation using permeated yeast cells treated with various kinds of surfactants has already been previously reported [45]. However, cell growth was strongly inhibited by addition of the surfactants used to achieve glutathione permeation through the cells, likely because of non-specific leakage of essential cellular components for cell growth. Therefore, specific export of glutathione from the cells is a key step for successful development of extracellular glutathione fermentation. Recently, the glutathione/proton exchangers, Gex1 and Gex2, have been identified in *S. cerevisiae* [46]. It has also been shown that the human ATP-binding cassette protein ABCG2 had glutathione exporting activity when expressed in *S. cerevisiae* [47]. However, these activities were not enough for extracellular glutathione production. Kiriyama et al. identified Gxa1 (Adp1), a novel *S. cerevisiae* glutathione export ABC protein, and overexpressed it in *S. cerevisiae* to develop efficient extracellular glutathione fermentation [48]. In that study, metabolic engineering of glutathione biosynthesis, degradation, and transport was carried out and the final engineered strain showed a 17.1-fold increase in extracellular glutathione production compared to the parental strain [48]. On the other hand, glutathione transport to the vacuoles is another strategy for solving the spatial problems, enabling host cells to avoid feedback regulation by GSH to GCS. Morgan et al. found that vacuolar membrane protein Ycf1 transports glutathione into the vacuoles and overexpression of the *YCF1* gene in *S. cerevisiae* cells increases total glutathione production in cells [49].

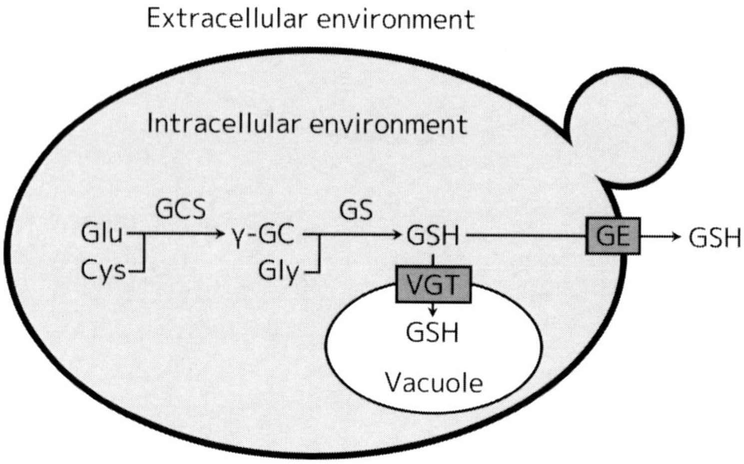

Figure 5. Glutathione transport to extracellular environment and the vacuole. Cys: L-Cysteine; γ-GC: γ-Glutamylcysteine; GE: Glutathione exporter; Glu: L-Glutamate; Gly: Glycine; GSH: Reduced glutathione; GCS: γ-Glutamylcysteine synthetase; GS: Glutathione synthetase; VGT: Vacuolar glutathione transporter.

2.6. Direct Biomass Utilization

2.6.1. Carbon Source

At present, glucose can be extracted from sugarcane *via* the glutathione fermentation process and it is generally used as a carbon source. Thus, both the extraction step for glucose and the glutathione fermentation step by *S. cerevisiae* are required for glutathione production. As such, production of glutathione from starchy materials using a conventional yeast fermentation process is complicated, time-consuming and costly. Hence, direct glutathione fermentation by *S. cerevisiae* from biomass could provide an alternative production process which is simple and efficient. Normally, glucose extraction from starchy materials is carried out by two consecutive steps: high temperature (140–180°C) cooking for conversion of the starchy materials to starch and a subsequent hydrolytic step to convert starch to glucose by amylases [50]. Thus, the glucose extraction process from starchy materials requires large amount of energy and amylases. The high costs associated with extensive energy consumption and enzyme requirement makes production of glutathione from starchy materials difficult to apply to industrial production. Yoshida et al. developed an *S. cerevisiae* strain expressing both α-amylase and glucoamylase genes and achieved an efficient and direct production of glutathione from reagent-grade starch [51]. These enzymes have the potential to reduce the amount of energy and the costs associated with the individual production of the enzymes normally needed for the hydrolysis step. Efforts have also been made to reduce the energy consumption at the cooking step. Low-temperature (75–85°C) cooking fermentation systems have been reported to be successful in reducing the energy required for the cooking step to approximately 40% of that required for high-temperature cooking (140–180°C) [50]. Thus, a combination of a low-temperature cooking system and the amylase-expressing yeast could make it possible to develop an energy-saving glutathione production method from starchy materials. Hara et al. directly produced glutathione from agricultural raw rice pretreated using a low-temperature cooking process and amylase-expressing *S. cerevisiae* [52].

S. cerevisiae can also utilize mannose, the main constituent of mannan-based bioresources [53, 54] and the third-most abundant polymer-based structure in hemicellulose. However, there is only one example of mannan fermentation using *S. cerevisiae* for ethanol production in the past because *S. cerevisiae* cannot hydrolyze mannan oligosaccharides [50]. A yeast-based mannan fermentation process for bioethanol is generally time-consuming and costly because it involves complicated mannose extraction steps. Mannose is normally extracted from mannan-based biomaterials by a two-step process: high temperature (170–200°C) cooking of mannan-based materials and subsequent hydrolysis by mannanase/mannosidase, which converts mannan to mannose [55]. Thus, extraction of mannose from mannan-based biomaterials requires large amount of energy and mannanase/mannosidase. Since such high-energy consumption and enzyme addition is costly, fermentation of mannan from mannan-based bioresources is non-economical for industrial production of bulk chemicals such as ethanol and value-added chemicals such as glutathione. An *S. cerevisiae* strain that expresses both β-mannanase and β-mannosidase on its cell surface has been developed. This strain produces glutathione directly from reagent-grade β-1,4-mannan and a mannan-based raw biomaterial as carbon sources [56]. Coexpression of β-mannanase and β-mannosidase can potentially reduce the energy and the costs of the individually produced enzymes required for the mannose extraction step.

2.6.2. Proteinaceous Bioresource

Some studies reported that supply of several kinds of amino acids during glutathione fermentation increased glutathione productivity in *S. cerevisiae* [19, 20]. Biomass is a renewable source for fermentations; biofuels are currently produced from carbohydrates and lipids in feedstock. By contrast, proteins have not been used to synthesize fuels because of the difficulties of deaminating protein hydrolysates. Recently, however, focus has been given to proteins as a resource for biofuels and fine chemicals [57]. Thus, glutathione fermentation using protein biomass resources as amino acid sources is an attractive approach. Although proteins cannot be transported directly into the *S. cerevisiae* cells, low-molecular weight peptides or amino acids can be internalized by transporters. Therefore, degradation of proteins is required to utilize proteins as amino acid sources. Protease surface display engineering on yeast cells has been used in analysis of the processing system of proteases [58] and in bioconversion of proteins to valuable peptides [59, 60]. Hara et al. have utilized this strategy to generate a protease-displaying yeast [21]. This yeast was then applied to production of glutathione from proteinaceous biomass resources, keratin, and soy protein [21].

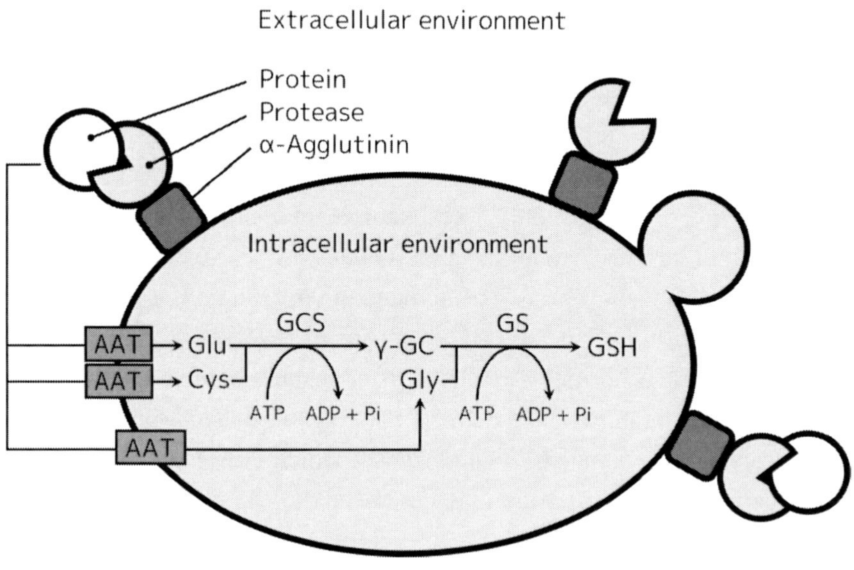

Figure 6. Glutathione production from proteinaceous resource by displaying protease on cell surface. AAT: Amino acid transporter; ATP: Adenosine triphosphate; ADP: Adenosine diphosphate; Cys: L-Cysteine; γ-GC: γ-Glutamylcysteine; Glu: L-Glutamate; Gly: Glycine; GSH: Reduced glutathione; GCS: γ-Glutamylcysteine synthetase; GS: Glutathione synthetase; Pi; Phosphate ion.

2.7. Other Genes

CYS4 and *DEF1* genes have been previously reported to promote glutathione synthesis by overexpression in *S. cerevisiae* cells [12]. The *CYS4* gene encodes the enzyme cystathionine β-synthase (CBS) (EC 4.2.1.22), which catalyzes synthesis of cystathionine from L-serine and homocysteine in the L-Cys synthetic pathway. As mentioned above, L-Cys is a rate-limiting precursor for GSH biosynthesis, therefore, Cys4 catalysis may be a rate-limiting step in the L-Cys biosynthesis. On the other hand, the *DEF1* gene encodes

RNAPII degradation factor, which forms a complex with the protein Rad26 in chromatin and facilitates the ubiquitination and proteolysis of RNAPII [61]. Intriguingly, the *DEF1* gene has been found to increase intracellular glutathione, GSSG in particular, by not only being overexpressed but also being deleted. This implies that both overexpression and deletion of the *DEF1* gene can induce stress in a host and consequently promote GSSG production in cells, avoiding feedback inhibition by GSH. As such, it is possible that genes seemingly unrelated to glutathione synthesis may affect and increase glutathione production, as seen with the *DEF1* gene, because glutathione synthesis is closely related to a variety of stressors. Formation of GSSG, which is induced by oxidative stress, also improves glutathione production.

3. Glutathione Production in Other Microorganisms

Among the scientific and industrial tools for bioproduction, *Escherichia coli* (*E. coli*) is also a good candidate due to its rapid growth rates and our accumulated knowledge for use of it in biological and genetic engineering. Therefore, *E. coli* has been used as a model microorganism for synthetic biology and as a host strain for bioproduction by metabolic engineering in various fields, from basic levels to industrial production. In *E. coli*, GSH is synthesized through consecutive reactions of GCS (encoded by *gshA*) and GS (encoded by *gshB*), in a similar way as in the case of *S. cerevisiae* (Figure 1).

Glutathione production by *E. coli* was first reported in 1976. Langer et al. enzymatically synthesized GSH employing an ATP regeneration system catalyzed by acetate kinase extracted from *E. coli* cells [62]. Since the early 1980s, the *E. coli* B strain - a mutant strain by which it is hard to produce glutathione - was further mutated resulting in the *E. coli*, RC912, which showed high glutathione production when screened [63]. The *E. coli* B strain was also used to improve glutathione production by the overexpression of the *gshA* and *gshB* genes [64]. Although expression levels of both genes were about 10-fold higher in the recombinant *E. coli* cells compared to those of the wild-type strain, the amount of intracellular GSH increased only 1.3-fold compared to the productivity of the wild-type strain. The glutathione yield reached 5.0 g/L by adding precursor amino acids and ATP.

As mentioned above, a novel bifunctional glutathione synthetase (GshF) was recently found in several bacterium stains including *Listeria monocytogenes* and *Streptococcus agalactiae* [65, 66]. Overexpression of its gene (*GshF*) was also applied for glutathione production using *E. coli*. Li et al. reported glutathione production using recombinant *E. coli* strain JM109, which overexpressed the *GshF* gene from *Streptococcus thermophiles* [67]. In that study, the recombinant *E. coli* strain was grown in presence of precursor amino acids and ATP, and produced 10.3 mM GSH in 5 h. Furthermore, the authors studied the same production system employing *E. coli* and overexpression of *GshF* from *S.*

thermophiles using four different vector systems (pET28a, pUC18, pUC19-P32, and pUC19-Pabb). In the fed-batch fermentation, the GSH accumulated by BL21(pUC18-gshF) reached 15.21 g/L, which was the highest level of GSH biosynthesis ever reported [68]. However, studies on GSH production by *E. coli* have peaked and the research interests related to glutathione and *E. coli* have shifted to the fundamental roles of GSH, such as oxidative damage tolerance, immune function, defense mechanism against various diseases, among others, whereas glutathione production using *E. coli* was much higher than those of various genetically engineered *S. cerevisiae* strains (Table 1). At present, the number of research papers and patents related to GSH production by recombinant *E. coli* are decreasing rapidly.

As another candidate for glutathione production, use of *Lactococcus lactis* has been previously reported [69]. In that report, the *GCS* and *GS* genes from *E. coli* were heterologously overexpressed in *L. lactis*, and the recombinant *L. lactis* strain produced 358 ± 17 nmol/mg of protein in the presence of precursor amino acids, whereas glutathione production in the control strain was not detected. However, volumetric production of glutathione (i.e., g/L) was not shown in this study. Therefore, this glutathione yield cannot be compared to the yields of other reports, and usefulness of *L. lactis* as a glutathione producer is unclear.

Various kinds of yeasts were also used as host microorganisms of glutathione production by genetic engineering. Fei et al. reported that *Pichia pastoris* GS115, a methylotrophic yeast, produced 0.92 g/L of glutathione by overexpressing the *GSH1* and *GSH2* from *S. cerevisiae* with no precursor amino acids in the medium [70]. Furthermore, when 20 mM of precursor amino acids were added, the engineered *P. pastoris* strain produced 4.15 g/L of glutathione. Another yeast, methylotrophic yeast *Hansenula polymorpha*, was also the focus of one study [71]. It was reported that the *GCS* gene from *H. polymorpha* was overexpressed in *H. polymorpha,* and that this engineered *H. polymorpha* strain produced 2.257 g/L of glutathione. The glutathione yields of the *P. pastoris* and *H. polymorpha* strains were apparently higher than those of the various genetically engineered *S. cerevisiae* strains (Table 1). Therefore, these yeasts show great potential in terms of microbial glutathione production. However, there are much fewer studies on glutathione production using these yeasts compared to studies using *S. cerevisiae*, most likely because of a lack of understanding for the genetic engineering of these yeasts. Recently, however, *Pichia* sp. have begun to receive attention as a host microorganism for bioproduction because of their high-density growth compared to *S. cerevisiae*, and because they don't produce ethanol [72]. Therefore, these other types of yeast may soon become the new protagonists of glutathione production studies in the future.

Table 2. Gene expression in other microorganisms for improving glutathione production

Host microorganisms (collective name)	Modified gene	Maximal yield (Relative yield to control strain)	Reference
Escherichia coli JM109	Expressing: *gshF* (Bifunctional glutathione synthetase) from *Streptococcus thermophilus* SIIM B218	11.1 g/L[a] (N.D.)	[67]
Escherichia coli JM109	Expressing: *gshF* from *S. thermophilus* SIIM B218	15.21 g/L[b] (N.D.)	[68]
Lactococcus lactis (Lactic acid bacteria)	Expressing: *gshA* (Bifunctional glutathione synthetase) from *E. coli* TG1 and *gshB* (Glutathione synthetase) from *E. coli* TG1	358 ± 17 nmol/mg-protein(N.D.)	[69]
Pichia pastoris GS115 (Methylotrophic yeast)	Expressing: *gshF* (Bifunctional glutathione synthetase) from *Listeria monocytogenes*	84.7 ± 10.7 mg/L (180%)	[71]
Pichia pastoris GS115 (Methylotrophic yeast)	Expressing: *GSH1* (Gamma glutamylcysteine synthetase) from S. cerevisiae and *GSH2* (Glutathione synthetase) from *S. cerevisiae*	0.92[c] g/L (N.D.)	[70]
		4.15[d] g/L (N.D.)	
Hansenula polymorpha (Methylotrophic yeast)	Expressing: *HpGSH2* (Gamma glutamylcysteine synthetase) from *H. polymorpha*	2.257 g/L (N.D.)	[72]
Candida utilis NBRC0988 (Torula yeast)	Expressing: *GSH1* (Gamma glutamylcysteine synthetase) from *C. utilis* NBRC0988	1.17 ± 0.33 ($nmol/OD_{600}$)	[12]
	Expressing: *CYS4* (Cystathionine beta-synthase) from *C. utilis* NBRC0988	0.47 ± 0.06 ($nmol/OD_{600}$)	[12]
	Expressing: *DEF1* (RNAPII degradation factor) from *C. utilis* NBRC0988	0.46 ± 0.03 ($nmol/OD_{600}$)	[12]

[a]Precursor amino acids and ATP were added to the medium. [b]Lac promoter was employed for the expression of the *gshF* gene. [c]No amino acids were added to the medium. [d]20 mM each precursor amino acids were added to the medium.

REFERENCES

[1] Meister, A.; Andersen, M. E. Glutathione. *Annu. Rev. Biochem.* 1983, 52, 711-760.

[2] Arjinpathana, N.; Asawanonda, P. Glutathione as an oral whitening agent: a randomized, double-blind, placebo-controlled study. *J. Dermatolog. Treat.* 2012, 23, 97-102.

[3] Li, Y.; Wei, G.; Chen, J. Glutathione: a review on biotechnological production. *Appl. Microbiol. Biotechnol.* 2004, 66, 233-242.

[4] Dröge, W.; Breitkreutz, R. Glutathione and immune function. *Proc. Nutr. Soc.* 2000, 59, 595-600.

[5] Penninckx, M. J. An overview on glutathione in *Saccharomyces versus* nonconventional yeasts. *FEMS Yeast Res.* 2002, 2, 295-305.

[6] Vartanyan, L. S.; Gurevich, S.; Kozachenko, A. I.; Nagler, L. G.; Lozovskaya, E. L.; Burlakova, E. B. Changes in superoxide production rate and in superoxide dismutase and glutathione peroxidase activities in subcellular organelles in mouse liver under exposure to low doses of low-intensity radiation. *Biochem (Mosc).* 2000, 65, 442-446.

[7] Ray, S.; Watkins, D. N.; Misso, N. L.; Thompson, P. J. Oxidant stress induces gamma-glutamylcysteine synthetase and glutathione synthesis in human bronchial epithelial NCI-H292 cells. *Clin. Exp. Allergy.* 2002, 32, 571-577.

[8] Rolseth, V.; Djurhuus, R.; Svardal, A. M. Additive toxicity of limonene and 50% oxygen and the role of glutathione in detoxification in human lung cells. *Toxicology.* 2002, 170, 75-88.

[9] Singh, R. J. Glutathione: a marker and antioxidant for aging. *J. Lab. Clin. Med.* 2002, 140, 380-381.

[10] Hara, K. Y.; Kiriyama, K.; Inagaki, A.; Nakayama, H.; Kondo, A. Improvement of glutathione production by metabolic engineering the sulfate assimilation pathway of *Saccharomyces cerevisiae*. *Appl. Microbiol. Biotechnol.* 2012, 94, 1313-1319.

[11] Ask, M.; Mapelli, V.; Höck, H.; Olsson, L.; Bettiga, M. Engineering glutathione biosynthesis of *Saccharomyces cerevisiae* increases robustness to inhibitors in pretreated lignocellulosic materials. *Microb. Cell Fact.* 2013, 12, 87.

[12] Suzuki, T.; Yokoyama, A.; Tsuji, T.; Ikeshima, E.; Nakashima, K.; Ikushima, S.; Kobayashi, C.; Yoshida, S. Identification and characterization of genes involved in glutathione production in yeast. *J. Biosci. Bioeng.* 2011, 112, 107-113.

[13] Xu, W.; Jia, H.; Zhang, L.; Wang, H.; Tang, H.; Zhang, L. Effects of *GSH1* and *GSH2* gene mutation on glutathione synthetases activity of *Saccharomyces cerevisiae*. *Protein J.* 2017, 36, 270-277.

[14] Biterova, E. I.; Barycki, J. J. Structural basis for feedback and pharmacological inhibition of *Saccharomyces cerevisiae* glutamate cysteine ligase. *J. Biol. Chem.* 2010, 285, 14459-14466.

[15] Vergauwen, B.; De, Vos, D.; Van, Beeumen, J. J. Characterization of the bifunctional gamma-glutamate-cysteine ligase/glutathione synthetase (GshF) of *Pasteurella multocida*. *J. Biol. Chem.* 2006, 281, 4380-4394.

[16] Qiu, Z.; Tan, H.; Zhou, S.; Cao, L. Surface display of a bifunctional glutathione synthetase on *Saccharomyces cerevisiae* for converting chicken feather hydrolysate into glutathione. *Mol. Biotechnol.* 2014, 56, 726-730.

[17] Qiu, Z.; Deng, Z.; Tan, H.; Zhou, S.; Cao, L. Engineering the robustness of *Saccharomyces cerevisiae* by introducing bifunctional glutathione synthase gene. *J. Ind. Microbiol. Biotechnol.* 2015, 42, 537-542.

[18] Tang, L.; Wang, W.; Zhou, W.; Cheng, K.; Yang, Y.; Liu, M.; Cheng, K.; Wang, W. Three-pathway combination for glutathione biosynthesis in *Saccharomyces cerevisiae*. Microb. *Cell Fact.* 2015, 14, 139.

[19] Alfafara, C. G.; Kanda, A.; Shioi, T.; Shimizu, H.; Shioya, S.; Suga, K. Effect of amino acids on glutathione production by *Saccharomyces cerevisiae*. *Appl. Microbiol. Biotechnol.* 1992, 36, 538-540.

[20] Wen, S.; Zhang, T.; Tana, T. Utilization of amino acids to enhance glutathione production in *Saccharomyces cerevisiae*. *Enz. Microbiol. Technol.* 2004, 35, 501-507.

[21] Hara, K. Y.; Kim, S.; Yoshida, H.; Kiriyama, K.; Kondo, T.; Okai, N.; Ogino, C.; Fukuda, H.; Kondo, A. Development of a glutathione production process from proteinaceous biomass resources using protease-displaying *Saccharomyces cerevisiae*. *Appl. Microbiol. Biotechnol.* 2011, 93, 1495-1502.

[22] Izawa, S.; Inoue, Y.; Kimura, A. Oxidative stress response in yeast: effect of glutathione on adaptation to hydrogen peroxide stress in *Saccharomyces cerevisiae*. *FEBS Lett.* 1995, 368, 73-76.

[23] Pócsi, I.; Prade, R. A.; Penninckx, M. J. Glutathione, altruistic metabolite in fungi. *Adv. Microb. Physiol.* 2004, 49, 1-76.

[24] Thorpe, C.; Hoober, K. L.; Raje, S.; Glynn, N. M.; Burnside, J.; Turi, G. K.; Coppock, D. L. Sulfhydryl oxidases: emerging catalysts of protein disulfide bond formation in eukaryotes. *Arch. Biochem. Biophys.* 2002, 405, 1-12.

[25] Mesecke, N.; Terziyska, N.; Kozany, C.; Baumann, F.; Neupert, W.; Hell, K.; Herrmann, J. M. A disulfide relay system in the intermembrane space of mitochondria that mediates protein import. *Cell.* 2005, 121, 1059-1069.

[26] Toledano, M. B.; Delaunay-Moisan, A.; Outten, C. E.; Igbaria, A. Functions and cellular compartmentation of the thioredoxin and glutathione pathways in yeast. *Antioxid. Redox Signal.* 2013, 18, 1699-1711.

[27] Gerber, J.; Mühlenhoff, U.; Hofhaus, G.; Lill, R.; Lisowsky, T. Yeast ERV2p is the first microsomal FAD-linked sulfhydryl oxidase of the Erv1p/Alrp protein family. *J. Biol. Chem.* 2001, 276, 23486-23491.

[28] Frand, A. R.; Kaiser, C. A. The *ERO1* gene of yeast is required for oxidation of protein dithiols in the endoplasmic reticulum. *Mol. Cell.* 1998, 1, 161-170.

[29] Pollard, M. G.; Travers, K. J.; Weissman, J. S. Ero1p: a novel and ubiquitous protein with an essential role in oxidative protein folding in the endoplasmic reticulum. *Mol. Cell.* 1998, 1, 171-182.

[30] Frand, A. R.; Kaiser, C. A. Ero1p oxidizes protein disulfide isomerase in a pathway for disulfide bond formation in the endoplasmic reticulum. *Mol. Cell.* 1999, 4, 469-477.

[31] Kim, S.; Sideris, D. P.; Sevier, C. S.; Kaiser, C. A. Balanced Ero1 activation and inactivation establishes ER redox homeostasis. *J. Cell Biol.* 2012, 196, 713-725.

[32] Santos, L. O.; Gonzales, T. A.; Ubeda, B. T.; Monte, Alegre, R. Influence of culture conditions on glutathione production by *Saccharomyces cerevisiae*. *Appl. Microbiol. Biotechnol.* 2007, 77, 763-769.

[33] Wang, M.; Sun, J.; Xue, F.; Shang, F.; Wang, Z.; Tan, T. The effect of intracellular amino acids on GSH production by high-cell-density cultivation of *Saccharomyces cerevisiae*. *Appl. Biochem. Biotechnol*. 2012, 168, 198-205.

[34] Fei, L.; Wang, Y.; Chen, S. Improved glutathione production by gene expression in *Pichia pastoris*. *Bioprocess Biosyst. Eng.* 2009, 32, 729-735.

[35] Sugimura, Y.; Yamamoto, K. Effect of orally administered reduced and oxidized-glutathione against acetaminophen-induced liver injury in rats. *J. Nutr. Sci. Vitaminol. (Tokyo)* 1998, 44, 613-624.

[36] Roxas, V. P.; Smith, R. K. Jr; Allen, E. R.; Allen, R. D. Overexpression of glutathione S-transferase/glutathione peroxidase enhances the growth of transgenic tobacco seedlings during stress. *Nat. Biotechnol.* 1997, 15, 988-991.

[37] Belmonte, M. F.; Yeung, E. C. The effects of reduced and oxidized glutathione on white spruce somatic embryogenesis. *In Vitro Cell Dev. Biol. Plant* 2004, 40, 61-66.

[38] Kiriyama, K.; Hara, K. Y.; Kondo, A. Oxidized glutathione fermentation using *Saccharomyces cerevisiae* engineered for glutathione metabolism. *Appl. Microbiol. Biotechnol.* 2013, 97, 7399-7404.

[39] Hara, K.Y.; Aoki, N.; Kobayashi, J.; Kiriyama, K.; Nishida, K.; Araki, M.; Kondo, A. Improvement of oxidized glutathione fermentation by thiol redox metabolism engineering in *Saccharomyces cerevisiae*. *Appl. Microbiol. Biotechnol.* 2015, 99, 9771-9778.

[40] Kobayashi, J.; Sasaki, D.; Hara, K. Y.; Hasunuma, T.; Kondo, A. Enzymatic improvement of mitochondrial thiol oxidase Erv1 for oxidized glutathione fermentation by *Saccharomyces cerevisiae*. *Microb. Cell Fact.* 2017, 16, 44.

[41] Tate, S.; Meister, A. γ-Glutamyltranspeptidase: catalytical, structural and functional aspects. *Mol. Cell Biochem.* 1981, 39, 357-368.

[42] Li, W.; Li, Z.; Ye, Q. Enzymatic synthesis of glutathione using yeast cells in two-stage reaction. *Bioprocess Biosyst. Eng.* 2010, 33, 675-682.

[43] Murata, K.; Tani, K.; Kato, J.; Chibata, I. Glycolytic pathway as an ATP generation system and its application to the production of glutathione and NADP. *Enzyme Microb. Technol.* 1981, 3, 233-242.

[44] Yoshida, H.; Hara, K. Y.; Kiriyama, K.; Nakayama, H.; Okazaki, F.; Matsuda, F.; Ogino, C.; Fukuda, H.; Kondo, A. Enzymatic glutathione production using

metabolically engineered *Saccharomyces cerevisiae* as a whole-cell biocatalyst. *Appl. Microbiol. Biotechnol.* 2011, 91, 1001-1006.

[45] Wei, G.; Li, Y.; Du, G.; Chen, J. Effect of surfactants on extracellular accumulation of glutathione by *Saccharomyces cerevisiae*. *Process Biochem.* 2003, 38, 1133-1138.

[46] Dhaoui, M.; Auchère, F.; Blaiseau, P. L.; Lesuisse, E.; Landoulsi, A.; Camadro, J. M.; Haguenauer-Tsapis, R.; Belgareh-Touzé, N. Gex1 is a yeast glutathione exchanger that interferes with pH and redox homeostasis. *Mol. Biol. Cell.* 2011, 22, 2054-2067.

[47] Brechbuhl, H. M.; Gould, N.; Kachadourian, R.; Riekhof, W. R.; Voelker, D. R.; Day, B. J. Glutathione transport is a unique function of the ATP-binding cassette protein ABCG2. *J. Biol. Chem.* 2010, 285, 16582-16587.

[48] Kiriyama, K.; Hara, K. Y.; Kondo, A. Extracellular glutathione fermentation using engineered *Saccharomyces cerevisiae* expressing a novel glutathione exporter. *Appl. Microbiol. Biotechnol.* 2012, 96, 1021-1027.

[49] Morgan, B.; Ezeriņa, D.; Amoako, T. N.; Riemer, J.; Seedorf, M.; Dick, T. P. Multiple glutathione disulfide removal pathways mediate cytosolic redox homeostasis. *Nat. Chem. Biol.* 2011, 9, 119-125.

[50] Matsumoto, N.; Fukushi, O.; Miyanaga, M.; Kakihara, K.; Nakajima, E.; Yoshizumi, H. Industrialization of a noncooking system for alcoholic fermentation from grains. *Agric. Biol. Chem.* 1982, 46, 1549-1558.

[51] Yoshida, H.; Arai, S.; Hara, K. Y.; Yamada, R.; Ogino, C.; Fukuda, H.; Kondo A. Efficient and direct glutathione production from raw starch using engineered *Saccharomyces cerevisiae*. *Appl. Microbiol. Biotechnol.* 2011, 89, 1417-1422.

[52] Hara, K. Y.; Kim, S.; Kiriyama, K.; Yoshida, H.; Arai, S.; Ishii, J.; Ogino, C.; Fukuda, H.; Kondo A. An energy-saving glutathione production method from low-temperature cooked rice using amylase-expressing *Saccharomyces cerevisiae*. *Biotechnol. J.* 2012, 7, 686-689.

[53] Ishii, J.; Okazaki, F.; Djohan, A. C.; Hara, K. Y.; Asai-Nakashima, N.; Teramura, H.; Andriani, A.; Tominaga, M.; Wakai, S.; Kahar, P.; Yopi, Prasetya, B.; Ogino, C.; Kondo, A. From mannan to bioethanol: cell surface co-display of β-mannanase and β-mannosidase on yeast *Saccharomyces cerevisiae*. *Biotechnol. Biofuels.* 2016, 9, 188.

[54] Malherbe, A. R.; Rose, S. H.; Viljoen-Bloom, M.; van, Zyl, W. H. Expression and evaluation of enzymes required for the hydrolysis of galactomannan. *J. Ind. Microbiol. Biot.* 2014, 41, 1201-1209.

[55] Hendriks, A. T.; Zeeman, G. Pretreatments to enhance the digestibility of lignocellulosic biomass. *Bioresour. Technol.* 2009, 100, 10-18.

[56] Prima, A.; Hara, K. Y.; Djohan, A. C.; Kashiwagi, N.; Kahar, P.; Ishii, J.; Nakayama, H.; Okazaki, F.; Prasetya, B.; Kondo, A.; Yopi; Ogino, C. Glutathione production

from mannan-based bioresource by mannanase/mannosidase expressing *Saccharomyces cerevisiae*. *Bioresour, Technol.* 2017, 245, 1400-1406.

[57] Huo, Y. X.; Cho, K. M.; Rivera, J. G.; Monte, E.; Shen, C. R.; Yan, Y.; Liao, J. C. Conversion of proteins into biofuels by engineering nitrogen flux. *Nat. Biotechnol.* 2011, 29, 346-351.

[58] Kato, M.; Kuzuhara, Y.; Maeda, H.; Shiraga, S.; Ueda, M. Analysis of a processing system for proteases using yeast cell surface engineering: conversion of precursor proteinase A to active proteinase A. *Appl. Microbiol. Biotechnol.* 2006, 72, 1229-1237.

[59] Ni, X.; Yue, L.; Chi, Z.; Li, J.; Wang, X.; Madzak, C. Alkaline protease gene cloning from the marine yeast *Aureobasidium pullulans* HN2–3 and the protease surface display on *Yarrowia lipolytica* for bioactive peptide production. *Mar. Biotechnol.* (NY) 2009, 11, 81-89.

[60] Yu, X. J.; Madzak, C.; Li, H. J.; Chi, Z. M.; Li, J. Surface display of acid protease on the cells of *Yarrowia lipolytica* for milk clotting. *Appl. Microbiol. Biotechnol.* 2010, 87, 669-677.

[61] Reid, J.; Svejstrup, J. Q. DNA damage-induced Def1-RNA polymerase II interaction and Def1 requirement for polymerase ubiquitylation *in vitro*. *J. Biol. Chem.* 2004, 279, 29875-29878.

[62] Langer, R. S.; Hamilton, B. K.; Gardner, C. R.; Archer, M. C.; Colton, C. K. Enzymatic regeneration of ATP. I. Alternative routes. *AIChE J.* 1976, 22, 1079-1090.

[63] Murata, K.; Kimura, A. Some properties of glutathione biosynthesis-deficient mutants of *Escherichia coli* B. *J. Gen. Microbiol.* 1982, 128, 1047-1052.

[64] Gushima, H.; Miya, T.; Murata, K.; Kimura, A. Construction of glutathione-producing strains of *Escherichia coli* B by recombinant DNA techniques. *J. Appl. Biochem.* 1983, 5, 43-52.

[65] Gopal, S.; Borovok, I.; Ofer, A.; Yanku, M.; Cohen, G.; Goebel, W.; Kreft, J.; Aharonowitz, Y. A multidomain fusion protein in *Listeria monocytogenes* catalyzes the two primary activities for glutathione biosynthesis. *J. Bacteriol.* 2005, 187, 3839-3847.

[66] Janowiak, B. E.; Griffith, O. W. Glutathione synthesis in *Streptococcus agalactiae*: One protein accounts for gamma-glutamylcysteine synthetase and glutathione synthetase activities. *J. Biol. Chem.* 2005, 280, 11829-11839.

[67] Li, W.; Li, Z.; Yang, J.; Ye, Q. Production of glutathione using a bifunctional enzyme encoded by *gshF* from *Streptococcus thermophilus* expressed in *Escherichia coli*. *J. Biotechnol.* 2011, 154, 261-268.

[68] Wang, C.; Zhang, J.; Wu, H.; Li, Z.; Ye, Q. Heterologous *gshF* gene expression in various vector systems in *Escherichia coli* for enhanced glutathione production. *J. Biotechnol.* 2015, 214, 63-68.

[69] Li, Y.; Hugenholtz, J.; Sybesma, W.; Abee, T.; Molenaar, D. Using *Lactococcus lactis* for glutathione overproduction. *Appl. Microbiol. Biotechnol.* 2005, 67, 83-90.

[70] Fei, L.; Wang, Y.; Chen, S. Improved glutathione production by gene expression in *Pichia pastoris*. *Bioprocess Biosyst. Eng.* 2009, 32, 729-735.

[71] Ubiyvovk, V. M.; Ananin, V. M.; Malyshev, A. Y.; Kang, H. A.; Sibirny, A. A. Optimization of glutathione production in batch and fed-batch cultures by the wild-type and recombinant strains of the methylotrophic yeast *Hansenula polymorpha* DL-1. *BMC Biotechnol.* 2011, 11, 8.

[72] Ge, S.; Zhu, T.; Li, Y. Expression of bacterial *GshF* in *Pichia pastoris* for glutathione production. *Appl. Environ. Microbiol.* 2012, 78, 5435-5439.

Biographical Sketch

Jyumpei Kobayashi

Affiliation:

Graduate School of Science, Technology and Innovation, Kobe University

Education

Doctor of Philosophy in Engineering (March 25, 2012)

Research and Professional Experience

Enzymology and Microbial engineering

Professional Appointments:

May 2012 - Postdoctoral fellow at Kyushu University, Japan
April 2014 - Project researcher at Tottori University, Japan
April 2015 - Postdoctoral fellow at Kobe University, Japan
December 2017 - Assistant professor at Kobe University, Japan.

Publications Last Three Years:

1. Hirokazu Suzuki, Tatsunari Taketani, Jyumpei Kobayashi, Takashi Ohshiro. *Geobacillus kaustophilus* HTA426 induces mutagenesis via *mfd* depression following growth inhibition by antibiotics. *J Antibiot*, 71:382–389 (2018)
2. Kiyotaka Y. Hara, Jyumpei Kobayashi, Ryosuke Yamada, Daisuke Sasaki, Yuki Kuriya, Yoko Hirono-Hara, Jun Ishii, Michihiro Araki, Akihiko Kondo. Transporter engineering in biomass utilization by yeast. *FEMS Yeast Res*, 17:fox061 (2017)

3. Jyumpei Kobayashi, Hirokazu Suzuki. Microbial responses and resistance to chloramphenicol, *Advances in Medicine and Biology* vol.123, Nova Science Publishers, 53–91 (2017)
4. Jyumpei Kobayashi, Daisuke Sasaki, Kiyotaka Y. Hara, Tomohisa Hasunuma, Akihiko Kondo. Enzymatic improvement of mitochondrial thiol oxidase Erv1 for oxidized glutathione fermentation by *Saccharomyces cerevisiae. Microb Cell Fact*, 16:44 (2017)
5. Yuta Mutaguchi, Jyumpei Kobayashi, Tadao Oikawa, Toshihisa Ohshima. D-Amino acids in fermentative foods. *D-Amino Acids: Physiology, Metabolism, and Application*, Springer Books 341–357 (2016)
6. Keisuke Wada, Jyumpei Kobayashi, Megumi Furukawa, Katsumi Doi, Takashi Ohshiro, Hirokazu Suzuki. A thiostrepton resistance gene and its mutants serve as selectable markers in *Geobacillus kaustophilus* HTA426. *Biosci. Biotechnol. Biochem.* 80, 368–375 (2016)
7. Jyumpei Kobayashi, Misaki Tanabiki, Shohei Doi, Akihiko Kondo, Takashi Ohshiro, Hirokazu Suzuki. Unique plasmids generated via pUC replicon mutagenesis in an error-prone thermophile derived from *Geobacillus kaustophilus* HTA426. *Appl. Environ. Microbiol.* 81, 7625–7632 (2015)
8. Shinya Hasegawa, Jyumpei Kobayashi, Tomoe Komoriya, Hideki Kohno, Kazuaki Yoshimue. Acetaldehyde dehydrogenase gene increases hydrogen production and acetate consumption by *Rhodobacter sphaeroides. Energ. Power Engin.* 7, 396–402 (2015)
9. Kiyotaka Y. Hara, Naoko Aoki, Jyumpei Kobayashi, Kentaro Kiriyama, Keiji Nishida, Michihiro Araki, Akihiko Kondo. Improvement of oxidized glutathione fermentation by thiol redox metabolism engineering in *Saccharomyces cerevisiae*. *Appl. Microbiol. Biotechnol.* 99, 9771–9778 (2015)
10. Jyumpei Kobayashi, Megumi Furukawa, Takashi Ohshiro, Hirokazu Suzuki. Thermoadaptation-directed evolution of chloramphenicol acetyltransferase in an error-prone thermophile using improved procedures. *Appl. Microbiol. Biotechnol.* 99, 5563–5572 (2015)
11. Jyumpei Kobayashi, Jotaro Yukimoto, Yasuhiro Shimizu, Taketo Ohmori, Hirokazu Suzuki, Katsumi Doi, Toshihisa Ohshima. Characterization of *Lactobacillus salivarius* alanine racemase: short-chain carboxylate-activation and the role of A131. *SpringerPlus* 4, 639 (2015)
12. Hirokazu Suzuki, Jyumpei Kobayashi, Keisuke Wada, Megumi Furukawa, Katsumi Doi. Thermoadaptation-directed enzyme evolution in an error-prone thermophile derived from *Geobacillus kaustophilus* HTA426. *Appl. Environ. Microbiol.* 81, 149–158 (2015)

In: Glutathione
Editor: Pál Perjési

ISBN: 978-1-53614-740-7
© 2019 Nova Science Publishers, Inc.

Chapter 4

REGULATING OF GLUTATHIONE SYNTHESIS IN YEAST USING COMBINATIONAL GENETIC STRATEGIES

Wenlong Zhou and Wei Wang*
Institute of Materia Medica, Peking Union Medical College and Chinese Academy of Medical Sciences, Beijing, China

ABSTRACT

Glutathione (GSH), a pivotal non-protein thiol, can be biosynthesized through different pathways in organisms: (1) two consecutive enzymatic reactions catalyzed by γ-glutamylcysteine synthetase (Gsh1 or GshA) and glutathione synthetase (Gsh2 or GshB); (2) a bifunctional γ-glutamylcysteine/glutathione synthetase (GshF); and (3) an alternative condensation of γ-glutamyl phosphate synthesized by γ-glutamyl kinase (Pro1 or ProB) with cysteine to form γ-glutamylcysteine, which was further conjugated with glycine by glutathione synthetase. The GSH level is strictly controlled by a complex regulatory system: biosynthesis, utilization, degradation, oxidation/reduction, and transport. Numerous enzymes and factors are involved in these processes. Based on the complex metabolic network, genetic engineering of one or several genes involving in the GSH biosynthesis can't largely improve the GSH producing capacity of the engineered cells. To break the bottleneck of GSH biosynthesis in *Saccharomyces cerevisiae*, an adaptation-based strategy combined with genetic engineering has been successfully employed. Here, we review the centre status of active

* Corresponding Author E-mail: wwang@imm.ac.cn.

antioxidant GSH in the oxidative stress and adapative bioprocess along with rational regulations of the target genes.

Keywords: glutathione, *Saccharomyces cerevisiae*, adaptive laboratory evolution, combinatorial biosynthesis, genetic engineering

INTRODUCTION

Glutathione (γ-L-glutamyl-L-cysteinyl-glycine, GSH), which is synthesized from glutamic acid (Glu), cysteine (Cys) and glycine (Gly), is an intracellular redox-active thiol found in living organisms. It has a wide variety of biological activities, such as antioxidant effect [1], detoxification [2] and immune regulation [3], and has been widely used in pharmaceutical, food and cosmetic industries.

GSH is primarily synthesized intracellularly through two consecutive reactions catalyzed by γ-glutamyl-cysteine synthetase (Gsh1 in eukaryotes and GshA in prokaryotes, encoded by *GSH1* and *gshA*, respectively) and glutathione synthetase (Gsh2 in eukaryotes and GshB in prokaryotes, encoded by *GSH2* and *gshB*, respectively). Gsh1 catalyzes formation of γ-glutamyl-cysteine (γ-GC), which is subsequently conjugated with glycine by Gsh2 to form GSH. The activity of Gsh1 could be physiologically feedback-inhibited by GSH to avoid over-accumulation of GSH [4]. However, many organisms without Gsh1 or Gsh2 were reported to have the ability to produce GSH, which suggests that they may have a different pathway for synthesis of GSH. A bifunctional enzyme GshF (encoded by *gshF*), found in some Gram-positive bacteria such as: *Listeria monocytogenes* [5], *Streptococcus agalactiae* [6] and *Pasteurella multocida* [7], was able to perform complete synthesis of GSH. In other studies, compensatory pathway for GSH synthesis was characterized in yeast and *Escherichia coli* that lack Gsh1 and GshA [8]. γ-Glutamyl kinase (GK) encoded by *PRO1* in eukaryotes or *proB* in prokaryotes, the first enzyme in the proline biosynthetic pathway, catalyzes formation of the intermediate γ-glutamyl phosphate, which - if accumulated - could partially react with cysteine to generate γ-GC and then Gsh2 adds glycine to γ-GC to form GSH.

To date, although yeast fermentation is the most common method for GSH production, several efforts have been done to obtain GSH overproducing strains using the strategies of metabolic engineering. These approaches mainly focus on improving the capacity of the GSH biosynthetic pathway [9], reducing GSH degradation [10], promoting GSH secretion [11], engineering the sulfate assimilation pathway [12], and increasing the efficiency of ATP utilization [13], resulting in 1-2% increase of GSH content of the modified strains. It is evident, however, that complexity of the cellular regulation restricts the possibility to rationally engineer it. GSH biosynthesis is strictly controlled by a complex regulatory system involving several factors including the feedback inhibition of Gsh1, substrate limitation, fermentation process and intracellular redox state [14].

Saccharomyces cerevisiae is an extremely well characterized model organism with a large variety of research tools and resources, including the complete genome sequence and detailed characterization of metabolic pathways. However, its metabolic system is too complex to accurately predict consequences of its metabolic modification. Although increase of GSH production was achieved based on a trial-and-error method, a rational strategy combined with adaptive evolution and metabolomics has been developed for radical and system-wide modification of metabolism [15, 16]. Here, in this section, we outline the current status of adaptive engineering for GSH production in the yeast *S. cerevisiae* and metabolomic-derived rational regulations of metabolic engineering.

Sources of Acrolein

Acrolein (CH_2=CH–CHO, ACR) is known as a volatile, mutagenic and carcinogenic compound with high reactivity and irritating properties. Being the simplest α,β-unsaturated aldehyde, it has by far the highest reactivity with various biomolecules including proteins, DNA, and glutathione [17, 18]. According to human exposure and toxicity, sources of ACR can be grouped into dietary, endogenous, and environmental origin [19]. ACR occurs as a ubiquitous pollutant in the environment, e.g., incomplete combustion of petrol, wood, and plastic materials, smoking of tobacco products, frying of foods in oils. As a secondary product generated in lipid peroxidation, ACR, along with other products such as 4-hydroxynonenal (HNE), malondialdehyde (MDA) and glyoxal, is generally formed endogenously during free radical-initiated lipid peroxidation of polyunsaturated fatty acids in cell membranes [18, 20]. ACR is also known as a metabolite formed in biotransformation of allyl compounds, (e.g., allyl alcohol (AA)) and the widely used anticancer drug cyclophosphamide [20, 21]. Moreover, ACR is regarded as an endogenous product from myeloperoxidase-mediated degradation of threonine in the presence of hydrogen peroxide and chloride, amine oxidase-mediated degradation of spermine and spermidine [22].

Targets of ACR

Toxicity of ACR reflects the pronounced reactivity of its electrophilic unsaturated bond - which is readily attacked by nucleophilic centers of proteins and DNA to form Michael adducts - disrupting the function of proteins, DNA, and lowering intracellular glutathione (GSH) levels [23]. The biological effects of acrolein are the consequence of its reactivity towards biological nucleophiles such as guanine in DNA and cysteine, lysine, histidine, and arginine residues in critical regions of nuclear factors, proteases, and other proteins. Acrolein adduction disrupts the function of these biomacromolecules which may result in mutations, altered gene transcription, and modulation of apoptosis.

The corresponding adducts reviewed by Stevens and Maier [19] are shown in Figure 1. Among these sites, the cysteine residues of proteins are believed to be the predominant targets of ACR reactivity [24]. ACR can form cyclic adduct with the guanidine group of arginine [25]. Addition of lysine onto ACR can generate two bis-adducts, N^{ε}-(3-formyl-3,4-dehydropiperidino)lysine (FDP-lysine) and N^{ε}-(3-methylpyridinium)lysine (MP-lysine) [26]. The histidine adducts of ACR are formed *via* Michael addition of the histidine residues (Figure 1) [27].

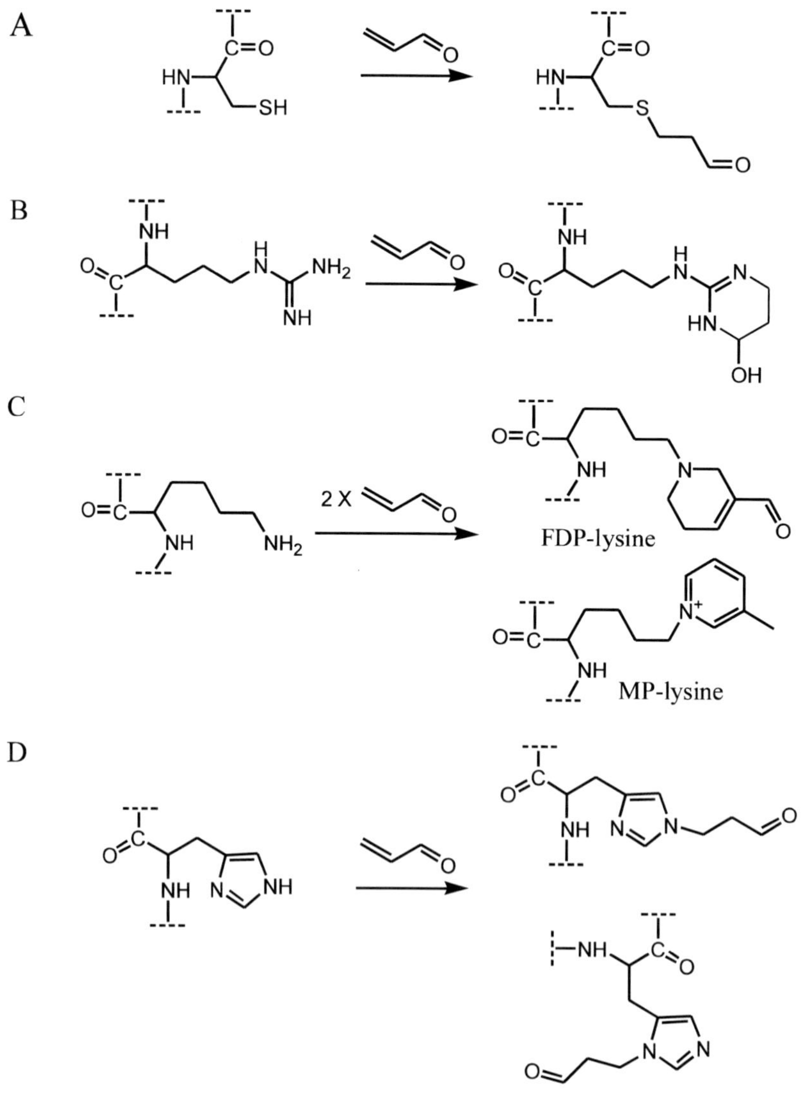

Figure 1. Reaction of amino acid residues with ACR [19, 23, 26, 27]. A: cysteine and its adduct; B: arginine and its adduct; C: lysine and its adducts; D: histidine and its adducts.

The most likely mode of action of ACR-induced mutagenicity is based on formation of DNA-adducts. The *in vitro* reaction of deoxyguanosine (dG) with ACR yielded isomeric 1,N^2-propanodeoxyguanosine (PdG) adducts. Adduct formation is initiated by Michael addition of either N-1 or N^2 of dG onto C-3 of ACR and the primary adducts yield two cyclic regioisomers, α- (α-HOPdG), and γ-hydroxy-PdGs (γ-HOPdG) (Figure 2).

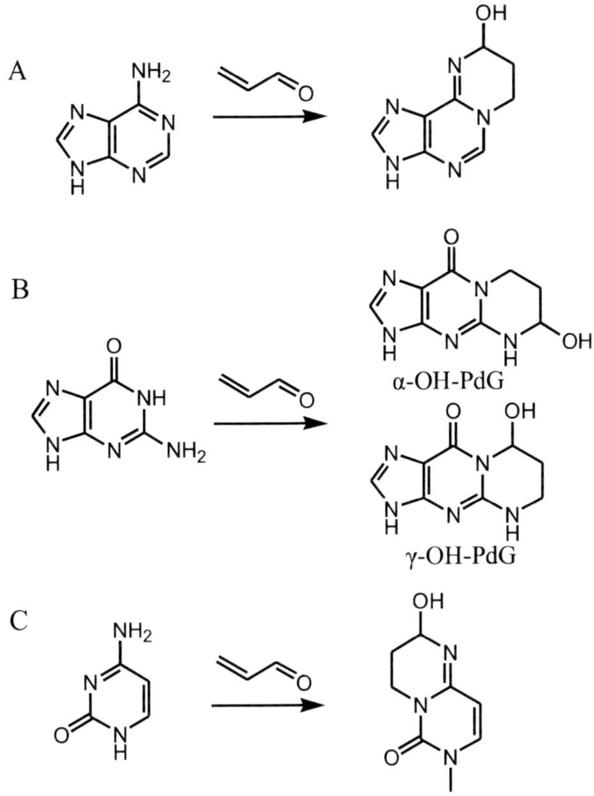

Figure 2. Addition reactions of ACR with adenine (A), guanosine (B), and cytosine (C) [28, 29, 30].

Incubation of ACR with dG or calf thymus DNA yielded the γ-hydroxy isomer as the major product. Both isomeric adducts were shown to induce base substitutions with predominating G→T transversions. α-HOPdG gave rise to a higher miscoding frequency when incorporated into DNA of human xeroderma pigmentosum A cells [19]. γ-HOPdG presents a block to normal replication not only by replicative DNA polymerases (Pols), but also by Y-family Pols such as η and κ. It was found that yeast Rev1 incorporates the correct nucleotide C opposite a permanently cyclic form of γ-HOPdG with nearly the same efficiency as opposite an undamaged G.

γ-HOPdG can exist both in cyclic and open-chain structures, and the open-chain base can participate in the standard Watson-Crick base pairing by adopting a regular *anti* orientation around the glycosidic bond, with the N^2-propyl chain in the minor groove pointing toward the solvent. Studies with model adducts showed that whereas the open-chain form of γ-HOPdG is not inhibitory to synthesis by human Pols η, ι, or κ, only Pol ι is able to incorporate nucleotides opposite the cyclic form, which is known to adopt a *syn* conformation in DNA [28]. Yeast Pol Rev1 contributes to both error-free and mutagenic bypass of γ-HOPdG [29, 30] along with the interaction of Pol ζ (zeta), which provides accurate and efficient synthesis through this potentially carcinogenic DNA lesion [29]. In addition, a direct reversal DNA repair enzyme, *Escherichia coli* AlkB reported by Singh et al. can efficiently repair a wide range of DNA and RNA alkylated lesions [31]. As an Fe(II)- and α-ketoglutarate-dependent dioxygenase, AlkB uses molecular oxygen to oxidize and remove alkyl groups from both N^1 and N^2 positions of guanine. In addition to the guanine adducts, 1,N^6-α-hydroxypropanoadenine (HPA) and 3,N^4-α-hydroxypropano-cytosine (HPC) are minor cyclic ACR adducts found in DNA (Figure 2). *E. coli* AlkB also efficiently removes the cyclic form of the ACR adducts HPA and HPC [32, 33]

Antitoxic Defense Systems of *S. cerevisiae* Exposed to ACR

The yeast *S. cerevisiae* appears to be an excellent model for studying toxicity of exogenous reactive aldehydes since yeast cells do not produce ω-6 polyunsaturated fatty acids and thus are not susceptible to lipid peroxidation [34]. However, the exogenous exposure of *S. cerevisiae* cells to 13-L-hydroperoxylinoleic acid (LOOH) caused their death, the degree of which was dependent on the growth phase of the cells [35]. The studied exogenous reactive aldehydes in yeast are thus not influenced by endogenous lipid peroxidation products. Based on the growing experimental findings with ACR or its precursor allyl alcohol (AA), we can define antitoxic capacity of yeast as a result of enzymatic and non-enzymatic antitoxic defense systems.

The major pathway of ACR detoxication in the cells is conjunction of ACR with GSH to form adducts of lower toxicity. Conjugation of ACR with GSH, however, may lead to depletion of cellular GSH in a concentration-dependent manner, which may cause damaging effects. Once the GSH level has reached a certain threshold, protein thiol groups could be progressively modified while a series of molecular effects may emerge such as cell proliferation, apoptosis and changes in gene/protein expression [36]. Besides this, ACR can also react with other thiols like cysteine, homocysteine, and N-acetylcysteine [16, 37, 38]. When engineered stain W303-1b/FGP^{PT} with high GSH production was exposed to ACR, accumulation of the ACR-GSH adduct showed a narrow liner formation prior to induced expression of the enzymatic defense system [16]. It is consistent with a rapid, dose-dependent decrease in the GSH content of the cells treated with AA, which is converted to ACR by endogenous alcohol dehydrogenase (ADH) [37]. The direct scavenging is to diminish the intracellular availability of ACR, thereby reducing alkylation of intracellular targets such as proteins and DNA. Depletion of GSH, however, alters the cellular redox homeostasis and may be the reason for increased production of reactive oxygen species (ROS) of the cells [38, 39]

Eukaryotic cells can adapt oxidative stress conditions, becoming resistant to a relatively high dose of ROS after exposure to a low, nonlethal dose of them. Adaptation depends on gene activation and *de novo* protein synthesis, typically altering global gene expression programs, including genes encoding antioxidants and other protective molecules. Trotter et al. [40] firstly performed a genome-wide screen to identify sensitive mutants to ACR and characterized two old yellow enzymes (*OYE2* and *OYE3*), which was originally identified as flavin-dependent oxidoreductases. The deletion screen proved that the *OYE2* mutant presented growth inhibition on plates containing 3 mM ACR, while disruption of *OYE3* did not affect ACR tolerance. Furthermore, *in vitro* analysis showed that *OYE2* and *OYE3* reduced toxicity of ACR and that of the α,β-unsaturated ketone, methyl vinyl ketone (MVK). Then, expression of *OYE2* and *OYE3* in response to ACR stress was tested. While expression of *OYE2* was enhanced approximately 15-fold after 0.1 mM ACR treatment, *OYE3* expression was particularly responsive and induced approximately 10-fold following treatment with 0.05 mM ACR and several hundredfolds following a 0.1 mM treatment [40]. It could be concluded that *OYE2* is required for basal levels of tolerance, whereas *OYE3* expression is induced following ACR stress.

Golla et al. [39] investigated the molecular cytotoxicity mechanisms of AA and revealed that yeast cells treated with a sublethal dose of AA (0.4 mM) showed differential regulation of approximately 30% of the yeast genome. The affected genes function in diverse cellular processes including the cell cycle, DNA damage repair, metal homeostasis, stress response genes, ribosomal biogenesis, metabolism, meiosis, ubiquitination, cell morphogenesis, and transport. Among them, the candidate genes *GSH1*, *SOD1*, *SOD2*, *YAP1*, *YBP1*, and *OYE3* were directly up-regulated to counteract AA induced oxidative stress [39].

Two *Yap1*-Activation Mechanisms Lead to Distinct Adaptive Responses

In yeast, the basic leucine-zipper transcription factor *Yap1* regulates most of the known cellular antioxidant genes and contributes to the adaptive response to diverse oxidative stress conditions. *Yap1* has six cysteine residues located in two cysteine-rich domains (CRDs): Cys303, Cys315, and Cys320 in the n-CRD and Cys598, Cys620, and Cys629 in the c-CRD. A nuclear export sequence (Y^{611}SD**<u>I</u>**D**<u>V</u>**D**<u>GL</u>**CSELMAKACSE) of *Yap1* is located within c-CRD, and can be recognized by the β-karyopherin-like nuclear exporter Crm1p. These cysteine residues function as a redox sensor under different cellular redox conditions. Under normal redox conditions, the maintenance of *Yap1* in the cytoplasm might rely on rapid Crm1p-dependent export of *Yap1*, rather than on inhibition of its nuclear import. When exposed to the oxidative conditions, disulfide formation of two cysteine residues between n-CRD and c-CRD or within c-CRD makes the nuclear export sequence unavailable to Crm1p, thus inhibiting nuclear exportation of *Yap1* and allowing it to accumulate in the nucleus [41, 42].

Ouyang et al. [41] had performed the comparative quantitative survival analyses and patch assays to investigate the *YAP1*-mediated adaptation to H_2O_2 and ACR along with N-ethylmaleimide (NEM), a cysteine-specific modifier. The results showed that the wild-type strain pretreated with 0.3 mM H_2O_2 acquired higher resistance to increasing concentrations of H_2O_2, showing fairly strong growth after treatment with 10 mM H_2O_2. In contrast, non-pretreated cells were mostly killed after treatment with 4 or 6 mM H_2O_2. This adaptive response to H_2O_2 was absent in the yeast strain lacking Yap1 (*Δyap1*). When the wild-type strain was pretreated with 0.3 mM H_2O_2, which was sufficient to activate Yap1 for an adaptive response to H_2O_2, no cross-protection against NEM was observed. In the same way, pretreatment of the BY4741 strain with either 0.016 mM NEM or 0.05 mM ACR (nonlethal doses) caused increased tolerance to both chemicals compared to the non-pretreated control. Moreover, the 0.3 mM H_2O_2 pretreatment, which did not trigger protection to NEM, also failed to induce higher tolerance against ACR. Accordingly, ACR or NEM pretreatment did not induce a significant protective response against H_2O_2. These data indicate that ACR and NEM induce similar Yap1-mediated adaptive responses that are distinct from the response induced by H_2O_2 [41].

H_2O_2 activates *Yap1 via* a Gpx3-dependent mechanism. *Yap1* is not oxidized directly by H_2O_2. In this adaptive response, a glutathione peroxidase (GPx)-like enzyme, Gpx3, serves as the H_2O_2 sensor and transducer of the oxidative signal to *Yap1* by forming an intermolecular disulfide bond; Gpx3 Cys36 bridges *Yap1* Cys598 by a disulfide bond. This disulfide bond is later converted into an intramolecular disulfide bond between the *Yap1* n-CRD Cys303 and c-CRD Cys598, and then a second intramolecular disulfide bond between Cys310 and Cys629 is formed for stabilizing the activated form of *Yap1* [43]. Consequently, the two disulfide bonds cause an n-terminal α-helix to mask the nuclear export sequence in the c-CRD, resulting in nuclear accumulation of *Yap1* [44].

Unlike H_2O_2, activation of *Yap1* by the model electrophile NEM *via* a Gpx3-independent mechanism has been raised by Azevedo et al. [45]. This c-CRD activation mechanism may not require any disulfide bond formation, because the chemical was found to form direct adducts on all three c-CRD cysteines of *Yap1*. Thus, by c-CRD cysteine adduction, *Yap1* may be activated through a conformational change, which is different from that one activated by H_2O_2 [45]. As for ACR, a thiol-reactive chemical, the same adaptation mechanism has been confirmed based on the observation that the ACR adaptive response in these two double mutants GPF-*Yap1*$^{C589S,\ C620S}$ or GFP-*Yap1*$^{C598S,\ C629T}$ was weaker than that one in the wild type; these mutants maintained sufficient inducible protection. Yeast cells carrying the double mutant GFP-*Yap1*$^{C620S,\ C629T}$ did not show increased resistance to ACR or NEM after pretreatment, indicating loss of an adaptive response. These results showed that single mutations of Cys620 or Cys629 did allow an adaptive response after pretreatment with either NEM or ACR. Cys598 is dispensable for this effect [41].

To identify the *Yap1*-targeted genes involving in the adaptive response to ACR and H_2O_2, microarray analysis revealed 43 genes specifically responding to H_2O_2, and 239 genes responding to both NEM and ACR. For the H_2O_2-up-regulated genes, gene ontology (GO) analysis grouped into (a) iron homeostasis as an enriched biological process and catalase activity, (b) iron ion transmembrane transporter activity, and (c) heme binding as enriched molecular functions. For the NEM- and ACR-up-regulated genes, GO analysis identified ribosome biogenesis, ribonucleoprotein complex biogenesis, RNA processing, nitrogen compound metabolic process, and sulfur metabolic process as enriched biological processes and no enriched molecular functions. Further deletion analysis proved that the candidate genes *CTA1* and *CTT1* from the H_2O_2-specific response group exerted selective protection against H_2O_2 toxicity, and the gene *YDR042C* showed protection against both NEM and ACR [41]. In addition, Golla et al. [39] also investigated comparative transcriptome of yeast cells treated with a sublethal dose of AA (0.4 mM) and revealed 22 up-regulated stress response genes including transcription factor *YAP1* and GSH biosynthetic genes GSH1 and GLR1. Based on these experimental evidences, Yap1-mediated adaptive responses were reviewed as shown in Figure 3 [41].

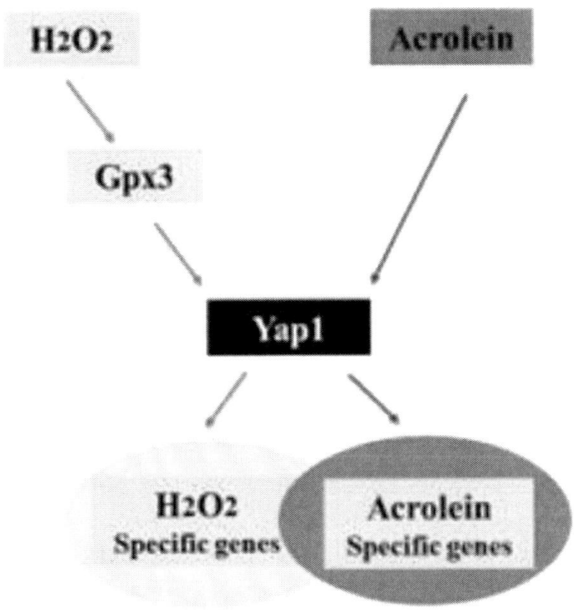

Figure 3. The *Yap1*-mediated adaptive responses to H_2O_2 and ACR [41].

Yeast Oxidative Stress Response Underlies Adaptive Evolution

Adaptability of cells was crucial during evolution. In recent years the adaptive stress response has been extensively studied in yeast cells. This possess is a biphasic dose-response to H_2O_2 and adapt to survive in media containing higher toxic concentrations of

H_2O_2 if they are pre-cultured in a non-lethal H_2O_2-containing medium [46]. The phenomenon, also known as hormesis, has been observed in various organisms: from bacteria to humans [47]. The stress response depends on the nature of the stressor and the downstream targets involved. In this adaptive process, the most critical transcription factor is *YAP1*. Sixty-five unique H_2O_2 and 327 NEM and ACR *Yap1*-dependent responsive genes were identified [41]. H_2O_2, the most abundant reactive oxygen species (ROS), directly causes oxidative stress in living organisms when cells are exposed to excess H_2O_2. It triggers metal loss and enzyme inactivation in iron-sulfur cluster- and iron-containing proteins, DNA damage, oxidation of cysteine (Cys) and methionine residues, as well as lipid peroxidation. As a classical model oxidant, it has been used to monitor a range of biological activities under both physiologic and pathogenic conditions. To counteract oxidative stress, intracellular GSH plays an important role in the adaptive response in yeast to oxidative damage. Moreover *Yap1*-mediated stress response controls changes in the expression of one or more growth- or stress-related genes or transcription factors secondary to a reduction in GSH, which is rapidly depleted following ACR treatment [48, 49, 50].

Most studies concerning oxidative stress response to H_2O_2 have focused on stress-inducible proteins and gene expression. Genome-wide surveys have identified several genes that are activated or repressed in response to oxidants. Until now, there is no effort to carry out systemic adaptive evolution based on H_2O_2-stressed response for GSH production.

ACR-Mediated Adaptation Laboratory Evolution (ALE) for GSH Production

Compared to the large number of studies on H_2O_2 adaptive responses, there are only a few contributions reported on AA toxicity in yeast. These report on hampering growth, inducing cell death, inducing ROS production that derives from depletion of the antioxidant GSH and changes of the GSH/GSSG redox potential, and activation of YAP1 [37, 38, 39, 40]. In these studies, however, AA, the reduced precursor of ACR was used, which possess much less direct toxicity to the yeast cells than ACR. Toxicity of AA mainly depends on the endogenous ADH activity, which converts it to ACR. Although these findings can't provide constructive instructions how to bio-engineer yeast, Patzschke et al. had pioneeringly carried out adaptive laboratory evolution (ALE) to enhance glutathione production of *Saccharomyces cerevisiae* strains [15]. The authors initially compared intracellular GSH levels and growth tolerance concentrations of ACR toward two basic strains, and then constructed a shake flask method of an evolutionary experiment starting with 0.2 mM ACR concentration and increasing by 0.01 mM increment in each transfer. After approximately 100 generations, the ACR concentration had been increased from 0.2 to 0.4 mM and all evolved populations were able to grow on this elevated ACR concentration. The fitness of the evolved populations was analyzed at single colony level and the better evolved isolates were selected for secondary evolution. To accelerate the selection process, the ACR concentration was increased by 0.02, 0.04, and 0.06 mM increments in parallel experiments. After 150 additional generations, the ACR concentration had been increased from 0.4 to 1.34 mM. As a result, an isolate (A4-19-13) with 5.9% GSH content was reached [15]. In this successful experiment the increased GSH level evidently contributed to counteract the ACR toxicity. This result is consistent with that we define non-enzymatic antitoxic defense systems against ACR. Intracellular GSH as a direct scavenger exerts a pivotal physiological role in ACR tolerance in the yeast.

Evolution Engineering of the Engineered Strain

Metabolic engineering strategy has overwhelmingly boosted the GSH production. In our experiments the engineered strain W303-1b/FGP with three biosynthetic pathways, constructed previously in our laboratory was used [9]. Considering the complexity of the metabolic networks, to further improve the inherited capacity, ALE was employed on the prototrophic engineered strain W303-1b/FGP^{PT}, which was derived from the W303-1b/FGP strain by reverse mutation *via* the CRISPR/Cas9-mediated method [16]. Prior to undertaking the ALE experiments, GSH production, biomass (DCW) and stability of the W303-1b/FGP and W303-1b/FGP^{PT} strains were systemically analyzed. The two strains had almost the same DCW (approximately 9.3 g/L), and GSH production (216 mg/L). However, it took the auxotrophic strain 96 h to reach the stationary phase, while the prototrophic W303-1b/FGP^{PT} grew to the stationary phase in 48 h.

Usually, continuous addition of N-methyl-N'-nitro-N-nitrosoguanidine (MNNG) - a chemical mutagen strategy to increase the genetic diversity of the strains - might be more suitable to shorten the adaptation period [51]. Therefore, 5 μg/ml of MNNG was added during adaptation, and an initial concentration of 0.2 mM ACR to inhibit the growth of both strains was chosen to initiate the adaptation assay. Four populations W303-1b/FGP^{PT} AM, W303-1b/FGP^{PT}m AM, W303-1b/FGP^{PT} A and W303-1b/FGP^{PT}m A ('AM' and 'A' referring to ALE assay using ACR as a selection agent with or without addition of MNNG, respectively) were monitored by measuring their GSH content in set intervals. After approximately 100 generations, concentration of ACR increased to 0.4 mM. The pools of W303-1b/FGP^{PT} A-100 and W303-1b/FGP^{PT}m A-75 showed GSH productions of 145.3±4.8 mg/L and 135.3±2.2 mg/L, respectively. Though the presence of MNNG can markedly increase the rate of mutations, it appears that the efficiency of ALE was not associated with genetic diversity of the parental strain, and neither the MNNG treatment of a single high dosage nor a continuous addition of the mutagen contributed to the GSH accumulation [16].

After the first round of adaptive evolution, only three colonies from W303-1b/FGP^{PT} A-100 showed significantly higher GSH production than their parental strain, while a consistent change in the GSH production was found in the isolates of W303-1b/FGP^{PT}m

A-75. To select the preponderant strain used for the second round of direct evolution, six isolates were repeatedly cultured in WMVIII medium [15] and nutrient-rich YPD medium [16]. The highest GSH production (252.3±7.7 mg/L) of isolate W303-1b/FGP^{PT}-8 from the W303-1b/FGP^{PT} A-100 pool was selected for the second round of evolution. After another 100-generation evolution, the GSH production of the evolved pool reached 242.3±10.1 mg/L in WMVIII medium and the corresponding ACR doses raised to 1.24 mM. Subsequently, the GSH productions of forty single isolates from this population were analyzed. Surprisingly, the amounts of GSH of the isolates showed almost identical GSH production. Finally, the isolate W303-1b/FGP^{PT}-8-12 with the maximum GSH production of 245.4±5.2 mg/L at 48 h was selected as the finally evolved strain due to a relatively higher biomass (7.0±0.3 g/L) [16].

Threshold Diffusion of ACR Across Yeast Membrane

According to the two successful ALE, it was observed that after two-round adaptive evolution, the tolerance concentration of the two evolved populations to ACR reached almost the same level, i.e., 1.34 mM (A4-19-13) and 1.24 mM (FGP^{PT}-8-12) [15, 16]. Moreover, after the second round of selection, forty isolates from the secondary evolved pool presented almost identical GSH titer. It is thus concluded that a membrane diffusion threshold of ACR crosses the cell membrane and a neutralization mechanism attenuating cellular toxicity induced by stress pressure of ACR might be operating. Before this observation, there was only one literature reporting that ACR might cross the cell membrane by means of aquaglyceroporin-mediated passive diffusion [52]. Therefore, we assumed that there was a limit not to form a selection pressure during the process of ACR transmembrane transport in the evolved strain, while the endogenous GSH rapidly detoxified ACR to form the less-toxic ACR-GSH adduct in enzymatic and non-enzymatic reactions. Zhou et al. [16] performed two supporting experiments. In the growth tolerance test the W303-1b/FGP^{PT}-8-12 cells were treated by higher concentrations of ACR (1.6, 1.8 and 2.0 mM ACR) than that one (1.24 mM) used in the adaptation. Though the three treatments greatly exceeded the final 1.24 mM of adaptation, the survival rates of W303-1b/FGP^{PT}-8-12 cells showed much less difference. In the other assay, accumulation of adduct of ACR-GSH was determined by LC-MS/MS using synthetic ACR-GSH conjugate as a reference standard. It was found that in case of single treatment with a higher dose of ACR, accumulation of the ACR-GSH adduct was nearly linear in the early 60 min. After the 90 min. timepoint, however, the ACR-GSH concentration became steeply declined. Based on the two results, it could be considered that ACR has a constant threshold diffusion coefficient across the biomembrane and the GSH production capacity of the evolved strain nearly counteracted the stress pressure of ACR of each transfer, leading to no severe growth inhibitions. Another reason can be related to loss of transport capacity of aquaglyceroporin

FPS1, which showed to confer resistance to ACR in *S. cerevisiae* [52]. Moreover, when the evolved yeast cells were treated for longer period of time than 90 min., there might be several antitoxic genes induced through the YAP1-mediated stress response. Thus, the ACR-GSH adduct can be accumulated before the enzyme systems exert its significant oxidative or reductive degradation [39, 40, 53]. Though, during the second round of selection, a higher concentration gradient between the extracellular and intracellular ACR can occur, the cellular homeostasis between the permeated ACR and biosynthesized thiols (e.g., GSH) forms an identical selection trait, leading to no beneficial phenotype or fitness to improve GSH production. Therefore, it is concluded that ACR is a threshold selection agent for GSH production in ALE [16].

Metabolomic Difference Exhibits the Evolved Inherited Traits

To observe the accumulated fitness by ALE and explain the molecular and metabolic bases of GSH over-production, Zhou et al. [16] performed a comparative metabolite analysis of the parental and the evolved strains, and identified 62 comparable metabolites including 27 amino acids, 11 organic acids, 3 sugars, 5 vitamins, 12 nucleotides, and 4 miscellaneous compounds. Among them, most amino acids exhibited significantly higher levels in the evolved strain W303-1b/FGP^{PT}-8-12. Especially, the relative concentration ratios of cysteine (Cys) and glycine (Gly) showed more than 11- and 5-fold increases, respectively. The concentration of glutamic acid (Glu) was also slightly higher in the evolved strain. It might be that Glu itself remained at a high concentration during the fermentation process. It was noted that the levels of intermediates in the Cys biosynthetic pathway, such as L-aspartate, homoserine, homocysteine and cystathionine, greatly varied. The ratio of these intermediates at 48 h, when the accumulation of GSH reached its maximum, were 0.3, 2.0, 3.9 and 10.3, respectively. The continuously declined ratio of L-aspartate from 2.2 at 12 h to 0.3 at 48 h accompanied by the increase of GSH accumulation, indicated that the evolved W303-1b/FGP^{PT}-8-12 had a higher capacity of sulfur assimilation. That was, the accumulated homoserine converted to homocysteine synthesis. Consequently, homocysteine and cystathionine contributed to a higher Cys accumulation and further leaded to an increased GSH production [16].

In addition to the increased intermediates of GSH production, the adequate ATP supplement is another element that triggers GSH overproduction in the evolved strain. In the engineered W303-1b/FGP^{PT}-8-12 cells, the ratios of glyceric acid, isocitric acid, pyruvic acid and citric acid, involving in glycolysis and citrate cycle, were also >1. This indicated that more ATP regeneration promoted two consecutive ATP-dependent GSH synthesis in the evolved strain, resulting in the enhancement of the GSH productivity [54].

Besides the two main concerns mentioned above, the evolved strain also presents a much 'stronger' metabolic capacity: the higher accumulation of the four nucleotides along with their intermediates would promote cellular DNA synthesis; the enhancements of vitamins nicotinic acid, pyridoxal, riboflavin and tocopherol acetate might improve catalytic efficiencies of the metabolic enzymes. The increase of *O*-phosphoethanolamine and 2-aminoethanol might promote the phospholipid biosynthesis. Thus, these metabolite enhancements indicate that the resulting strain has been evolved at a multivariate fitness, and the corresponding metabolic processes may be distributed among most of the major biological processes [40].

Non-Enzymatic Detoxication Mechanism of the Evolved Strain

According to the non-enzymatic detoxication mechanism described above, the major pathway of ACR detoxification in the evolved cells is conjunction of ACR with GSH. Besides the enhanced accumulation of GSH and the precursor Cys along with its biosynthetic intermediates (e.g., homocysteine and cystathionine), it was clearly observed that the amount of N-acetylcysteine (NAC), a powerful antioxidant amino acid, showed a10-fold accumulation at 48 h compared to that of the parental strain. To confirm whether NAC accumulation converted to GSH production, Zhou et al. employed another supplemental assay. The result showed that the different dosages of NAC added into the cultured W303-1b/FGP^{PT} did not promote the GSH production. Moreover, all the thiols enhanced in the adaptation process can directly trap ACR through formation of the less-toxic ACR-thiol adducts [55]. Hence, it is evident that, in the ACR-stress evolution, a forward selective pressure enables the sulfur flux of the evolved strain to be diverted to produce more thiols, directly neutralizing the intracellular toxicity of ACR. Once the ACR-stress selection is removed, the positively selected strain still inherits a stronger capacity of sulfur assimilation and accumulate a higher amount of NAC along with other thiol intermediates including homocysteine, cystathionine and Cys, which further contribute to GSH production. Thus, it is concluded that the mechanism of action of the stress-induced thiols is direct detoxification of ACR upon exposure to the cells.

Rational Regulation of the GSH Biosynthetic Pathway

Based on the metabolic difference, Cys, the major limiting precursor for GSH production, can be converted to the target synthesis. GSH is synthesized from the amino acids Glu, Cys and Gly, of which an extra addition of Cys could greatly increase the GSH content at the special fermentation stage [56, 57]. In general, the use of a stable supply of intracellular Cys, *via* modulation of the Cys biosynthetic pathway, is a promising strategy for GSH production. To divert the Cys precursors (i.e., homocysteine and cystathionine) accumulated in the evolved strain to Cys synthesis and then increase GSH production, the two key genes encoding cystathionine γ-lyase (γ-CTLase, *cys3*) and cystathionine β-synthase (β-CTSase, *cys4*) of the Cys biosynthetic pathway were independently or simultaneously overexpressed in the evolved strain under the control of a constitutive promoter. GSH productions of the engineered strains W303-1b/FGP^{PT}-8-12/cys3, W303-1b/FGP^{PT}-8-12/cys4, and W303-1b/FGP^{PT}-8-12/cys3/cys4 were increased up to 264 mg/L, 260 mg/L and 272 mg/L, respectively. The results indicated that up-regulation of the Cys biosynthetic pathway might change the flux to Cys and improved GSH production; the DCWs of the two strains W303-1b/FGP^{PT}-8-12/cys3 and W303-1b/FGP^{PT}-8-12/cys4 were not changed [16]. Earlier, Suzuki et al. had overexpressed *cys4* and *cys3* in *S. cerevisiae* DBY7286, which had less effect on GSH production [58]. It can be assumed that, due to the poor Cys precursor supply and conversion to GSH compared to those of the evolved strain, the two regulated genes can't effeciently function. In addition to the strategy described above, production of Cys can be enhanced by disrupting the Cys degradation genes [59] or by metabolic engineering of the sulfur metabolism [60]. It is evident that Cys overproduction depends on the stronger sulfur assimilation, i.e., accumulation of the intermediates homocysteine and cystathionine boosting the GSH production.

The metabolic difference also showed the marked levels of both homocysteine and cystathionine accumulated in the evolved strain. To prompt their accumulation and redirection of the sulfur flux further to improve synthesis of cystathionine and enhance the GSH production, cystathionine beta-lyase *str3* was disrupted. Under normal physiological situation, it can reversely cleave cystathionine *via* an α,β-elimination reaction to produce homocysteine and the by-products pyruvate and ammonia [61], and maintain the physiological cycle between cystathionine and homocysteine. The resulting strain W303-1b/FGP^{PT}-8-12/Δstr3 produced higher GSH levels of 261.0±3.8 mg/L than the control (132.5±7.2 mg/L). Although, previous research showed that yeast cells in the absence of *str3* were unable to grow on Cys or cystathionine as the sole sulfur source [62], this culture condition did not occur in the evolved strain [16]. Thus, the gene deletion of *str3* is a feasible strategy to boost GSH production by redirecting the sulfur flux.

In the secondary ALE case, the basic strain W303-1b/FGP^{PT} had been integrating three recombinant genes along with the *cys3* and *cys4* genes involving in Cys precursor synthesis under the uniformly constitutive promoter, which can guide the target genes to higher express. Multi-protein overexpression may cause the cellular metabolic burden. Especially many of the protein folding maturation processes taking place in the endoplasmic reticulum (ER) are regulated by a mechanism referred to as the unfolded protein response (UPR) under conditions of protein miss-folding or unfolded stress [63]. During the maturation of proteins, the correct disulfide formations play a central role. In this process, protein disulfide isomerase (PDI) as a molecular chaperone is essential for formation of disulfide bonds in the protein folding. The formed oxidized PDIs will be re-reduced by intrinsic GSH-mediated oxidation-reduction reaction, which maintains the balance of the oxidized and reduced PDIs [64]. Considering this UPR-induced effect, Zhou et al. further up-regulated the PDI expression under the control of the phosphoglycerate kinase 1 (PGK1) promotor of *S. cerevisiae*. Firstly, that of the evolved strain W303-1b/FGP^{PT}-8-12 along with overexpression of *pdi1*; its intracellular GSH yield increased to 276.5±3.0 mg/L, and its DCW was increased to 7.7±0.1 g/L. This result showed that overexpression of the additional Pdi1 attenuated their growth inhibition induced by the UPR effects of protein misfolding, along with the increased DCW and GSH contents compared to the parental strains [16].

To obtain a higher GSH production, combinatorial up-regulation of *pdi* and *cys3* and *cys4* were further investigated. The engineered strain W303-1b/FGP^{PT}-8-12/Δstr3/cys3/cys4/pdi showed a particularly high GSH production of 296.3±5.6 mg/L, 1.21-fold higher than that of the control strain W303-1b/FGP^{PT}-8-12, and its DCW reached 6.5±0.2 g/L. It could be inferred that the *pdi1* overexpression partially restored the growth inhibition due to the co-expression of *cys3* and *cys4*. These results indicated that deletion of *str3* and overexpression of *cys3*, *cys4* and *pdi1* function dramatically increased intracellular GSH production. It is evident that multiple protein over-expressions can saturate the cell's capacity to properly fold proteins and expressing additional PDIs can compensate for the defects in disulfide formation and isomerization, resulting in enhanced capacity of the engineered yeast cells to GSH production. Although several efforts focusing on modulation of the GSH biosynthetic pathway have been made to develop the engineered yeasts for GSH production [9], very little attention has been paid to the fundamental cellular processes, such as folding of proteins. The positive regulation of additional PDIs indicates that a complex metabolic process, especially in biosynthetic industry, requires systemic manipulation to counteract the UPR effects induced by the overexpressed misfolding proteins.

CONCLUSION

Although metabolic engineering has successfully been employed to improve the capacity of GSH production in yeast, considering the complex metabolic network, genetic engineering of only several candidate genes involving in the GSH biosynthesis can't largely improve capacity of the engineered cells. The engineered strain should further be conducted metabolomic analysis due to the higher GSH accumulation, which may inevitability induce the unexpected physiological imbalance. Besides this, H_2O_2 as a classical model oxidant not only induces oxidant stress in the living organism, but also functions as signal molecule to regulate normal physiological actions. But, until now there haven't been performed studies on sublethal dose adaptive evolution of H_2O_2 and the corresponding metabolomic analysis. As the increasing experimental evidence indicates adaptive response to H_2O_2 and ACR has a close overlap such as YAP1-mediated activation of down-stream genes and accumulation of effective molecules. Whether a synergistic effect occurs in combination of different stressors, needs to be explored. In summary, ALE combined with metabolomic analysis offers a constructive approach of rational metabolic engineering to further boost GSH production in *S. cerevisiae.*

REFERENCES

[1] Jones, D. P. Redefining oxidative stress. *Antioxid. Redox Signal.* 2006, 8, 1865-1879.

[2] Ketterer, B.; Coles, B.; Meyer, D. J. The role of glutathione in detoxication. *Environ. Health Perspect.* 1983, 49, 59-69.

[3] Dröge, W.; Breitkreutz, R. Glutathione and immune function. *Proc. Nutr. Soc.* 2000, 59, 595-600.

[4] Richman, P.; Meister, A. Regulation of gamma-glutamyl-cysteine synthetase by nonallosteric feedback inhibition by glutathione. *J. Biol. Chem.* 1975, 250, 1422-1426.

[5] Gopal, S.; Borovok, I.; Ofer, A.; Yanku, M.; Cohen, G.; Goebel, W.; Kreft, J.; Aharonowitz, Y. A multidomain fusion protein in *Listeria monocytogenes* catalyzes the two primary activities for glutathione biosynthesis. *J. Bacteriol.* 2005, *187*, 3839-3847.

[6] Janowiak, B. E.; Griffith, O. W. Glutathione synthesis in *Streptococcus agalactiae* one protein accounts for γ-glutamylcysteine synthetase and glutathione synthetase activities. *J. Biol. Chem.* 2005, 280, 11829-11839.

[7] Vergauwen, B.; De Vos, D.; Van Beeumen, J. J. Characterization of the bifunctional γ-glutamate-cysteine ligase/glutathione synthetase (GshF) of *Pasteurella multocida. J. Biol. Chem.* 2006, 281, 4380-4394.

[8] Veeravalli, K.; Boyd, D.; Iverson, B. L.; Beckwith, J.; Georgiou, G. Laboratory evolution of glutathione biosynthesis reveals natural compensatory pathways. *Nat. Chem. Biol.* 2011, 7, 101-105.

[9] Tang, L.; Wang, W.; Zhou, W.; Cheng, K.; Yang, Y.; Liu, M.; Cheng, K.; Wang, W. Three-pathway combination for glutathione biosynthesis in *Saccharomyces cerevisiae*. *Microb. Cell Fact.* 2015, 14, 139.

[10] Zhang, J.; Quan, C.; Wang, C.; Wu, H.; Li, Z.; Ye, Q. Systematic manipulation of glutathione metabolism in *Escherichia coli* for improved glutathione production. *Microb. Cell Fact.* 2016, 15, 38.

[11] Kiriyama, K.; Hara, K. Y.; Kondo, A. Extracellular glutathione fermentation using engineered *Saccharomyces cerevisiae* expressing a novel glutathione exporter. *Appl. Microbiol. Biotechnol.* 2012, 96, 1021-1027.

[12] Hara, K. Y.; Kiriyama, K.; Inagaki, A.; Nakayama, H.; Kondo, A. Improvement of glutathione production by metabolic engineering the sulfate assimilation pathway of *Saccharomyces cerevisiae*. *Appl. Microbiol. Biotechnol.* 2012, 94, 1313-1319.

[13] Suzuki, T.; Yokoyama, A.; Tsuji, T.; Ikeshima, E.; Nakashima, K.; Ikushima, S.; Kobayashi, C.; Yoshida, S. Identification and characterization of genes involved in glutathione production in yeast. *J. Biosci. Bioeng.* 2011, 112, 107-113.

[14] Semchyshyn, H. M.; Lozinska, L. M.; Miedzobrodzki, J.; Lushchak, V. I. Fructose and glucose differentially affect aging and carbonyl/oxidative stress parameters in *Saccharomyces cerevisiae* cells. *Carbohydr. Res.* 2011, 346, 933-938.

[15] Patzschke, A.; Steiger, M. G.; Holz, C.; Lang, C.; Mattanovich, D.; Sauer, M. Enhanced glutathione production by evolutionary engineering of *Saccharomyces cerevisiae* strains. *Biotech. J.* 2015, 10, 1719-1726.

[16] Zhou, W.; Yang, Y.; Tang, L.; Cheng, K.; Li, C.; Wang, H.; Liu, M.; Wang, W. Acrolein-stressed threshold adaptation alters the molecular and metabolic bases of an engineered *Saccharomyces cerevisiae* to improve glutathione production. *Sci. Rep.* 2018, 8, 4506.

[17] Kehrer, J. P.; Biswal, S. S. The molecular effects of acrolein. *Toxicol. Sci.* 2000, 57, 6-15.

[18] Zhu, Q.; Sun, Z.; Jiang, Y.; Chen, F.; Wang. M. Acrolein scavengers: reactivity, mechanism and impact on health. *Mol. Nutr. Food Res.* 2011, 55, 1375-1390.

[19] Stevens, J. F.; Maier, C. S. Acrolein: sources, metabolism, and biomolecular interactions relevant to human health and disease. *Mol. Nutr. Food Res.* 2008, 52, 7-25.

[20] Uchida K. Current status of acrolein as a lipid peroxidation product. *Trends Cardiovasc. Med.* 1999, 9, 109-113.

[21] Cannon, J.; Linke, C. A.; Cos, L. R. Cyclophosphamide-associated carcinoma of urothelium: modalities for prevention, *Urology* 1991, 38, 413-416.

[22] Anderson, M. M.; Hazen, S. L.; Hsu, F. F.; Heinecke, J. W. Human neutrophils employ the myeloperoxidase-hydrogen peroxide-chloride system to convert hydroxy-amino acids into glycolaldehyde, 2-hydroxypropanal, and acrolein. A mechanism for the generation of highly reactive alpha-hydroxy and alpha,beta-unsaturated aldehydes by phagocytes at sites of inflammation. *J. Clin. Invest.* 1997, 99, 424-432.

[23] Uchida, K.; Kanematsu, M.; Sakai, K.; Matsuda, T.; Hattori, N.; Mizuno, Y.; Suzuki, D.; Miyata, T.; Noguchi, N.; Niki, E.; Osawa T. Protein-bound acrolein: potential markers for oxidative stress. *Proc. Natl. Acad. Sci. USA* 1998, 95, 4882-4887.

[24] Cai, J.; Bhatnagar, A,; Pierce, W. M.; Jr. Protein modification by acrolein: formation and stability of cysteine adducts. *Chem Res Toxicol.* 2009, 22, 708-716.

[25] Lambert, C.; Li, J.; Jonscher, K.; Yang, T. C.; Reigan, P.; Quintana, M.; Harvey, J.; Freed, B. M. Acrolein inhibits cytokine gene expression by alkylating cysteine and arginine residues in the NF-kappaB1 DNA binding domain. *J. Biol. Chem.* 2007, 282, 19666-19675.

[26] Furuhata, A.; Ishii, T.; Kumazawa, S.; Yamada, T.; Nakayama, T.; Uchida, K. N(ε)-(3-methylpyridinium)lysine, a major antigenic adduct generated in acrolein-modified protein. *J. Biol. Chem.* 2003, 278, 48658-48665.

[27] Baba, S. P.; Hoetker, J. D.; Merchant, M.; Klein, J. B.; Cai, J.; Barski, O. A.; Conklin, D. J.; Bhatnagar, A. Role of aldose reductase in the metabolism and detoxification of carnosine-acrolein conjugates. *J. Biol. Chem.* 2013, 288, 28163-28179.

[28] Wolfle W. T.; Johnson R. E.; Minko I. G.; Lloyd R. S.; Prakash S.; Prakash L. Human DNA polymerase ι promotes replication through a ring-closed minor-groove adduct that adopts a syn conformation in DNA. *Mol. Cell. Biol.* 2005, 25, 8748–8754.

[29] Minko, I. G.; Kozekov, I. D.; Harris, T. M.; Rizzo, C. J.; Lloyd, R. S.; Stone, M. P. Chemistry and biology of DNA containing 1,N(2)-deoxyguanosine adducts of the alpha,beta-unsaturated aldehydes acrolein, crotonaldehyde, and 4-hydroxynonenal. *Chem. Res. Toxicol.* 2009, 22, 759-78.

[30] Nair, D. T.; Johnson, R. E.; Prakash, L.; Prakash, S.; Aggarwal, A. K. Protein-template-directed synthesis across an acrolein-derived DNA adduct by yeast Rev1 DNA polymerase. *Structure*. 2008, 16, 239-245.

[31] Singh, V.; Fedeles, B. I.; Li, D.; Delaney, J. C.; Kozekov, I. D.; Kozekova, A.; Marnett, L. J.; Rizzo, C, J.; Essigmann, J. M. Mechanism of repair of acrolein- and malondialdehyde-derived exocyclic guanine adducts by the α-ketoglutarate/Fe(II) dioxygenase AlkB. *Chem. Res. Toxicol.* 2014, 27, 1619-1631.

[32] Dylewska, M.; Kuśmierek, J. T.; Pilżys, T.; Poznański, J.; Maciejewska, A. M. 1,N^6-α-hydroxypropanoadenine, the acrolein adduct to adenine, is a substrate for AlkB dioxygenase. *Biochem. J.* 2017, 474, 1837-1852.

[33] Maciejewska, A.M.; Poznanski, J.; Kaczmarska, Z.; Krowisz, B.; Nieminuszczy, J.; Polkowska-Nowakowska, A. Alkb dioxygenase preferentially repairs protonated substrates: specificity against exocyclic adducts and molecular mechanism of action. *J. Biol. Chem.* 2013, 288, 432–441.

[34] Do, T. Q., J. R. Schultz, and C. F. Clarke. Enhanced sensitivity of ubiquinone-deficient mutants of *Saccharomyces cerevisiae* to products of autoxidized polyunsaturated fatty acids. *Proc. Natl. Acad. Sci. USA* 1996, 93,7534–7539.

[35] Aoshima, H.; Kadoya, K.; Taniguchi, H.; Satoh, T.; Hatanaka, H. Generation of free radicals during the death of *Saccharomyces cerevisiae* caused by lipid hydroperoxide. *Biosci. Biotechnol. Biochem.* 1999, 63, 1025-1031.

[36] Liao, X. Y.; Shen, W.; Chen, J.; Li, Y.; Du, G. C. Improved glutathione production by gene expression in *Escherichia coli*. *Lett. Appl. Microbiol.* 2006, 43, 211-214.

[37] Kwolek-Mirek, M.; Bednarska, S.; Bartosz, G.; Biliński, T. Acrolein toxicity involves oxidative stress caused by glutathione depletion in the yeast *Saccharomyces cerevisiae*. *Cell Biol. Toxicol.* 2009, 25, 363-378.

[38] Kwolek-Mirek, M.; Zadrąg-Tęcza, R.; Bednarska, S.; Bartosz, G. Acrolein-Induced Oxidative Stress and Cell Death Exhibiting Features of Apoptosis in the Yeast *Saccharomyces cerevisiae* Deficient in SOD1. *Cell Biochem. Biophys.* 2015, 71, 1525-1536.

[39] Golla, U.; Bandi, G.; Tomar, R. S. Molecular cytotoxicity mechanisms of allyl alcohol (acrolein) in budding yeast. *Chem. Res. Toxicol.* 2015, 28, 1246-1264.

[40] Trotter, E. W.; Collinson, E. J.; Dawes, I. W.; Grant, C. M. Old yellow enzymes protect against acrolein toxicity in the yeast *Saccharomyces cerevisiae*. *Appl. Environ. Microbiol.* 2006, 72, 4885-4892.

[41] Ouyang, X.; Tran, Q. T.; Goodwin, S.; Wible, R. S.; Sutter, C. H.; Sutter, T. R. Yap1 activation by H_2O_2 or thiol-reactive chemicals elicits distinct adaptive gene responses. *Free Radic. Biol. Med.* 2011, 50, 1-13.

[42] Yan, C.; Lee, L. H.; Davis L. I. Crm1p mediates regulated nuclear export of a yeast AP-1-like transcription factor. *EMBO J.* 1998, 17, 7416-7429.

[43] Delaunay, A,; Pflieger, D.; Barrault, M. B.; Vinh, J.; Toledano, M. B. A thiol peroxidase is an H_2O_2 receptor and redox-transducer in gene activation. *Cell*, 2002. 111, 471-481.

[44] Wood, M. J.; Storz, G.; Tjandra, N. Structural basis for redox regulation of Yap1 transcription factor localization. *Nature*, 2004, 430, 917-921.

[45] Azevedo, D.; Tacnet, F.; Delaunay, A.; Rodrigues-Pousada, C.; Toledano, M. B. Two redox centers within Yap1 for H_2O_2 and thiol-reactive chemicals signaling. *Free Radic. Biol. Med.* 2003, 35, 889-900.

[46] Ludovico, P.; Burhans, W. C. Reactive oxygen species, ageing and the hormesis police. *FEMS Yeast Res.* 2014, 14, 33-39.

[47] Musa, M,; Perić, M.; Bou Dib, P.; Sobočanec, S.; Šarić, A.; Lovrić, A.; Rudan, M.; Nikolić, A.; Milosević, I.; Vlahoviček, K.; Raimundo, N.; Kriško, A. Heat-induced longevity in budding yeast requires respiratory metabolism and glutathione recycling. *Aging* (Albany NY). 2018, 10, 2407-2427.

[48] Ng, C. H.; Tan, S. X.; Perrone, G. G,; Thorpe, G. W.; Higgins, V. J.; Dawes, I.W. Adaptation to hydrogen peroxide in *Saccharomyces cerevisiae*: the role of NADPH-generating systems and the SKN7 transcription factor. *Free Radic. Biol. Med.* 2008, 44, 1131-1145.

[49] Fernandes, P. N.; Mannarino, S. C.; Silva, C. G.; Pereira, M. D.; Panek, A. D.; Eleutherio, E. C. Oxidative stress response in eukaryotes: effect of glutathione, superoxide dismutase and catalase on adaptation to peroxide and menadione stresses in *Saccharomyces cerevisiae*. *Redox Rep.* 2007, 12, 236-244.

[50] Maris, A. F.; Kern, A. L.; Picada, J. N.; Boccardi, F.; Brendel, M.; Henriques, J. A. Glutathione, but not transcription factor Yap1, is required for carbon source-dependent resistance to oxidative stress in *Saccharomyces cerevisiae*. *Curr. Genet.* 2000, 37, 175-182.

[51] Lee, D.-H.; Feist, A. M.; Barrett, C. L.; Palsson, B. Ø. Cumulative number of cell divisions as a meaningful timescale for adaptive laboratory evolution of *Escherichia coli*. *PLoS One*. 2011, 6, e26172.

[52] Shepherd, A.; Piper, P. W. The Fps1p aquaglyceroporin facilitates the use of small aliphatic amides as a nitrogen source by amidase-expressing yeasts. *FEMS Yeast Res*. 2010, 10, 527-534.

[53] Mitchell, D.; Petersen, D. Metabolism of the glutathione-acrolein adduct, S-(2-aldehydo-ethyl) glutathione, by rat liver alcohol and aldehyde dehydrogenase. *J. Pharmacol. Exp. Ther*. 1989, 251, 193-198.

[54] Yoshida, H.; Hara, K. Y.; Kiriyama, K.; Nakayama, H.; Okazaki, F.; Matsuda, F.; Ogino, C.; Fukuda, H.; Kondo, A. Enzymatic glutathione production using metabolically engineered *Saccharomyces cerevisiae* as a whole-cell biocatalyst. *Appl. Microbiol. Biotech*. 2011, 91, 1001-1006.

[55] Nunoshiba, T.; Yamamoto, K. Role of glutathione on acrolein-induced cytotoxicity and mutagenicity in *Escherichia coli*. *Mut. Res*. 1999, 442, 1-8.

[56] Alfafara, C.; Miura, K.; Shimizu, H.; Shioya, S.; Suga, K.-i. Cysteine addition strategy for maximum glutathione production in fed-batch culture of *Saccharomyces cerevisiae*. *Appl. Microbiol. Biotech*. 1992, 37, 141-146.

[57] Liang, G.; Liao, X.; Du, G.; Chen, J. Elevated glutathione production by adding precursor amino acids coupled with ATP in high cell density cultivation of *Candida utilis*. *J. Appl. Microbiol*. 2008, 105, 1432-1440.

[58] Suzuki, T.; Yokoyama, A.; Tsuji, T.; Ikeshima, E.; Nakashima, K.; Ikushima, S.; Kobayashi, C.; Yoshida, S. Identification and characterization of genes involved in glutathione production in yeast. *J. Biosci. Bioeng*. 2011, 112, 107-113.

[59] Zhang, J.; Quan, C.; Wang, C.; Wu, H.; Li, Z.; Ye, Q. Systematic manipulation of glutathione metabolism in *Escherichia coli* for improved glutathione production. *Microb. Cell Fact*. 2016, 15, 38.

[60] Hara, K. Y.; Kiriyama, K.; Inagaki, A.; Nakayama, H.; Kondo, A. Improvement of glutathione production by metabolic engineering the sulfate assimilation pathway of *Saccharomyces cerevisiae*. *Appl. Microbiol. Biotech*. 2012, 94, 1313-1319.

[61] Holt, S.; Cordente, A. G.; Curtin, C. *Saccharomyces cerevisiae* STR3 and yeast cystathionine β-lyase enzymes: The potential for engineering increased flavor release. *Bioeng Bugs*. 2012, 3, 178-180.

[62] Hansen, J.; Johannesen, P. F. Cysteine is essential for transcriptional regulation of the sulfur assimilation genes in *Saccharomyces cerevisiae*. *Mol. Gen. Genet*. 2000, 263, 535-542.

[63] Bernales, S.; Papa, F. R.; Walter, P. Intracellular signaling by the unfolded protein response. *Annu. Rev. Cell Dev. Biol*. 2006, 22, 487-508.

[64] Khan, H. A.; Mutus, B. Protein disulfide isomerase a multifunctional protein with multiple physiological roles. *Front. Chem*. 2014, 2, 70.

BIOGRAPHICAL SKETCH

Wei Wang

Present Position

Professor of Institute of Materia Medica, Peking Union Medical College, Beijing, China

Affiliation

Institute of Materia Medica, Peking Union Medical College & Chinese Academy of Medical Sciences, Beijing, China

Education

PhD, the Studies on Metabolic Engineering of Natural Medicines from Medicinal Plant, Peking Union Medical College (PUMC), 1999 – 2002.

MS, Molecular Biology of Plant, Capital Normal University, 1996-1997; Institute of Botany, Chinese Academy of Science, 1997-1999.

BS, Biology, QuFu Normal University, 1989 – 1993.

Research and Professional Experience

Wei Wang gained his PhD in pharmacognosy from Peking Union Medical College in 2002. After postdoctoral research in Chemical Biology at Nanyang Technology University (2006–2008), he joined the Institute of Materia Medica, Chinese Academy of Medical Sciences working on enzyme catalysis and synthetic biology, where he became a group leader in 2008. His current research focuses on biocatalysis for biosynthesis of natural drugs and pharmaceutical manufacturing, as well as biobased chemical production.

Publications Last Three Years

1. Zhou, W.; Yang, Y.; Tang, L.; Cheng, K.; Li, C.; Wang, H.; Liu, M.; Wang, W. Acrolein-stressed threshold adaptation alters the molecular and metabolic bases of an engineered Saccharomyces cerevisiae to improve glutathione production. *Sci. Rep.* 2018, 8, 4506.
2. Tang, L.; Wang, W.; Zhou, W.; Cheng, K.; Yang, Y.; Liu, M.; Cheng, K.; Wang, W. Three-pathway combination for glutathione biosynthesis in *Saccharomyces cerevisiae*. *Microb. Cell Fact.* 2015, 14, 139.

3. Wang, H., Yang, Y., Lin, L., Zhou, W., Liu, M., Cheng, K., Wang, W. Engineering *Saccharomyces cerevisiae* with the deletion of endogenous glucosidases for the production of flavonoid glucosides. *Microb Cell Fact,* 2016, 15, 134.
4. Yang, Y., Wang, H., Tong, Y., Liu, M., Cheng, K., Wu, S., Wang, W. Systems metabolic engineering of *Escherichia coli* to enhance the production of flavonoid Glucuronides. *RSC Adv*, 2016, 6, 33622–33630.

Session II. Functions-Malfunctions

In: Glutathione
Editor: Pál Perjési

ISBN: 978-1-53614-740-7
© 2019 Nova Science Publishers, Inc.

Chapter 5

GLUTATIONE AND GLUTATHIONE-DEPENDENT ENZYMES IN REDOX REGULATION OF CELLULAR PROCESSES

Elena. V. Kalinina**, *Nikolai. N. Chernov and Maria. D. Novichkova
T.T. Berezov Department of Biochemistry, Peoples' Friendship University of Russia (RUDN University), Moscow, Russian Federation

ABSTRACT

The chapter summarizes the data on the role of glutathione, glutathione transferase, and glutaredoxin in the regulation of cellular redox-dependent processes and includes the following sections: glutathione and cellular redox homeostasis, the role of glutathione in the redox-dependent regulation of apoptosis, glutaredoxin and redox regulation of cellular viability, the family of glutathione transferases in the regulation of redox-dependent processes. The role of glutathione in redox-dependent processes in the cytoplasm, nucleus and mitochondria including its role in redox-dependent regulation of MAPK-induced apoptotic signaling is discussed. It is considered the important role of the superfamily of glutathione S-transferase isoenzymes in redox-dependent processes. A separate section is devoted to the value of glutaredoxin in redox regulation including S-glutathionylation and protein-protein interaction.

Keywords: glutathione, glutathione transferase, mitochondrial GSH, glutaredoxin, cellular redox-dependent processes

* Corresponding Author E-mail: elena.v.kalinina@gmail.com.

Glutathione and Cellular Redox Homeostasis

The state of redox status determines the most vital functions of cells, including proliferation, differentiation, apoptosis and autophagy, and depends on the balance of oxidation–reduction processes. It is controlled by the redox-dependent regulation based, in great part, on the thiol-disulfide exchange, which is highly sensitive to changes in the ROS/antioxidants balance, providing a cellular response to a change in the ROS level [1-4].

Maintenance of the GSH/GSSG ratio is important for control of the thiol-disulfide exchange. Among the several functions of glutathione, its participation in the defense of the cell from products of oxidative stress should be mentioned as the most important one. Hydrogen peroxide is reduced by glutathione peroxidase using GSH as the cosubstrate. Organic hydroperoxides can be reduced to the corresponding alcohols in the reactions catalyzed by glutathione peroxidases as well as due to the peroxidase activity of Se-independent glutathione S-transferases that also use GSH as the cosubstrate. Glutathione can reduce oxidized glutaredoxin that is formed during reduction of disulfides [5].

Glutathione is a low molecular weight antioxidant that can also take part in the nonenzymatic antioxidant defense, playing a role as an efficient scavenger of free radicals [6]. Oxidative stress leads to damage of carbohydrates, lipids, and nucleic acids, resulting in cell dysfunctions and death. Oxidative stress and/or change in cellular redox status can affect the state of the nuclear chromatin and alter gene expressions. Progression of oxidative stress can result in single- or double-strand breaks in DNA molecules. Mitochondrial damage is followed by a decrease in the transmembrane potential, changes in membrane permeability, and accelerated release of apoptotic factors, which leads to cell death [7]. Under physiological conditions, reactive oxygen and nitrogen species (ROS and RNS, respectively) are involved in processes of redox-signaling, which are fast, specific, and reversible reactions regulating the activity of proteins that are important for cell functioning. The processes involved in redox-signaling can occur in various cell compartments at certain times with participation of different redox pairs, such as GSH/GSSG or $NADH(H^+)/NAD^+$ [8]. Special attention is paid to the GSH/GSSG ratio as the main marker of redox status and an important factor of signal transduction [9].

Functionally important cysteine residues of many proteins are subjected to posttranslational modifications including oxidation. Under physiological conditions, the thiol group of free cysteine is mainly protonated ($pK_a = 8.5$), which excludes its oxidative modification. Within proteins, however, cysteine can be activated, i.e., exists as the thiolate anion. This is due to numerous factors including hydrogen bonding, the impact of neighboring amino acid residues, the microenvironment of cysteine residues, and binding with substrates [10]. The cysteine residue of GSH can interact with cysteine residues of proteins yielding a disulfide bond (protein-SSG). This process called S-glutathionylation defends proteins from oxidative stress and makes a significant contribution to redox signaling and regulation of protein activities [11]. S-glutathionylation can affect ability of the protein to form disulfide bonds and to correct its folding, which influences the functional state of the protein. Also, S-glutathionylation protects sulfenic acid derivatives (Cys-SOH), formed during oxidation of cysteine residues, from oxidation to sulfinic acid (Cys-SO_2H) and then further to sulfonic acid (Cys-SO_3H), which cannot be reduced under physiological conditions. The reversibility of oxidation of cysteine residues is of great importance for the normal functioning of proteins and their ability to participate in signal transduction cascades. Glutathione is the main substrate for the reduction of oxidized cysteine residues.

Overproduction of glutathionylated proteins indicates the progression of oxidative stress leading to cell death. Change in sulfhydryl homeostasis of the cell, especially the steady state of glutathionylation of specific regulatory proteins, modulates various pathways of signal transduction, shifting the cell state from survival to death. For example, the functioning of the cell actin is regulated by reversible S-glutathionylation, and disturbance in S-glutathionylation changes the structural organization of stress fibrils of the actin cytoskeleton [12]. Under normal conditions, modifications such as protein-SSG are transient and reversible. If the cysteine residue is essential, its S-glutathionylation can affect the functioning of the protein. For example, S-glutathionylation of subunits p65 and p50 of the transcriptional factor NF-κB inhibits their binding with DNA [13], while S-glutathionylation of the β-subunit of IκB kinase suppresses the activation of NF-κB [14].

The state of the thiol–disulfide system is determined by the cellular redox status, which is characterized by the GSH/GSSG ratio. Under physiological conditions, GSH/GSSG is about 100:1, which minimizes the oxidative action of ROS/RNS. Disturbance of this ratio significantly affects, in context of the redox regulation, processes of signal transduction, control of gene expression, cell proliferation, differentiation, state of cell metabolism, and vital functions overall [15, 16]. Glutathione is synthesized only in the cytoplasm and then transferred to the mitochondria, peroxisomes, endoplasmic reticulum, and nucleus. More than 70% of the total GSH remains in the cytoplasm, while the nucleus and mitochondria can accumulate up to 10% and 30% of the total intracellular GSH content, respectively [17].

The content of GSH in mitochondria (mGSH) is approximately the same as in the cytoplasm (10-14 mM) [18]. Glutathione is not synthesized in mitochondria, but it is transported from the cytoplasm. GSH easily penetrates through the porin channels of the outer mitochondrial membrane. Since under physiological conditions GSH exists as an anion, it is unable to diffuse into the matrix through the inner mitochondrial membrane that has a high negative transmembrane potential. GSH is transferred into the mitochondrial matrix by transporters in the inner mitochondrial membrane working against the electrochemical gradient. In mitochondria of liver and kidneys, this role is played by carriers of 2-oxoglutarate (OGC) and dicarboxylates (DIC), which transfer GSH into the mitochondrial matrix by the antiport mechanism in exchange with 2-oxoglutarate and inorganic phosphate, respectively. Since these transporters provide liver mitochondria only 45-50% of their total GSH, some additional mechanism must exist.

GSSG does not come out from mitochondria to the cytoplasm, being instead reduced by glutathione reductase yielding GSH. This process depends on the presence of a sufficient amount of NAPDH(H^+). It should be noted that accumulation of GSSG affects glutathionylation of mitochondrial proteins, thus changing their functioning. For example, the activity of NADH-ubiquinone reductase (complex I of the mitochondrial respiratory chain) depends on the GSH/GSSG ratio [19].

In an experimental model with mitochondrial membranes from rat heart, it was found that addition of GSSG after the action of Grx2 resulted in glutathionylation of complex I. In contrast, addition of GSH and Grx2 caused its deglutathionylation [20]. For mitochondria from bovine heart, the sites of glutathionylation in complex I were found to be the cysteine residues Cys531 and Cys704 of the 75-kDa NDUSF1 subunit [21]. The role of S-glutathionylation is of importance for the defense of NADH-ubiquinone reductase from irreversible oxidation and for the control of ROS production by the mitochondria in response to changes in the local redox environment. In this case, the important condition is the simultaneous S-glutathionylation of complex I and the α-ketoglutarate dehydrogenase complex, since the latter supplies NADH(H^+) for oxidation by complex I, i.e., both protein complexes contribute to generation of ROS by the mitochondria [22]. Under conditions of oxidative stress, glutathionylation of these two enzyme complexes decreases their activities and production of ROS [23, 24]. After the return of $O_2^{\cdot-}$ and H_2O_2 to the normal level, the α-ketoglutarate dehydrogenase complex and NADH dehydrogenase are deglutathionylated by Grx2, and oxidative phosphorylation is restored.

The role of mGSH is of great interest. Functions of mitochondria are closely related to the maintenance of cellular redox balance. Mitochondria are the main consumers of oxygen and the main source of ROS, which are mainly generated during functioning of the electron transfer chain. Under physiological conditions, incomplete one-electron reduction of molecular oxygen results in formation of superoxide anion radical $O_2^{\cdot-}$, which gives rise to another ROS. The concentration of $O_2^{\cdot-}$ in the mitochondrial matrix under steady-state conditions exceeds 5-10-fold its concentration in the cytoplasm [25]. Besides, the action of various toxins and some pathological states affecting mitochondrial functions can increase production of ROS. The presence of the antioxidant defense system in mitochondria prevents disturbances in mitochondrial functioning. The main component of this system is mGSH, which prevents or repairs damage occurring under normal aerobic metabolism.

In mitochondria Mn-dependent superoxide dismutase converts $O_2^{\cdot-}$ to H_2O_2, which can be neutralized by GSH-dependent systems with the participation of glutathione transferase and glutathione peroxidase. Glutathione peroxidase GPx1, the most active isoform towards H_2O_2, is mostly localized in the cytoplasm, but a small amount is also present in the mitochondrial matrix [18]. In mitochondria, GST catalyzes formation of GSH-conjugates and reduction of organic hydroperoxides using GSH as a cosubstrate. In contrast to Se-containing GPx, GST does not interact with H_2O_2, but it efficiently reduces hydroperoxides of polyunsaturated (linoleic and arachidonic) fatty acids, phospholipids, mononucleotides and DNA. The Se-independent GPx4 plays an important role in detoxication of lipid hydroperoxides in mitochondria. Recently, it was shown that GPx4 prevents development of cell apoptosis in presence of apoptosis inducing factor (AIF), and supports oxidative phosphorylation in the intestinal epithelial cells [26]. Due to its ability to reduce hydroperoxides of cardiolipin, GPx4 takes part in regulation of release of apoptogenic proteins from mitochondria [27]. In the mitochondria of human cells, peroxidase activity is exhibited by GST isoenzymes hGSTA4-4, hGSTA1-1, hGSTA2-2, and hGSTP1-1, among which isoform hGSTA4-4 is the most active [28].

Grx2 is the mitochondrial isoform of glutaredoxin which also occurs in the nucleus. Interestingly, the oxidized form of Grx2 can be reduced by both thioredoxin reductase (TrxR) and GSH, providing functional activity of Grx2 in the oxidized microenvironment that is common for mitochondria.

In mitochondria, Grx2 plays a significant role in the antioxidant defense system and in redox-dependent signaling *via* interaction of the GSH pool with thiol groups of proteins [28-30]. The thioredoxin-dependent system in mitochondria is represented by thioredoxin 2 (Trx2), thioredoxin reductase (TrxR2) and Prx3, the most important peroxiredoxin isoform. The mGSH/GPx and Prx3/Trx2 systems, that defend against H_2O_2, are interrelated. For example, it was shown that decrease in mGSH resulted in oxidation of Trx2 [31]. Thus, the role of the redox-cycle of mGSH in maintenance of an efficient antioxidant system and homeostasis of hydrogen peroxide in mitochondria is evident.

Some investigations have shown that GSH is accumulated in the nucleus at the beginning of the G1 phase [32], so it could play an important role in maintenance of the redox status of the nucleus during the cell cycle [33]. During mitosis, the nuclear membrane breaks down and appears again around the daughter DNA molecules packed into the chromosomes. During cell division, a high pool of GSH is maintained near the chromatin, which is consistent with reported data on high redox potential of the dividing cell. The nuclear glutathione (nGSH) pool is resistant to the factors decreasing the cellular and the mitochondrial GSH levels such as the action of the thiol-binding compounds N-ethylmaleimide and diethyl maleate, as well as that of the GSH synthesis inhibitor buthionine sulfoximine (BSO). It has been found that cells with a high level of nGSH are more resistant to apoptosis under oxidative stress conditions. However, the role of nGSH in oxidative stress has been little studied [33].

The mechanisms of transport and deposition of nGSH are still little studied. A significant amount of nGSH presumably comes from the parental nucleus to the daughter nuclei during the telophase due to the high concentration of GSH in proximity to the replicating genetic material [33].

Nuclear pores allow various ions and small molecules, including GSH, to penetrate the nucleus [34]. At the same time, an ATP-dependent mechanism of GSH transport into the nucleus was also demonstrated [35]. Currently, the role of protein Bcl-2 in transmembrane transport of GSH is being discussed. It has been found that the nGSH content is significantly increased in tumor cells exhibiting overexpression of the *bcl-2* gene [36]. It should be noted that the BH-3 domain of Bcl-2 protein can bind to GSH. Participation of Bcl-2 in the maintenance of GSH level in mitochondria has been reported [37]. These data together with the fact that the antiapoptotic protein Bcl-2 is involved in formation of pores in the membrane indicate the ability of members of the Bcl-2 protein family to serve as mediators of GSH translocation into the nucleus [32, 36].

Several studies on plant cells have demonstrated that gene expression is sensitive to accumulation of GSH in the nucleus and to its decrease in the cytoplasm [36]. Decrease in redox potential in the cytoplasm and its growth in the nucleus affects not only gene expressions, but the ability of proteins to bind with their targets in the nucleus as well. It was shown that at the beginning of the G1 phase of the cell cycle in animal cells, activation of oxidative processes in the cytoplasm caused by epidermal growth factor resulted in accumulation of ROS, which activated phosphorylation cascades and DNA replication, and induced cell division [38]. The decrease in the cytoplasmic GSH level in the G1 phase may cause accumulation of ROS. It is suggested that changes in the nuclear redox status may act as the trigger element for the other components that are essential for transcription. Such a mechanism was shown for the proteins NF-κB, AP-1, and p53 [39]. For example, for the interaction of NF-κB with DNA, the cysteine residue in the DNA-binding region of NF-κB must be reduced. Similar behavior was reported for the transcriptional factors Fos, Jun, and Nrf2 [36].

GSH content and its changes in the nucleus can modulate structural organization of chromatin [40]. The extent of glutathionylation of nuclear proteins increases at the beginning of cell proliferation [32]. In the animal cells, while going from proliferation through differentiation to cell death, the cellular redox status changes towards a more oxidized state. Thus, the GSH/GSSG ratio is a kind of switch from proliferation to differentiation and then to programmed cell death [41]. The nuclear GSH is related to the synthesis of DNA, presumably being the redox sensor for the beginning of DNA synthesis. GSH maintains the necessary organization of the nucleus at the expense of the optimal redox status for replication of DNA and maintenance of its intact structure. Also, it was found that nGSH affects the proteasomal degradation of nuclear proteins [32].

Activity of Se-dependent glutathione peroxidase was found in the nuclear fraction of Wistar rat hepatocytes [35]. The nucleus-specific isoform of glutathione peroxidase GPx4/snGPx was shown to maintain stability of the chromatin structure in sperm [42]. The nuclear localization of this enzyme emphasizes the important role of glutathione in the regulation of the cell cycle and the chromosomal organization, since nuclear proteins, mainly histones and other chromatin-binding proteins, presumably should be maintained in the reduced state for optimal functioning [33].

The role of GSH in the repair of DNA damage should also be pointed out. Glutathione is not the most efficient protector of DNA from X-radiation, but it controls repair mechanisms of damaged DNA molecules [43]. An important component of the repair mechanism of oxidative damages of DNA is poly(ADP-ribose)polymerase (PARP), which catalyzes the growth of the polymer chains from the ADP-ribose molecules on target proteins (particularly on histones). This process proceeds virtually in all eukaryotic cells in response to DNA damage [44]. Expression of genes and activity of PARP proteins are related to nGSH level during the cell cycle. For plant cells, it was demonstrated that mRNA of PARP1 and PARP2 was accumulated with the growth of the GSH pool in the nucleus [45]. Similar character of changes in polyribosylation activity was observed in NIH3T3 fibroblast cells: polyribosylation of histones in their nuclei grew during proliferation, when the nGSH level was maximal [32, 46]. It should be mentioned, however, that several details concerning the mechanisms of GSH transport into the nucleus and the role of nGSH in various genetic and epigenetic processes remain unclear [32].

ROLE OF GLUTATHIONE IN REDOX-DEPENDENT REGULATION OF APOPTOSIS

The protective role of GSH in the mechanism of apoptosis is widely considered. According to a contemporary concept, decrease of GSH level below a critical value results in appearance of a signal for apoptosis, which is initiated by activation of the death receptor or by mitochondrial apoptotic signaling. In contrast, an increase in the GSH level provides defense of cells from Fas-induced apoptosis [47]. Numerous data indicate the crucial role of GSH in cell defense from various apoptotic stimuli, since disturbances of redox homeostasis of the cell caused by GSH oxidation or GSH export facilitate development of apoptosis [48, 49]. In studies with various types of cells, it has been shown that disturbance of the GSH/GSSG balance caused by an oxidant precedes induction of mitochondrial apoptotic signaling [50, 51]. Restoration of the GSH/GSSG ratio to the normal level after the action of the oxidant does not prevent progress of the apoptosis, suggesting that it is induced in the early stage of the GSH/GSSG disbalance. Use of the thiol antioxidant N-acetylcysteine before the action of the compounds resulting in the oxidative (*tert*-butyl hydroperoxide) or carbonyl (methylglyoxal) stress prevents induction of apoptosis. These data are consistent with other reports and indicate that the signal for apoptotic death is triggered at the very beginning of the decrease of the GSH/GSSG ratio [50-52].

Special attention has been given to investigation of the mitochondrion as the cell organelle involved in activation of apoptosis. The GSH/GSSG ratio is considered to be the main redox system maintaining redox homeostasis of the mitochondrial matrix and defending mitochondrial proteins and DNA from the action of ROS. Using various cell models, it has been shown that selective decrease in the mGSH content resulted in a reduction in activity of the respiratory chain complexes, growth in production of ROS, decrease in transmembrane potential (ΔΨ), and release of apoptogenic factors from mitochondria. For example, in diabetic cardiomyocytes, stress-induced oxidation of mitochondrial, but not cytoplasmic, GSH resulted in a decrease in the ΔΨ value and activation of caspase-9 and caspase-3 [53, 3]. In human B-cell lymphoma cells, ROS-induced decrease in mGSH level initiated apoptosis that was accompanied by drop in the ΔΨ value, release of cytochrome *c*, and activation of caspase-3 [54]. A direct relation between decrease in the mGSH content and activation of apoptosis has been demonstrated in different cell types. In hepatocytes, decrease in mGSH content was a necessary condition for TNF-α-induced apoptosis, which was preceded by tBid/Bax-initiated permeabilization of the mitochondrial membrane, release of cytochrome *c*, assembly of apoptosomes, and activation of caspase-3 [55]. In large intestine cells, oxidation of mGSH was the main factor in development of menadione-induced mitochondrial dysfunction and cytochrome *c*-dependent activation of apoptosis [56].

The precise mechanism of mitochondrial dysfunction caused by a decrease in the mGSH content is not completely understood. However, it was found that cisplatin-induced apoptosis is related to disbalance in the mGSH/GSSG ratio, decrease in NADPH(H^+) content, and oxidative damage of cardiolipin and aconitase, which disturbs functioning of mitochondria and activates caspase-3 [57]. Later works demonstrated that a sharp drop in the mGSH content induced generation of ROS/RNS, leading to apoptosis in HL-60 and Raji cells. In this case, apoptosis was caused by breakdown of complex I of the respiratory chain due to destabilization of the Fe-S cluster of the NDUGS3 subunit of the complex, resulting in inhibition of respiration and drop in the ΔΨ value [58]. Noteworthy, a slight decrease in the mGSH content of hepatocytes caused by moderate hypoxia did not lead to apoptosis. This fact demonstrates that the content of mGSH must drop to a critical level to induce apoptosis [17].

Permeability of the mitochondrial membrane may be controlled by the mGSH content. In early works, decrease in mGSH was attributed to changes in mitochondrial permeability that were caused by redox modulation of adenine nucleotide translocase and subsequent release of apoptogenic factors such as cytochrome *c* and AIF from the mitochondria to the cytoplasm [59, 60]. Later investigations showed that change in redox balance of mGSH was a crucial factor in control of mitochondrial membrane permeability [61]. Drop in the mGSH/GSSG ratio from 300:1 to 20:1 leads to the opening of the anion channels in the inner mitochondrial membrane and the mitochondrial pores. If the mGSH/GSSG ratio lies in the region from 150:1 to 100:1, instability of the ΔΨ value is observed. In the case of pronounced oxidation, when the ratio is less than 50:1, irreversible depolarization of the mitochondrial membrane takes place, accompanied by opening channels and breakdown of the mitochondria [61]. Accelerated transport of glutathione into mitochondria suppressed the menadione-induced increase of the mGSSG level, preventing decrease in the ATP content, drop in the $\Delta\Psi$ value, release of cytochrome *c* into the cytoplasm, and activation of caspase-3 and caspase-9 [56].

Induction of apoptosis under oxidative stress is caused by activation of mitogen-activated protein kinases (MAPKs) [62]. There are three classes of MAPKs: ERK (extracellular signal-regulated kinase), JNK (c-Jun N-terminal kinase), and p38. The signal transduction cascade includes consecutive phosphorylation steps resulting in activation of a specific MAPKKK (MAP3K, kinase of MAPK kinase) that activates MAPK kinase (MAP2K) which in turn, activates MAPK. Stress-induced apoptosis is related to activation of JNK and p38 MAPK, and this can be triggered by the kinase cascade involving apoptosis signal-regulating kinase 1 (ASK1), MAPK kinase 4/7 (MEK4/7), and JNK, or through the cascade of ASK1, MAPK kinase-3/6 (MEK 3/6), and p38 [62, 1, 63].

The role of GSH in redox-dependent regulation of MAPK-induced apoptotic signaling has been little studied to date. Since GSH is a key factor in maintenance of the intracellular redox homeostasis and plays an important role in antioxidant defense of the cell, it might be a modulator of MAPK-dependent signaling pathways. In some cell models it was shown that disbalance of GSH/GSSG activated the MAPK-signaling pathway and facilitated apoptosis. For example, induction of ROS formation by aloe-emodin resulted in disbalance of the GSH/GSSG ratio and redox-dependent activation of the GSTP1/JNK-signaling pathway in hepatoma cells [64].

Other works report that inhibition of the *de novo* GSH synthesis by BSO promotes redox-activation of MAPK and the apoptotic signaling pathway. Treatment of breast cancer cells with the antitumor agent aplidine after treatment with BSO, resulted in activation of the JNK- and p38-dependent signaling pathways and development of apoptosis [65]. Apoptosis of HepG2 cells treated with BSO was induced by andrographolide through activation of the ASK1/MEK4/JNK signaling cascade [66]. Addition of thiols (N-acetyl cysteine and GSH) prevented toxin-induced activations of MAPK, which suggested participation of GSH in functioning of MAPK and cellular response to stress [65].

While investigating the role of GSSG in initiation of apoptosis, it was found that extracellular GSSG can selectively activate the MAP kinase cascade ASK1/MEK3/6/p38 through the mechanism of GSSG-induced thiol-disulfide exchange on the cellular membrane and formation of mixed protein disulfides. Such redox stress can in turn result in breakdown of the Trx1/ASK1 complex and activation of the p38-dependent pathway of apoptosis [67]. The fact that each of these events can be prevented by GSH is consistent with the idea of the protective role of glutathione [67]. In this connection, it should be noted that in SH-SY5Y neuroblastoma cells, that are resistant to GSSG-induced apoptosis, apoptosis was activated after preliminary treatment with BSO, which was accompanied by increased ROS production and activation of JNK [68]. Such data point to the assumption that drop in GSH concentration below a critical level is a necessary condition for the activating effect of GSSG on the MAPK-signaling pathway and induction of apoptosis. GSH regulates the redox state of Trx1 and the Trx1-dependent ASK1 signaling cascade inducing apoptosis. For example, the action of agents oxidizing GSH (diamide and dithionitrobenzoate) on stomach adenocarcinoma cells initiated the mitochondrial pathway of apoptosis [69]. In this case, redox activation of the Trx1/ASK1/p38 signaling cascade was triggered by increase in the GSSG content. Resistance of the cells towards systems (paraquat and xanthine/xanthine oxidase) producing H_2O_2 and ROS correlated with the Nrf2-dependent increase of the GSH content and protein S-glutathionylation [69]. Another response was observed in SH-SY5Y neuroblastoma cells, where H_2O_2 activated the Trx1/p38/p53 cascade and cellular apoptosis, while diamide activated the ERK-signaling pathway, Nrf2-dependent increase in GSH content, and expression of the heme oxygenase-1 gene, which assisted cell survival.

Concerning regulation of the GSH-dependent post-translational modifications of cysteine residues of the proteins involved in the MAPK-signaling pathways, it can be noted that under the oxidative stress caused by menadione (2-methyl-1,4-naphthoquinone), S-glutathionylation of Cys1238 in the ATP-binding domain of MEKK1 inhibits activity of the kinase [70]. However, a question is still open concerning the specific relation between the oxidative stress and S-glutathionylation in terms of activation/inactivation of specific MAPK-dependent signaling pathways and induction of apoptosis.

The presented data show that GSH plays an important role in the redox regulation of MAPK-dependent pathways of signal transduction. However, the difference between the induction of different signaling pathways that are responsible for cell death or survival is likely determined by not only the cellular content of GSH, but also by the cell type [69, 71].

Glutaredoxin and Redox Regulation of Cellular Viability

Glutaredoxin (Grx, EC 1.20.4.1) is one of the most important enzymes involved in processes of disulfide reduction and deglutathionylation. Isoenzymes of Grx are thermoresistant and low molecular weight (10-16 kDa) proteins functioning as GSH-dependent oxidoreductases. According to their structure, they belong to the thioredoxin superfamily, and together with Trx they play an important role in redox-dependent processes in cells. The active site sequence Cys-X-X-Cys/Ser is located in the N-terminal region of Grx, while the conserved glutathione-binding domain is at the C-terminal part of the molecule. Isoenzymes of Grx are found in virtually all living organisms except for some types of bacteria and archaea [72, 73].

Grx isoenzymes are classified in terms of presence of a cysteine residue in the second position of the active site sequence. Isoenzymes with the sequence Cys-Cys-X-Cys/Ser in the active site are called Grx of C-C type. Originally, it was suggested that the main functions of Grx are reduction of disulfide bonds and deglutathionylation of proteins. However, later it was found that certain Grx isoenzymes rather serve as transfer proteins for iron-sulfur clusters [FeS] using GSH as ligand [74]. The oxidized form of Grx formed after the reduction of protein disulfides and glutathionylated thiols is reduced by GSH. However, some Grx isoenzymes are reduced by ferredoxin- or NADPH-dependent thioredoxin reductase, for example Grx4 of *E. coli* and human Grx2 [29]. Depending on the active site structure, Grx isoenzymes can be dithiol or monothiol (active site sequences Cys-X-X-Cys and Cys-X-X-Ser, respectively) [75]. The binding of the Fe-S clusters can lead to formation of dimers and tetramers. In these interactions, alternative protein-protein contact sites are possible in mono- and dithiol Grx isoenzymes, providing existence of both mono- and multidomain forms of Grx [76].

Concerning classification of the Grx isoenzymes, it should be noted that since bacteria, yeast, and mammals have a limited number of these proteins, their classification into mono- and dithiol isoenzymes is sufficient. For photosynthesizing organisms containing a wide range of Grx isoenzymes, a new classification is used [77] dividing the isoenzymes into six classes based the homology of their amino acid sequences. Dithiol Grx isoenzymes belong to class I, monothiol Grx isoenzymes occur in both classes I and II, and glutaredoxins of the C-C type belong to class III. Grx isoenzymes of classes I and II are found virtually in all organisms. Isoenzymes of class III are present in higher plants, where they control the functional activity of plants, for example, flowering [5, 78]. Isoenzymes of class IV are found in photosynthesizing eukaryotes. Glutaredoxins of class V occur in cyanobacteria and proteobacteria, while Grx of class VI are present only in cyanobacteria [79].

Grx isoenzymes use two catalytic mechanisms: monothiol and dithiol. The monothiol mechanism is characteristic for the reactions of deglutathionylation. In this case, only the catalytic cysteine residue (the first of two active site cysteines at the N-terminus) participates in the catalysis. Reduction of a glutathionylated substrate starts from nucleophilic attack of the thiol group of the Grx CysA residue. The substrate is released with formation of the intermediate glutathionylated product Grx-SSG. Further, glutaredoxin is regenerated by GSH, yielding $Grx(SH)_2$ and GSSG. The monothiol mechanism is used by both monothiol and dithiol Grx isoenzymes [5]. The dithiol mechanism, besides the catalytic cysteine residue, requires another cysteine residue (so-called recycling cysteine) that can be either the second cysteine residue of the Grx active site (CysB) or a cysteine residue apart from the active site (CysC). If the substrate is deglutathionylated, the first stage proceeds by the monothiol mechanism, but the glutathionylated intermediate Grx-SSG then releases GSH yielding the intramolecular disulfide bond $Grx(S_2)$ between the catalytic cysteine residue and one of the recycling cysteines. Furthermore, the disulfide bond is reduced using two GSH molecules or by thioredoxin reductase. If the substrate of Grx requires the reduction of an intra- or intermolecular disulfide bond, the CysA residue of Grx forms a transient disulfide bond with one of the substrate cysteines, and then the reduced substrate is released, while the disulfide bond is formed between the CysA and CysB or CysC residues of Grx. Finally, the disulfide bond in Grx(S_2) is reduced by two molecules of GSH or by thioredoxin reductase [80]. All dithiol isoforms of Grx investigated so far are capable of functioning by the monothiol mechanism, but not all of them have been tested for their ability for dithiol catalysis. However, all dithiol Grx isoforms that have the ability for dithiol catalysis can also function using the monothiol mechanism.

Four Grx isoenzymes have been found in mammals: Grx1, Grx2, Grx3 (also known as protein interacting cousin of Trx (PICOT)), and Grx5 [81]. The dithiol isoenzyme Grx1 is localized mainly in the cytoplasm, but also can be translocated into the nucleus, secreted from the cell, and localized in the intermembrane space of mitochondria [75]. The dithiol Grx2 was originally found in mitochondria, but later it was found in the cytoplasm and nucleus of testes and in a number of tumor cells [82]. The monothiol isoenzyme Grx3 is a multidomain protein that is present in the nucleus as well as in the cytoplasm, while the monothiol Grx5 is found in mitochondria [81].

Although the amino acid sequences of Grx1 and Grx2 exhibit only 34% homology, these isoenzymes use the same catalytic mechanism [83, 84]. Grx1 is approximately 10-fold more active than Grx2, but the content of Grx2 in the intermembrane space of mitochondria is higher than that of Grx1, which presumably compensates the difference in their catalytic activity. Oxidized Grx1 is only reduced by GSH, while Grx2 can be reactivated using either glutathione or thioredoxin reductase, which suggests that this protein exhibits properties of both Grx and Trx [29] and implies the connection between the metabolic pathways controlled by glutathione and thioredoxin in mitochondria. Besides, the possibility of reduction of Grx2 by thioredoxin reductase allows Grx2 to function over a wide range of GSH/GSSG values and rather strong oxidative stress in mitochondria [20].

The thiol groups of the active site of some Grx isoenzymes can form complexes with iron-sulfur clusters. These enzymes include a limited number of dithiol Grx isoenzymes of humans, plants, and trypanosomes and virtually all monothiol Grx isoenzymes [85-87,161-163]. Most of such complexes were found in mitochondria. The cluster $[2Fe2S]^{2+}$ is located between two monomers of Grx, forming coordinate bonds with two active site cysteine residues at the N-termini and with two non-covalently bound GSH molecules. The GSH comes from the free GSH pool, indicating the important role of GSH in stabilizing Fe-S clusters [88]. Since the cofactor $[2Fe2S]^{2+}$within the holo-Grx complex interacts with the cysteine residues involved in the catalysis, such complex is catalytically inactive [85]. Degradation of the cluster and dissociation of the holo-complex restore activity of Grx. Slow degradation of the complex under aerobic conditions is efficiently prevented by GSH. In contrast, GSSG facilitates degradation of the cluster and activation of Grx [85]. Two GSH molecules in the complex successfully screen the iron atoms from the environment. Thus, the iron of the $[2Fe2S]^{2+}$ cluster has no possibility to interact with oxidants that require direct molecular interaction, particularly H_2O_2. It has been shown that release of Grx monomers is caused by $O_2^{\bullet-}$ [89]. Presumably, the breakdown of the cluster as a response to the action of the oxidant, is related to formation of GSSG. It is suggested that the human Grx2/Fe-S complex is a kind of a redox sensor: under high GSH/GSSG values, Grx is bound to the complex in the inactive state, while changes in the cellular redox status result in release of the catalytically active Grx [85]. Besides, Grx2 is resistant to oxidative inactivation [90] and successfully functions as an alternative reducing system of Trx1 in the cytosol and Trx2 in mitochondria, under the action of inhibitors of thioredoxin reductase. Overexpression of the Grx2 gene in mitochondria protects Trx2 against oxidation, significantly decreasing development of apoptosis caused by the production of ROS in mitochondria [90].

Some studies have shown importance of the yeast cytosolic multidomain monothiol isoenzymes Grx3 and Grx4 in the intracellular distribution of iron [91]. A simultaneous decrease in Grx3 and Grx4 decelerates all iron-dependent processes in the cytoplasm, mitochondria, and nucleus, which is caused by iron deficiency in the organelles and insufficiency of its incorporation into proteins despite sufficient amount of iron in the cytoplasm. The ability of Grx to bind Fe-S complexes is necessary for bioavailability of iron in the cell [91]. For a better understanding functions of various human Grx isoenzymes, their specificity towards different disulfides must be investigated considering different cellular localization. Besides, their contribution to the processes of iron transport and maintenance of Fe-S clusters are also of importance. In this connection, recent investigation of Grx5 seems to be interesting. It was demonstrated that the gene of Grx5 had high expression level in bone tissue and played an antiapoptotic regulatory role in osteoblasts [92]. However, it remains unclear, what precisely affected development of the oxidative stress-induced apoptosis while changing the level of Grx5 (overexpression or knockout of the gene): alterations of the thiol–disulfide homeostasis or homeostasis of the iron–sulfur clusters.

Thiol–disulfide exchange influences not only the substrate structure, but also the structure of Grx. Comparison of the oxidized and reduced forms of Grx1 from *E. coli* and T4 bacteriophage demonstrated that the structures of the two forms of the isoenzymes are very similar, although there are some differences [93]. In the presence of mixed disulfide bond with GSH, the Grx1 of *E. coli* exhibits properties of the oxidized protein [94]. The structural changes involve the region of the active site, increasing its flexibility in the reduced form of the enzyme. Besides, in the oxidized Grx, the surface of the molecule involved in protein–protein interactions is masked. Therefore, affinity of Grx to the substrate decreases as soon as the substrate is reduced, resulting in dissociation of the Grx–substrate complex.

Grx1 and monomeric Grx2 can catalyze both deglutathionylation and the reverse reaction of S-glutathionylation. The direction of the reaction depends on the relative concentrations of the participants protein-SSG, protein-SH, GSH, and GSSG. The redox potential of the GSH/GSSG pair is most important for determining the cellular redox potential. The value of the GSH/GSSG redox potential substantially depends on the functional state of the cell. During cell proliferation, this value is approximately –240 mV, during cell differentiation it reaches – 200 mV, and apoptosis results in further growth of this value to – 170 mV [95]. It has been ascertained that Grx acts as the GSH-dependent reductase at – 240 mV, while at – 170 mV it acts as the GSSG-dependent oxidase [96]. Under conditions when GSH/GSSG value is decreased, i.e., under the action of oxidizing factors, Grx can catalyze the S-glutathionylation reaction, while under weakening oxidative stress, Grx catalyzes deglutathionylation [97]. Grx facilitates S-glutathionylation of proteins *via* reaction of the disulfide bond with the GS· radical yielding the intermediate anion radical Grx-SSG, which then gives the mixed disulfide P-SSG. Reversion of S-glutathionylation depends on the extent and duration of the initiating stress, the removal of which usually results in deglutathionylation. The half-reduction time of the glutathionylated bonds is 2-3 h [98]. Evidently, Grx contributes to the control of signal transduction, regulating the processes of glutathionylation and deglutathionylation.

Glutathionylation in cells of humans and other mammals participates in regulation of several key proteins and processes in response to redox signals. More than 200 mammalian proteins are known to be involved in thiol–disulfide exchange. For example, S-glutathionylation has been shown to inhibit phosphofructokinase, carboanhydrase III, nuclear factor NF1, glyceraldehyde-3-phosphate dehydrogenase, protein tyrosine phosphatase 1B, protein kinase Cα, creatine kinase, actin, protein phosphatase 2A, protein kinase A, tyrosine hydroxylase, complex I of the mitochondrial respiratory chain, NR-κB transcription factor, and IκB kinase (IKK). In contrast, such proteins as microsomal S-glutathione transferase (GST), phosphatase of carbonic anhydrase III, HIV-1 protease, matrix metalloproteinase, HRAS GTPase, sarcoplasmic calcium ATPase, and complex II of the mitochondrial respiratory chain are activated by S-glutathionylation. Progression of oxidative stress and change in functioning of Grx disturb regulation of S-glutathionylation, which may facilitate several pathophysiological changes observed in diabetes, disorders of the lungs and heart, oncological diseases, and different neurodegenerative processes. For example, disturbance of glutathionylation of cytoskeletal elements promotes pathological changes in heart and skeletal muscles in ischemia and in neurons in Friedreich's ataxia [99]. Glutathionylation of actin prevents its polymerization, so the redox-dependent reversible glutathionylation of actin regulates the cytoskeleton structure, which is of special importance for functioning of such cells as thrombocytes, in which actin is the main protein [100]. Besides, glutathionylation of actin was shown to be necessary for dissociation of the actin–myosin complexes during cell adhesion [101].

The specific regulation of protein activities through glutathionylation/deglutathionylation processes are important for many aspects of cell functioning including regulation of the apoptotic signaling cascades. It was found that apoptosis induced by TNF-α and FasL is highly sensitive to S-glutathionylation. Thus, in epithelial lung cells, the Fas receptor is glutathionylated on Cys294 during degradation of Grx1 involving caspase-8 and/or caspase-3, this resulting in the acceleration of apoptosis [102]. The results indicate existence of feedback between caspase-3 and Grx, since Grx activates procaspase-3, which causes degradation of Grx. At the same time, *in vitro* studies demonstrated inactivation of caspase-3 by glutathionylation [103]. For a better understanding of the interactions between Grx and caspase-3, it is necessary to determine the specific cysteine residues of caspase-3 that are glutathionylated/deglutathionylated. This is of importance for ascertaining the relation between the redox status of caspase-3 and the mechanism of activation of apoptosis. It is assumed that the deglutathionylation catalyzed by Grx can play a crucial role in the mechanism of the redox regulation of the processes from proliferation to apoptosis that is specific for different cell types [104].

In Alzheimer's disease, change in metabolism is partially associated with a decrease in the activity of α-ketoglutarate dehydrogenase. Activity of this enzyme decreased with glutathionylation under conditions of oxidative stress, which may take place in brain cells in Alzheimer's disease [105]. Besides, selective glutathionylation of protein p53 in brain cells was found in Alzheimer's disease, which also may facilitate progression of the oxidative stress [106]. In patients with type 2 diabetes, glutathionylated hemoglobin (HbSSG) was found, and its level correlated with development of diabetic microangiopathy [107]. At the same time, in a rat diabetes model an increased expression of Grx1 gene was revealed, facilitating translocation of NF-κB into the nucleus and activation of the cell adhesion molecules ICAM-1. Both processes make a significant contribution to the development of retinopathy. Disturbance in their regulation, in which glutathionylation plays an important role, is observed under the activation of Grx1 [108].

The crystalline lens contains high concentrations of GSH (6 mM) that acts as an antioxidant, maintaining transparency of the lens [109]. During progression of cataract, the GSH/GSSG ratio decreases and the lens proteins undergo structural changes resulting in unfolding of the protein globules and exposure of the buried cysteine residues, which increases disulfide bonding and S-glutathionylation [109]. Presumably, the maintenance of proper GSH level prevents or decelerates the progression of cataract. In the rat diabetes model, N-acetylcysteine and glutathione ethyl ester that are easily converted into GSH *in vivo*, successfully suppressed development of cataract in early stages [110].

Glutathionylation of the transcription factor p53 significantly decreased its ability to bind to DNA. Consequently, glutathionylation inhibits p53 that suppresses development of malignant tumors, which may influence oncogenesis [111]. It is supposed that inactivation of p53 through its glutathionylation provides the mechanism for cell adaptation that suppresses development of the apoptotic response in the early stage of oxidative stress and allows the cell to avoid immediate death.

Of note is that age is a risk factor for many diseases, since different damages accumulate with age, while repair systems slow down their activity. With age, mitochondrial functions can also be affected by negative changes facilitating ROS production, which is observed simultaneously with a decrease in activity of the antioxidant enzymes. Such disorders in redox-regulation affect the S-glutathionylation of proteins, which makes the cell more sensitive to apoptosis and promotes development of pathologic changes [104].

It should be noted that Grx is capable of protein–protein interactions. For example, Grx participates in regulation of ASK1 protein kinase activating JNK1- and p38-dependent signaling pathways of apoptosis [112]. Using different cell lines, it was shown that ASK1 is activated by ROS, by H_2O_2 in particular, due to the breakdown of the complex with Grx1. The reduced Grx1 binds to the C-terminal domain of ASK1, resulting in inactivation of the kinase. In contrast, oxidation of Grx1 leads to dissociation of the complex, activation of ASK1, and induction of apoptosis [113]. This dissociation is prevented by catalase or N-acetylcysteine. The decrease in GSH content using BSO inhibits binding of Grx1 to ASK1. Presumably, GSH is necessary for reduction of the intermolecular disulfide bonds between the adjacent cysteine residues in Grx1, which enables the protein to bind ASK1 [113]. These data suggest that Grx1 can be considered as a redox-sensitive factor involved in regulation of signaling cascades of JNK1 and p38 MAP-kinases.

The Family of Glutathione Transferases in the Regulation of Redodox-Dependent Processes

Glutathione S-transferase (GST, EC 2.5.1.18) is represented by a superfamily of isoenzymes catalyzing conjugation of glutathione (GSH) with a wide range of nonpolar compounds of exogenous and endogenous origin containing electrophilic atoms of carbon, sulfur, nitrogen, and phosphorous, which assists in defense of the cell against possible toxic action of these compounds [114-117]. The GST superfamily includes three subfamilies of isoenzymes: cytosolic, mitochondrial, and microsomal. Isoenzymes of GST have now been found in most living organisms including aerobic bacteria, yeast, plants, insects, and vertebrates. In mammals, GST is present in virtually all organs and tissues, but the highest content of the enzyme is found in the liver.

The cytosolic isoenzymes of GST account for approximately 90% of the total GST activity in the cell. Based on the amino acid sequence homology, mammalian cytosolic GST isoenzymes are grouped into seven classes (α, μ, π, θ, ζ, ω, and σ) that comprise 17 isoenzymes [114, 115]. In humans and rodents, the cytosolic isoenzymes within the same class exhibit more than 40% homology (sometimes more than 90%), while the homology between the enzymes in different classes is less than 25%. Special attention in the contemporary classification is given to the primary structure of the conserved N-terminus of the polypeptide chain containing catalytic residues of tyrosine, cysteine, or serine [115].

In species other than mammals, GST isoenzymes of β, δ, ε, ϕ, λ, τ, and ν classes have been found [118].

A mitochondrial isoenzyme of human GST is GSTK1-1, which belongs to the κ-class [119]. The same isoenzyme was found in human peroxisomes. GSTK1-1 of rodents and humans is active towards several traditional GST substrates, 1-chloro-2,4-dinitrobenzol in particular. In *Caenorhabditis elegans*, GST is involved in metabolism of lipids [120].

Microsomal GSTs are integral membrane proteins that are now called membrane-associated proteins in eicosanoid and glutathione metabolism (MAPEG) [121, 122]. Isoenzymes of the MAPEG microsomal subfamily are divided into four subgroups (I-IV), the amino acid sequence homology between the subgroups constituting less than 20%. In humans, six isoenzymes have been found that belong to subgroups I, II, and IV [121]. Likewise the cytosolic and mitochondrial GST isoenzymes, the microsomal isoenzymes also catalyze conjugation of GSH with electrophilic compounds, but they also participate in reactions of isomerization of unsaturated compounds and biosynthesis of leukotrienes and prostaglandins [121, 123].

Cytosolic and mitochondrial isoforms of GST are homo- or heterodimers, and the subunits within heterodimers belong to the same class. Although the cytosolic GSTs are dimers, representatives of the microsomal subfamily can be mono-, di-, and trimers, and multienzyme complexes also occur [115]. Each subunit is composed of two domains linked with a small irregular region.

The N-terminal domain is the GSH-binding site (G-site). It exhibits topology that is similar to thioredoxin and contains four β-sheets (β1, β2, β3, and β4), three of which are antiparallel, and three α-helices. These elements of the secondary structure are composed into the sequences βαβαββα. The C-terminal domain is the cosubstrate-binding site (H-site), a completely α-helical region composed of either five or six α-helices (α4-8 or α4-9). In contrast to other classes, the isoenzymes of the α, θ, and ω classes have an additional α9 helix. The isoenzymes of the μ-class have a unique μ-loop at the C-terminus, while the isoenzymes of the ω-class contain an additional 19 amino acid sequence at the N-terminus. The θ-isoenzymes have a large loop between the α4- and α5-helices [114, 124, 125]. The differences in the structures of the representatives of various GST isoenzymes provide their wide substrate specificity and diversity of functions. Detailed comparative analysis of the amino acid sequence and structure of cytosolic GST isoenzymes considering the presence of certain amino acid residues in their active site allowed their division into two subgroups: Y-GST, isoenzymes using a tyrosine residue for the activation of GSH (α, μ, π, and σ classes), and S/C-GST, isoenzymes binding GSH through a serine (ϕ, τ, θ, and ζclasses) or a cysteine (β and ω classes) residue [125]. In GST of both subgroups and in mitochondrial GST, these amino acid residues being essential for GSH activation are relocated in the so-called catalytic loop that is next to the first β-sheet in the thioredoxin-like domain. As mentioned above, structure of the H-site exhibits significant variability in representatives of different classes. While forming the dimeric structure, domain I of one of the subunits and domain II of the other interact with each other by the lock-and-key principle. Certain aromatic amino acid residues of the loop between the α3-helix and β2-sheet of the first monomer play a role as the key. They are located in the hydrophobic "lock" formed by the cavity between the α4- and α5-helices of the second monomer.

The main function of GST is its participation in the antioxidant system through its ability to reduce organic hydroperoxides to alcohols using GSH as the co-substrate [115]. *Via* Se-independent glutathione-peroxidase activity, GST reduces hydroperoxides of polyunsaturated higher fatty acids, phospholipids, and cholesterol [126]. Among the substrates of GSTA4-4 are the lipid peroxidation pruducts acrolein and 4-hydroxynonenal (4-HNE) [115, 127]. Conjugation of these compounds with glutathione protects proteins and DNA against covalent modification. As a result of oxidative stress, nucleotides can be oxidized to propenals and hydroperoxides, which are substrates of GSTP1-1. Oxidation of catecholamines also results in formation of compounds (aminochrome, dopachrome, noradrenochrome, and adrenochrome) that are substrates of GST isoforms. Conjugation of such compounds with GSH contributes to the cellular antioxidant defense system, since they contain a quinone structure producing $O_2^{\bullet-}$, and, consequently, promoting oxidative stress [115]. The cytosolic GSTM2-2 was shown to detoxify *o*-quinones of dopamine, which can protect dopaminergic systems of the brain against degenerative processes [115]. GSTP1-1 mediates defense against oxidative stress, recovering the peroxidase activity of the oxidized peroxiredoxin Prx6 [128].

As discussed above, GST was shown to be involved in the process of S-glutathionylation. Originally, it was considered that growth in ROS production leads to S-glutathionylation to prevent irreversible oxidation of protein cysteine residues and disturbance of protein functions [11]. Later it was found that S-glutathionylation plays an important role in the mechanisms of the cell signaling, changing the sensitivity of cysteine residues towards redox modification. The list of proteins whose structure and functions are modulated by S-glutathionylation is large: proteins involved in metabolism, proteins forming the cytoskeleton and ion channels, signal proteins (kinases and phosphatases), transcription factors, ras-proteins, and heat shock proteins [129]. The process of S-glutathionylation can proceed both non-enzymatically and with participation of enzymes, one of which is GSTP1-1. The ability of GSTP1-1 for S-glutathionylation is based on the catalytic activity of the enzyme. Under oxidative stress, GSTP1-1 is auto-S-glutathionylated at residues Cys47 and Cys101, each of which affects the catalytic activity of the enzyme and its ability to bind target proteins. Besides, specific S-glutathionylation causes oligomerization of GSTP1-1, which presumably has significant consequences for other components of the cellular stress response.

S-Glutathionylation of the GSTP1-1 monomer decreases the number of α-helical regions, i.e., alters the secondary structure of the enzyme, which subsequently leads to a change in the tertiary and quaternary structures [98], affecting ability of the GSTP1-1 to interact with proteins. An example of such regulation is the complex between GSTP1-1 and JNK1. S-glutathionylation of GSTP1-1 at residues Cys47 and/or Cys101 results in dissociation of the complex between GSTP1-1 and JNK1, activation of JNK1, and aggregation of GSTP1-1 [98, 130].

Of special importance is the role of GST in regulation of cell signaling due to the protein–protein interactions with kinases that are activated by oxidative stress. Under physiological conditions, some GSTP1-1 is bound to kinase JNK1, resulting in its inactivation, which regulates the level of active JNK1. Under conditions increasing ROS level, which is observed, for example, under the action of a number of antitumor drugs, the complex between GSTP1-1 and JNK1 dissociates, and GSTP1-1 associates into oligomers. The release of JNK1 induces a cascade of processes, starting from the phosphorylation of Jun-c and resulting in apoptosis. The enhanced expression of the GSTP1-1 gene observed in some tumors can significantly inhibit JNK1, and consequently, suppress the signaling pathway leading to apoptosis, which contributes to formation of drug resistance of the tumor cells [131, 132]. A similar interaction of GSTP1-1 with TRAF2 (factor 2 bound to the TNF-α receptor) blocks action of the kinases JNK1, p38, and ASK1 in the case of the signaling cascade induced by TNF-α. Dissociation of the complex between GSTP1-1 and TRAF2 activates proliferation under gentle oxidative stress, while a prolonged and strong oxidative stress leads to apoptosis [133]. It should be noted that catalytic activity of GSTP1-1 does not change during protein–protein interactions, suggesting that the active sites of the enzyme are not involved in this process [116].

The GSTA1-1 isoenzyme also participates in regulation of apoptotic signaling pathways through protein–protein interactions with JNK1. Enhanced expression of the GSTA1-1 gene significantly decreases the number of cells subjected to apoptosis due to inhibition of JNK1-dependent phosphorylation of Jun-c and activation of caspase-3 [134]. GSTM1-1 exhibits regulatory functions that are similar to those of GSTP1-1. The complex between GSTM1-1 and ASK1 is important for maintenance of the normal level of phosphorylation of p38. Under stress conditions of heat shock or increased ROS level, the complex dissociates, and the GSTM1-1 associates into oligomers, while ASK1 is activated [135]. Since ASK1 is a kinase of MAPK kinase activating the JNK1- and p38-dependent signaling pathways, dissociation of this complex results in cytokine- and stress-induced apoptosis [136].

Ability of the homodimers of GSTP1-1 (sometimes GSTM1-1) to dissociate and form heterodimers with other monomeric proteins underlies its ability to provide these proteins with glutathione [137]. The cytosolic isoforms of GST are catalytically active in the dimeric form, the surface of the dimer being the site of non-catalytic binding of ligands. Several studies report that the isoforms of mammalian GSTP1-1 and GSTM1-1 in the monomeric form can interact with ASK1, JNK1, or with peroxiredoxin 6 (Prx6) [128, 133, 135]. Investigations of the GSTP1-1 molecule have shown that structural features of its C-terminus promote dissociation of the homodimer into monomers. At the same time, the Trx-like domain at the N-terminus promotes formation of heterodimers between GSTP1-1 monomers with other proteins, especially with those containing a Trx-like domain [138]. Recovery of the peroxidase activity of Prx6 can be an example of the protein–protein interaction of GSTP1-1 with simultaneous reduction of the protein by glutathione. The Prx6 molecule has one catalytically active cysteine residue, Cys47, at the N-terminus. Oxidation of this residue yields a sulfenic acid derivative, inactivating peroxidase activity of Prx6 towards H_2O_2 and hydroperoxides of phospholipids. It has been found that GSTP1-1, as a complex with GSH, forms a heterodimer with Prx6 and reduces the Cys47 residue. Binding of GSH induces conformational changes allowing formation of the heterodimer between GSTP1-1 and Prx6 [139]. Then Prx6 is S-glutathionylated at the oxidized Cys47 residue with subsequent disulfide bonding between Cys47 of Prx6 and Cys47 of GSTP1-1, followed by reduction of the disulfide bond by GSH.

GST takes part in regulation of the serine/threonine AMP-activated protein kinase (AMPK), which controls the energy balance of the cell [140]. A diversity of AMPK functions are involved in the control of various metabolic pathways and physiological processes such as proliferation and cell motility. AMPK is activated by ROS and RNS through AMP-dependent and AMP-independent mechanisms and can be involved in cellular redox regulation [141]. *In vitro* studies demonstrated that, under close to physiological conditions, the mammalian isoenzymes GSTM1-1 and GSTP1-1 promoted S-glutathionylation of AMPK at the same cysteine residues that were glutathionylated during the nonenzymatic H_2O_2-dependent process, which also increased the kinase activity [142, 143]. The interaction with AMPK activates GSTM1-1 and GSTP1-1, which in turn results in the S-glutathionylation and activation of AMPK. These data illustrate well the role of AMPK as an important element in redox-dependent signal transduction [144-146]. The activated AMPK activates the transcriptional factor FOXO3 that affects such processes as cell proliferation, gluconeogenesis, and defense against oxidative stress through the activation of the PI3K/AKT signaling pathway [141, 147]. The contribution of FOXO3 to antioxidant defense is accounted for the enhanced expression of Mn-superoxide dismutase, catalase, thioredoxin [148, 149], metallothioneins [150], mitochondrial uncoupling protein UCP2 [151], γ-glutamylcysteine synthetase [148], glutathione peroxidase [149], and GSTM1-1 [150]. The GST-mediated S-glutathionylation and activation of AMPK can be considered as an additional mechanism of regulation of AMPK as a redox sensor of energetic stress and antioxidant defense [142].

As mentioned above, one of the main products of lipid peroxidation is 4-hydroxynonenal (4-HNE), which forms adducts with proteins and nucleic acids. 4-HNE is involved in the MAPK-dependent signaling pathways of cellular stress response, particularly by facilitation of phosphorylation of JNK and p38, which results in their activation [152]. A dose-dependent regulation of cellular signaling pathways by 4-HNE has been demonstrated: at concentrations above 10 μM, 4-HNE exerted cytotoxic effect, while at concentrations below 10 μM (physiological range) 4-HNE modulated cell growth, i.e., affected cell proliferation [153]. Besides, 4-HNE inhibits expression of cyclins D1,

D2, and A and, consequently, activity of the cyclin-dependent kinases 4/6 (Cdk4/6) and Cdk2 [154]. Furthermore, it increases expression of p21waf1, which inhibits the functioning of some cyclin-dependent complexes [155]. Thus, 4-HNE can simultaneously affect the expression of different genes involved in the control of cell proliferation. Undoubtedly, the intracellular content of 4-HNE must be regulated to protect the cell from damage and/or to control stress-dependent signaling pathways. 4-HNE is metabolized by GST, forming the HNE-GSH conjugates. The most specific isoform towards 4-HNE is GSTA4-4 [156].

It has been found that under conditions of oxidative stress, phosphorylation of GSTA4 subunits in the cytoplasm increases, facilitating their binding to the Hsp70 protein, fast dimerization, and subsequent translocation into mitochondria. If the subunits are not hyperphosphorylated, they do not exhibit high affinity to Hsp70. In this case, the formed dimers remain in the cytoplasm [157]. Thus, the oxidative stress-activated import to mitochondria of the GSTA4-4 isoform exhibiting high specificity towards 4-HNE protects mitochondria from oxidative stress and modulates signaling pathways that are affected by 4-hydroxynonenal [157]. It should be noted that TNFα, IL-6, and epidermal growth factor enhance the GSTA4-4 content in mitochondria *in vivo* [158]. Decrease in the GSTA4-4 level results in an increase of ROS production and disturbs mitochondrial functions, which promotes development of the insulin resistance and type 2 diabetes [159]. In whole, different studies show that concentration of 4-HNE in the cell is important for activation of the cell cycle and the signaling cascades regulating cell differentiation, proliferation, and apoptosis. The level of 4-HNE is strongly dependent on the activity of GSTA4-4 both in the cytoplasm and the mitochondria.

There are several specific features of the impact of GSTP1-1 on the redox-dependent pathways regulating cell signaling and metabolism. It is supposed that some of the numerous changes of regulatory proteins observed under acute or chronic cocaine injections could be related to S-glutathionylation catalyzed by GSTP1-1. For example, actin, JNK and AMP-dependent protein kinase are regulated through S-glutathionylation under the action of cocaine [98, 160]. Presumably, it is the enhanced S-glutathionylation that results in neuroadaptation under cocaine-induced oxidative stress. GSTP1-1 was shown to directly inhibit the cyclin-dependent kinase Cdk5, interacting with its regulatory p25/p35 subunit [160]. Stimulation of Cdk5 results in generation of ROS, which leads to cell death due to a feedback mechanism. Under neurotoxic conditions, introduction of the GSTP1-1 gene provides successful neuroprotection due to the ability of GSTP1-1 to modulate the Cdk5-dependent signaling, which protects the cell from oxidative stress and prevents neurodegeneration [161].

GSTP1-1 was shown to prevent the origin and progression of Parkinson's disease, suppressing activation of Jun-c [162]. GSTP1-1 gene knockout mice were more sensitive to the neurotoxin 1-methyl-4-phenyl-1,2,3,6-tetrahydropyridine, which led to the early degeneration of dopaminergic neurons and corpus striatum fibers.

Expression levels of GST isoforms differ in normal and tumor tissues. High expression of the GSTP1-1 gene often correlates with drug resistance that is observed in tumor tissues of the ovaries, lungs, mammary glands, large intestine, and in some oncological diseases of the blood [163, 164]. Ability of GSTP1-1 as the inhibitor of JNK1, ASK1, and TRAF2 to regulate kinase signaling pathways determining the cell fate can provide the resistance of tumor cells to antitumor drugs including those with prooxidant action [165, 166]. Transfection of the GSTA1-1 gene into H69 cells (small-cell lung cancer) led to resistance towards doxorubicin, exhibiting prooxidant action [134]. Overexpression of the GSTA1-1 gene protected the cells from the decrease in GSH level caused by doxorubicin, decelerating lipid peroxide oxidation. Besides, overexpression of the GSTA1-1 gene significantly decreased the number of apoptotic cells due to the inhibition of JNK1-dependent phosphorylation of both Jun-c and caspase-3 [134].

The mechanisms regulating the work of genes of the GST isoforms are still not completely understood. It was shown that exogenous or endogenous compounds of different structure are inducers of GST. Some of them activate transcription of the GST genes, acting on the antioxidant-responsive (ARE), xenobiotic-responsive (XRE), or glucocorticoid-responsive (GRE) elements of the promoter region [132, 167, 168]. The presence of ARE in the promoter region is characteristic for genes whose products are involved in the defense of the cell from oxidative stress or xenobiotics. GST isoforms are also encoded by genes that are often called ARE genes, and the corresponding proteins are called ARE proteins. Expression of ARE genes is controlled by the transcription factor Nrf2. Normally, Nrf2 is located in the cytoplasm in a complex with Keap1 protein, which provides ubiquitinylation of Nrf2 and its proteasomal degradation. The mechanism that was supposed to be responsible for activation of Nrf2 involves oxidation of cysteine residues of Keap1 under oxidative stress, resulting in the dissociation of the Keap1–Nrf2 complex and translocation of Nrf2 to the nucleus, where it forms a dimeric complex with small Maf protein. This complex activates expression of genes whose products are involved in cell defense [169]. However, some data suggest that the idea of direct dissociation of the Keap1–Nrf2 complex is incorrect, since the affinity of this interaction is rather high. It is suggested that stress conditions do not affect affinity of the complex, but rather decrease the ability of Keap1 to ubiquitinylate Nrf2, which finally allows the transcription factor to be accumulated in the nucleus and to stimulate expression of the ARE genes [170].

Conclusion

The data, which have been reviewed in the chapter, show significance of GSH and GSH-related enzymes (glutathione transferase and glutaredoxin) in supporting redox homeostasis and consequently, in redox regulation of vital cellular functions such as proliferation, apoptosis, protein folding, and cell signaling. A special role in maintenance of the cellular redox status due to participation in thiol–disulfide exchange and antioxidant defense belongs to reduced glutathione through the suppression of free radical damage as well as regulation of numerous cellular redox-dependent processes from enzyme activity to gene expression. The crucial conditions for cellular viability are supporting the optimal level in GSH/GSSG ratio. Its decrease is related to pathological anomalies including cardiovascular diseases and cancer. Disruption of GSH/GSSG level causes an alteration in the including cardiovascular mechanism of cellular redox signaling that is controlled both nonenzymatically and enzymatically with participation of glutathione transferases and glutaredoxins

ACKNOWLEDGMENT

The publication has been prepared with the support of the «RUDN University Program 5 - 100».

REFERENCES

[1] Sies, H; Berndt, C; Jones, DP. Oxidative Stress. *Ann. Rev. Biochem.* 2017, 86, 715-748.

[2] Yamamoto, M; Kensler, TW; Motohashi, H. The KEAP1-NRF2 System: A thiol-based sensor-effector apparatus for maintaining redox homeostasis. *Physiol. Rev.* 2018, 98, 1169-1203.

[3] Go, YM; Fernandes, J; Hu, X; Uppal, K; Jones, DP. Mitochondrial network responses in oxidative physiology and disease. *Free Rad. Biol. Med.* 2018, 116, 31-40.

[4] Ursini, F; Maiorino, M; Forman, HJ. Redox homeostasis: The golden mean of healthy living. *Redox Biol.* 2016, 8, 205-215.

[5] Deponte, M. Glutathione catalysis and the reaction mechanisms of glutathione-dependent enzymes. *Biochim. Biophys. Acta.* 2013, 1830, 3217-3266.

[6] Galano, A; Alvarez-Idaboy, JR. Glutathione: mechanism and kinetics of its non-enzymatic defense action against free radicals. *RSC Adv.* 2011, 1, 1763-1771.

[7] Ortega, AL; Mena, S; Estrela, JM. Glutathione in cancer cell death. *Cancers (Basel).* 2011, 3, 1285-1310.

[8] Go, YM; Jones, DP. Redox compartmentalization in eukaryotic cells. *Biochim. Biophys. Acta.* 2008, 1780, 1273-1290.

[9] Jones, DP. Redefining oxidative stress. *Antiox. Redox Signal.* 2006, 8, 1865-1879.

[10] Cai, Z; Yan, LJ. Protein oxidative modifications: Beneficial roles in disease and health. *J. Biochem. Pharmacol. Res.* 2013, 1, 15-26.

[11] Zhang, J; Ye, ZW; Singh, S1; Townsend, DM; Tew, KD. An evolving understanding of the S-glutathionylation cycle in pathways of redox regulation. *Free Rad. Biol. Med.* 2018, 120, 204-216.

[12] Fiaschi, T; Cozzi, G; Raugei, G; Formigli, L; Ramponi, G; Chiarugi, P. Redox regulation of beta-actin during integrin-mediated cell adhesion. *J. Biol. Chem.* 2006, 281, 22983-22991.

[13] Qanungo, S; Starke, DW; Pai, HV; Mieyal, JJ; Nieminen, AL. Glutathione supplementation potentiates hypoxic apoptosis by S-glutathionylation of p65-NFkappaB. *J. Biol. Chem.* 2007, 282, 18427-18436.

[14] Reynaert, NL; Van der Vliet, A; Guala, AS; McGovern, T; Hristova, M; Pantano, C; Heintz, NH; Heim, J; Ho, YS; Matthews, DE; Wouters, EF; Janssen-Heininger, YM. Dynamic redox control of NF-kappaB through glutaredoxin-regulated S-glutathionylation of inhibitory kappaB kinase beta. *Proc. Natl. Acad. Sci. USA.* 2006, 103, 13086-13091.

[15] Fitzpatrick, AM; Jones, DP; Brown, LA. Glutathione redox control of asthma: from molecular mechanisms to therapeutic opportunities. *Antiox. Redox Signal.* 2012, 17, 375-408.

[16] He, L; He, T; Farrar, S; Ji, L; Liu, T; Ma, X. Antioxidants maintain cellular redox homeostasis by elimination of reactive oxygen species. *Cell. Physiol. Biochem.* 2017, 44, 532-553.

[17] Lluis, JM; Morales, A; Blasco, C; Colell, A; Mari, M; Garcia-Ruiz, C; Fernandez-Checa, JC. Critical role of mitochondrial glutathione in the survival of hepatocytes during hypoxia. *J. Biol. Chem.* 2005, 280, 3224-3232.

[18] Mari, M; Morales, A; Colell, A; Garcia-Ruiz, C; Fernandez-Checa, JC. Mitochondrial glutathione, a key survival antioxidant. *Antiox. Redox Signal.* 2009, 11, 2685-2700.

[19] Taylor, ER; Hurrell, F; Shannon, RJ; Lin, TK; Hirst, J; Murphy, MP. Reversible glutathionylation of complex I increases mitochondrial superoxide formation. *J. Biol. Chem.* 2003, 278, 19603-19610.

[20] Beer, SM; Taylor, ER; Brown, SE; Dahm, CC; Costa, NJ; Runswick, MJ; Murphy, MP. Glutaredoxin 2 catalyzes the reversible oxidation and glutathionylation of mitochondrial membrane thiol proteins: implications for mitochondrial redox regulation and antioxidant defence. *J. Biol. Chem.* 2004, 279, 47939-47951.

[21] Hurd, TR; Requejo, R; Filipovska, A; Brown, S; Prime, TA; Robinson, AJ; Fearnley, IM; Murphy, MP. Complex I within oxidatively stressed bovine heart mitochondria is glutathionylated on Cys-531 and Cys-704 of the 75-kDa subunit: potential role of CYS residues in decreasing oxidative damage. *J. Biol. Chem.* 2008, 283, 24801-24815.

[22] Mailloux, RJ; Jin, X; Willmore, WG. Redox regulation of mitochondrial function with emphasis on cysteine oxidation reactions. *Redox Biol.* 2014, 2, 123-139.

[23] Tretter, L; Adam-Vizi, V. Inhibition of Krebs cycle enzymes by hydrogen peroxide: A key role of [alpha]-ketoglutarate dehydrogenase in limiting NADH production under oxidative stress. *J. Neurosci.* 2000, 20, 8972-8979.

[24] Gibson, GE; Blass, JP; Beal, MF; Bunik, V. The alpha-ketoglutarate-dehydrogenase complex: a mediator between mitochondria and oxidative stress in neurodegeneration. *Mol. Neurobiol.* 2005, 31, 43-63.

[25] Cadenas, E; Davies, KJ. Mitochondrial free radical generation, oxidative stress, and aging. *Free Rad. Biol. Med.* 2000, 29, 222-230.

[26] Cole-Ezea, P; Swan, D; Shanley, D; Hesketh, J. Glutathione peroxidase 4 has a major role in protecting mitochondria from oxidative damage and maintaining oxidative phosphorylation complexes in gut epithelial cells. *Free Rad. Biol. Med.* 2012, 53, 488-497.

[27] Liang, H; Ran, Q; Jang, YC; Holstein, D; Lechleiter, J; McDonald-Marsh, T; Musatov, A; Song, W; Van Remmen, H; Richardson, A. Glutathione peroxidase 4 differentially regulates the release of apoptogenic proteins from mitochondria. *Free Rad. Biol. Med.* 2009, 47, 312-320.

[28] Mari, M; Morales, A; Colell, A; Garcia-Ruiz, C; Kaplowitz, N; Fernandez-Checa, JC. Mitochondrial glutathione: features, regulation and role in disease. *Biochim. Biophys. Acta.* 2013, 1830, 3317-3328.

[29] Johansson, C; Lillig, CH; Holmgren, A. Human mitochondrial glutaredoxin reduces S-glutathionylated proteins with high affinity accepting electrons from either glutathione or thioredoxin reductase. *J. Biol. Chem.* 2004, 279, 7537-7543.

[30] Knoops, B; Goemaere, J; Van der Eecken, V; Declercq, JP. Peroxiredoxin 5: structure, mechanism, and function of the mammalian atypical 2-Cys peroxiredoxin. *Antiox. Redox Signal.* 2011, 15, 817-829.

[31] Zhang, H; Go, YM; Jones, DP. Mitochondrial thioredoxin-2/peroxiredoxin-3 system functions in parallel with mitochondrial GSH system in protection against oxidative stress. *Arch. Biochem. Biophys.* 2007, 465, 119-126.

[32] Markovic, J; Borras, C; Ortega, A; Sastre, J; Vina, J; Pallardo, FV. Glutathione is recruited into the nucleus in early phases of cell proliferation. *J. Biol. Chem.* 2007, 282, 20416-20424.

[33] Garcia-Gimenez, JL; Markovic, J; Dasi, F; Queval, G; Schnaubelt, D; Foyer, CH; Pallardo, FV. Nuclear glutathione. *Biochim. Biophys. Acta.* 2013, 1830, 3304-3316.

[34] Bellomo, G; Palladini, G; Vairetti, M. Intranuclear distribution, function and fate of glutathione and glutathione-S-conjugate in living rat hepatocytes studied by fluorescence microscopy. *Microsc. Res. Tech.* 1997, 36, 243-252.

[35] Ho, YF; Guenthner, TM. Uptake and biosynthesis of glutathione by isolated hepatic nuclei. *Toxicologist.* 1994, 14, 178.

[36] Voehringer, DW; McConkey, DJ; McDonnell, TJ; Brisbay, S; Meyn, RE. Bcl-2 expression causes redistribution of glutathione to the nucleus. *Proc. Natl. Acad. Sci. USA*. 1998, 95, 2956-2960.

[37] Zimmermann, AK; Loucks, FA; Schroeder, EK; Bouchard, RJ; Tyler, KL; Linseman, DA. Glutathione binding to the Bcl-2 homology-3 domain groove: a molecular basis for Bcl-2 antioxidant function at mitochondria. *J. Biol. Chem.* 2007, 282, 29296-29304.

[38] Carpenter, G; Cohen, S. Epidermal growth factor. *J. Biol. Chem.* 1990, 265, 7709-7712.

[39] Jang, JH; Surh, YJ. Potentiation of cellular antioxidant capacity by Bcl-2: implications for its antiapoptotic function. *Biochem. Pharmacol.* 2003, 66, 1371-1379.

[40] Gadea, J; Gumbao, D; Matas, C; Romar, R. Supplementation of the thawing media with reduced glutathione improves function and the *in vitro* fertilizing ability of boar spermatozoa after cryopreservation. *J. Androl.* 2005, 26, 749-756.

[41] Schafer, FQ; Buettner, GR. Redox environment of the cell as viewed through the redox state of the glutathione disulfide/glutathione couple. *Free Rad. Biol. Med.* 2001, 30, 1191-1212.

[42] Conrad, M; Moreno, SG; Sinowatz, F; Ursini, F; Kolle, S; Roveri, A; Brielmeier, M; Wurst, W; Maiorino, M; Bornkamm, GW. The nuclear form of phospholipid hydroperoxide glutathione peroxidase is a protein thiol peroxidase contributing to sperm chromatin stability. *Mol. Cell. Biol.* 2005, 25, 7637-7644.

[43] Pujari, G; Berni, A; Palitti, F; Chatterjee, A. Influence of glutathione levels on radiation-induced chromosomal DNA damage and repair in human peripheral lymphocytes. *Mutat. Res.* 2009, 675, 23-28.

[44] Berger, F; Ramirez-Hernandez, MH; Ziegler, M. The new life of a centenarian: signalling functions of NAD(P). *Trends Biochem. Sci.* 2004, 29, 111-118.

[45] Pellny, TK; Locato, V; Vivancos, PD; Markovic, J; De Gara, L; Pallardo, FV; Foyer, CH. Pyridine nucleotide cycling and control of intracellular redox state in relation to poly(ADP-ribose) polymerase activity and nuclear localization of glutathione during exponential growth of Arabidopsis cells in culture. *Mol. Plant.* 2009, 2, 442-456.

[46] Garcia-Gimenez, JL; Ledesma, AM; Esmoris, I; Roma-Mateo, C; Sanz, P; Vina, J; Pallardo, FV. Histone carbonylation occurs in proliferating cells. *Free Rad. Biol. Med.* 2012, 52, 1453-1464.

[47] Cazanave, S; Berson, A; Haouzi, D; Vadrot, N; Fau, D; Grodet, A; Letteron, P; Feldmann, G; El-Benna, J; Fromenty, B; Robin, MA; Pessayre, D. High hepatic glutathione stores alleviate Fas-induced apoptosis in mice. *J. Hepatol.* 2007, 46, 858-868.

[48] Aoyama, K; Watabe, M; Nakaki, T. Modulation of neuronal glutathione synthesis by EAAC1 and its interacting protein GTRAP3-18. *Amino Acids*. 2012, 42, 163-169.

[49] Thompson, JA; Franklin, CC. Enhanced glutathione biosynthetic capacity promotes resistance to As^{3+}-induced apoptosis. *Toxicol. Lett.* 2010, 193, 33-40.

[50] Wang, TG; Gotoh, Y; Jennings, MH; Rhoads, CA; Aw, TY. Lipid hydroperoxide-induced apoptosis in human colonic CaCo-2 cells is associated with an early loss of cellular redox balance. *FASEB J.* 2000, 14, 1567-1576.

[51] Ekshyyan, O; Aw, TY. Decreased susceptibility of differentiated PC12 cells to oxidative challenge: relationship to cellular redox and expression of apoptotic protease activator factor-1. *Cell. Death Differ.* 2005, 12, 1066-1077.

[52] Okouchi, M; Okayama, N; Aw, TY. Differential susceptibility of naive and differentiated PC-12 cells to methylglyoxal-induced apoptosis: influence of cellular redox. *Curr. Neurovasc. Res.* 2005, 2, 13-22.

[53] Ghosh, S; Pulinilkunnil, T; Yuen, G; Kewalramani, G; An, D; Qi, D; Abrahani, A; Rodrigues, B. Cardiomyocyte apoptosis induced by short-term diabetes requires mitochondrial GSH depletion. *Am. J. Physiol. Heart Circ. Physiol.* 2005, 289, H768-776.

[54] Armstrong, JS; Steinauer, KK; Hornung, B; Irish, JM; Lecane, P; Birrell, GW; Peehl, DM; Knox, SJ. Role of glutathione depletion and reactive oxygen species generation in apoptotic signaling in a human B lymphoma cell line. *Cell. Death Differ.* 2002, 9, 252-263.

[55] Mari, M; Colell, A; Morales, A; Caballero, F; Moles, A; Fernandez, A; Terrones, O; Basanez, G; Antonsson, B; Garcia-Ruiz, C; Fernandez-Checa, JC. Mechanism of mitochondrial glutathione-dependent hepatocellular susceptibility to TNF despite NF-kappaB activation. *Gastroenterology*. 2008, 134, 1507-1520.

[56] Circu, ML; Rodriguez, C; Maloney, R; Moyer, MP; Aw, TY. Contribution of mitochondrial GSH transport to matrix GSH status and colonic epithelial cell apoptosis. *Free Rad. Biol. Med.* 2008, 44, 768-778.

[57] Martins, NM; Santos, NA; Curti, C; Bianchi, ML; Santos, AC. Cisplatin induces mitochondrial oxidative stress with resultant energetic metabolism impairment, membrane rigidification and apoptosis in rat liver. *J. Appl. Toxicol.* 2008, 28, 337-344.

[58] Chen, G; Chen, Z; Hu, Y; Huang, P. Inhibition of mitochondrial respiration and rapid depletion of mitochondrial glutathione by beta-phenethyl isothiocyanate: mechanisms for anti-leukemia activity. *Antiox. Redox Signal*. 2011, 15, 2911-2921.

[59] Chernyak, BV; Bernardi, P. The mitochondrial permeability transition pore is modulated by oxidative agents through both pyridine nucleotides and glutathione at two separate sites. *Eur. J. Biochem.* 1996, 238, 623-630.

[60] Costantini, P; Chernyak, BV; Petronilli, V; Bernardi, P. Modulation of the mitochondrial permeability transition pore by pyridine nucleotides and dithiol oxidation at two separate sites. *J. Biol. Chem.* 1996, 271, 6746-6751.

[61] Aon, MA; Cortassa, S; Maack, C; O'Rourke, B. Sequential opening of mitochondrial ion channels as a function of glutathione redox thiol status. *J. Biol. Chem.* 2007, 282, 21889-21900.
[62] Son, Y; Cheong, YK; Kim, NH; Chung, HT; Kang, DG; Pae, HO. Mitogen-activated protein kinases and reactive oxygen species: How can ROS activate MAPK pathways? *J. Signal Transduct.* 2011, 2011, 792639.
[63] Circu, ML; Aw, TY. Reactive oxygen species, cellular redox systems, and apoptosis. *Free Rad. Biol. Med.* 2010, 48, 749-762.
[64] Lu, GD; Shen, HM; Chung, MC; Ong, CN. Critical role of oxidative stress and sustained JNK activation in aloe-emodin-mediated apoptotic cell death in human hepatoma cells. *Carcinogenesis.* 2007, 28, 1937-1945.
[65] Cuadrado, A; Garcia-Fernandez, LF; Gonzalez, L; Suarez, Y; Losada, A; Alcaide, V; Martinez, T; Fernandez-Sousa, JM; Sanchez-Puelles, JM; Munoz, A. Aplidin induces apoptosis in human cancer cells via glutathione depletion and sustained activation of the epidermal growth factor receptor, Src, JNK, and p38 MAPK. *J. Biol. Chem.* 2003, 278, 241-250.
[66] Ji, L; Shen, K; Jiang, P; Morahan, G; Wang, Z. Critical roles of cellular glutathione homeostasis and jnk activation in andrographolide-mediated apoptotic cell death in human hepatoma cells. *Mol. Carcinog.* 2011, 50, 580-591.
[67] Filomeni, G; Rotilio, G; Ciriolo, MR. Glutathione disulfide induces apoptosis in U937 cells by a redox-mediated p38 MAP kinase pathway. *FASEB J.* 2003, 17, 64-66.
[68] Filomeni, G; Aquilano, K; Civitareale, P; Rotilio, G; Ciriolo, MR. Activation of c-Jun-N-terminal kinase is required for apoptosis triggered by glutathione disulfide in neuroblastoma cells. *Free Rad. Biol. Med.* 2005, 39, 345-354.
[69] Piccirillo, S; Filomeni, G; Brune, B; Rotilio, G; Ciriolo, MR. Redox mechanisms involved in the selective activation of Nrf2-mediated resistance versus p53-dependent apoptosis in adenocarcinoma cells. *J. Biol. Chem.* 2009, 284, 27721-27733.
[70] Cross, JV; Templeton, DJ. Oxidative stress inhibits MEKK1 by site-specific glutathionylation in the ATP-binding domain. *Biochem. J.* 2004, 381, 675-683.
[71] Filomeni, G; Piccirillo, S; Rotilio, G; Ciriolo, MR. p38(MAPK) and ERK1/2 dictate cell death/survival response to different pro-oxidant stimuli via p53 and Nrf2 in neuroblastoma cells SH-SY5Y. *Biochem. Pharmacol.* 2012, 83, 1349-1357.
[72] Fernandes, AP; Holmgren, A. Glutaredoxins: glutathione-dependent redox enzymes with functions far beyond a simple thioredoxin backup system. *Antiox. Redox Signal.* 2004, 6, 63-74.
[73] Berndt, C; Lillig, CH. Glutathione, Glutaredoxins, and Iron. *Antiox. Redox Signal.* 2017, 27, 1235-1251.

[74] Rouhier, N. Plant glutaredoxins: pivotal players in redox biology and iron-sulphur centre assembly. *New Phytol*. 2010, 186, 365-372.

[75] Lillig, CH; Berndt, C; Holmgren, A. Glutaredoxin systems. *Biochim. Biophys. Acta*. 2008, 1780, 1304-1317.

[76] Johansson, C; Roos, AK; Montano, SJ; Sengupta, R; Filippakopoulos, P; Guo, K; von Delft, F; Holmgren, A; Oppermann, U; Kavanagh, KL. The crystal structure of human GLRX5: iron-sulfur cluster co-ordination, tetrameric assembly and monomer activity. *Biochem. J*. 2011, 433, 303-311.

[77] Couturier, J; Jacquot, JP; Rouhier, N. Evolution and diversity of glutaredoxins in photosynthetic organisms. *Cell. Mol. Life Sci*. 2009, 66, 2539-2557.

[78] Hofmann, NR. Glutaredoxin functions in floral development. *Plant Cell*. 2009, 21, 363.

[79] Benyamina, SM; Baldacci-Cresp, F; Couturier, J; Chibani, K; Hopkins, J; Bekki, A; de Lajudie, P; Rouhier, N; Jacquot, JP; Alloing, G; Puppo, A; Frendo, P. Two Sinorhizobium meliloti glutaredoxins regulate iron metabolism and symbiotic bacteroid differentiation. *Environ. Microbiol*. 2013, 15, 795-810.

[80] Xing, S; Lauri, A; Zachgo, S. Redox regulation and flower development: a novel function for glutaredoxins. *Plant Biol (Stuttg)*. 2006, 8, 547-555.

[81] Hanschmann, EM; Godoy, JR; Berndt, C; Hudemann, C; Lillig, CH. Thioredoxins, glutaredoxins, and peroxiredoxins--molecular mechanisms and health significance: from cofactors to antioxidants to redox signaling. *Antiox. Redox Signal*. 2013, 19, 1539-1605.

[82] Lonn, ME; Hudemann, C; Berndt, C; Cherkasov, V; Capani, F; Holmgren, A; Lillig, CH. Expression pattern of human glutaredoxin 2 isoforms: identification and characterization of two testis/cancer cell-specific isoforms. *Antiox. Redox Signal*. 2008, 10, 547-557.

[83] Gallogly, MM; Starke, DW; Mieyal, JJ. Mechanistic and kinetic details of catalysis of thiol-disulfide exchange by glutaredoxins and potential mechanisms of regulation. *Antiox. Redox Signal*. 2009, 11, 1059-1081.

[84] Stroher, E; Millar, AH. The biological roles of glutaredoxins. *Biochem. J*. 2012, 446, 333-348.

[85] Lillig, CH; Berndt, C; Vergnolle, O; Lonn, ME; Hudemann, C; Bill, E; Holmgren, A. Characterization of human glutaredoxin 2 as iron-sulfur protein: a possible role as redox sensor. *Proc. Natl. Acad. Sci. USA*. 2005, 102, 8168-8173.

[86] Riondet, C; Desouris, JP; Montoya, JG; Chartier, Y; Meyer, Y; Reichheld, JP. A dicotyledon-specific glutaredoxin GRXC1 family with dimer-dependent redox regulation is functionally redundant with GRXC2. *Plant Cell. Environ*. 2012, 35, 360-373.

[87] Feng, Y; Zhong, N; Rouhier, N; Hase, T; Kusunoki, M; Jacquot, JP; Jin, C; Xia, B. Structural insight into poplar glutaredoxin C1 with a bridging iron-sulfur cluster at the active site. *Biochemistry*. 2006, 45, 7998-8008.

[88] Berndt, C; Hudemann, C; Hanschmann, EM; Axelsson, R; Holmgren, A; Lillig, CH. How does iron-sulfur cluster coordination regulate the activity of human glutaredoxin 2? *Antiox. Redox Signal*. 2007, 9, 151-157.

[89] Mitra, S; Elliott, SJ. Oxidative disassembly of the [2Fe-2S] cluster of human Grx2 and redox regulation in the mitochondria. *Biochemistry*. 2009, 48, 3813-3815.

[90] Zhang, H; Du, Y; Zhang, X; Lu, J; Holmgren, A. Glutaredoxin 2 reduces both thioredoxin 2 and thioredoxin 1 and protects cells from apoptosis induced by auranofin and 4-hydroxynonenal. *Antiox. Redox Signal.* 2014, 21, 669-681.

[91] Muhlenhoff, U; Molik, S; Godoy, JR; Uzarska, MA; Richter, N; Seubert, A; Zhang, Y; Stubbe, J; Pierrel, F; Herrero, E; Lillig, CH; Lill, R. Cytosolic monothiol glutaredoxins function in intracellular iron sensing and trafficking via their bound iron-sulfur cluster. *Cell. Metab*. 2010, 12, 373-385.

[92] Linares, GR; Xing, W; Govoni, KE; Chen, ST; Mohan, S. Glutaredoxin 5 regulates osteoblast apoptosis by protecting against oxidative stress. *Bone*. 2009, 44, 795-804.

[93] Wang, Y; Amegbey, G; Wishart, DS. Solution structures of reduced and oxidized bacteriophage T4 glutaredoxin. *J. Biomol. NMR*. 2004, 29, 85-90.

[94] Bushweller, JH; Billeter, M; Holmgren, A; Wuthrich, K. The nuclear magnetic resonance solution structure of the mixed disulfide between Escherichia coli glutaredoxin(C14S) and glutathione. *J. Mol. Biol.* 1994, 235, 1585-1597.

[95] Watson, WH; Chen, Y; Jones, DP. Redox state of glutathione and thioredoxin in differentiation and apoptosis. *Biofactors*. 2003, 17, 307-314.

[96] Aslund, F; Berndt, KD; Holmgren, A. Redox potentials of glutaredoxins and other thiol-disulfide oxidoreductases of the thioredoxin superfamily determined by direct protein-protein redox equilibria. *J. Biol. Chem.* 1997, 272, 30780-30786.

[97] Starke, DW; Chock, PB; Mieyal, JJ. Glutathione-thiyl radical scavenging and transferase properties of human glutaredoxin (thioltransferase). Potential role in redox signal transduction. *J. Biol. Chem.* 2003, 278, 14607-14613.

[98] Townsend, DM; Manevich, Y; He, L; Hutchens, S; Pazoles, CJ; Tew, KD. Novel role for glutathione S-transferase pi. Regulator of protein S-Glutathionylation following oxidative and nitrosative stress. *J. Biol. Chem.* 2009, 284, 436-445.

[99] Sparaco, M; Gaeta, LM; Santorelli, FM; Passarelli, C; Tozzi, G; Bertini, E; Simonati, A; Scaravilli, F; Taroni, F; Duyckaerts, C; Feleppa, M; Piemonte, F. Friedreich's ataxia: oxidative stress and cytoskeletal abnormalities. *J. Neurol. Sci.* 2009, 287, 111-118.

[100] Johansson, M; Lundberg, M. Glutathionylation of beta-actin via a cysteinyl sulfenic acid intermediary. *BMC Biochem.* 2007, 8, 26. https://doi.org/10.1186/1471-2091-8-26 (accessed: August 28, 2018).

[101] Wang, J; Boja, E. S; Tan, W; Tekle, E; Fales, HM; English, S; Mieyal, JJ; Chock, PB. Reversible glutathionylation regulates actin polymerization in A431 cells. *J. Biol. Chem.* 2001, 276, 47763-47766.

[102] Anathy, V; Aesif, SW; Guala, AS; Havermans, M; Reynaert, NL; Ho, YS; Budd, RC; Janssen-Heininger, YM. Redox amplification of apoptosis by caspase-dependent cleavage of glutaredoxin 1 and S-glutathionylation of Fas. *J. Cell. Biol.* 2009, 184, 241-252.

[103] Huang, Z; Pinto, JT; Deng, H; Richie, JP. Jr. Inhibition of caspase-3 activity and activation by protein glutathionylation. *Biochem. Pharmacol.* 2008, 75, 2234-2244.

[104] Allen, EM; Mieyal, JJ. Protein-thiol oxidation and cell death: regulatory role of glutaredoxins. *Antiox. Redox Signal.*, 2012, 17, 1748-1763.

[105] Shi, Q; Xu, H; Kleinman, WA; Gibson, GE. Novel functions of the alpha-ketoglutarate dehydrogenase complex may mediate diverse oxidant-induced changes in mitochondrial enzymes associated with Alzheimer's disease. *Biochim. Biophys. Acta.* 2008, 1782, 229-238.

[106] Di Domenico, F; Cenini, G; Sultana, R; Perluigi, M; Uberti, D; Memo, M; Butterfield, DA. Glutathionylation of the pro-apoptotic protein p53 in Alzheimer's disease brain: implications for AD pathogenesis. *Neurochem. Res.* 2009, 34, 727-733.

[107] Sampathkumar, R; Balasubramanyam, M; Sudarslal, S; Rema, M; Mohan, V; Balaram, P. Increased glutathionylated hemoglobin (HbSSG) in type 2 diabetes subjects with microangiopathy. *Clin. Biochem.* 2005, 38, 892-899.

[108] Shelton, MD; Kern, TS; Mieyal, JJ. Glutaredoxin regulates nuclear factor kappa-B and intercellular adhesion molecule in Muller cells: model of diabetic retinopathy. *J. Biol. Chem.* 2007, 282, 12467-12474.

[109] Craghill, J; Cronshaw, AD; Harding, JJ. The identification of a reaction site of glutathione mixed-disulphide formation on gammaS-crystallin in human lens. *Biochem. J.* 2004, 379, 595-600.

[110] Zhang, S; Chai, FY; Yan, H; Guo, Y; Harding, JJ. Effects of N-acetylcysteine and glutathione ethyl ester drops on streptozotocin-induced diabetic cataract in rats. *Mol. Vis.* 2008, 14, 862-870.

[111] Velu, CS; Niture, SK; Doneanu, CE; Pattabiraman, N; Srivenugopal, KS. Human p53 is inhibited by glutathionylation of cysteines present in the proximal DNA-binding domain during oxidative stress. *Biochemistry.* 2007, 46, 7765-7780.

[112] Ichijo, H; Nishida, E; Irie, K; ten Dijke, P; Saitoh, M; Moriguchi, T; Takagi, M; Matsumoto, K; Miyazono, K; Gotoh, Y. Induction of apoptosis by ASK1, a mammalian MAPKKK that activates SAPK/JNK and p38 signaling pathways. *Science.* 1997, 275, 90-94.

[113] Song, JJ; Rhee, JG; Suntharalingam, M; Walsh, SA; Spitz, DR; Lee, YJ. Role of glutaredoxin in metabolic oxidative stress. Glutaredoxin as a sensor of oxidative stress mediated by H_2O_2. *J. Biol. Chem.* 2002, 277, 46566-46575.

[114] Wu, B; Dong, D. Human cytosolic glutathione transferases: structure, function, and drug discovery. *Trends Pharmacol. Sci.* 2012, 33, 656-668.

[115] Hayes, JD; Flanagan, JU; Jowsey, IR. Glutathione transferases. *Ann. Rev. Pharmacol. Toxicol.* 2005, 45, 51-88.

[116] Tew, KD; Townsend, DM. Glutathione-s-transferases as determinants of cell survival and death. *Antiox. Redox Signal.* 2012, 17, 1728-1737.

[117] Allocati, N; Masulli, M; Di Ilio, C; Federici, L. Glutathione transferases: substrates, inihibitors and pro-drugs in cancer and neurodegenerative diseases. *Oncogenesis.* 2018, 7, 8. https://www.nature.com/articles/s41389-017-0025-3 (accessed: August 28, 2018).

[118] Krajewski, MP; Kanawati, B; Fekete, A; Kowalski, N; Schmitt-Kopplin, P; Grill, E. Analysis of Arabidopsis glutathione-transferases in yeast. *Phytochemistry.* 2013, 91, 198-207.

[119] Gallagher, EP; Gardner, JL; Barber, DS. Several glutathione S-transferase isozymes that protect against oxidative injury are expressed in human liver mitochondria. *Biochem. Pharmacol.* 2006, 71, 1619-1628.

[120] Morel, F; Aninat, C. The glutathione transferase kappa family. *Drug Metab. Rev.* 2011, 43, 281-291.

[121] Morgenstern, R; Zhang, J; Johansson, K. Microsomal glutathione transferase 1: mechanism and functional roles. *Drug Metab. Rev.* 2011, 43, 300-306.

[122] Brautigam, L; Zhang, J; Dreij, K; Spahiu, L; Holmgren, A; Abe, H; Tew, KD; Townsend, DM; Kelner, MJ; Morgenstern, R; Johansson, K. MGST1, a GSH transferase/peroxidase essential for development and hematopoietic stem cell differentiation. *Redox Biol.* 2018, 17, 171-179.

[123] Mandal, AK; Skoch, J; Bacskai, BJ; Hyman, BT; Christmas, P; Miller, D; Yamin, TT; Xu, S; Wisniewski, D; Evans, JF; Soberman, RJ. The membrane organization of leukotriene synthesis. *Proc. Natl. Acad. Sci. USA.* 2004, 101, 6587-6592.

[124] Mohana, K; Achary, A. Human cytosolic glutathione-S-transferases: quantitative analysis of expression, comparative analysis of structures and inhibition strategies of isozymes involved in drug resistance. *Drug Metab. Rev.* 2017, 49, 318-337.

[125] Oakley, A. Glutathione transferases: a structural perspective. *Drug Metab. Rev.* 2011, 43, 138-151.

[126] Prabhu, KS; Reddy, PV; Jones, EC; Liken, AD; Reddy, CC. Characterization of a class alpha glutathione-S-transferase with glutathione peroxidase activity in human liver microsomes. *Arch. Biochem. Biophys.* 2004, 424, 72-80.

[127] Ronis, MJ; Mercer, KE; Gannon, B; Engi, B; Zimniak, P; Shearn, CT; Orlicky, DJ; Albano, E; Badger, TM; Petersen, DR. Increased 4-hydroxynonenal protein adducts

in male GSTA4-4/PPAR-alpha double knockout mice enhance injury during early stages of alcoholic liver disease. *Am. J. Physiol. Gastrointest. Liver Physiol.* 2015, 308, G403-415.

[128] Manevich, Y; Feinstein, SI; Fisher, AB. Activation of the antioxidant enzyme 1-CYS peroxiredoxin requires glutathionylation mediated by heterodimerization with pi GST. *Proc. Natl. Acad. Sci. USA.* 2004, 101, 3780-3785.

[129] Lindahl, M; Mata-Cabana, A; Kieselbach, T. The disulfide proteome and other reactive cysteine proteomes: analysis and functional significance. *Antiox. Redox Signal.* 2011, 14, 2581-2642.

[130] Xiong, Y; Uys, JD; Tew, KD; Townsend, DM. S-glutathionylation: from molecular mechanisms to health outcomes. *Antiox. Redox Signal.* 2011, 15, 233-270.

[131] Adler, V; Yin, Z; Fuchs, SY; Benezra, M; Rosario, L; Tew, KD; Pincus, MR; Sardana, M; Henderson, CJ; Wolf, CR; Davis, RJ; Ronai, Z. Regulation of JNK signaling by GSTp. *EMBO J.* 1999, 18, 1321-1334.

[132] Bartolini, D; Galli, F. The functional interactome of GSTP: A regulatory biomolecular network at the interface with the Nrf2 adaption response to oxidative stress. *J. Chromatogr. B. Anal. Technol. Biomed. Life Sci.* 2016, 1019, 29-44.

[133] Wu, Y; Fan, Y; Xue, B; Luo, L; Shen, J; Zhang, S; Jiang, Y; Yin, Z. Human glutathione S-transferase P1-1 interacts with TRAF2 and regulates TRAF2-ASK1 signals. *Oncogene.* 2006, 25, 5787-5800.

[134] Sharma, A; Patrick, B; Li, J; Sharma, R; Jeyabal, PV; Reddy, PM; Awasthi, S; Awasthi, YC. Glutathione S-transferases as antioxidant enzymes: small cell lung cancer (H69) cells transfected with hGSTA1 resist doxorubicin-induced apoptosis. *Arch. Biochem. Biophys.* 2006, 452, 165-173.

[135] Dorion, S; Lambert, H; Landry, J. Activation of the p38 signaling pathway by heat shock involves the dissociation of glutathione S-transferase Mu from Ask1. *J. Biol. Chem.* 2002, 277, 30792-30797.

[136] Matsuzawa, A; Nishitoh, H; Tobiume, K; Takeda, K; Ichijo, H. Physiological roles of ASK1-mediated signal transduction in oxidative stress- and endoplasmic reticulum stress-induced apoptosis: advanced findings from ASK1 knockout mice. *Antiox. Redox Signal.* 2002, 4, 415-425.

[137] Ralat, LA; Misquitta, SA; Manevich, Y; Fisher, AB; Colman, RF. Characterization of the complex of glutathione S-transferase pi and 1-cysteine peroxiredoxin. *Arch. Biochem. Biophys.* 2008, 474, 109-118.

[138] Tew, KD; Manevich, Y; Grek, C; Xiong, Y; Uys, J; Townsend, DM. The role of glutathione S-transferase P in signaling pathways and S-glutathionylation in cancer. *Free Rad. Biol. Med.* 2011, 51, 299-313.

[139] Ralat, LA; Manevich, Y; Fisher, AB; Colman, RF. Direct evidence for the formation of a complex between 1-cysteine peroxiredoxin and glutathione S-transferase pi with activity changes in both enzymes. *Biochemistry.* 2006, 45, 360-372.

[140] Hardie, DG; Ross, FA; Hawley, SA. AMPK: a nutrient and energy sensor that maintains energy homeostasis. *Nat. Rev. Mol. Cell. Biol*. 2012, 13, 251-262.

[141] Zhao, Y; Hu, X; Liu, Y; Dong, S; Wen, Z; He, W; Zhang, S; Huang, Q; Shi, M. ROS signaling under metabolic stress: cross-talk between AMPK and AKT pathway. *Mol. Cancer*. 2017, 16, 79.

[142] Klaus, A; Zorman, S; Berthier, A; Polge, C; Ramirez, S; Michelland, S; Seve, M; Vertommen, D; Rider, M; Lentze, N; Auerbach, D; Schlattner, U. Glutathione S-transferases interact with AMP-activated protein kinase: Evidence for S-glutathionylation and activation *in vitro*. *PLoS One*. 2013, 8, e62497. https://doi.org/10.1371/journal.pone.0062497 (accessed: August 28, 2018).

[143] Zmijewski, JW; Banerjee, S; Bae, H; Friggeri, A; Lazarowski, ER; Abraham, E. Exposure to hydrogen peroxide induces oxidation and activation of AMP-activated protein kinase. *J. Biol. Chem.* 2010, 285, 33154-33164.

[144] Zou, MH; Kirkpatrick, SS; Davis, BJ; Nelson, JS; Wiles, WG; IV. Schlattner, U; Neumann, D; Brownlee, M; Freeman, MB; Goldman, MH. Activation of the AMP-activated protein kinase by the anti-diabetic drug metformin *in vivo*. Role of mitochondrial reactive nitrogen species. *J. Biol. Chem.* 2004, 279, 43940-43951.

[145] Xie, Z; Dong, Y; Zhang, M; Cui, MZ; Cohen, RA; Riek, U; Neumann, D; Schlattner, U; Zou, MH. Activation of protein kinase C zeta by peroxynitrite regulates LKB1-dependent AMP-activated protein kinase in cultured endothelial cells. *J. Biol. Chem.* 2006, 281, 6366-6375.

[146] Trewin, AJ; Berry, BJ; Wojtovich, AP. Exercise and mitochondrial dynamics: Keeping in shape with ROS and AMPK. *Antioxidants (Basel)*. 2018, 7, 7. https://doi.org/10.3390/antiox7010007 (accessed: August 28, 2018).

[147] Tikhanovich, I; Cox, J; Weinman, SA. Forkhead box class O transcription factors in liver function and disease. *J. Gastroenterol. Hepatol*. 2013, 28 Suppl. 1, 125-131.

[148] Colombo, SL; Moncada, S. AMPKalpha1 regulates the antioxidant status of vascular endothelial cells. *Biochem. J.* 2009, 421, 163-169.

[149] Wang, S; Dale, GL; Song, P; Viollet, B; Zou, MH. AMPKalpha1 deletion shortens erythrocyte life span in mice: role of oxidative stress. *J. Biol. Chem*. 2010, 285, 19976-19985.

[150] Greer, EL; Oskoui, PR; Banko, MR; Maniar, JM; Gygi, MP; Gygi, SP; Brunet, A. The energy sensor AMP-activated protein kinase directly regulates the mammalian FOXO3 transcription factor. *J. Biol. Chem.* 2007, 282, 30107-30119.

[151] Xie, Z; Zhang, J; Wu, J; Viollet, B; Zou, MH. Upregulation of mitochondrial uncoupling protein-2 by the AMP-activated protein kinase in endothelial cells attenuates oxidative stress in diabetes. *Diabetes*. 2008, 57, 3222-3230.

[152] Jang, EJ; Kim, DH; Lee, B; Lee, EK; Chung, KW; Moon, KM; Kim, MJ; An, HJ; Jeong, JW; Kim, YR; Yu, BP; Chung, HY. Activation of proinflammatory signaling

by 4-hydroxynonenal-Src adducts in aged kidneys. *Oncotarget*. 2016, 7, 50864-50874.

[153] Zarkovic, N; Ilic, Z; Jurin, M; Schaur, RJ; Puhl, H; Esterbauer, H. Stimulation of HeLa cell growth by physiological concentrations of 4-hydroxynonenal. *Cell. Biochem. Funct*. 1993, 11, 279-286.

[154] Pizzimenti, S; Barrera, G; Dianzani, MU; Brusselbach, S. Inhibition of D1, D2, and A-cyclin expression in HL-60 cells by the lipid peroxydation product 4-hydroxynonenal. *Free Rad. Biol. Med.* 1999, 26, 1578-1586.

[155] Barrera, G; Pizzimenti, S; Laurora, S; Moroni, E; Giglioni, B; Dianzani, MU. 4-Hydroxynonenal affects pRb/E2F pathway in HL-60 human leukemic cells. *Biochem. Biophys. Res. Commun.* 2002, 295, 267-275.

[156] McElhanon, KE; Bose, C; Sharma, R; Wu, L; Awasthi, YC; Singh, SP. Gsta4 null mouse embryonic fibroblasts exhibit enhanced sensitivity to oxidants: Role of 4-hydroxynonenal in oxidant toxicity. *Open J. Apoptosis*. 2013, 2, 1-11. http://dx.doi.org/10.4236/ojapo.2013.21001 (accessed: August 28, 2018).

[157] Raza, H. Dual localization of glutathione S-transferase in the cytosol and mitochondria: implications in oxidative stress, toxicity and disease. *FEBS J.* 2011, 278, 4243-4251.

[158] Desmots, F; Rissel, M; Gilot, D; Lagadic-Gossmann, D; Morel, F; Guguen-Guillouzo, C; Guillouzo, A; Loyer, P. Pro-inflammatory cytokines tumor necrosis factor alpha and interleukin-6 and survival factor epidermal growth factor positively regulate the murine GSTA4 enzyme in hepatocytes. *J. Biol. Chem*. 2002, 277, 17892-17900.

[159] Curtis, JM; Grimsrud, PA; Wright, WS; Xu, X; Foncea, RE; Graham, DW; Brestoff, JR; Wiczer, BM; Ilkayeva, O; Cianflone, K; Muoio, DE; Arriaga, EA; Bernlohr, DA. Downregulation of adipose glutathione S-transferase A4 leads to increased protein carbonylation, oxidative stress, and mitochondrial dysfunction. *Diabetes*. 2010, 59, 1132-1142.

[160] Uys, JD; Knackstedt, L; Hurt, P; Tew, KD; Manevich, Y; Hutchens, S; Townsend, DM; Kalivas, PW. Cocaine-induced adaptations in cellular redox balance contributes to enduring behavioral plasticity. *Neuropsychopharmacol*. 2011, 36, 2551-2560.

[161] Sun, KH; Chang, KH; Clawson, S; Ghosh, S; Mirzaei, H; Regnier, F; Shah, K. Glutathione-S-transferase P1 is a critical regulator of Cdk5 kinase activity. *J. Neurochem.* 2011, 118, 902-914.

[162] Davis, W; Jr. Ronai, Z; Tew, KD. Cellular thiols and reactive oxygen species in drug-induced apoptosis. *J. Pharmacol. Exp. Ther*. 2001, 296, 1-6.

[163] Board, PG; Menon, D. Glutathione transferases, regulators of cellular metabolism and physiology. *Biochim. Biophys. Acta*. 2013, 1830, 3267-3288.

[164] Pathania, S; Bhatia, R; Baldi, A; Singh, R; Rawal, RK. Drug metabolizing enzymes and their inhibitors' role in cancer resistance. *Biomed. Pharmacother.* 2018, 105, 53-65.

[165] Tew, KD; Townsend, DM. Regulatory functions of glutathione S-transferase P1-1 unrelated to detoxification. *Drug Metab. Rev.* 2011, 43, 179-193.

[166] Castro-Caldas, M; Carvalho, AN; Rodrigues, E; Henderson, C; Wolf, CR; Gama, MJ. Glutathione S-transferase pi mediates MPTP-induced c-Jun N-terminal kinase activation in the nigrostriatal pathway. *Mol. Neurobiol.* 2012, 45, 466-477.

[167] Hayes, JD; Pulford, DJ. The glutathione S-transferase supergene family: regulation of GST and the contribution of the isoenzymes to cancer chemoprotection and drug resistance. *Crit. Rev. Biochem. Mol. Biol.* 1995, 30, 445-600.

[168] Higgins, LG; Hayes, JD. Mechanisms of induction of cytosolic and microsomal glutathione transferase (GST) genes by xenobiotics and pro-inflammatory agents. *Drug Metab. Rev.* 2011, 43, 92-137.

[169] Bryan, HK; Olayanju, A; Goldring, CE; Park, BK. The Nrf2 cell defence pathway: Keap1-dependent and -independent mechanisms of regulation. *Biochem. Pharmacol.* 2013, 85, 705-717.

[170] McMahon, M; Thomas, N; Itoh, K; Yamamoto, M; Hayes, JD. Dimerization of substrate adaptors can facilitate cullin-mediated ubiquitylation of proteins by a "tethering" mechanism: a two-site interaction model for the Nrf2-Keap1 complex. *J. Biol. Chem.* 2006, 281, 24756-24768.

BIOGRAPHICAL SKETCH

Elena. V. Kalinina

Present Position

Professor of Biochemistry

Affiliation

Peoples' Friendship University of Russia (RUDN University), Moscow, Russian Federation

Education

Voronezh State University

Research and Professional Experience

Redox-dependent regulation of multidrug resistance in cancer cells

Oxidative stress as pathogenesis factor in development of diseases

Glutathione and glutathione-related enzymes in adaptive antioxidant response of cancer cells

Short Biography

1980 –	Biological Faculty, Voronezh State University
1988 –	Ph.D., Biochemistry, Peoples' Friendship University of Russia (RUDN University), Moscow
1980-2005 –	Researcher, Senior Researcher, Head of Laboratory, Institute of Biotechnology, Center for Theoretical Problems of Physico-Chemical Pharmacology, Russian Academy of Sciences, Moscow
2003-2009 –	Assistant professor of Biochemistry, Biochemistry Department, RUDN University
2009 –	D.Sc., Biochemistry, RUDN University, Moscow
2010-present –	Professor of Biochemistry, Biochemistry Department, RUDN University

Professional Appointments

1988 –	Ph.D. (Biochemistry), Peoples' Friendship University of Russia (RUDN University)
2009 –	D.Sc. (Biochemistry), Peoples' Friendship University of Russia (RUDN University)

Publications from the Last 3 Years

1. Kalinina, E. V., Andreev, Y. A.; Petrova, A. S.; Lubova, K. I.; Shtil', A. A.; Chernov, N. N.; Novichkova, M. D.; Nurmuradov, N. K. Redox-Dependent Expression of Genes Encoding NADPH Oxidase 5 and the Key Antioxidant Enzymes during Formation of Drug Resistance of Tumor Cells to Cisplatin. *Bull. Exp. Biol. Med.* 2018, 165, 678-681.
2. Kalinina, E. V.; Chernov, N. N. Why does chemotherapy stop affecting the cells of ovarian and breast tumors? *Future Oncol.* 2018, 14, 1137-1140.
3. Kalinina, E. V.; Andreev, Y. A.; Brodsky, I. B.; Chernov, N. N.; Novichkova, M. D. Role of peroxiredoxins in the development of drug resistance in cancer cells. *Free Rad. Biol. Med.* 2018, 120(S1), S49.
4. Kalinina, E. V.; Chernov, N. N.; Novichkova, M. D.; Nurmuradov, N. K. Thioredoxins, glutaredoxins and peroxiredoxins in redox-dependent formation of cancer cell resistance. *Free Rad. Biol. Med.* 2017, 108, 34-35.

In: Glutathione
Editor: Pál Perjési

ISBN: 978-1-53614-740-7
© 2019 Nova Science Publishers, Inc.

Chapter 6

DIVERSE EFFECTS OF GLUTATHIONE IN THE PATHOPHYSIOLOGY OF SOME ORGAN DISEASES

Janka Vašková*

Department of Medical and Clinical Biochemistry, Faculty of Medicine,
Pavol Jozef Šafárik University in Košice, Košice, Slovak Republic

ABSTRACT

Reduced glutathione (GSH) is the most important low molecular weight antioxidant synthesized in cells. Since its level is higher than that of the two other significant redox systems, the GSH/GSSH system is also considered as the principal redox buffer. The reducing properties of GSH are applied in the GSH-dependent enzyme system in reduction of peroxides, including glutathione peroxidases and glutathione reductase. Although reduction of peroxides alone is relatively slow, direct reduction of other reactive species is effective (such as hydroxyl radicals, HOCl, carbonate radicals and hydrogen peroxide). Consumption of GSH as a reducing substrate underlines its importance, especially in organs whose specific metabolism is directly related to formation of reactive oxygen and nitrogen species (ROS, RNS). In fact, organ distribution and efficacy of the eight glutathione peroxidase isoforms differ with different substrate peroxides. This chapter will therefore be focused on description of mechanisms of formation of GSH deficiency, effectiveness of cellular GSH transport mechanisms resulting from increased production of ROS and RNS in the main organs of xenobiotic metabolism, such as the liver and the kidneys, and their impact on development of pathological conditions in these organs. About 10-15% of the total GSH in the liver is found in the mitochondria, and more than 30% of the total kidney GSH in the proximal tubules. The relatively alkaline pH of mitochondria, along with higher production of ROS due to metabolism of certain substances, creates conditions leading to stress-induced apoptosis or necrosis of cells of these tissues. GSH loss from lipid peroxide formation and low activity of GPx1 was also associated with progression of heart failure and cardiac remodeling. In nerve cells, oxytosis-induced GSH

* Corresponding Author E-mail: janka.vaskova@upjs.sk.

deficiency generally affects neurotransmission and behavior. Finally, it can lead to a nerve cell death in a form distinct from apoptosis. Thus, GSH deficiency prevents tissues from overcoming oxidative stress conditions and leads to structural damage and functional impairment. Due to the interconnectivity of organs in its distribution and precursor supply, GSH deficiency may result in local as well as systemic damage.

Keywords: glutathione, glutathione reductase, glutamate cysteine ligase, glutathione peroxidase, glutathione synthetase, parenchymal organs

1. Introduction

The most abundant non-protein thiol present in all mammalian cells is the tripeptide γ-L-glutamyl-L-cysteinylglycine, glutathione. Occurring in both reduced (GSH) and oxidized (GSSG) forms, its concentration in tissues ranges from 1-10 mM, with the highest concentration found in the liver [1]. The reduced form (GSH) is prevalent, accounting for more than 98% total GSH [2]. GSH is found in three main sites within the cell: in the cytosol (80 - 85%), in mitochondria (10 - 15%) and the endoplasmic reticulum [3].

Synthesis of GSH is controlled and takes place in the cytosol in two ATP-dependent steps: formation of γ-glutamylcysteine from glutamate and cysteine and formation of GSH from γ-glutamylcysteine and glycine. The first step of GSH biosynthesis is rate limiting and catalyzed by glutamate cysteine ligase (GCL) (formerly γ-glutamylcysteine synthetase, GCS) (EC 6.3.2.2). Formation of GSH from γ-glutamylcysteine and glycine is catalyzed by GSH synthetase (GS, EC 6.3.2.3) [1]. Their need for formation and proper function of GSH and survival of the organism has also been confirmed [4, 5]. An unusual amide bond is formed between glutamate and cysteine, which can only be hydrolyzed by γ-glutamyltranspeptidase (GGT, EC 2.3.2.2). GGT is also essential for hydrolysis of glutathione S-conjugates [6]. As GGT is present only on the outer surface of certain cells, such as lymphoid cells and especially epithelia, which have secretory and uptake capabilities. Therefore, GSH is resistant to intracellular degradation and is only metabolized extracellularly by cells that express GGT. GGT transmits the γ-glutamyl moiety to another amino acid or peptide forming a γ-glutamyl amino acid or peptide derivative. The γ-glutamyl amino acid can be transported back into the cell where it can be further metabolized by means of γ-glutamyl cyclotransferase (EC 2.3.2.4) to release the amino acid and 5-oxoproline. Subsequent cleavage of 5-oxo-L-proline to glutamate is catalyzed by 5-oxoprolinase. The cysteinyl-glycine residue is further degraded by the membrane-bound Cys-Gly dipeptidase (MBD, EC 3.4.13.6) to cysteine and glycine, both of which may be re-absorbed and re-incorporated into GSH in the cell (γ-glutamyl cycle) [7]. Knockout of a single GGT gene is lethal in mice [8].

Cysteine is a non-essential amino acid and is formed from methionine, which is essential and should be taken in as part of a diet. In the organism, however, cysteine is mostly transported in the form of GSH, hence the inactivity of GSH hydrolysis leads to a lack of cysteine in cells [7]. In addition to acting as a cysteine reservoir, GSH has several other vital functions. It forms the main redox system and is involved in anti-oxidative and detoxification processes, storage and transport of nitric oxide, cell cycle regulation, apoptosis, immunological events, fibrogenesis, metabolism of estrogens, leukotrienes, prostaglandins, reduction of ribonucleotides, maturation of ferrous sulphur clusters, and the activity of some transcription factors [1, 9].

The ubiquity and importance of the events in which GSH is involved outline its irreplaceable role. Therefore, influencing its transcription, formation, hydrolysis, or reduction of its oxidized form directly affects its concentration. There could be a particularly severe GSH deficiency in tissues directly related to biotransformation of xenobiotics and higher burden of oxidative stress.

2. Role of Liver in GSH Synthesis, Transport and Utilization

Liver plays a central role in inter-organ GSH homeostasis [10]. GSH turnover in the liver is necessary for inter-organelle transport and cellular utilization. However, the main mechanism of hepatocyte GSH turnover is its efflux into the bile, as no significant activity of GGT was determined in the liver [11]. GSH thus serves to supply its synthetic precursors to other tissues that express GGT, particularly the kidney and the intestine. However, disruption of GSH synthesis in the liver has an overall impact on GSH homeostasis in the body. Decrease of GSH synthesis is the basis of all pathogenesis of liver disorders such as cholestasis, endotoxemia, alcoholic and non-alcoholic fatty liver diseases [1]. In addition, systemic oxidative stress arising from liver diseases can also cause damage to extra-hepatic organs such as the brain and the kidneys. Systemic oxidative stress was suggested as a significant “first hit”, acting synergistically with ammonia to induce brain edema in chronic liver failure [12, 13]. Due to its central role in GSH homeostasis, it is necessary to mention the specifics of GSH synthesis and transport in the liver.

2.1. GSH Synthesis in the Liver

Due to the lack of GGT, ensuring a sufficient concentration of GSH for redox, conjugation reactions and other tissue supply in the liver is possible only by synthesis and reduction of oxidized glutathione *via* glutathione reductase (GR, EC 1.8.1.7).

The first and the rate-limiting step of GSH biosynthesis is catalyzed by GCL, which is composed of a heavy or catalytic (GCLC,) and light or modifier (GCLM) subunits. GCL regulation occurs both by allosteric regulation and through altered expression of its catalytic (GCLC) and modulatory (GCLM) subunits [14]. GCLC exhibits all the catalytic activity of the isolated enzyme and feedback inhibition by GSH. The smaller GCLM modulates the catalytic activity of GCLC by reducing the feedback inhibition by GSH and decreasing the K_m for glutamate. The K_m values of GCL for glutamate and cysteine are 1.8 and 0.1 – 0.3 mM, respectively. The intracellular glutamate concentration is 10-fold higher than the K_m value, but cysteine concentration gives the apparent K_m value [15, 16]. Activity of the subunits is also regulated by phosphorylation [17]. The catalytic complex is formed involving both subunits and elevated GCLM/GCLC ratio leads to an increase in GCL activity. Increased GCLC expression elevates GSH and increasing GCLM/GCLC will further increase GSH [18]. Expression is regulated by several mechanisms. Electrophiles including hydroperoxides increase transcription and stability of mRNA of both subunits [19], and the electrophile-response elements (antioxidant-response element, ARE) and TPA-response element (palindromic AP-1 binding site) are present in both the GCLC and GCLM promoter sequences [20, 21]. In hepatocytes, a series of processes are necessary for cysteine availability such as membrane transport of cysteine *via* the ASC system (Alanine Serine Cysteine system), cysteine *via* the xc^- system (glutamate/cysteine antiporter) which is induced under oxidative stress, and methionine *via* the L system (leucine-preferring system) [22]. While liver catabolizes up to half the daily intake of methionine, and the transsulfuration (cystathionine) pathway is particularly active in hepatocytes, there is little or no evidence of transsulfuration outside the liver. The pathway is also absent in the fetus and newborn infant [19, 23].

GS is composed of two identical subunits (Mr~118 kDa) and does not exhibit feedback inhibition by GSH. There is accumulating evidence of its importance in overall GSH synthesis in certain tissues and under certain conditions [19]. Specific activity of GS is 2 to 4 times higher than that of the GCL activity in normal liver [24]. Research has shown, however, that GS expression can be increased by the effects of substances that have no effect on GCL expression, suggesting that GS regulation may also be important under certain conditions, especially in non-hepatic tissues. GS deficiency can result in dramatic metabolic consequences (metabolic acidosis, hemolytic anemia or central nervous system damage) due to accumulation of γ-glutamylcysteine, which is subsequently converted to pyroglutamic acid (5-oxoproline) [25]. According to Lu et al. [1] it is therefore possible to assume that GS activity is the limiting factor in GCL induction.

2.2. GSH Transport from the Liver

Hepatocytes export GSH into the plasma to maintain inter-organ GSH homeostasis. The overall sinusoidal efflux of GSH is electrogenic and is stimulated by uncharged thiols and inhibited by disulphides. Extracellular methionine was also found to trans-inhibit the efflux of GSH [26]. Export of GSH occurs *via* anion transporter polypeptides (OATPs), in an electroneutral exchange of anions HCO_3^-, GSH, GSSH, and glutathione-S-conjugates. Although other organic anion transporters may mediate GSH efflux in human hepatocytes [27].

Almost half of rat hepatic GSH efflux passes into bile through the hepatic canalicular membrane. Canalicular GSH efflux involves multidrug resistance proteins (Mrps), particularly Mrp2 [28]. Mrp2 also exports GSH in the form of glutathione S-conjugate and facilitates GSSG export.

GSH efflux is an important event in Fas-induced apoptosis [29]. It is speculated that GSH export may facilitate apoptosis by enhancing cell susceptibility to oxidative stress and the apoptotic signaling pathway. In support of this hypothesis, blocking of GSH export was found to delay or suppress apoptosis [30].

2.3. Liver Pathologies Associated with Lack of Glutathione

Cholestatic liver injury is a major cause of chronic liver disease. Chronic retention of toxic bile acids is a major mechanism of cell toxicity induced by oxidative stress, apoptosis, and fibrosis, leading to cirrhosis [31]. In several studies, a decrease in GSH level has been reported. While a decrease has been measured in GCL activity and in the mRNA levels of GCLC and GCLM, a decrease has also been found in GSL activity [32, 33]. Yang et al. [34] found changes with transient increase followed by a significant decline of expression of GSH synthetic enzymes induced by bile duct ligation during experimental cholestasis. This was the result of a decrease in the Nrf2 nuclear binding activity of ARE. Ursodeoxycholic acid and S-adenosylmethionine can prevent this fall by inducing the expression of Nrf2 and suppressing the expression of Maf proteins repressing ARE [34].

The fundamental difference between chronic alcoholic liver disease and the physiological state is the 45-60% selective depletion of the mitochondrial GSH pool in the liver [26]. As GSH is synthesized only in the cytosol, it must be transported into the mitochondria. Normally, GSH passes easily through the external mitochondrial membrane *via* the voltage-dependent anion channel (VDAC). It then passes through the inner membrane *via* the dicarboxylate carrier protein (DIC) or the oxoglutarate carrier (OGC). By transporting GSH through the internal mitochondrial membrane, the mitochondria lose citrate cycle intermediates, which must be replaced by anaplerotic reactions [35]. DIC acts as an oxaloacetate shuttle *via* an internal mitochondrial membrane after conversion to malate in exchange for phosphate, sulphate or thiosulfate. In the cytosol, malate is converted to oxaloacetate following conversion into phoshoenolpyruvate by a carboxykinase-catalyzed reaction. The rate of reaction is limited by gluconeogenesis. Reduced DIC expression results in significantly reduced levels of glutathione in the mitochondria and subsequently alters redox conditions [36]. The OGC provides for the transfer of oxoglutarate by replacing dicarboxylate. Recently, however, this GSH transmission *via* the internal mitochondrial membrane has been questioned by the DIC or OGC, and the exact GSH transmitter is not known [37]. However, binding of succinate to the matrix side of the carrier increases the affinity for malate, while phenylsuccinate, pyridoxalphosphate, retinoic acid or alcohol have been found to inhibit OGC [38, 39]. Inhibition of the transporter thus leads not only to reduced energy generation but also to the lack of GSH, and subsequent development of oxidative stress. It has been confirmed that administration of an excess of retinyl palmitate can lead to fibrosis and hepatocellular dysfunction [40]. In addition to fibrosis, oxidative stress has also been implicated in pathogenesis of a wide range of liver injuries including steatosis, steatohepatitis, fibrosis and cirrhosis, known as non-alcoholic fatty liver disease (NAFLD). A diminished antioxidant capacity was found in the patients with steatosis and steatohepatitis, including a decrease of GSH and superoxide dismutase (SOD) levels [41]. In a study of a classical fatty liver model, complex I and cardiolipin oxidation were identified [42]. Cardiolipin provides stability and fluidity of the internal mitochondrial membrane as well as an effective phosphorylation coupling. Under physiological conditions, GSH protects cardiolipin from oxidative damage as its high content of polyunsaturated fatty acids makes it susceptible to oxidation. GSH thus prevents changes in membrane properties and dissociation of cytochrome c, and induction of apoptosis. Increased production of reactive oxygen species (ROS) also leads to increased Ca^{2+} permeability of the internal mitochondrial membrane, which can lead to cell death by apoptosis or necrosis [35].

Despite the easy transfer of GSH through the external mitochondrial membrane, it has been found that VDAC closure caused by aldehyde formation and lipid peroxidation is a common feature leading to liver pathologies as was pointed out on almost indistinguishable histopathological manifestations in alcoholic liver disease, non-alcoholic fatty liver diseases, and toxicant-associated steatohepatitis. The closure of VDAC by aldehydes is explained by Lemaster et al. [43] on the metabolism of ethanol. Ethanol is metabolized predominantly by the liver in two consecutive oxidation steps. First, oxidation to acetaldehyde by alcohol dehydrogenase (ALD), second, oxidation to acetate by aldehyde dehydrogenase (ALDH). Acetaldehyde formation also takes place in the cytosol and the peroxisomes by CYP2E1 and catalase, respectively. Although oxidation by ALD is prevalent, consequence of cytochrome P450 metabolism is overwhelmingly negative due to formation of ROS, e.g., hydroxyethyl radicals. Of 19 known mammalian ALDH genes, mitochondrial ALDH2 with high affinity for acetaldehyde (K_m < 1 μM) is the most important in oxidation of acetaldehyde to acetate [44]. In both reactions, by oxidation of 1 mole of ethanol, 2 moles of NADH are formed, further requiring oxidation in the respiratory chain. In 2.5 h after a single bolus dose of ethanol ingestion, a swift increase of alcohol metabolism (SIAM) occurs as an adaptive change in hepatic ethanol metabolism [45]. Mitochondrial respiration nearly doubles along with NADH oxidation, but it does not lead to an increase in ATP generation. On the contrary, hepatic ATP decreases, glycolysis is stimulated, and glycogen stores are depleted. Furthermore, mitochondrial β-oxidation of fatty acids becomes inhibited, promoting fat accumulation within hepatocytes-steatosis [46]. This is probably the result of a decrease in mitochondrial outer membrane permeability by VDAC closure, promoting selective oxidation of acetaldehyde, because VDAC closure blocks mitochondrial ATP release, respiratory substrate uptake and the uptake of fatty acids for β-oxidation [43]. Adrenergic hormones release free fatty acids from adipose tissue, which serve as substrates for long-chain fatty acid peroxisomal β-oxidation. The ensuing peroxisomal H_2O_2 formation can then promote catalase-dependent alcohol metabolism [43]. Acetaldehyde is toxic to mitochondria and increases oxidative stress by binding to GSH and promoting GSH leakage [47]. Moreover, as mentioned before, ethanol metabolism and NADH overproduction cause formation of ROS, lipid peroxidation, onset of mitochondrial permeability transition and apoptosis [48]. However, as has been shown, short- and medium-chain fatty acids can cross the mitochondrial membrane freely using carnitine shuttle or other transport systems [49], demonstrating that dietary nutrients have the capacity to prevent development of steatosis. Also, activity of the hepatic transsulfuration pathway is markedly impaired in cirrhotic patients. Part of the mechanism relates to decreased cofactor availability (such as B vitamins). In addition, cirrhotic patients have decreased methionine-adenosyltransferase activity and diminished S-adenosylmethionine [1].

Endotoxins or lipopolysaccharides (LPS) are found on the outer part of the Gram-negative bacteria membrane. They enter the bloodstream when integrity of the intestinal wall is disrupted or during intestinal dysfunctions and cause inflammatory processes. From there, they are picked up by Kupfer cells in the liver. The first activity of hepatic sinusoidal cells is therefore disrupted, which subsequently affects the ability to eliminate commonly innocuous amounts of LPS from the intestine. A marked increase in sensitivity to LPS (10-to- 1000-fold) leads to further hepatocytic damage and prevents the enteric endotoxins from entering the systemic circulation, resulting in extrahepatic manifestations associated with liver injury [50]. Endotoxemia worsens alcoholic liver disease, non-alcoholic steatohepatitis and lowers GSH level in the liver [51, 52].

A decrease in GSH level was also observed in septic patients and found to be related to LPS-induced mortality and increased TNFα secretion [53]. GSH supplementation decreased mortality and led to suppression of the inflammatory response [54]. Payabvash et al. [51] showed the fall in GCLC mRNA levels and GCL activity as the main consequence of GSH depletion in the liver. There is a 50-60% fall in the mRNA and protein levels of GCLC and GCLM, up to 40% in GS expression and an over 50% in GSH concentration [55]. Moreover, in septic patients and patients with multi-organ failure, significant increases were found in the glutathione-S-transferase (GST, EC 2.5.1.18) activity [56]. GST belongs to a family of phase II xenobiotic detoxifying enzymes, catalyzing conjugation of GSH with electrophiles, including the secondary products of lipid peroxidation [57]. No statistically significant increase in serum GST activity was found in hepatocellular damage caused secondarily to various diseases [56].

3. Cardiac Dysfunction from GSH Deficiency

Because of the previously described liver functions on synthesis of glutathione, this organ is responsible for maintaining GSH concentration in other tissues, including heart. An important function of glutathione in heart muscle is its protection against oxidative stress. Heart muscle requires a continuous supply of energy in form of ATP, coming from oxidative phosphorylation. Energy is primarily obtained from oxidation of fatty acids and, to a lesser extent, lactate and glucose [58]. Oxidation of fatty acids is associated with formation of hydrogen peroxide, and reactive molecules are formed within the respiratory chain. Since catalase activity is low in the heart, peroxides are mainly degraded by glutathione peroxidase and glutathione itself [59]. Glutathione peroxidase (GPx) catalyzes conversion of hydrogen peroxide and substrate peroxides to water and the corresponding alcohols using GSH, which is itself oxidized in turn. Its reduction back to GSH is catalyzed by GR. Some studies have shown that excessive oxidation of fatty acids and oxidation of intracellular lipids cause an increase in oxidative stress and result in cardiac dysfunction [60]. These deleterious effects include intracellular acidosis, electrochemical dysfunction, alteration in cardiac action potential, and a decrease in contractility [61, 62]. Ghosh et al. [63] revealed that cardiolipotoxicity is the result of impaired function of cystathionine β-synthase (Cbs, EC 4.2.1.22). In heterozygous mice with targeted disruption of the gene for Cbs (+/-), normal GSH concentrations were found in the liver after feeding with high fatty diets. In the plasma, however, cysteine concentrations were lowered. Lower GSH concentrations, and increased triglyceride and apoptosis markers were found in the heart. About 50% of cysteine is formed by the transsulfuration pathway that occurs almost exclusively in the liver. Hyperhomocysteinemia due to the lack of Cbs activity has been associated with disruption of lipid metabolism in the liver [64]. However, expression of Cbs (and cystathionase, CSE) is also present in the heart to a lesser extent [65]. Heterozygous Cbs activity was therefore enough to facilitate GSH synthesis in the liver but was no longer sufficient for proper supply to the heart under conditions of increased oxidative stress. Increased concentrations of homocysteine and decreased GSH concentrations were present in adolescent children of parents with premature coronary heart disease (CHD) and were even predictive of unknown or suspected CHD in parents [66]. A study by Damy et al. [67] on patients undergoing cardiac surgery for dilated cardiomyopathy also confirmed that systemic GSH deficiency was related to abnormalities of the functional status and structural hart of patients. Significant GSH depletion occurred prior to detectable elevation of the marker of symptomatic heart failure severity in the blood, sTNFR1 (cleaved extracellular domain of tumor necrosis factor-alpha type-1 receptor). Both parameters, homocysteine and GSH, are therefore suitable independent predictors of CVD development.

In Cbs(+/-) mice, low GR activity, GPx-1 expression and altered NF-κB signaling were also found in the heart [63]. Activation of NF-κB signaling occurs with moderate levels of GSH and is inhibited by GSH depletion [68]. Reduction or absence of GPx-1 activity, which is localized in the cytosol and the mitochondria, was directly related to development of left ventricular hypertrophy and dysfunction or vascular medial hypertrophy [69]. Reduction in GPx-1 is already considered to be a risk factor for cardiovascular events in patients [70].

It has also been demonstrated that polymorphisms in GST genes may modify the risk of toxicant exposure associated with coronary heart diseases [71]. A study by Li et al. [72] recently confirmed that maternal smoke exposure during pregnancy is associated with an increased risk of heart disease because of GST polymorphisms and consequently modified metabolism of GSH-conjugates.

4. Role of GSH in Neurological Disorders

As the brain contains a large amount of polyunsaturated lipids, which are particularly prone to oxidation, ROS production poses a particularly high potential threat of increased oxidative stress conditions. This occurs for three main reasons; first, the brain utilizes up to about 20% of the oxygen consumed by the whole body, although it accounts for only 2% of the total body mass. Second, some areas of the brain contain high levels of iron. Finally, the levels of antioxidant enzymes are much lower in the brain than in the liver or the kidneys. These antioxidant enzymes include superoxide dismutase (SOD), catalase, and GPx, which also acts upon substrate peroxides. The importance of GSH in maintaining the redox state and conversion of peroxides in nerve cells is unquestionable. The amount of GSH thus affects susceptibility of cells to damage due to oxidative stress, which can cause neurodegenerative diseases [73].

The amount of GSH in each cell type in the brain depends on availability of its precursors from the extracellular space. GSH content is higher in astrocytes than in neurons, but less is known about GSH metabolism in oligodendroglial and microglial cells [73]. Cultured astroglial cells were found to have released approximately 10% of the synthesized GSH within 1 hour [74]. GGT is expressed on the surface of astrocytes; therefore, after hydrolysis of GSH, astrocytes are capable of resynthesizing GSH in the γ-glutamyl cycle. Cys-Gly is transported back into the cells *via* the PepT2 transporter [75]. In addition to astrocytes, the PepT2 transporter is present in epithelial cells of the choroid plexus [76]. As found in neuron cultures, GSH content is independent of glutamine or glycine but depends on availability of cysteine. Thus, neurons can use Cys-Gly, γGlu-Cys and N-acetylcysteine as cysteine donors. Cys-Gly formed in the GGT reaction on the surface of astrocytes is used mostly by neurons [77]. The probable transfer mechanism is PepT2 transport or that Cys and Gly which are received by neurons after hydrolysis by dipeptidase. Cysteine is also transferred by the glutamate/cysteine antiporter in a 1:1 ratio (xc^- system), which provides cysteine transfer to the cells instead of glutamate. Glutamate is taken up by astrocytes that release glutamine. As neurons are unable to synthesize the crucial excitatory neurotransmitter, glutamate, they must receive glutamate synthesized in astrocytes (glutamate/GABA-glutamine cycle). Localization of this antiporter was found in mice and humans, particularly in peripheral areas including vascular endothelial cells, ependymal cells, choroid plexus and leptomeninges [78]. Astrocytes supplying the neurons with intermediates for GSH synthesis are very important for the maintenance of redox homeostasis [73], while neurotransmission and behavioral processes are also affected by sufficient amount of glutamate.

Inhibition of cysteine import through the xc^- system leads to GSH depletion, increased oxidative stress due to accumulation of peroxides, and induction of cell death by the mechanism of oxytosis [78]. In the case of glutamate toxicity, glutamate receptors are activated by glutamate, or structurally similar amino acids, leading to the loss of post-synaptic structures including dendrites and cell bodies. Sensitivity of nerve cells to glutamate and its agonists differ depending on receptor expression and tend to change with age [79]. Acute toxicity arises in response to a variety of severe insults including cerebral ischemia, traumatic brain injury (TBI), hypoglycemia, and status epilepticus [80, 81]. While chronic toxicity plays a role in neurological disorders such as amyotrophic lateral sclerosis (ALS), Alzheimer's disease (AD) and Huntington's disease (HD) [79].

The lack of GSH in nerve cells, due to inhibition of other enzymes associated with its synthesis or regeneration, also leads to increased oxidative stress that results in disruption of mitochondrial and cellular functions and consequently development of neurodegenerative diseases. Laboratory, preclinical and clinical studies of AD have shown a link between low levels of GSH and temporal and parietal amyloid levels [82-84]. Low levels of GSH have been associated with mitochondrial dysfunction and increased levels of cognitive impairment [85, 86]. Elevated GSH concentrations in earlier AD stages are considered a consequence of the compensatory mechanism, and these individuals were only affected with mild cognitive impairment.

The study by Chiang et al. [87], however, confirmed association between low GSH concentrations and amyloidosis, even with mild cognitive impairment. Negative association was also found between low concentrations of GSH in neurons and body mass index. Significantly reduced GCLC and GS activities were found in tissue lysates from autopsied human brains of patients with AD, suggesting inadequate GSH synthesis [88]. In addition, increased levels of GR have been confirmed as a compensatory mechanism for GSH deficiency. Increased activity of GR in Parkinson's disease and other neurodegenerative disorders were also detected [89, 90]. In contrast to neurodegenerative disorders, no change in GCL and GR levels was measured, although a reduction in GPx levels was found in post-mortem prefrontal cortex biopsies from patients with schizophrenia, major depressive disorder, and bipolar disorder [91]. In addition, patients with HIV-1 infection are susceptible to development of a severe form of tuberculosis, tuberculous meningitis. Interestingly, decreased concentrations of GSH, S-adenosylmethionine, Cys-Gly, and cysteine were found in autopsied brains of patients with HIV-1, although homocysteine was not affected. Patients with low CD4+ T cell counts had significantly reduced levels of both GCLS and GS, but patients with high CD4+ T cell counts had reduced levels of only GS [88]. The likely mechanism may be the high level of TGF-β (transforming growth factor β) found in plasma and macrophage supernatants, which inhibits GCLC synthesis [92, 93]. The decrease in GS enzyme activity is also affected by the action of HIV-1 transactivator protein (TAT), which is essential for the efficient transcription of the viral genome [94]. Moreover, HIV-1 related proteins also down-regulate Nrf2 expression and activity. Nrf2 is involved in the regulation of antioxidant and phase II metabolizing enzymes [95].

5. Kidney Injury and Impairment of GSH Synthesis

It has already been emphasized that liver is a key organ for GSH synthesis, supplying other tissues with effective redox-active agents, and acting as a way of transporting cysteine into tissues (particularly kidney and gut). Expressing GGT on their surface. Several studies have confirmed presence of oxidative stress conditions in chronic renal disease, with severity rising along with the progression of renal failure [96-98]. Ryu et al. [99] found a positive correlation between increased serum GGT activities and an increased risk of chronic kidney disease. The likely mechanism behind the increase in GGT activity in association with kidney failure was not presented; however, it can be assumed that it is an adaptation mechanism leading to an increase in the amount of substrates for GSH synthesis in the kidney. Interestingly, however, a link was found between increased GGT activity, the incidence of chronic kidney disease and a high body mass index [100]. Similarly, this parameter was found to be significantly associated with GSH depletion, even in the case of nerve cell damage [88]. Several studies have also confirmed association between insulin resistance, GGT activity, and an increased risk of chronic kidney disease [101-103]. Impaired systemic GSH synthesis has been clarified as a reason for low GSH content in acute kidney injury. A common cause of acute kidney injury is ischemia-reperfusion e.g., in surgery procedures, renal transplantation or as comorbidity in critically ill patients. The most important mechanism in its pathogenesis is increased oxidative stress often leading to a distant organ injury - multi-organ dysfunction [104, 105]. Rats subjected to ischemia-reperfusion showed a decrease in expression of GCL, CSE and Nrf2 (nuclear factor-erythroid 2-related factor-2) accompanied by a marked decrease in plasma and hepatic glutathione levels [106]. GSH depletion could be directly related to Nrf2-mediated inhibition of GCL expression. Increased homocysteine levels found in the kidney are also related to down-regulation of the transsulfuration pathway and consequently increased oxidative stress conditions in the organ due to lack of GSH synthesis [107, 108]. As mentioned above, the transsulfuration pathway mostly takes place in the liver, and -to a lesser extent- in the kidney, intestine, heart, aorta, brain, and the lungs [109, 110]. GSH depletion in acute kidney injury pointed to induction of the enzyme expression having an impact on GSH synthesis in the liver, which then further affects the redox state in other organs.

CONCLUSION

This chapter summarizes the current knowledge on the effects of glutathione in the pathophysiology of some organ diseases. Its main tasks include facilitating reduction of compounds changed by oxidation, preventing their damage, and transport of cysteine from the site of its dominant synthesis, the liver, to individual organs. The peculiarities of its metabolism are described in individual organs, and in terms of its deficiency, which leads not only to local organ failure, but also to systemic organ failure.

REFERENCES

[1] Lu, S. C. Glutathione synthesis. *Biochim. Biophys. Acta*. 2013, 1830, 3143–3153.

[2] Forman, H. J.; Zhang, H.; Rinna, A. Glutathione: overview of its protective roles, measurement, and biosynthesis. *Mol. Aspects Med.* 2009, 30, 1–12.

[3] Yuan, L.; Kaplowitz, N. Glutathione in liver diseases and hepatotoxicity. *Mol. Aspects Med.* 2009, 30, 29–41.

[4] Kim, B. J.; Choi, C. H.; Lee, C. H.; Jeong, S. Y.; Kim, J. S.; Kim, B. Y.; Yim, H. S.; Kang, S. O. Glutathione is required for growth and prespore cell differentiation in Dictyostelium. *Dev. Biol.* 2005, 284, 387–398.

[5] Cairns, N. G.; Pasternak, M.; Wachter, A.; Cobbett, C. S.; Meyer, A. J. Maturation of rabidopsis seeds is dependent on glutathione biosynthesis within the embryo. *Plant Physiol.* 2006, 141, 446–455.

[6] Martin, M. N.; Saladores, P. H.; Lambert, E.; Hudson, A. O.; Leustek, T. Localization of members of the γ-glutamyl transpeptidase family identifies sites of glutathione and glutathione S-conjugate hydrolysis. *Plant Physiol.* 2007, 144, 1715–1732.

[7] Meister, A.; Tate, S. S.; Thompson, G. A. *On the function of the γ-glutamyl cycle in the transport of amino acids and peptides*. 1. Elliott K, Maeve O'Connor M, editors.

[8] *Ciba Foundation Symposium 50* - Peptide Transport and Hydrolysis: Nova Foundation Symposia, 2008.

[9] Lieberman, M. W.; Wiseman, A. L.; Shi, Z. Z.; Carter, B. Z.; Barrios, R.; Ou, C. N.; Chevez-Barrios, P.; Wang, Y.; Habib, G. M.; Goodman, J. C.; Growth retardation and cysteine deficiency in γ-glutamyl transpeptidase deficient mice. *Proc. Natl. Acad. Sci. USA*. 1996, 93, 7923–7926.

[10] Halliwell, B.; Gutteridge, J. M. C. *Free Radicals in Biology and Medicine*. Claredon Press: Oxford: UK, 1989.

[11] Ookhtens, M.; Kaplowitz, N. Role of the liver in interorgan homeostasis of glutathione and cyst(e)ine, Semin. *Liver Dis*. 1998, 18, 313–329.

[12] Kaplowitz, N. The importance and regulation of hepatic glutathione. *Yale J. Biol. Med.* 1981, 54, 497-502.

[13] Bosoi, C. R.; Yang, X.; Huynh, J.; Parent-Robitaille, C.; Jiang, W.; Tremblay, M.; Rose, C. F. Systemic oxidative stress is implicated in the pathogenesis of brain edema in rats with chronic liver failure. *Free Rad. Biol. Med.* 2012, 52, 1228–1235.

[14] Wang, H.; Chen, X.; Su, Y.; Paueksakon, P.; Hu, W.; Zhang, M. Z.; Harris, R. C; Blackwell, T. S.; Zent, R.; Pozzi, R. p47(phox) contributes to albuminuria and kidney fibrosis in mice. *Kidney Int.* 2015, 87, 948–962.

[15] Forman, H. J. Glutathione - From antioxidant to post-translational modifier. *Arch. Biochem. Biophys.* 2016, 595, 64-67.

[16] Dalton, T. P.; Chen, Y.; Schneider, S. N.; Nebert, D. W.; Shertzer, H. G. Genetically altered mice to evaluate glutathione homeostasis in health and disease. *Free Rad. Biol. Med.* 2004, 37, 1511–1526.

[17] Huang, C. S.; Anderson, M. E.; Meister, A. The function of the light subunit of γ-glutamylcysteine synthetase (rat kidney). *FASEB J.* 1993, 7, A1102.

[18] Sun, W. M.; Huang, Z. Z.; Lu, S. C. Regulation of γ-glutamylcysteine synthetase by protein phosphorylation. *Biochem. J.* 1996, 320, 321-328.

[19] Krzywanski, D. M.; Dickinson, D. A.; Iles, K. E.; Wigley, A. F.; Franklin, C. C.; Liu, R. M.; Kavanagh, T. J.; Forman, H. J. Variable regulation of glutamate cysteine ligase subunit proteins affects glutathione biosynthesis in response to oxidative stress. *Arch. Biochem. Biophys.* 2004, 423, 116-125.

[20] Lu, S. C. Regulation of glutathione synthesis. *Mol. Aspects Med.* 2009, 30, 42-59.

[21] Dickinson, D. A.; Iles, K. E.; Watanabe, N.; Iwamoto, T.; Zhang, H.; Krzywanski, D. M.; Forman H. J. 4-hydroxynonenal induces glutamate cysteine ligase through JNK in HBE1 cells. *Free Rad. Biol. Med.* 2002, 33, 974-987.

[22] Wild, A. C.; Gipp, J. J.; Mulcahy, R. T. Overlapping antioxidant response element and PMA response element sequences mediate basal β-naphthoflavone induced expression of the human γ-glutamylcysteine synthetase catalytic subunit gene. *Biochem. J.* 1998, 332, 373-381.

[23] Bannai, S.; Tateishi, N. Role of membrane transport in metabolism and function of glutathione in mammals. *J. Membr. Biol.* 1986, 89, 1–8.

[24] Finkelstein, J. D. Methionine metabolism in mammals. *J. Nutr. Biochem.* 1990, 1, 228–237.

[25] Wu, Y.; Zhang, X.; Robel, R. C. V.; Chen, L. X.; Zeng, Y.; Aguilo, J.; Bardag-Gorce, F.; Hwang, K.; French, S. W.; Lu, S. C.; Wan, Y. Y. Retinoid X receptor a regulates amino acid metabolism, glutathione homeostasis and xenobiotic detoxification processes in mouse liver. *Mol. Pharmacol.* 2004, 65, 550–557.

[26] Shi, Z. Z.; Habib, G. M.; Rhead, W. J.; Gahl, W. A.; He, X.; Sazer, S.; Lieberman, M. W. Mutations in the glutathione synthetase gene cause 5-oxoprolinuria. *Nat. Genet.* 1996, 14, 361–365.

[27] Yuan, L.; Kaplowitz, N. Glutathione in liver diseases and hepatotoxicity. *Mol. Aspects Med.* 2009, 30, 29-41.

[28] Mahagita, C.; Grassl, S. M.; Piyachaturawat, P.; Ballatori, N. Human organic anion transporter 1B1 and 1B3 function as bidirectional carriers and do not mediate GSH-bile acid cotransport. *Am. J. Physiol.* 2007, 293, G271-G278.

[29] Ballatori, N.; Rebbeor, J. F. Roles of MRP2 and oatp1 in hepatocellular export of reduced glutathione. *Semin. Liver Dis.* 1998, 18, 377–387.

[30] Franco, R.; Panayiotidis, M. I.; Cidlowski, J. A. Glutathione depletion is necessary for apoptosis in lymphoid cells independent of reactive oxygen species formation. *J. Biol. Chem.* 2007, 282, 30452–30465.

[31] Hammond, C. L.; Marchan, R.; Krance, S. M.; Ballatori, N. Glutathione export during apoptosis requires functional multidrug resistance-associated proteins. *J. Biol. Chem.* 2007, 282, 14337–14347.

[32] Guicciardi, M. E.; Gores, G. J. Bile acid-mediated hepatocyte apoptosis and cholestatic liver disease. *Dig. Liver Dis.* 2002, 34, 387–392.

[33] Serviddio, G.; Pereda, J.; Pallardo, F. V.; Carretero, J.; Borras, C.; Cutrin, J.; Vendemiale, G.; Poli, G.; Vina, J.; Sastre, J. Ursodeoxycholic acid protects against secondary biliary cirrhosis in rats by preventing mitochondrial oxidative stress. *Hepatology*. 2004, 39, 711–720.

[34] Gumpricht, E.; Devereaux, M. W.; Dahl, R. H.; Sokol, R. J. Glutathione status of isolated rat hepatocytes affects bile acid-induced cellular necrosis but not apoptosis. *Toxicol. Appl. Pharmacol.* 2000, 164, 102–111.

[35] Yang, H.; Ramano, K.; Xia, M.; Ko, K. S.; Li, T. W. H; Oh, P.; Vendemiale, G.; Poli, G.; Viña, J.; Sastre, J. Dysregulation of glutathione synthesis during cholestasis in mice: Molecular mechanisms and therapeutic implications. *Hepatology*. 2009, 49, 1982-1991.

[36] Lushchak, V. I. Glutathione homeostasis and functions: Potential targets for medical intervention. *J. Amino Acids*. 2012, ID 736837, doi:10.1155/2012/736837.

[37] Kamga, C. K.; Zhang, S. X.; Wang, Y. Dicarboxylate carrier-mediated glutathione transport is essential for reactive oxygen species homeostasis and normal respiration in rat brain mitochondria *Am. J. Physiol. Cell Physiol.* 2010, 299C, 497-505.

[38] Booty, L. M.; King, M. S.; Thangaratnarajah, C.; Majd, H.; James, A. M.; Kunji, E. R. S.; Murphy, M. P. The mitochondrial dicarboxylate and 2-oxoglutarate carriers do not transport glutathione. *FEBS Lett.* 2015, 589, 621–628.

[39] Cione, E.; Pingitore, A.; Perri, M.; Genchi, G. Influence of all-trans-retinoic acid on oxoglutarate carrier via retinoylation reaction. *Biochim. Biophys. Acta.* 2009, 1791, 3-7.

[40] Coll, O.; Colell, A.; García-Ruiz, C.; Kaplowitz, N.; Fernández-Checa, J. C. Sensitivity of the 2-oxoglutarate carrier to alcohol intake contributes to mitochondrial glutathione depletion. *Hepatology*. 2003, 38, 692-702.

[41] Choudhary, P. R.; Swami, K. K. Hypervitaminosis-A is associated with increased risk of hepatotoxicity in swiss albino mice. *Int. J. Biol. Res.* 2012, 3, 1311-1314.

[42] Videla, L. A.; Rodrigo, R.; Orellana, M.; Fernandez, V.; Tapia, G.; Quinones, L.; Varela, N.; Contreras, J.; Lazarte, R.; Csendes, A.; Rojas, J.; Maluenda, F.; Burdiles, P.; Diaz, J. C.; Smok, G.; Thielemann, L.; Poniachik, J. Oxidative stress-related parameters in the liver of non-alcoholic fatty liver disease patients. *Clin. Sci.* (Lond), 2004, 106, 261–268.

[43] Petrosillo, G.; Portincasa, P.; Grattagliano, I.; Casanova, G.; Matera, M.; Ruggiero, F. M.; Ferri, D.; Paradies, G. Mitochondrial dysfunction in rat with nonalcoholic fatty liver Involvement of complex I, reactive oxygen species and cardiolipin. *Biochim. Biophys. Acta.* 2007, 1767, 1260–1267.

[44] Lemasters, J. J.; Holmuhamedov, E. L.; Czerny, C.; Zhong, Z.; Maldonado, E. N. Regulation of mitochondrial function by voltage dependent anion channels in ethanol metabolism and the Warburg effect. *Biochim. Biophys. Acta* 2012, 1818, 1536-1544.

[45] Marchitti, S. A.; Brocker, C.; Stagos, D.; Vasiliou, V. Non-P450 aldehyde oxidizing enzymes: the aldehyde dehydrogenase superfamily. *Expert Opin. Drug Metab. Toxicol* 2008, 4(6), 697-720.

[46] Yuki, T.; Thurman, R. G. The swift increase in alcohol metabolism. Time course for the increase in hepatic oxygen uptake and the involvement of glycolysis *Biochem J* 1980, 186119-186126.

[47] Bradford, B. U.; Rusyn, I. increase in alcohol metabolism (SIAM): understanding the phenomenon of hypermetabolism in liver. *Alcohol.* 2005, 35, 13-17.

[48] Lieber, C. S. Alcoholic fatty liver: its pathogenesis and mechanism of progression to inflammation and fibrosis *Alcohol.* 2004, 34, 9-19.

[49] Adachi, M.; Ishii, H. Role of mitochondria in alcoholic liver injury *Free Rad. Biol. Med* 2002, 32, 487-491.

[50] Bartlett, K.; Eaton, S. *Mitochondrial beta-oxidation* 2004, 271, 462-469.

[51] Nolan, J. P. The role of intestinal endotoxin in liver injury: A long and evolving history. *Hepatology*. 2010, 52, 1829-1835.

[52] Su, G. L. Lipopolysaccharides in liver injury: molecular mechanisms of Kupffer cell activation. *Am. J. Physiol. Gastrointest. Liver Physiol*. 2002, 283, G256–G265.

[53] Payabvash, S.; Ghahremani, M. H.; Goliaei, A.; Mandegary, A.; Shafaroodi, H.; Amanlou, M.; Dehpour, A. R. Nitric oxide modulates glutathione synthesis during endotoxemia. *Free Rad. Biol. Med*. 2006, 41, 1817–1828.

[54] Thimmulappa, R. K.; Lee, H.; Rangasamy, T.; Reddy, S. P.; Yamamoto, M.; Kensler, T. W.; Biswal, S. Nrf2 is a critical regulator of the innate immune response and survival during experimental sepsis. *J. Clin. Invest*. 2006, 116, 984–995.

[55] Sun, S.; Zhang, H.; Xue, B.; Wu, Y.; Wang, J.; Yin, Z.; Luo, L. Protective effect of glutathione against lipopolysaccharide-induced inflammation and mortality in rats. *Inflamm. Res*. 2006, 55, 504–510.

[56] Ko, K. S.; Yang, H. P.; Noureddin, M.; Iglesia-Ara, A.; Xia, M.; Wagner, C.; Luka, Z.; Mato, J. M.; Lu, S. C. Changes in S-adenosylmethionine and glutathione homeostasis during endotoxemia in mice. *Lab. Invest*. 2008, 88, 1121–1129.

[57] Muttigi, M. S.; Kedage, V. K.; Suvarna, R.; Rao, S. S.; Joshi, C.; Shetty, M. S.; Prakas, M. Serum GST acitivty and total thiols status in patients with liver disease secondary to various disorders. *The Open Hepatol. J*. 2009, 1, 5-8.

[58] Gemma, S.; Vichi, S.; Testai, E. Individual susceptibility and alcohol effects: biochemical and genetic aspects. *Ann. Ist. Super Sanita*. 2006, 42, 8-16.

[59] Stanley, W. C.; Chandler, M. P. Energy metabolism in the normal and failing heart: potential for therapeutic interventions. *Heart Fail. Rev*. 2002, 7, 115-130.

[60] Antunes, F.; Han, D.; Cadenas, E. Relative contributions of heart mitochondria glutathione peroxidase *and* catalase to H_2O_2 detoxification *in vivo* conditions. *Free Rad. Biol. Med*. 2002, 33, 1260–1267.

[61] Keaney, J. F.; Jr. Larson, M. G.; Vasan, R. S.; Wilson, P. W.; Lipinska, I.; Corey, D.; Massaro, J. M.; Sutherland, P.; Vita, J. A.; Benjamin, E. J. Obesity and systemic oxidative stress clinical correlates of oxidative stress in the Framingham study. *Arterioscler. Thromb. Vasc. Biol*. 2003, 23, 434–439.

[62] Azevedo, A. M.; Martins, V. C.; Prazeres, D. M.; Vojinović, V.; Cabral, J. M.; Fonseca, L. P. Horseradish peroxidase: a valuable tool in biotechnology. *Biotechnol. Annu Rev*. 2003, 9, 199-249.

[63] Ferdinandy, P.; Danial, H.; Ambrus, I.; Rothery, R. A.; Schulz, R. Peroxynitrite is a major contributor to cytokine-induced myocardial contractile failure. *Circ Res*. 2000, 87, 241–247.

[64] Ghosh, S.; Sulistyoningrum, D. C.; Glier, M. B.; Bruce Verchere, C. B.; Devlin, A. M. Altered glutathione homeostasis in heart augments cardiac lipotoxicity associated with diet-induced obesity in mice. *J. Biol. Chem*. 2011, 286, 42483-42493.

[65] Devlin, A. M.; Singh, R.; Wade, R. E.; Innis, S. M.; Bottiglieri, T.; Lentz, S. R. Hypermethylation of Fads2 and altered hepatic fatty acid and phospholipid metabolism in mice with hyperhomocysteinemia. *J. Biol. Chem.* 2007, 282, 37082–37090.

[66] Brosnan, J. T.; Brosnan, M. E. The sulfur-containing amino acids: an overview. *J. Nutr.* 2006, 136, 1636S–1640S.

[67] Morrison, J. A.; Jacobsen, D. W.; Sprecher, D. L.; Robinson, K.; Khoury, P.; Daniels, S. R. Serum glutathione in adolescent males predicts parental coronary heart disease. *Circulation*. 1999, 100, 2244-2247.

[68] Damy, T.; Kirsch, M.; Khouzami, L.; Caramelle, P.; LeCorvoisier, P.; Roudot-Thoraval, F.; Dubios-Rande, J. L.; Hittinger, L.; Pavione, C.; Pecker, C. Glutathione deficiency in cardiac patients is related to the functional status and structural cardiac abnormalities. *PLoS One*. 2009, 4, e4871.

[69] Lou, H.; Kaplowitz, N. Glutathione depletion down-regulates tumor necrosis factor alpha-induced NF-kappaB activity via IkappaB kinase-dependent and -independent mechanisms. *J. Biol. Chem.* 2007, 282, 29470–29481.

[70] Ardanaz, N.; Yang, X. P.; Cifuentes, M. E.; Huarani, M. J.; Jackson, K. W.; Liao, T. D.; Carretero, O. A.; Pagano, P. J. Lack of glutathione peroxidase 1 accelerates cardiac-specific hypertrophy and dysfunction in angiotensin II hypertension. *Hypertension*. 2010, 55, 116-123.

[71] Blankenberg, S.; Rupprecht, H. J.; Bickel, C.; Torzewski, M.; Hafner, G.; Tiret, L.; Smieja, M.; Cambien, F.; Meyer, J.; Lackner, K. J. Glutathione peroxidase 1 activity and cardiovascular events in patients with coronary artery disease. *N. Engl. J. Med.* 2003, 349, 1605-1613.

[72] Cresci, M.; Foffa, I.; Ait-Ali, L.; Pulignani, S.; Gianicolo, E. A.; Botto, N.; Picano, E.; Andreassi, M. G. Maternal and paternal environmental risk factors, metabolizing GSTM1 and GSTT1 polymorphisms, and congenital heart disease. *Am. J. Cardiol.* 2011, 108, 1625–1631.

[73] Li, X.; Liu, Z.; Deng, Y.; Li, S.; Mu, D.; Tian, X.; Lin, Y.; Yang, J.; Li, J.; Li, N.; Wang, Y.; Chen, X.; Deng, K.; Zhu, J. Modification of the association between maternal smoke exposure and congenital heart defects by polymorphisms in glutathione S-transferase genes. *Sci. Rep.* 2015, 5, 14915.

[74] Dringen, R. Metabolism and functions of glutathione in brain. *Prog. Neurobiol.* 2000, 62, 649-671.

[75] Dringen, R.; Kranich, O.; Hamprecht, B. The γ-glutamyl transpeptidase inhibitor acivicin preserves glutathione released by astroglial cells in culture. *Neurochem. Res.* 1997, 22, 727-733.

[76] Dringen, R.; Hamprecht, B.; Bröer, S. The peptide transporter PepT2 mediates the uptake of the glutathione precursor CysGly in astroglia-rich primary cultures. *J. Neurochem.* 1998, 71, 388-393.

[77] Kamal, M. A.; Keep, R. F.; Smith, D. E. Role and relevance of PEPT2 in drug disposition, dynamics, and toxicity. *Drug Metab. Pharmacokinet.* 2008, 23, 236-242.

[78] Dringen, R.; Pfeiffer, B.; Hamprecht, B. Synthesis of the antioxidant glutathione in neurons: supply by astrocytes of CysGly as precursor for neuronal glutathione. *J. Neurosci.* 1999, 19, 562-569.

[79] Albrecht, P.; Lewerenz, J.; Dittmer, S.; Noack, R.; Maher, P.; Methner, A. Mechanisms of oxidative glutamate toxicity: the glutamate/cystine antiporter system xc- as a neuroprotective drug target. *CNS Neurol. Disord. Drug Targets.* 2010, 9, 373-382.

[80] Lewerenz, J.; Maher, P. Chronic glutamate toxicity in neurodegenerative diseases—What is the evidence? *Front. Neurosci.* 2015, 9, 469.

[81] Meldrum, B.; Garthwaite, J. Excitatory amino acid neurotoxicity and neurodegenerative disease. *Trends Pharmacol. Sci.* 1990, 11, 379-387.

[82] Meldrum, B. S. The role of glutamate in epilepsy and other CNS disorders. *Neurology.* 1994, 44, S14-S23.

[83] Pratico, D.; Uryu, K.; Leight, S.; Trojanoswki, J. Q.; Lee, V. M. Increased lipid peroxidation precedes amyloid plaque formation in an animal model of Alzheimer amyloidosis. *J. Neurosci*. 2001, 21, 4183–4187.

[84] Komatsu, H.; Liu, L.; Murray, I. V.; Axelsen, P. H. A mechanistic link between oxidative stress and membrane mediated amyloidogenesis revealed by infrared spectroscopy. *Biochim. Biophys. Acta*. 2007, 1768, 1913–1922.

[85] Bonda, D. J.; Wang, X.; Lee, H. G.; Smith, M. A.; Perry, G.; Zhu, X. Neuronal failure in Alzheimer's disease: a view through the oxidative stress looking-glass. *Neurosci. Bull.* 2014, 30, 243–252.

[86] Venkateshappa, C.; Harish, G.; Mahadevan, A.; Srinivas Bharath, M. M.; Shankar, S. K. Elevated oxidative stress and decreased antioxidant function in the human hippocampus and frontal cortex with increasing age: implications for neurodegeneration in Alzheimer's disease. *Neurochem. Res.* 2012, 37, 1601–1614.

[87] Gibson, G. E.; Starkov, A.; Blass, J. P.; Ratan, R. R.; Beal, M. F. Cause and consequence: mitochondrial dysfunction initiates and propagates neuronal dysfunction, neuronal death and behavioral abnormalities in age-associated neurodegenerative diseases. *Biochim. Biophys. Acta.* 2010, 1802, 122–134.

[88] Chiang, G. C.; Mao, X. X.; Kang, X. G.; Chang, X. E.; Pandya, X. S.; Vallabhajosula, X. S.; Isaacson, R.; Radvin, L. D.; Shungu, D. C. Relationships among cortical glutathione levels, brain amyloidosis, and memory in healthy older adults investigated *in vivo* with ^{1}H MRS and PiB PET. *Am. J. Neuroradiol.* 2017, 38, 1130-1137.

[89] Saing, T.; Lagman, M.; Castrillon, J.; Gutierrez, E.; Guilford, F. T; Venketaraman, V. Analysis of glutathione levels in the brain tissue samples from HIV-1-positive

individuals and subject with Alzheimer's disease and its implication in the pathophysiology of the disease process. *Biochim. Biophys. Acta Clin.* 2016, 29, 38-44.

[90] Martin, H. L.; Teismann, P. Glutathione-review on its role and significance in Parkinson's disease, *FASEB J.* 2008, 23, 3263–3272.

[91] Drechsel, D. A.; Liang, L.; Patel, M. 1-Methyl-4-phenylpyridinium-induced alterations of glutathione status in immortalized rat dopaminergic neurons, *Toxicol. Appl. Pharmacol.* 2007, 220, 341–348.

[92] Gawryluk, J. W.; Wang, J. F.; Andreazza, A. C.; Shao, L.; Young, L. T. Decreased levels of glutathione, the major brain antioxidant, in post-mortem prefrontal cortex from patients with psychiatric disorders. *Int. J. Neuropsychopharmacol.* 2011, 14, 123-130.

[93] Morris, D.; Guerra, C.; Donohue, C.; Oh, H.; Khurasany, M.; Venketaraman, V. Unveiling the mechanisms for decreased glutathione in individuals with HIV-1 infection. *Clin. Dev. Immunol.* 2012, 2012, 734125.

[94] Venketaraman, V.; Dayaram, Y. K.; Talaue, M. T.; Connell, N. D. Glutathione and nitrosoglutathione in macrophage defense against Mycobacterium tuberculosis. *Infect. Immun.* 2005, 73, 1886–1889.

[95] Choi, J.; Liu, R. M.; Kundu, R. K.; Sangiorgi, F.; Wu, W.; Maxson, R.; Forman, H. J. Molecular mechanism of decreased glutathione content in human immuno-deficiency virus type 1 transgenic mice. *J. Biol. Chem.* 2000, 275, 3693–3698.

[96] Fan, X.; Staitieh, B. S.; Jensen, J. S.; Mould, K. J.; Greenberg, J. A.; Joshi, P. C.; Koval, M.; Guidot, D. M. Activating the Nrf2- mediated antioxidant response element restores barrier function in the alveolar epithelium of HIV-1-1 transgenic rats. *Am. J. Phys. Lung Cell. Mol. Phys.* 2013, 305, L267–L277.

[97] Annuk, M.; Fellström, B.; Akerblom, O.; Zilmer, K.; Vihalemm, T.; Zilmer, M. Oxidative stress markers in pre-uremic patients. *Clin. Nephrol.* 2001, 56, 308–314.

[98] Nguyen-Khoa, T.; Massy, Z. A.; De Bandt, J. P.; Kebede, M.; Salama, L.; Lambrey, G.; *Witko-Sarsat, V.; Drueke, T. B.; Lacour, B.; Thevenin, M.* Oxidative stress and haemodialysis: role of inflammation and duration of dialysis treatment. *Nephrol. Dial. Transplant.* 2001, 16, 335–340.

[99] Tarng, D. C.; Wen Chen, T.; Huang, T. P.; Chen, C. L.; Liu, T. Y.; Wei, Y. H. Increased oxidative damage to peripheral blood leukocyte DNA in chronic peritoneal dialysis patients. *J. Am. Soc. Nephrol.* 2002, 13, 1321–1330.

[100] Ryu, S; Chang, Y; Kim, DI; Kim, WS; Suh, BS. γ-Glutamyltransferase as a predictor of chronic kidney disease in nonhypertensive and nondiabetic Korean men. *Clin. Chem.* 2007, 53, 71-77.

[101] Gelber, R. P.; Kurth, T.; Kausz, A. T.; Manson, J. E.; Buring, J. E.; Levey, A. S.; Gaziano, J. M. Association between body mass index and CKD in apparently healthy men. *Am. J. Kidney Dis.* 2005, 46, 871-880.

[102] Lonardo, A.; Lombardini, S.; Scaglioni, F.; Carulli, L.; Ricchi, M.; Ganazzi, D.; Adinolfi, L. E.; Ruggiero, G.; Carulli, N.; Loria, P. Hepatic steatosis and insulin resistance: does etiology make a difference? *J. Hepatol.* 2006, 44, 190-196.

[103] Chen, J.; Muntner, P.; Hamm, L. L.; Fonseca, V.; Batuman, V.; Whelton, P. K.; He, J. Insulin resistance and risk of chronic kidney disease in nondiabetic US adults. *J. Am. Soc. Nephrol.* 2003, 14, 469-477.

[104] Chen, J.; Muntner, P.; Hamm, L. L.; Jones, D. W.; Batuman, V.; Fonseca, V.; Whelton, P. K.; He, J. The metabolic syndrome and chronic kidney disease in U.S. adults. *Ann. Intern. Med.* 2004, 140, 167-174.

[105] Grams, M. E.; Rabb, H. The distant organ effects of acute kidney injury. *Kidney Int.* 2012, 81, 942–948.

[106] Yap, S. C.; Lee, H. T. Acute kidney injury and extrarenal organ dysfunction: new concepts and experimental evidence. *Anesthesiology.* 2012, 116, 1139–1148.

[107] Shang, Y.; Siow, Y. L.; Isaak, C. K.; Karmin, C. Downregulation of glutathione biosynthesis contributes to oxidative stress and liver dysfunction in acute kidney injury. *Oxid. Med. Cell Longev.* 2016, 2016, 9707292.

[108] Prathapasinghe, G. A.; Siow, Y. L., O. K. Detrimental role of homocysteine in renal ischemia-reperfusion injury. *Am. J. Physiol. Renal Physiol.* 2007, 292, F1354–F1363.

[109] Prathapasinghe, G. A.; Siow, Y. L.; Xu, Z., O. K. Inhibition of cystathionine-*β*-synthase activity during renal ischemia-reperfusion: role of pH and nitric oxide. *Am. J. Physiol. Renal Physiol.* 2008, 295, F912–F922.

[110] Mosharov, E.; Cranford, M. R.; Banerjee, R. The quantitatively important relationship between homocysteine metabolism and glutathione synthesis by the transsulfuration pathway and its regulation by redox changes. *Biochemistry.* 2000, 39, 13005–13011.

[111] Brosnan, J. T.; Brosnan, M. E. The sulfur-containing amino acids: an overview. *J. Nutr.* 2006, 136, 1636S–1640S.

In: Glutathione
Editor: Pál Perjési

ISBN: 978-1-53614-740-7
© 2019 Nova Science Publishers, Inc.

Chapter 7

GENETIC POLYMORPHISM OF GLUTATHIONE S-TRANSFERASE

Mariana Chantre-Justino and Gilda Alves*
Department of Pathology, Faculty of Medical Sciences,
State University of Rio de Janeiro, Rio de Janeiro, Brazil

ABSTRACT

Macromolecules (nucleic acids, lipids, and proteins) are constantly exposed to a wide variety of chemical agents that can affect their integrity and thus contribute to increased cellular instability. However, the cells present different enzymatic systems specialized in minimizing the toxic and/or mutagenic effects of the different chemical compounds through their continuous removal from the organism. The glutathione S-transferase (GST) gene family encodes crucial metabolic enzymes of phase II involved in xenobiotic detoxification processes by promoting the conjugation of an electrophilic compound to the reduced glutathione (GSH), thereby minimizing the presence of oxidizing species in the intracellular environment. Consequently, inter-individual genetic variability in GST genes may influence their activity. This chapter focuses on the clinical significance of GST polymorphisms in humans and on the consequence of gene-gene and gene-environment interactions on the susceptibility for several pathologies.

Keywords: Glutathione S-transferase (GST), genetic variability, GST-polymorphism, gene-gene interactions, gene-environment interaction

GST POLYMORPHISMS

Single nucleotide polymorphisms (SNPs) are variations of a single base pair in a specific position in the DNA sequence and appear in frequencies of more than 1% of the

* Corresponding Author E-mail: galvesbrown@gmail.com.

population. SNPs may occur in non-coding as well as coding DNA sequences, and in this latter case can promote amino acid substitution in the polypeptide sequence and present clinical significance [1]. Therefore, glutathione S-transferase (GST) genetic polymorphisms may modulate their enzymatic activities and influence the individual predisposition to different pathologies and in the ability to metabolize different chemical compounds [2].

According to the cellular localization, GSTs have been described as cytosolic, mitochondrial, and microsomal. In humans, cytosolic GSTs are divided into seven distinct classes, namely alpha (GSTA), mu (GSTM), omega (GSTO), pi (GSTP), theta (GSTT), sigma (GSTS), and zeta (GSTZ). Polymorphisms in the isoenzymes GSTA1, GSTM1, GSTT1, GSTM3, and GSTP1 have shown clinical relevance and are therefore the most commonly studied. The *GSTA1* gene presents SNPs in the promoter region: −567T>G, −69C>T, and −52G>A, in which the variant GSTA1*B allele (−567G, −69T, and −52A) shows lower transcriptional activity than the GSTA1*A allele (−567T, −69C, and −52G). The homozygous deletions (null genotypes) in *GSTM1* and *GSTT1* genes result in deficiency of the detoxification activities. In the GSTM3 gene, the variant GSTM3*B allele differs from the GSTM3*A allele in intron 6 by a 3 bp deletion, creating a recognition motif (-aagata-) for the multifunctional transcription factor YY1. Finally, two functional SNPs have been described in the coding sequence for the *GSTP1* gene, the c.313A>G and the c.341C>T, which determine the amino acid substitutions Ile105Val and Ala114Val, respectively. These changes in GSTP1 also result in lower enzyme activity.

Environmental exposure to some carcinogens plays a critical role in the susceptibility to several pathological abnormalities, such as cancer and hematological and neurological disorders. Evidence indicates that polymorphisms in the GST genes increase the inter-individual susceptibility to these malignancies, showing the gene–environment interaction. Furthermore, some studies suggest that GST polymorphisms also play a role in effectiveness of chemotherapeutic agents, modulating the drug resistance process. Thus, individuals carrying GST polymorphisms with an impaired ability to eliminate carcinogenic compounds may be defined as at-risk groups.

Clinical Significance of GST Polymorphisms in Mental Disorders and Brain Tumors

In the human brain, GST enzymes confer neurological protection by regulating the neuronal exposure to potentially toxic substances derived from blood or cerebrospinal fluid (CSF) [3]. Furthermore, aging is characterized by increase of oxidative stress due to reduced activity of the antioxidant machinery, favoring the excess of molecular damage throughout life [4]. Parkinson’s disease (PD) and Alzheimer's disease (AD) are the most prevalent age-related neurodegenerative diseases associated with excessive generation of oxidative damages in the brain [5]. Therefore, polymorphisms in GST genes may contribute to susceptibility to neurological disorders and brain tumors. A meta-analysis in China revealed that the GSTM1 null genotype and GSTP1 (Ile105Val) polymorphism were associated with increased risk of AD [6]. A meta-analysis conducted by Wang in the USA, however, suggested no associations between polymorphisms of GSTM1, GSTT1, GSTM3, GSTP1, GSTO1, and AD risk [7]. Perez-Pastene et al. found a significant association of the GSTM1 null genotype with PD in a Chilean population, being strongest in the earlier age range [8]. Golbe et al. screened genetic variations in GST for an effect on the age of onset (AO) of PD development in affected individuals of Italian or Greek origin. The authors observed that subjects carrying the GG genotype of GSTP1 (Ile105Val) (A313G) had a mean AO acceleration of 15.2 years, while a 9.7-year acceleration of AO was found for those carrying the CT heterozygote relative to the CC homozygote genotype of GSTP1 Ala114Val (C341T) [9].

Physical illnesses are more prevalent among people with severe mental disorders compared to the general population. Furthermore, people with mental illness appear to give less priority to their physical health needs [10]. This specific behavior may contribute to increased infection susceptibility. Tuberculosis (TB) and mental illness have been identified as closely related diseases [11]. In addition, the reactive metabolites of anti-TB drugs can induce serious side effects, including mental disorders such as schizophrenia. Schizophrenia (SCZ) is a severe psychiatric disorder with the involvement of environmental factors and genetic predisposition characterized by emotional disruptions and altered perception of some aspects of reality [12]. Some evidence suggests that chronic oxidative stress has a critical role in the pathophysiology of the disease, in which reduction of plasma glutathione levels has been demonstrated in patients with SCZ, consequently leading to brain injury [13, 14]. Moreover, an additional adverse effect of anti-TB therapy is the anti-tuberculosis drug-induced hepatotoxicity (ATDH), also known as anti-tuberculosis drug-induced liver injury (ATDILI). Therefore, GST polymorphisms may modulate the individual susceptibility to SCZ and ATDH.

A meta-analysis revealed that GSTM1-polymorphisms may be a genetic risk factor for SCZ in the East Asian population [15]. Cai et al. demonstrated in a meta-analysis that ATDH and SCZ may be co-occurring in subjects with the GSTM1 null genotype [16]. Another meta-analysis reported that the GSTM1 null genotype may be associated with an increased risk of ATDH, particularly among the Chinese population [17]. The drug resistance is observed in many individuals carrying polymorphic GST genotypes and may significantly affect the treatment. In a Brazilian study, patients diagnosed with SCZ carrying combination of GSTT1 and GSTM1 deletion polymorphisms had a 4.6-fold increased risk for developing treatment-resistant SCZ, revealing contribution of these genes to SCZ development and its refractory manifestation [18].

Brain tumors represent neoplasms arising from within the brain or from metastases. Tumors of the central nervous system (CNS) are the second most common childhood and adolescent cancers after hematological malignancies and account for 20 to 25% of all primary pediatric tumors [19]. A study in the USA observed an association between increased risk of childhood brain tumor and maternal consumption of cured meat during pregnancy among children with the GSTT1 null genotype or with potentially reduced levels of GSTM3 (-63C allele), modifying fetal ability to deactivate nitrosoureas [20]. Another study highlighted significant associations in pediatric brain tumor susceptibility: functional GSTM1 allele and high-grade pediatric astrocytomas; GSTP1 variant genotype (Val114/Val114) and pediatric astrocytomas; and GSTP1 variant genotype (Val114/Val114) among pediatric tumors with microsatellite instability (MSI) [21].

A meta-analysis involving GST variants and the risk of adult brain tumors revealed that only the GSTT1 null genotype was significantly associated with a risk of meningioma [22]. However, other associations have been reported in different studies. In a Turkish study, only the GSTM1 null genotype was associated with adult brain tumor incidence. Nevertheless, no significant association was observed between any of the GST variants analyzed and histological types of brain tumor [23]. A study in the San Francisco Bay Area also found no significant difference in GST genotype frequencies among cases with different histological types of adult brain tumors; however, the authors reported that the GSTT1 null genotype was less common among patients with astrocytoma or anaplastic astrocytoma, while the GSTP1 A114V polymorphic genotypes (Ala/Val or Val/Val) were more common among patients categorized with other histology. Additionally, the GSTT1 null genotype was near-significantly more common among glioblastoma multiforme (GBM) cases whose tumors had p53 mutation [24]. In contrast, another study conducted in the United States showed that the GSTP1 105 Val/Val genotype was associated with an 80% increase of glioma incidence. Moreover, association between the GSTT1 null genotype and meningioma was also reported. However, the study found no association of the GSTM1 or GSTP1 I114V variants with the risk of any tumor type [25].

Exposure to Environmental Carcinogens, GST Polymorphisms and Cancer Risk

Epidemiologic studies strongly suggest that human exposure to some environmental agents is associated with congenital diseases and carcinogenesis process. One of these xenobiotic agents is represented by polycyclic aromatic hydrocarbons (PAHs). PAH exposure occurs by inhalation, ingestion and topical absorption, with subsequent formation of PAH-DNA adducts that can concentrate in different human tissues and increase susceptibility of developing cancer [26]. The major sources of PAH exposure include tobacco smoke and air pollution. Tobacco contains approximately 4,000 compounds and several carcinogens, including PAHs, benzene, aromatic amines, and N-nitroso compounds [27].

Maternal smoking during pregnancy is a risk factor for congenital heart diseases (CHDs), an important cause of death in childhood. Therefore, association between maternal PAH exposure and GST polymorphisms has been investigated regarding CHD susceptibility. A study conducted by Li's group in China evaluated the effect of maternal cigarette smoke exposure and GST variants on CHD risk. Among the pregnant women with deletions in GSTM1 and/or GSTT1, the lowest hair nicotine concentration (HNC) range associated with an increased risk of any CHD in the fetuses was 0.213–0.319 ng/mg; in contrast, that range was 0.319–0.573 ng/mg among the mothers with normal copies of GSTM1 and GSTT1. Additionally, the mothers with an HNC of >0.573 ng/mg and carrying the GSTP1 AG or GG genotype showed a much higher risk of CHDs than those with the GSTP1 AA genotype. These findings suggest that the combined effect of maternal genetic variations in GSTs and smoking during pregnancy, may seriously affect fetal heart development [28].

Cigarette consumption is a well-known risk factor for lung cancer. Since GST enzymes detoxify inhaled carcinogens, several reports have evaluated the role of GST polymorphisms on susceptibility of lung cancer in smokers. Sreeja et al. reported that individuals carrying the null GSTT1 genotype had a 2.4-fold increase in the risk of developing lung cancer in a South Indian population. Also, a significant 2.2-fold risk associated with the GSTT1 null genotype was observed in individuals who had smoked. The study also indicated that the risk was 4-fold higher in those individuals who had <20 pack years and with the GSTT1 null genotype [29]. The GSTP1 (Ile/Val) heterozygous genotype and the GSTM1-null genotype were also associated with increased lung cancer risk in a Greek smoking population. Additionally, in the non-small cell lung cancer group, the GSTP1 homozygous variant was significantly associated with increased lung cancer risk and shorter survival [30]. In contrast, a German study found no relevant modifying effect of polymorphisms of the GSTM1, GSTT1, or GSTP1 on lung cancer risk and cumulative smoking dose [31]. Similarly, no significant association was found between genetic polymorphisms in GSTM1, GSTT1, and GSTP1, and lung cancer risk in a Spanish study. Moreover, negative results were also observed in the gene-gene, gene-tobacco smoking, or gene-occupational exposure interactions analysis relative to lung cancer susceptibility [32]. Similar findings were obtained by Honma's group, in which the genetic polymorphisms in GSTM1, GSTT1, and GSTP1 (Ile105Val), isolated or combined, had no influence on the risk of lung cancer in Brazilian patients, even when stratified by clinical pathological features such as stage, histology, and smoking status [33].

A meta-analysis including 7,415 lung cancer cases and 6,084 controls identified significant association between the GSTT1 null genotype and lung cancer among the East Asian populations. Furthermore, the risk significantly increased in smokers carrying the

GSTT1 null genotype compared with non-smokers [34]. Another meta-analysis, including 31,802 subjects, also showed that the GSTT1 null genotype was associated with lung cancer in overall populations and in Asians, but not in Caucasians, Brazilian population, and Africans [35]. Likewise, the combined effects of GSTM1 null and GSTT1 null polymorphisms were associated with increased lung cancer risk in Asians, Caucasians, and Indians in a meta-analysis including 13,706 cases and 13,093 controls [36]. Following these findings, an increased risk of lung cancer with the GSTM1 and GSTT1 null genotypes, especially with the dual null genotype, was found in the Chinese population. In addition, a significant relationship was observed between squamous carcinoma (SC), adenocarcinoma (AC) or small cell lung carcinoma (SCLC) and the GSTM1 null genotype, as well as SC or AC and the GSTT1 null genotype. Additionally, smokers with the GSTM1 null genotype had a higher lung cancer risk than non-smokers [37]. Sharma et al. revealed that the GSTM1 null genotype was associated with a twofold increased risk for adenocarcinoma (ADCC) in the Indian population. A positive and significant association between the GSTT1 null and lung cancer was found for ADCC only, while individuals carrying the GSTP1 genotype showed a threefold risk for SCLC. Unlike other findings, Sharma's study observed that non-smokers with the null GSTM1 genotype had an increased lung cancer risk as compared to smokers. The authors believe that these never-smoker patients carrying the null GSTM1 genotype might be exposed to environmental tobacco smoke or exposure to occupational hazards, suggesting that genetic factors may play a more important role in cancer development [38].

Chronic obstructive pulmonary disease (COPD) is another lung function impairment caused by chronic cigarette smoking. Symptoms usually include airflow obstruction due to chronic bronchitis, emphysema, and/or small airways disease. In addition to smoking raising the oxidative stress levels, genetic factors also play a role in modulating this susceptibility. In Tunisia, Lakhdar et al. showed that the combination of the null-GSTM1/GSTP1 (105Val/Val) genotypes significantly increased the risk of COPD. Also, the null-GSTM1/null-GSTT1, GSTP1 (105Val/Val)/null GSTT1, and null-GSTM1/GSTP1 (105Val/Val) genotypes were related to the development of emphysema [39]. Cao's group in China also demonstrated that GSTM1 and GSTT1 gene deletion may be a risk factor for the pathogenesis of COPD. In addition, the serum total antioxidant capacity in the GST gene deletion group was lower than observed in the wild-type group, revealing a protective function of *GSTM1* and *GSTT1* genes in the lungs [40].

Exposure to chemical carcinogenes through tobacco use (smoking/chewing) and alcohol consumption represent important risk factors in etiology of head and neck squamous cell carcinoma (HNSCC), in which development occurs in the upper aerodigestive epithelium. Therefore, genetic polymorphism in the GST genes may modulate the effects of these environmental carcinogens and thus influence the susceptibility to HNSCC. A meta-analysis presented evidence of association between the null genotypes of GSTM1 and GSTT1 with HNSCC risk. According to ethnicity, an increased risk of HNSCC was observed in carriers of the GSTM1 null genotype in Asia, the GSTT1 null genotype in South America, and the dual null genotype in Europe and Asia. Also, a significant association of the GSTM1 null genotype with HNSCC risk was observed in smokers in this study [41]. In India, Singh's group highlighted importance of GST polymorphism and the gene-environment interaction in modulating the risk of HNSCC. The authors reported that combination of both the GSTM1 and GSTT1 deletion genotypes showed an increased risk of HNSCC. Also, an increased risk was observed in the patients carrying a genotype combination of GSTM1 null, GSTT1 null, and GSTP1 (Ile/Ile). Moreover, interaction between tobacco chewing and the null genotype of GSTM1 or GSTT1 showed a 3.5- and a 2.2-fold increase in the risk, respectively. Similarly, the study also revealed that alcohol consumption resulted in an increased risk in patients with the null genotype of GSTM1 or the null genotype of GSTT1 as compared to the non-drinkers [42].

Development of second primary malignancy (SPM) post-treatment is estimated to occur in approximately 15% of HNSCC patients and may be influenced by genetic susceptibility. In the USA, Zafereo et al. observed that patients with the GSTP1 (Ile105Val) variant genotype had a 1.7-fold elevated risk for developing SPM compared to patients with the wild-type genotype. Moreover, the combined effects of a panel of four GST polymorphisms (GSTM1 null, GSTT1 null, GSTP1 Ile/Val + Val/Val, and GSTP1 Ala/Ala) revealed that patients with 3 or 4 risk genotypes had a short SPM-free survival compared to those with 0 or 1 or 2 risk genotypes, thus identifying biomarkers of susceptibility for second primary head and neck malignancies [43].

Interesting to note, none of these previously described studies evaluated presence of human papillomavirus (HPV) as an additional risk factor for HNSCC or even for development of second primary cancers. The HPV infection is an important and known risk factor for HNSCC, with prevalence of the high-risk types HPV16 and HPV18 [44]. HPV infects mucosal and/or cutaneous skin *via* micro wounds and causes various types of lesions such as common and genital warts and anal condyloma, as well as lesions in the aero-digestive tract that can evolve with epithelial dysplasia and cancer. High-risk HPV infections are associated with malignant transformation through viral DNA integration into the host genome, resulting in higher expression of the oncoproteins E6 and E7 in the basal layer. These oncoproteins are critical for cell cycle dysregulation in the host cell, with degradation of pRB by E7 and degradation of p53 by E6, promoting cellular instability [45].

HPV infection is also an important risk factor for cervical carcinogenesis. The interaction between the HPV-16 E7 oncoprotein and GSTP1 has been demonstrated [46]. Similarly, interaction of GSTM3 with the E7 oncoprotein from HPV18 was observed [47]. In this way, some studies analyzed the combined role of GST polymorphisms and HPV infections in cervical cancer. The GSTP1 (A>G; Ile105Val) G allele was associated with a lower risk for cervical cancer in Thailand. Among Thai women with the high-risk type of

HPV infection, a significantly lower risk for cervical cancer was observed in heterozygous carriers [48]. In contrast, a Brazilian study found no significant association between the GSTP1(A>G; Ile105Val)-polymorphism and any HPV-related cervical abnormalities. Nevertheless, the findings revealed a synergic effect of use of oral contraceptives, the GSTP1 AG genotype, and high-risk HPV infection, with higher susceptibility to development of cervical lesions. The study also highlighted that synthetic estrogens have been associated with increased cancer risk and are able to enhance HPV E6/E7 expression in specific sequences. Thus, since GSTs are involved in estrogen metabolism, the long-term exposure to synthetic estrogens associated with polymorphic GSTs could be a key factor for development of HPV-related cervical cancer, particularly the high-risk HPV type [49].

A meta-analysis provided strong evidence that HPV infection associated with the GSTM1 null genotype significantly increased the risk of cervical cancer, especially in Chinese and Indian populations [50]. An Indian study found that GSTM1 and GSTT1 deletions were prominent among cervical cancer cases compared to controls, with a higher proportion of both deletions detected in HPV-16-positive subjects. Moreover, a significantly increased risk was found between the GSTM1 deleted genotype and the HPV status [51]. Another Indian study showed a significant association in cervical cancer cases, with an almost four-fold increased risk, in women who were HPV-positive smokers with the GSTT1 null genotype and the GSTM1/T1 null genotypes compared to HPV-positive non-smokers [52]. Lee's work also reported that the GSTM1 null genotype in women with high-risk HPV infection was significantly associated with an increased risk of cervical cancer in South Korea. Also, the combination of the GSTT1 null and p21 Arg/Arg genotypes showed a significant interactive effect on the risk of cervical cancer. In addition to mutations in p21 which affect cellular proliferation, the authors suggested modulating effect of the GSTM1 genotype on the cervical carcinogenesis caused by HPV infection [53].

Besides its detoxification activity, GSTP1 plays an important regulatory role. It has been demonstrated that GSTP1 can interact with tumor necrosis factor receptor-associated factor 2 (TRAF2) and thus decrease apoptosis activation by reducing signal transduction from tumor necrosis factor-alpha (TNF-alpha) and c-Jun NH(2)-terminal kinase (JNK) pathways [54, 55]. In a recent study, GSTM3 and GSTP1 were reported as players driving tumor progression in cervical cancer. Firstly, the authors have demonstrated that GSTM3 interacts with TRAF6. Under UV stress, a synergistic effect between GSTs (GSTM3 and GSTP1) and HPV-18 E7 oncoprotein was observed, contributing to cell survival. In addition, longer cell survival post-treatment with cisplatin was observed in cell lines that co-expressed the GSTs and HPV E7. Finally, the loss of GSTs inhibited tumor progression in cervical cancer, suggesting that GSTM3 and GSTP1 expression might significantly influence the survival of patients with cervical cancer [47].

In addition to cervical carcinogenesis, breast cancer also affects women. Breast cancer is the most common type of cancer diagnosed in women worldwide and has a significant impact on the overall cancer incidence and mortality. It is characterized as a multifactorial and heterogeneous disease with tumor subtypes stratified according to gene expression profiles presenting variable prognostics [56]. In addition to genetic predisposition, epidemiological studies have reported interactions between environmental chemical exposure and breast cancer risk [57].

Tobacco smoke has been associated with breast cancer risk, even in women who had never smoked but who had been exposed to environmental tobacco smoke in utero or throughout their childhood [58]. Moreover, it has been shown that breast tissues are susceptible to the effects of aromatic DNA adducts [59]. Firozi et al. demonstrated that the GSTM1 null genotype affected the level of DNA adducts in normal breast tissues of women having breast cancer, especially in smokers [60]. McCarty's group reported a noteworthy, though not statistically significant, finding of a 56% increase in breast cancer risk among women with detectable PAH–DNA adducts and with at least three GST variants [GSTT1 null variant, GSTM1 null variant, GSTP1 AG/GG genotype, GSTA1 common (A*/B* or B*/B*) genotypes] compared to those with all common genotypes and no detectable PAH–DNA adducts [61].

Breast cancer is a known estrogen-related cancer since the hormone can induce proliferation of the breast tissue cells and release toxic metabolites [62]. GSTs may detoxify reactive intermediates involved in the steroid metabolism. Mitrunen et al. evaluated the menopausal status in Finnish women and reported increased risk of breast cancer among postmenopausal women with the GSTM1 null genotype. However, a substantially increased risk of breast cancer was obtained for premenopausal women concurrently carrying the GSTM3*B (variant allele) genotype, the GSTT1 null genotype, and the GSTP1 Ile/Ile genotype, who had a 10-fold risk of breast cancer. These observations suggest that etiology of breast cancer may differ by menopausal status and the risk may be modified by GST genotypes [63]. Martínez-Ramírez et al. observed that the premenopausal group carrying the GSTP1Val/Val genotype had a higher risk of breast cancer than the postmenopausal group with the same genotype in a cohort of Mexican women [64].

People infected with hepatitis B and/or hepatitis C virus (HBV/HCV) are at a higher risk of developing chronic liver disease and hepatocellular carcinoma (HCC). Chronic HBV and HCV infections induce oxidative stress and trigger the neoplastic transformation process in the liver [65]. Martínez et al. reported that the GSTT1 null genotype, associated with the GSTM1 null genotype, may facilitate HCV infection becoming chronic by a significant deficit in the hepatic antioxidant function, with more reactive oxygen species (ROS) and electrophilic compounds generated [66]. In addition to hepatitis viruses, environmental exposure to some chemical agents also plays a role in hepatocarcinogenesis. One of the best-known agents is aflatoxin B1 (AFB1), a potent carcinogenic mycotoxin produced by the *Aspergillus* genus and commonly found in peanuts, corn, and other grains. AFB1-DNA adducts increase the risk of hepatocarcinogenesis, which may be modulated by GST activity since they are involved in detoxifying chemical carcinogens. The effect of aflatoxin B1-related HCC risk was more pronounced among chronic hepatitis B virus carriers with the null genotypes of GSTT1 and GSTM1 in the Taiwanese population [67,68].

Chronic liver injuries are frequently observed in individuals with a history of alcohol abuse since ethanol promotes oxidative stress. A Spain study noted that the GSTT1 null genotype was more frequent among patients with advanced alcoholic liver disease (ALD) and that the simultaneous occurrence of the GSTM1 and GSTT1 null genotypes was four times more likely in the patients than in healthy controls [69]. Roy et al. reported the GSTM1 null genotype as an independent risk factor for development of ALD in Indian alcoholic patients [70]. Also, in India, Khan et al. observed an increased risk to alcoholic cirrhosis in alcoholic cirrhotic patients (ACPs) with the GSTM1 null genotype. Likewise, the authors also reported a significantly increased risk in ACPs with the variant genotype of GSTP1 when compared with non-alcoholic controls [71]. Together, these findings show the genetic modifier role that GST polymorphisms play in the susceptibility to alcoholic liver damage.

Regardless the environmental exposure, GST polymorphisms also play a crucial role in HCC susceptibility. A meta-analysis revealed that both the GSTM1 and GSTT1 null genotypes were associated with an increased HCC risk in the Chinese population [72]. Taiwanese individuals aged $\leq$ 57 years and carrying the AG or GG genotype of GSTP1 (Ile105Val; A>G) had a 2.18-fold risk and 5.64-fold risk, respectively, of developing HCC compared to individuals with the AA genotype [73]. A study conducted in China investigated association of GSTM-, GSTT- and GSTP-polymorphisms with HCC. The results of the study showed that individuals carrying the GSTM1 null genotype had a 1.64-fold risk of HCC compared to those with the non-null genotype, while individuals carrying the GSTP1 Val/Val variant genotype had a 93% increased risk over those with the GSTP1 Ile/Ile genotype. According to the survival outcome of HCC, individuals with the GSTM1 null genotype and with the GSTP1 Val/Val genotype showed better survival rates. Among patients who received chemotherapy, individuals carrying the GSTM1 null genotype and the GSTP1 Val/Val genotype had a lower risk of death from HCC compared with those without chemotherapy. Together, all these findings highlight the influence of the null GSTM1 and GSTP1 Val/Val genotypes on the susceptibility and the survival rates from HCC [74]. Another Chinese study investigated the impact of 12 SNPs in GST genes on the overall survival of 469 patients with HCC. Only the rs4147581 (G>C) in the GSTP1 gene exhibited a significant association with survival of HCC patients, in which patients carrying the mutant allele had a significantly lower risk of death compared to those with the homozygous wide-type [75].

Urine contributes to the homeostasis of the body fluids. It is produced by filtration of the blood in the kidneys and temporarily stored in the bladder. However, urine also plays an important role in carcinogenesis of the urinary system by transporting different substances to be excreted, including carcinogens and harmful metabolites. Thus, prolonged exposure to mutagenic and carcinogenic substances, associated with genetic polymorphisms, may result in preneoplastic changes and trigger the urinary system cancer process. A meta-analysis performed with 63,876 subjects revealed that the GSTT1 null genotype increased the risk of bladder cancer and prostate cancer, but not renal cell carcinoma (RCC) [76]. Another meta-analysis reported that none of the GSTM1, GSTT1 and GSTP1-polymorphisms investigated had a significant association with the RCC risk [77]. In contrast, Coric's work showed that individuals with the GSTM1-null genotype were at a 2.07-fold higher risk of developing clear cell renal cell carcinoma (cRCC), while the risk for individuals carrying the GSTP1 variant genotype (IleVal+ValVal) was 3.14-fold [78]. In Iran, Safarinejad et al. observed a significantly increased risk for transitional cell carcinoma (TCC) of the bladder in individuals carrying combinations of the GSTT1 positive and GSTP1 Ile/Val or Val/Val genotypes, and in those with the GSTM1 positive and GSTP1 Ile/Val or Val/Val combinations. Moreover, combination of the GSTM1 null genotype with the GSTP1 Ile/Val or Val/Val genotype resulted in a 4.76-fold increased risk for bladder TCC. According to tumor stage/grade, patients carrying the GSTP1 Val/Val variant genotype had a significantly increased risk of developing high-grade and muscle invasive tumors of bladder cancer [79].

Prostate cancer is the most frequently diagnosed cancer in men and epidemiological studies have associated the role of chronic oxidative stress in the development and progression of this malignancy [80]. Therefore, GST activity is crucial in protecting cells against oxidative stress to avoid DNA mutations. Thus, GST polymorphisms are important biomarkers to evaluate the prostate cancer susceptibility. A meta-analysis reported that the GSTM1 null genotype conferred an increased risk of prostate cancer on a wide population basis including Caucasians and Asians, except for Africans. No relationship was found between GSTT1 and GSTP1 (Ile105Val) status and the risk of prostate cancer [81]. Nonetheless, different findings have been observed in other populations. Safarinejad's group found that the GSTM1 null, GSTT1 null, and GSTP1 (Ile105Val) Ile/Val or Val/Val genotypes were associated with an increased risk of prostate cancer in Iran. Combination of both null genotypes of GSTM1 and GSTT1 was associated with a 6.82-fold increased risk of advanced stage of prostate cancer. Also, a higher frequency of both null genotypes was observed in patients with Gleason score >7. In addition, the GSTP1 Val allele was more prevalent in patients having advanced-stage cancers [82].

Unlike Safarinejad's findings, Medeiros's work noted no increased prostate cancer risk in Portuguese men carrying any of the GSTM1 or GSTT1 genotypes. The authors also observed that the GSTT1 null was overrepresented in individuals with advanced prostate cancer. Interestingly, men carrying the GSTM3 BB variant genotype had a 5-fold increased risk of prostate cancer [83]. A Brazilian study reported the influence of GSTA1-polymorphism in the prostate cancer risk, with a protective role of the genotypes AB and BB compared to the AA genotype [84]. An Argentine study evaluated the effect of GST polymorphisms on the risk of prostate cancer biochemical relapse, defined as a rise in serum PSA above 0.2 ng/ml after radical retropubic prostatectomy (RRP). Patients carrying the GSTP1 GG variant genotype (p. 105 Val/Val) had shorter biochemical-relapse free survival (BRFS) and a 3-fold higher risk for recurrence compared to those with the AA and AG genotypes. Thus, the authors suggested that GSTP1 (Ile105Val, A>G) polymorphism might be a relevant predictor of biochemical relapse of prostate cancer in the Argentine population, supporting the implementation of GST genotyping for personalized therapies [85].

Sperms are also susceptible to oxidative damage and this condition can affect spermatogenesis and spermatozoa function, leading to reproductive problems. In this way, several studies have evaluated the effect of GST polymorphisms on susceptibility of male infertility. A meta-analysis encompassing 6,934 subjects in a Han Chinese population showed that the GSTM1 null genotype was significantly associated with idiopathic oligozoospermia, while the null genotype of GSTT1 was significantly associated with normozoospermia and azoospermia. Moreover, an elevated risk of male infertility was observed in both Asian and Caucasian groups carrying the GSTM1 null genotype, increasing the vulnerability to oxidative DNA damage. Some limitations of this meta-analysis include not performing gene-gene and gene-environmental factor interactions [86].

Significantly higher levels of oxidative stress and damage biomarkers were found in seminal plasma and spermatozoa of idiopathic infertile men with the GSTM1 null genotype in Turkey [87]. In India, Tirumala et al. also reported that subjects with the GSTM1 null genotype had lower sperm concentrations and motility [88]. Like these findings, Safarinejad's work revealed that combination of the GSTM1 null and GSTT1 null genotypes showed higher risk of infertility in the Iranian population. In addition, a much higher risk for infertility was obtained in the patients carrying a genotype combination of GSTM1 null, GSTT1 null, and GSTP1 (Ile/Ile) [89]. Tang's group showed that the GSTM1 null genotype was significantly associated with susceptibility to idiopathic male infertility. In contrast to the aforementioned findings, no significantly increased risk of azoospermia was found in either the GSTM1 null or GSTT1 null genotypes when stratified by sperm concentration of the case group. Likewise, regarding ethnicity, no association was found between polymorphism and the risk of idiopathic male infertility in any of the GSTM1 null and GSTT1 null genotypes, neither in Asian nor in European populations [90].

PHARMACOGENETIC INFLUENCE OF GST POLYMORPHISMS ON CLINICAL OUTCOMES

Variations in chemotherapeutic response are observed among patients following the same protocol regimen. Pharmacogenetics has the potential to provide an important explanation for part of the observed inter-individual variability affecting pharmacokinetics and pharmacodynamics of medications. Thus, successful response to chemotherapeutic treatment can be genetically determined. GST polymorphisms are promising biological pathways for pharmacogenetic studies since variability in enzyme activity may affect drug availability. Therefore, identification of polymorphic GST genes in patients receiving chemotherapy is important to predict a possible process of therapeutic resistance and toxicities, thus showing a significant impact on clinical outcomes of these patients.

Docetaxel is a chemotherapeutic agent used in management of solid tumors. Influence of genetic polymorphisms on docetaxel pharmacokinetics and toxicity was investigated in a French study in patients with histologically confirmed solid tumors (breast cancer, prostate cancer, lung cancer, and other cancers) following docetaxel therapy. Febrile neutropenia, the major side effect observed during the treatment with docetaxel, occurred more frequently in patients carrying the GSTP1*A/*B genotype (GSTP1*A [Ile105/Ala114]; GSTP1*B [Val105/Ala114]) [91]. A study in Texas reported that patients with primary malignant glioma treated with nitrosoureas as regimens of chemotherapy carrying the combined GSTP1*A/*A (Ile at exon 5 and Ala at exon 6) and GSTM1 null genotypes were 5.7 times more likely to experience adverse events secondary to chemotherapy [92]. In France, Lecomte et al. observed that gastrointestinal cancer patients who received oxaliplatin-based chemotherapy and carried the Ile/Ile genotype of GSTP1 (Ile105Val) polymorphism had a significantly higher risk of developing a grade 3 cumulative peripheral neuropathy. This observation could be explained by diminished capacity of the GSTP1 105 Ile allele in detoxifying oxaliplatin [93].

A study conducted in North India investigated the effect of common GST polymorphisms on the treatment of anthracycline-based chemotherapy regimens in breast cancer patients. While analyzing the effect of these polymorphisms on myelotoxicity, a significant association was observed of the Ile/Val and Ile/Val+Val/Val genotypes of the GSTP1 (Ile105Val) polymorphism with grade 2–4 anemia. The GSTM1 null-GSTP1 Ile/Val genotype interaction was significantly associated with response to neo-adjuvant chemotherapy (NACT) [94]. In Poland, a study evaluated polymorphic variants in genes with known or potential role in activity on FAC drugs (combination of fluorouracil, doxorubicin, and cyclophosphamide) and the clinical risk factors of FAC chemotherapy-related toxicities in breast cancer patients. According to hematological toxicities, the risk of FAC-related anemia of any grade was elevated in the presence of both GSTT1 and GSTM1 genes. The presence of both GSTT1 and GSTM1 genes also contributed to nephrotoxicity of FAC treatment, with an elevated creatinine level. The common allele A of the GSTP1 (Ile105Val) polymorphism influenced nausea risk. The authors suggested that the full activity of GSTT1 and GSTM1 enzymes decrease intracellular GSH level, affecting intracellular redox balance and protein-thiol status, which in turn are responsible for tissue damage. Additionally, they described that the presence of allele A of GSTP1 (Ile105Val) reduces detoxification rate of GSTP1 substrates, with the subsequent slower removal of cytotoxic agents from the cells [95].

Hematopoietic stem cell transplantation (HSCT) has become the standard of treatment for many patients with congenital or acquired disorders of the hematopoietic system. Busulfan (BU), a bifunctional alkylating agent, is a key component of the conditioning regimen prior to HSCT. In Israel, Elhasid's work evaluated the association between BU pharmacokinetics and the GSTA1, GSTP1, GSTM1, and GSTT1 genotypes in 18 pediatric patients treated for congenital hemoglobinopathies and who received BU in their pre-HSCT conditioning regimen. The authors found that the maximal concentration (C_{max}) of BU was significantly lower in the GSTA1*A/*B heterozygote children compared to that of the GSTA1*A/*A and GSTA1*B/*B homozygote groups. In addition, the GSTM1-null individuals demonstrated lower BU-AUC(area under the concentration-time curve)/kg compared to the GSTM1-positive individuals. Among the 18 children, only 3 of them did not require dose adjustment and were GSTT1-positive, GSTM1-null, and GSTP1 heterozygotes. Finally, 5 patients of the study developed graft versus host disease (GVHD) and all of them had the GSTM1-null genotype [96].

Ansari's group investigated influence of polymorphisms in the *GSTA1* and *GSTM1* genes in pediatric patients with thalassemia who received BU in their pre-HSCT conditioning regimen in Switzerland. The authors observed that individuals carrying the T variant allele (*GSTA1* -69C > T) showed higher busulfan concentrations compared to those with the CC genotype, since the T variant allele is related to the reduction of enzyme activity. Moreover, the group also observed a higher occurrence of toxicity among patients with the GSTM1 null genotype [97]. Influence of GST gene polymorphisms on BU pharmacokinetics was also reported in adult Chinese patients who received allogeneic HSCT. The authors of this study detected lower clearance in the GSTA1 *A/*B genotype group compared with the GSTA1 *A/*A genotype group. In addition, the BU clearance in the GSTP1 *A/*A genotype was higher than the *A/*G and *G/*G genotypes [98].

Cyclophosphamide (CP) is an alkylating agent widely used in several therapy protocols. In a German study, Vester et al. reported influence of GST polymorphisms (GSTT1, GSTM1, and GSTP1) on the long-term remission rate after cyclophosphamide treatment of steroid-sensitive nephrotic syndrome in 26 children. The authors observed that patients with the GSTM1 null genotype showed a significantly better rate of sustained remission and longer median time to the first relapse than patients with the heterozygous or wild-type GSTM1 genotype. In contrast, children with the GSTP1 (Ile105Val) heterozygous or homozygous variant genotypes had a significantly worse rate of sustained remission and a shorter time to the first relapse than those with the wild-type genotype. The GSTT1 genotype did not influence the outcome after CP treatment. Also, patients with combination of the GSTM1 null and GSTP1 wild-type genotypes had the best-sustained remission rate - 50% compared to 6% in other children - revealing that pediatric patients with this genotype combination had the maximum benefit from the CP therapy [99]. In India, 37 (50%) out of 74 children evaluated with idiopathic nephrotic syndrome responded to CP therapy, in which 27 were steroid sensitive (SS) and 10 were steroid resistant (SR). In contrast to Vester's study, no significant association of individual GST genotypes with response to CP was found. However, a synergistic effect of three genotypic combinations showed a significant correlation with response to CP therapy among the SS group: (1) GSTP1 wild-type (I/I genotype) with GSTM1 null genotype, (2) GSTM1 null and GSTT1 null genotypes, and (3) GSTP1 wild-type together with GSTM1 and GSTT1 null genotypes. Lower GST enzyme activities may modulate drug effects for individuals following chemotherapeutic treatment by prolonging drug action in the body and enhancing CP exposure and thus its effectiveness to achieve long-term remission. Accordingly, these data also might be useful to select children with nephrotic syndrome who are likely to benefit from CP therapy [100].

Cyclophosphamide has also been used as an immunosuppressive agent in the treatment of several autoimmune diseases, such as systemic lupus erythematosus (SLE). However, severe side effects have been observed in patients following cyclophosphamide therapy. Zhong et al. observed that Chinese patients with SLE carrying the heterozygote and the homozygote variant genotypes of GSTP1 (Ile105Val) had an increased risk of myelotoxicity when treated with pulsed high-dose cyclophosphamide therapy, especially in patients younger than 30 years, or even in patients treated with a total cyclophosphamide dose greater than 1.0 g. Also, SLE patients with these GSTP1 genotypes had an increased risk of gastrointestinal toxicity when treated with an initial pulsed high-dose cyclophosphamide regimen. The GSTM1 and GSTT1 null genotypes did not alter the risks of these short-term side-effects of pulsed high-dose cyclophosphamide therapy in these SLE patients [101]. Renal involvement is frequent in SLE patients, being defined as lupus nephritis (LN). A French study investigated genetic polymorphisms of GST in 70 patients with proliferative LN treated with cyclophosphamide. The GSTP1 (Ile105Val) genotype was an independent factor of poor renal outcome and the GSTM1 null genotype influenced the occurrence of adverse drug reactions related to cyclophosphamide therapy [102].

GST Polymorphisms as Risk Modifiers for Occupational Exposure to Carcinogenic Compounds

Altered detoxification activity of toxic compounds due to SNPs in GST genes may increase the health risks previously mentioned, especially for those who are occupationally exposed to such compounds. Thus, it is important to detect individuals with a genetic profile indicating a higher risk of developing harmful effects of occupational exposure and provide them with constant biomonitoring and protection. Some studies indicate that pesticide exposure may result in genotoxic effects. The GSTM1 null genotype was significantly associated with an increased micronucleus (MN) frequency in pesticide-exposed agricultural workers in Turkey [103]. A study conducted in India found significantly higher DNA damage in workers occupationally exposed to organophosphate pesticides (OPs) with the GSTM1 null genotype and with the wild-type GSTP1 (Ile/Ile), suggesting the influence of the gene-environment interactions on DNA damage [104].

Some evidence points to the association between pesticide exposure and urological cancer. Buzio et al. investigated influence of the GSTM1- and GSTT1-polymorphisms on renal cell carcinoma (RCC) risk associated with occupational exposure to chemicals in individuals recruited from an Italian hospital. Interestingly, the authors found that individuals carrying the GSTM1-present genotype and occupationally exposed to metals or pesticides showed an increased risk for RCC. The GSTT1-present genotype also contributed to enhancing the risk of RCC among subjects exposed to organic solvents and pesticides, compared to those with the GSTT1-null genotype. Despite these findings, the authors emphasized that specific identification of the chemical agents and estimation of the levels of exposure to these compounds in the workplace were not performed in the study [105]. Like Buzio's work, a study conducted in four countries of Central and Eastern Europe also reported a significantly increased risk of RCC among pesticide-exposed subjects carrying the GSTM1 active and the GSTT1 active genotypes. Additionally, a higher risk of RCC was associated with duration and cumulative exposure to pesticides and no elevated risk was observed among pesticide individuals with the null GST genotypes [106]. Despite its central role in detoxification reactions, a possible explanation for association of the active GST with a higher risk of RRC would be its participation in bioactivation reactions of the substrate, such as halogenated pesticides, thus promoting formation of reactive intermediates with observed nephrotoxicity [107].

1,3-Butadiene (BD) is an important industrial chemical classified as a human carcinogen. Chinese BD-exposed workers from a rubber factory carrying GSTM1(-) or GSTT1(-) genotypes, and especially those who carried both null genotypes, exhibited significantly higher MN frequencies than those carrying the GSTM1(+) or GSTT1(+) genotypes or their combination. Moreover, incidence of BD exposure-induced chromosomal damage increased in proportion to the number of deleted alleles. It was also observed that individuals with the GSTP1 (Ile105Val) AG and GG genotypes developed higher MN frequencies than those with the GSTP1 AA genotype [108]. Influence of GST polymorphisms on BD metabolism was investigated in Italian workers occupationally exposed to low levels of BD during production of BD monomers and polymers. Subjects with the GSTM1 null and GSTT1 null genotypes showed higher levels of hemoglobin N-(2,3,4-trihydroxybutyl)valine (THBVal) adduct [109].

Benzene is an aromatic hydrocarbon also classified as carcinogenic to humans and its metabolites can accumulate in bone marrow and induce hematotoxicity [110]. Exposure to benzene leads to formation of benzene oxide (BO), a reactive intermediate, of which detoxification occurs *via* reaction with GSH catalyzed by GST activity [111]. Thus, genetic polymorphisms in GST genes may significantly contribute to increasing susceptibility of DNA damages and hematological risks associated with benzene exposure. Petrochemical workers from Bulgaria carrying GSTT1 gene deletion showed a consistent 35-40% rise in DNA single-strand break (DNA-SSB) levels [112]. In a study conducted in China by Ye's group, the authors observed that the benzene-exposed workers recruited from shoe factories showed a significant reduction in their white blood cell (WBC) counts compared with those of the unexposed control group, particularly in individuals carrying the null-GSTT1 and null-GSTM1 genotypes. WBC counts also declined with increasing cumulative exposure dose of benzene in individuals with the null-GSTT1 and the null-GSTM1 genotypes [113]. When analyzing genotoxic effects of benzene exposure in Italian traffic policemen, a gene-gender interaction was found for individuals carrying the GSTM1-null genotype, with a significantly higher median MN frequency observed only in men [114]. The GSTM1 null, GSTT1 null as well as the heterozygote and variant genotypes of GSTP1 (Ile105Val) were significantly associated with higher MN frequency in gasoline pump workers from India [115]. In Taiwan, workers of a chemical factory carrying the GSTT1-positive genotype showed higher levels of the urinary benzene metabolite S-phenylmercapturic acid (S-PMA) than those with the GSTT1-deficient genotype [116]. Similarly, Carrieri's group reported that reduced S-PMA excretion in the urine of Italian petrochemical workers was significantly influenced by the GSTT1 null genotype [117].

Parental occupational exposure to benzene and other toxic compounds may seriously affect development of the fetus and the newborns. An Italian study conducted by Cresci et al. revealed that maternal and paternal occupational/environmental exposure to toxic xenobiotics (such as cigarette smoke, x-rays, solvents, pesticides) increased the risk of congenital heart diseases (CHDs) in children carrying the combined GSTM1 and GSTT1 null genotypes. The effect for the combined null GST genotypes was significantly stronger when both parents were exposed, which might predispose the children to the harmful effects of these xenobiotics [118].

Conclusion

Study of GST polymorphisms is of utmost importance to evaluate influence of gene-gene and gene-environmental interactions on susceptibility for several pathological aberrations, such as congenital diseases, hematological and neurological disorders, and cancer. This relationship can also vary according to the different ethnicities. Furthermore, GST polymorphisms in pharmacogenetic studies highlight importance of evaluating the genetic profile of each patient to personalize the dose adjustment of the therapeutic protocol to avoid therapeutic resistance and related toxicities. In conclusion, inter-individual variability in the GST genes may modify susceptibility to harmful effects of endogenous and environmental toxic metabolites.

Table 1 summarizes the major GST polymorphisms associated with pathological abnormalities in some studies described in this chapter.

Table 1. Human pathological abnormalities and the related GST polymorphisms commonly associated

Pathological abnormalities	GST polymorphisms	References
Mental disorders	GSTM1 null, GSTP1 (Ile105Val), GSTT1 null	[6, 8, 9, 15, 18].
Brain tumors	GSTM3 (-63C allele), functional GSTM1, GSTP1 Ala114 Val; GSTT1 null; GSTP1 (Ile105Val)	[20, 21, 22, 23, 24, 25].
CHDs	GSTM1 null, GSTT1 null	[28]
Lung cancer	GSTT1 null, GSTM1 null, GSTP1 (Ile105Val)	[29, 30, 34, 35, 36, 37, 38].
HNSCC	GSTT1 null, GSTM1 null	[41, 42].
Cervical cancer	GSTP1 (Ile105Val), GSTM1 null, GSTT1 null	[49, 50, 51, 52, 53].
Breast cancer	GSTM1 null, GSTT1 null, GSTP1 (Ile105Val), GSTA1*B, GSTM3*B	[60, 61, 63, 64].
HCC	GSTM1 null, GSTT1 null, GSTP1 (Ile105Val)	[66, 67, 68, 69, 70, 71, 72, 73, 74].
RCC	GSTM1 null, GSTP1 (Ile105Val)	[78, 79]
Bladder cancer	GSTT1 null	[76]
Prostate cancer	GSTT1 null, GSTM1 null, GSTP1 (Ile105Val), GSTM3*B	[76, 81, 82, 83, 85].
Chemotherapy-related toxicity	GSTP1*A/*B, GSTP1*A/*A, GSTM1 null, GSTP1 (Ile105Val)	[91, 92, 93, 94, 97, 101, 102].

References

[1] Sachidanandam, R.; Weissman, D.; Schmidt, S. C.; Kakol, J. M.; Stein, L. D.; Marth, G.; Sherry, S.; Mullikin, J. C.; Mortimore, B. J.; Willey, D. L.; Hunt, S. E.; Cole, C.

G.; Coggill, P. C.; Rice, C. M.; Ning. Z.; Rogers, J.; Bentley, D. R.; Kwok, P. Y.; Mardis, E. R; Yeh, R. T.; Schultz, B.; Cook, L.; Davenport, R.; Dante, M.; Fulton, L.; Hillier, L.; Waterston, R. H.; McPherson, J. D.; Gilman, B.; Schaffner, S.; Van Etten, W. J.; Reich, D.; Higgins, J.; Daly, M. J.; Blumenstiel, B.; Baldwin, J.; Stange-Thomann, N.; Zody, M. C.; Linton, L.; Lander, E. S.; Altshuler, D. International SNP Map Working Group. A map of human genome sequence variation containing 1.42 million single nucleotide polymorphisms. *Nature*. 2001, 409, 928-933.

[2] Ahmadi, K. R.; Weale, M. E.; Xue, Z. Y.; Soranzo, N.; Yarnall, D. P.; Briley, J. D.; Maruyama, Y.; Kobayashi, M.; Wood, N. W.; Spurr, N. K; Burns, D. K.; Roses, A. D.; Saunders, A. M.; Goldstein D. B. A single-nucleotide polymorphism tagging set for human drug metabolism and transport. *Nat. Genet*. 2005, 37, 84-89.

[3] Carder, P. J.; Hume, R.; Fryer, A. A.; Strange, R. C.; Lauder, J.; Bell, J. E. Glutathione S-transferase in human brain. *Neuropathol. Appl. Neurobiol*. 1990, 16, 293-303.

[4] Kumawat, M.; Sharma, T. K.; Singh, I.; Singh, N.; Singh, S. K.; Ghalaut, V. S.; Shankar, V.; Vardey, S. K. Decrease in antioxidant status of plasma and erythrocytes from geriatric population. *Dis. Markers*. 2012, 33, 303-308.

[5] Kim, G. H.; Kim, J. E.; Rhie, S. J.; Yoon, S. The Role of Oxidative Stress in Neurodegenerative Diseases. *Exp. Neurobiol*. 2015, 24, 325-340.

[6] Wang, M.; Li, Y.; Lin, L.; Song, G.; Deng, T. GSTM1 Null genotype and GSTP1 Ile105Val polymorphism are associated with Alzheimer's disease: a meta-analysis. *Mol. Neurobiol.* 2016, 53, 1355-1364.

[7] Wang, T. Glutathione S-transferases variants as risk factors in Alzheimer’s disease. *Neurol. Sci*. 2015, 36, 1785-1792.

[8] Perez-Pastene, C.; Graumann, R.; Díaz-Grez, F.; Miranda, M.; Venegas, P.; Godoy, O. T.; Layson, L.; Villagra, R.; Matamala, J. M.; Herrera, L.; Segura-Aguilar, J. Association of GST M1 null polymorphism with Parkinson's disease in a Chilean population with a strong Amerindian genetic component. *Neurosci. Lett*. 2007, 418, 181-185.

[9] Golbe, L. I.; Di Iorio, G.; Markopoulou, K.; Athanassiadou, A.; Papapetropoulos, S.; Watts, R. L.; Vance, J. M.; Bonifati, V.; Williams, T. A.; Spychala, J. R.; Stenroos, E. S.; Johnson, W. G. Glutathione S-transferase polymorphisms and onset age in alpha-synuclein A53T mutant Parkinson's disease. *Am. J. Med. Genet. B Neuropsychiatr. Genet*. 2007, 144B, 254-258.

[10] Buhagiar, K.; Parsonage, L.; Osborn, D. P. Physical health behaviours and health locus of control in people with schizophrenia-spectrum disorder and bipolar disorder: a cross-sectional comparative study with people with non-psychotic mental illness. *BMC Psychiatry*. 2011, 11, 104.

[11] Doherty, A. M.; Kelly, J.; McDonald, C.; O'Dywer, A. M.; Keane, J.; Cooney, J. A review of the interplay between tuberculosis and mental health. *Gen. Hosp. Psychiatry.* 2013, 35, 398-406.

[12] Arneth, B. M. Multiple sclerosis and schizophrenia. *Int. J. Mol. Sci.* 2017, 18, 1760.

[13] Nucifora, L. G.; Tanaka, T.; Hayes, L. N.; Kim, M.; Lee, B. J.; Matsuda, T.; Nucifora, F. C. Jr.; Sedlak, T.; Mojtabai, R.; Eaton, W.; Sawa, A. Reduction of plasma glutathione in psychosis associated with schizophrenia and bipolar disorder in translational psychiatry. *Transl. Psychiatry.* 2017, 7, e1215.

[14] Zhang, X. Y.; Tan, Y. L.; Cao, L. Y.; Wu, G. Y.; Xu, Q.; Shen, Y.; Zhou, D. F. Antioxidant enzymes and lipid peroxidation in different forms of schizophrenia treated with typical and atypical antipsychotics. *Schizophr. Res.* 2006, 81, 291-300.

[15] Kim, S. K.; Kang, S. W.; Chung, J. H.; Park, H. J.; Cho, K. B.; Park, M. S. Genetic polymorphisms of glutathione-related enzymes (GSTM1, GSTT1, and GSTP1) and schizophrenia risk: A meta-analysis. *Int. J. Mol. Sci.* 2015, 16, 19602-19611.

[16] Cai, L.; Cai, M. H.; Wang, M. Y.; Xu, Y. F.; Chen, W. Z.; Qin, S. Y.; Wan, C. L.; He, L. Meta-analysis-based preliminary exploration of the connection between ATDILI and schizophrenia by GSTM1/T1 gene polymorphisms. *PLoS One.* 2015, 10, e0128643.

[17] Tang, N.; Deng, R.; Wang, Y.; Lin, M.; Li, H.; Qiu, Y.; Hong, M.; Zhou, G. GSTM1 and GSTT1 null polymorphisms and susceptibility to anti-tuberculosis drug-induced liver injury: a meta-analysis. *Int. J. Tuberc. Lung. Dis.* 2013, 17, 17-25.

[18] Pinheiro, D. S.; Santos, R. D. S.; de Brito, R. B.; Cruz, A. H. D. S.; Ghedini, P. C.; Reis, A. A. S. GSTM1/GSTT1 double-null genotype increases risk of treatment-resistant schizophrenia: A genetic association study in Brazilian patients. *PLoS One.* 2017, 12, e0183812.

[19] Mueller, S.; Chang, S. Pediatric brain tumors: current treatment strategies and future therapeutic approaches. *Neurotherapeutics.* 2009, 6, 570-586.

[20] Searles Nielsen, S.; Mueller, B. A.; Preston-Martin, S.; Farin, F. M.; Holly, E. A.; McKean-Cowdin, R. Childhood brain tumors and maternal cured meat consumption in pregnancy: differential effect by glutathione S-transferases. *Cancer Epidemiol. Biomarkers Prev.* 2011, 20, 2413-2419.

[21] Ezer, R.; Alonso, M.; Pereira, E.; Kim, M.; Allen, J. C.; Miller, D. C.; Newcomb, E. W. Identification of glutathione S-transferase (GST) polymorphisms in brain tumors and association with susceptibility to pediatric astrocytomas. *J. Neurooncol.* 2002, 59, 123-134.

[22] Lai, R.; Crevier, L.; Thabane, L. Genetic polymorphisms of glutathione S-transferases and the risk of adult brain tumors: a meta-analysis. *Cancer Epidemiol. Biomarkers Prev.* 2005, 14, 1784-1790.

[23] Pinarbasi, H.; Silig, Y.; Gurelik, M. Genetic polymorphisms of GSTs and their association with primary brain tumor incidence. *Cancer Genet. Cytogenet.* 2005, 156, 144-149.

[24] Wrensch, M.; Kelsey, K. T.; Liu, M.; Miike, R.; Moghadassi, M.; Aldape, K.; McMillan, A.; Wiencke, J. K. Glutathione-S-transferase variants and adult glioma. *Cancer Epidemiol. Biomarkers Prev.* 2004, 13, 461-467.

[25] De Roos, A. J.; Rothman, N.; Inskip, P. D.; Linet, M. S.; Shapiro, W. R.; Selker, R. G.; Fine, H. A.; Black, P M.; Pittman, G. S.; Bell, D. A. Genetic polymorphisms in GSTM1, -P1, -T1, and CYP2E1 and the risk of adult brain tumors. *Cancer Epidemiol. Biomarkers Prev.* 2003, 12, 14-22.

[26] Pratt, M. M.; John, K.; MacLean, A. B.; Afework, S.; Phillips, D. H.; Poirier, M. C. Polycyclic aromatic hydrocarbon (PAH) exposure and DNA adduct semi-quantitation in archived human tissues. *Int. J. Environ. Res. Public. Health.* 2011, 8, 2675–2691.

[27] IARC Working Group on the Evaluation of Carcinogenic Risks to Humans. Tobacco smoke and involuntary smoking. *IARC Monogr. Eval. Carcinog. Risks. Hum.* 2004, 83, 1-1438.

[28] Li, X.; Liu, Z.; Deng, Y.; Li, S.; Mu, D.; Tian, X.; Lin, Y.; Yang, J.; Li, J.; Li, N.; Wang, Y.; Chen, X.; Deng, K.; Zhu, J. Modification of the association between maternal smoke exposure and congenital heart defects by polymorphisms in glutathione S-transferase genes. *Sci. Rep.* 2015, 5, 14915.

[29] Sreeja, L.; Syamala, V.; Hariharan, S.; Madhavan, J.; Devan, S. C.; Ankathil, R. Possible risk modification by CYP1A1, GSTM1 and GSTT1 gene polymorphisms in lung cancer susceptibility in a South Indian population. *J. Hum. Genet.* 2005, 50, 618-627.

[30] Pliarchopoulou, K.; Voutsinas, G.; Papaxoinis, G.; Florou, K.; Skondra, M.; Kostaki, K.; Roussou, P.; Syrigos, K.; Pectasides, D. Correlation of CYP1A1, GSTP1 and GSTM1 gene polymorphisms and lung cancer risk among smokers. *Oncol. Lett.* 2012, 3, 1301-1306.

[31] Schneider, J.; Bernges, U.; Philipp, M.; Woitowitz, H. J. GSTM1, GSTT1, and GSTP1 polymorphism and lung cancer risk in relation to tobacco smoking. *Cancer Lett.* 2004, 208, 65-74.

[32] López-Cima, M. F.; Alvarez-Avellón, S. M.; Pascual, T.; Fernández-Somoano, A.; Tardón, A. Genetic polymorphisms in CYP1A1, GSTM1, GSTP1 and GSTT1 metabolic genes and risk of lung cancer in Asturias. *BMC Cancer.* 2012, 12, 433.

[33] Honma, H. N.; De Capitani, E. M.; Perroud, M. W. Jr.; Barbeiro, A. S.; Toro, I. F.; Costa, D. B.; Lima, C. S.; Zambon, L. Influence of p53 codon 72 exon 4, GSTM1, GSTT1 and GSTP1*B polymorphisms in lung cancer risk in a Brazilian population. *Lung Cancer.* 2008, 61, 152-162.

[34] Zhao, Y.; Wang, B.; Hu, K.; Wang, J.; Lu, S. U.; Zhang, Y.; Lu, W.; Zhao, E.; Yuan, L. Glutathione S-transferase θ1 polymorphism contributes to lung cancer susceptibility: A meta-analysis of 26 case-control studies. *Oncol. Lett.* 2015, 9, 1947-1953.

[35] Zhou, H. F.; Feng, X.; Zheng, B. S.; Qian, J.; He, W. A meta-analysis of the relationship between glutathione S-transferase T1 null/presence gene polymorphism and the risk of lung cancer including 31802 subjects. *Mol. Biol. Rep.* 2013, 40, 5713-5721.

[36] Gao, Y.; Gao, F.; Hu, T. T.; Li, G.; Sui, Y. X. Combined effects of glutathione S-transferase M1 and T1 polymorphisms on risk of lung cancer: Evidence from a meta-analysis. *Oncotarget.* 2017, 8, 28135-28143.

[37] Liu, K.; Lin, X.; Zhou, Q.; Ma, T.; Han, L.; Mao, G.; Chen, J.; Yue, X.; Wang, H.; Zhang, L.; Jin, G.; Jiang, J.; Zhao, J.; Zou, B. The associations between two vital GSTs genetic polymorphisms and lung cancer risk in the Chinese population: evidence from 71 studies. *PLoS One*. 2014, 9, e102372.

[38] Sharma, N.; Singh, A.; Singh, N.; Behera, D.; Sharma, S. Genetic polymorphisms in GSTM1, GSTT1 and GSTP1 genes and risk of lung cancer in a North Indian population. *Cancer Epidemiol*. 2015, 39, 947-955.

[39] Lakhdar, R.; Denden, S.; Knani, J.; Leban, N.; Daimi, H.; Hassine, M.; Lefranc, G.; Chibani, J. B.; Khelil, A. H. Combined analysis of EPHX1, GSTP1, GSTM1 and GSTT1 gene polymorphisms in relation to chronic obstructive pulmonary disease risk and lung function impairment. *Dis. Markers*. 2011, 30, 253-263.

[40] Cao, T.; Xu, N.; Wang, Z.; Liu, H. Effects of Glutathione S-Transferase Gene Polymorphisms and Antioxidant Capacity per Unit Albumin on the Pathogenesis of Chronic Obstructive Pulmonary Disease. *Oxid. Med. Cell. Longev*. 2017, 2017, 6232397.

[41] Zhang, Y.; Ni, Y.; Zhang, H.; Pan, Y.; Ma, J.; Wang, L. Association between GSTM1 and GSTT1 allelic variants and head and neck squamous cell cancinoma. *PLoS One*. 2012, 7, e47579.

[42] Singh, M.; Shah, P. P.; Singh, A. P.; Ruwali, M.; Mathur, N.; Pant, M. C.; Parmar, D. Association of genetic polymorphisms in glutathione S-transferases and susceptibility to head and neck cancer. *Mutat. Res*. 2008, 638, 184-194.

[43] Zafereo, M. E.; Sturgis, E. M.; Aleem, S.; Chaung, K.; Wei, Q.; Li, G. Glutathione S-transferase polymorphisms and risk of second primary malignancy after index squamous cell carcinoma of the head and neck. *Cancer Prev. Res*. (Phila). 2009, 2, 432-439.

[44] Psyrri, A.; Gouveris, P.; Vermorken, J. B. Human papillomavirus-related head and neck tumors: clinical and research implication. *Curr. Opin. Oncol*. 2009, 21, 201-205.

[45] Dok, R.; and Nuyts, S. HPV Positive Head and Neck Cancers: Molecular Pathogenesis and Evolving Treatment Strategies. *Cancers* (Basel). 2016, 8, 41.

[46] Mileo, A. M.; Abbruzzese, C.; Mattarocci, S.; Bellacchio, E.; Pisano, P.; Federico, A.; Maresca, V.; Picardo, M.; Giorgi, A.; Maras, B.; Schininà, M. E.; Paggi, M. G. Human papillomavirus-16 E7 interacts with glutathione S-transferase P1 and enhances its role in cell survival. *PLoS One*. 2009, 4, e7254.

[47] Checa-Rojas, A.; Delgadillo-Silva, L. F.; Velasco-Herrera, M. D. C.; Andrade-Domínguez, A.; Gil, J.; Santillán, O.; Lozano, L.; Toledo-Leyva, A.; Ramírez-Torres, A.; Talamas-Rohana, P.; Encarnación-Guevara, S. GSTM3 and GSTP1: novel players driving tumor progression in cervical cancer. *Oncotarget*. 2018, 9, 21696–21714.

[48] Phuthong, S.; Settheetham-Ishida, W.; Natphopsuk, S.; Ishida, T. Genetic Polymorphism of the Glutathione S-transferase Pi 1 (GSTP1) and Susceptibility to Cervical Cancer in Human Papilloma Virus Infected Northeastern Thai Women. *Asian Pac. J. Cancer Prev*. 2018, 19, 381-385.

[49] Chagas, B. S.; Gurgel, A. P. A. D.; Paiva Júnior, S. S. L.; Lima, R. C. P.; Cordeiro, M. N.; Moura, R. R.; Coelho, A. V. C.; Nascimento, K. C. G.; Silva Neto, J. C.; Crovella, S.; Freitas, A. C. Synergic effect of oral contraceptives, GSTP1 polymorphisms, and high-risk HPV infection in development of cervical lesions. *Genet. Mol. Res*. 2017, 16.

[50] Zhen, S.; Hu, C. M.; Bian, L. H. Glutathione S-transferase polymorphism interactions with smoking status and HPV infection in cervical cancer risk: an evidence-based meta-analysis. *PLoS One*. 2013, 8, e83497.

[51] Joseph, T.; Chacko, P.; Wesley, R.; Jayaprakash, P. G.; James, F. V.; Pillai, M. R. Germline genetic polymorphisms of CYP1A1, GSTM1 and GSTT1 genes in Indian cervical cancer: associations with tumor progression, age and human papillomavirus infection. *Gynecol. Oncol*. 2006, 101, 411-417.

[52] Sharma, A.; Gupta, S.; Sodhani, P.; Singh, V.; Sehgal, A.; Sardana, S.; Mehrotra, R.; Sharma, J. K. Glutathione S-transferase M1 and T1 Polymorphisms, Cigarette Smoking and HPV Infection in Precancerous and Cancerous Lesions of the Uterine Cervix. *Asian Pac. J. Cancer Prev*. 2015, 16, 6429-6438.

[53] Lee, S. A.; Kim, J. W.; Roh, J. W.; Choi, J. Y.; Lee, K. M.; Yoo, K. Y.; Song, Y. S.; Kang, D. Genetic polymorphisms of GSTM1, p21, p53 and HPV infection with cervical cancer in Korean women. *Gynecol. Oncol*. 2004, 93, 14-18.

[54] Adler, V.; Yin, Z.; Fuchs, S. Y.; Benezra, M.; Rosario, L.; Tew, K. D.; Pincus, M. R.; Sardana, M.; Henderson, C. J.; Wolf, C. R.; Davis, R. J.; Ronai, Z. Regulation of JNK signaling by GSTp. *EMBO J*. 1999, 18, 1321-1334.

[55] Wu. Y.; Fan, Y.; Xue, B.; Luo, L.; Shen, J.; Zhang, S.; Jiang, Y.; Yin, Z. Human glutathione S-transferase P1-1 interacts with TRAF2 and regulates TRAF2-ASK1 signals. *Oncogene*. 2006, 25, 5787-5800.

[56] Sørlie, T.; Perou, C. M.; Tibshirani, R.; Aas, T.; Geisler, S.; Johnsen, H.; Hastie, T.; Eisen, M. B.; van de Rijn, M.; Jeffrey, S. S.; Thorsen, T.; Quist, H.; Matese, J. C.; Brown, P. O.; Botstein, D.; Lønning, P. E.; Børresen-Dale, A. L. Gene expression patterns of breast carcinomas distinguish tumor subclasses with clinical implications. *Proc. Natl. Acad. Sci. USA*. 2001, 98, 10869-10874.

[57] Rodgers, K. M.; Udesky, J. O.; Rudel, R. A.; Brody, J. G. Environmental chemicals and breast cancer: An updated review of epidemiological literature informed by biological mechanisms. *Environ. Res*. 2018, 160, 152-182.

[58] White, A. J.; D'Aloisio, A. A.; Nichols, H. B.; DeRoo, L. A.; Sandler, D. P. Breast cancer and exposure to tobacco smoke during potential windows of susceptibility. *Cancer Causes Control.* 2017, 28, 667-675.

[59] Li, D.; Wang, M.; Dhingra, K.; Hittelman, W. N. Aromatic DNA adducts in adjacent tissues of breast cancer patients: clues to breast cancer etiology. *Cancer Res*. 1996, 56, 287-293.

[60] Firozi, P. F.; Bondy, M. L.; Sahin, A. A.; Chang, P.; Lukmanji, F.; Singletary, E. S.; Hassan, M M.; Li, D. Aromatic DNA adducts and polymorphisms of CYP1A1, NAT2, and GSTM1 in breast cancer. *Carcinogenesis.* 2002, 23, 301-306.

[61] [61] McCarty, K. M.; Santella, R. M.; Steck, S. E.; Cleveland, R. J.; Ahn, J.; Ambrosone, C. B.; North, K.; Sagiv, S. K.; Eng, S. M.; Teitelbaum, S. L.; Neugut, A. I.; Gammon, M. D. PAH-DNA adducts, cigarette smoking, GST polymorphisms, and breast cancer risk. *Environ. Health. Perspect*. 2009, 117, 552–558.

[62] Yager, J. D.; and Davidson, N. E. Estrogen carcinogenesis in breast cancer. *N. Engl. J. Med*. 2006, 354, 270-282.

[63] Mitrunen, K.; Jourenkova, N.; Kataja, V.; Eskelinen, M.; Kosma, V. M.; Benhamou, S.; Vainio, H.; Uusitupa, M.; Hirvonen, A. Glutathione S-transferase M1, M3, P1, and T1 genetic polymorphisms and susceptibility to breast cancer. *Cancer Epidemiol Biomarkers Prev*. 2001, 10, 229-236.

[64] Martínez-Ramírez, O. C.; Pérez-Morales, R.; Castro, C.; Flores-Díaz, A.; Soto-Cruz, K. E.; Astorga-Ramos, A.; Gonsebatt, M. E.; Casas, L.; Valdés-Flores, M.; Rubio, J. Polymorphisms of catechol estrogens metabolism pathway genes and breast cancer risk in Mexican women. *Breast.* 2013, 22, 335-343.

[65] Ivanov, A. V.; Valuev-Elliston, V. T.; Tyurina, D. A.; Ivanova, O. N.; Kochetkov, S. N.; Bartosch, B.; Isaguliants, M. G. Oxidative stress, a trigger of hepatitis C and B virus-induced liver carcinogenesis. *Oncotarget*. 2017, 8, 3895-3932.

[66] Martínez, C.; García-Martín, E.; Ladero, J. M.; Herráez, O.; Ortega, L.; Taxonera, C.; Suárez, A.; Díaz-Rubio, M.; Agúndez, J. A. GSTT1 and GSTM1 null genotypes may facilitate hepatitis C virus infection becoming chronic. *J. Infect. Dis*. 2007, 195, 1320-1323.

[67] Chen, C. J.; Yu, M. W.; Liaw, Y. F.; Wang, L. W.; Chiamprasert, S.; Matin, F.; Hirvonen, A.; Bell, D. A.; Santella, R. M. Chronic hepatitis B carriers with null

genotypes of glutathione S-transferase M1 and T1 polymorphisms who are exposed to aflatoxin are at increased risk of hepatocellular carcinoma. *Am. J. Hum. Genet.* 1996, 59, 128-134.

[68] Sun, C. A.; Wang, L. Y.; Chen, C. J.; Lu, S. N.; You, S. L.; Wang, L. W.; Wang, Q.; Wu, D. M.; Santella, R M. Genetic polymorphisms of glutathione S-transferases M1 and T1 associated with susceptibility to aflatoxin-related hepatocarcinogenesis among chronic hepatitis B carriers: a nested case-control study in Taiwan. *Carcinogenesis.* 2001, 22, 1289-1294.

[69] Ladero, J. M.; Martínez, C.; García-Martin, E.; Fernández-Arquero, M.; López-Alonso, G.; de la Concha, E. G.; Díaz-Rubio, M.; Agúndez, J. A. Polymorphisms of the glutathione S-transferases mu-1 (GSTM1) and theta-1 (GSTT1) and the risk of advanced alcoholic liver disease. *Scand. J. Gastroenterol.* 2005, 40, 348-353.

[70] Roy, N.; Dasgupta, D.; Mukhopadhyay, I.; Chatterjee, A.; Das, K.; Bhowmik, P.; Das, S.; Basu, P.; Santra, A. K.; Datta, S.; Dhali, G. K.; Chowdhury, A.; Banerjee, S. Genetic Association and Gene-Gene Interaction Reveal Genetic Variations in ADH1B, GSTM1 and MnSOD Independently Confer Risk to Alcoholic Liver Diseases in India. *PLoS One.* 2016, 11, e0149843.

[71] Khan, A. J.; Choudhuri, G.; Husain, Q.; Parmar, D. Polymorphism in glutathione-S-transferases: a risk factor in alcoholic liver cirrhosis. *Drug Alcohol Depend.* 2009, 101, 183-190.

[72] Liu, K.; Zhang, L.; Lin, X.; Chen, L.; Shi, H.; Magaye, R.; Zou, B.; Zhao, J. Association of GST genetic polymorphisms with the susceptibility to hepatocellular carcinoma (HCC) in Chinese population evaluated by an updated systematic meta-analysis. *PLoS One.* 2013, 8, e57043.

[73] Chen, Y. L.; Tseng, H. S.; Kuo, W. H.; Yang, S. F.; Chen, D. R.; Tsai, H. T. Glutathione S-Transferase P1 (GSTP1) gene polymorphism increases age-related susceptibility to hepatocellular carcinoma. *BMC Med. Genet.* 2010, 11, 46.

[74] Li, C. G.; Zhao, Z. M.; Hu, M. G.; Liu, R. Predictive role of glutathione-S-transferase gene polymorphisms in risk and prognosis of hepatocellular carcinoma. *Asian Pac. J. Cancer Prev.* 2012, 13, 3247-3252.

[75] Wang, Z.; Qu, K.; Niu, W.; Lin, T.; Xu, X.; Huang, Z.; Liu, S.; Liu, S.; Chang, H.; Liu, Y.; Dong, X.; Liu, C.; Zhang, Y. Glutathione S-transferase P1 gene rs4147581 polymorphism predicts overall survival of patients with hepatocellular carcinoma: evidence from an enlarged study. *Tumour Biol.* 2016, 37, 943-952.

[76] Wang, Y.; He, J.; Ma, T. J.; Lei, W.; Li, F.; Shen, H.; Shen, Z. Y. GSTT1 Null Genotype Significantly Increases the Susceptibility to Urinary System Cancer: Evidences from 63,876 Subjects. *J. Cancer.* 2016, 7, 1680-1693.

[77] Yang, X.; Long, S.; Deng, J.; Deng, T.; Gong, Z.; Hao, P. Glutathione S-transferase polymorphisms (GSTM1, GSTT1 and GSTP1) and their susceptibility to renal cell carcinoma: an evidence-based meta-analysis. *PLoS One.* 2013, 8, e63827.

[78] Coric, V. M.; Simic, T. P.; Pekmezovic, T. D.; Basta-Jovanovic, G. M.; Savic Radojevic, A. R.; Radojevic-Skodric, S. M.; Matic, M G.; Dragicevic, D. P.; Radic, T. M.; Bogdanovic, L. M.; Dzamic, Z. M.; Pljesa-Ercegovac, M. S. Combined GSTM1-null, GSTT1-active, GSTA1 low-activity and GSTP1-variant genotype is associated with increased risk of clear cell renal cell carcinoma. *PLoS One*. 2016, 11, e0160570.

[79] Safarinejad, M. R.; Safarinejad, S.; Shafiei, N.; Safarinejad, S. Association of genetic polymorphism of glutathione S-transferase (GSTM1, GSTT1, GSTP1) with bladder cancer susceptibility. *Urol. Oncol.* 2013, 31, 1193-1203.

[80] Oh, B.; Figtree, G.; Costa, D.; Eade, T.; Hruby, G.; Lim, S.; Elfiky, A.; Martine, N.; Rosenthal, D.; Clarke, S.; Back, M. Oxidative stress in prostate cancer patients: A systematic review of case control studies. *Prostate Int.* 2016, 4, 71–87.

[81] Mo, Z.; Gao, Y.; Cao, Y.; Gao, F.; Jian, L. An updating meta-analysis of the GSTM1, GSTT1, and GSTP1 polymorphisms and prostate cancer: a HuGE review. *Prostate*. 2009, 69, 662-688.

[82] Safarinejad, M. R.; Shafiei, N.; Safarinejad, S. H. Glutathione S-transferase gene polymorphisms (GSTM1, GSTT1, GSTP1) and prostate cancer: a case-control study in Tehran, Iran. *Prostate Cancer Prostatic Dis*. 2011, 14, 105-113.

[83] Medeiros, R.; Vasconcelos, A.; Costa, S.; Pinto, D.; Ferreira, P.; Lobo, F.; Morais, A.; Oliveira, J.; Lopes, C. Metabolic susceptibility genes and prostate cancer risk in a southern European population: the role of glutathione S-transferases GSTM1, GSTM3, and GSTT1 genetic polymorphisms. *Prostate.* 2004, 58, 414-420.

[84] Så, R. A.; Moreira Ados, S.; Cabello, P. H.; Ornellas, A. A.; Costa, E. B.; Matos Cda, S.; Alves, G.; Hatagima, A. Human glutathione S-transferase polymorphisms associated with prostate cancer in the Brazilian population. *Int. Braz. J. Urol*. 2014, 40, 463-473.

[85] Cotignola, J.; Leonardi, D. B.; Shahabi, A.; Acuña, A. D.; Stern, M. C.; Navone, N.; Scorticati, C.; De Siervi, A.; Mazza, O.; Vazquez, E. Glutathione-S-transferase (GST) polymorphisms are associated with relapse after radical prostatectomy. *Prostate Cancer Prostatic Dis.* 2013, 16, 28–34.

[86] Wu, W.; Lu, J.; Tang, Q.; Zhang, S.; Yuan, B.; Li, J.; Di, Wu; Sun, H.; Lu, C.; Xia, Y.; Chen, D.; Sha, J.; Wang, X. GSTM1 and GSTT1 null polymorphisms and male infertility risk: an updated meta-analysis encompassing 6934 subjects. *Sci. Rep.* 2013, 3, 2258.

[87] Aydemir, B.; Onaran, I.; Kiziler, A. R.; Alici, B.; Akyolcu, M. C. Increased oxidative damage of sperm and seminal plasma in men with idiopathic infertility is higher in patients with glutathione S-transferase Mu-1 null genotype. *Asian J. Androl.* 2007, 9, 108-115.

[88] Tirumala Vani, G.; Mukesh, N.; Siva Prasad, B.; Rama Devi, P.; Hema Prasad, M.; Usha Rani, P.; Pardhanandana Reddy, P. Role of glutathione S-transferase Mu-1

(GSTM1) polymorphism in oligospermic infertile males. *Andrologia.* 2010, 42, 213-217.

[89] Safarinejad, M. R.; Shafiei, N.; Safarinejad, S. The association of glutathione-S-transferase gene polymorphisms (GSTM1, GSTT1, GSTP1) with idiopathic male infertility. *J. Hum. Genet.* 2010, 55, 565-570.

[90] Tang, M.; Wang, S.; Wang, W.; Cao, Q.; Qin, C.; Liu, B.; Li, P.; Zhang, W. The glutathione-S-transferase gene polymorphisms (GSTM1 and GSTT1) and idiopathic male infertility risk: a meta-analysis. *Gene.* 2012, 511, 218-223.

[91] Tran, A.; Jullien, V.; Alexandre, J.; Rey, E.; Rabillon, F.; Girre, V.; Dieras, V.; Pons, G.; Goldwasser, F.; Tréluyer, J. M. Pharmacokinetics and toxicity of docetaxel: role of CYP3A, MDR1, and GST polymorphisms. *Clin. Pharmacol. Ther.* 2006, 79, 570-580.

[92] Okcu, M. F.; Selvan, M.; Wang, L. E.; Stout, L.; Erana, R.; Airewele, G.; Adatto, P.; Hess, K.; Ali-Osman, F.; Groves, M.; Yung, A. W.; Levin, V. A.; Wei, Q.; Bondy, M. Glutathione S-transferase polymorphisms and survival in primary malignant glioma. *Clin. Cancer Res.* 2004, 10, 2618-2625.

[93] Lecomte, T.; Landi, B.; Beaune, P.; Laurent-Puig, P.; Loriot, M. A. Glutathione S-transferase P1 polymorphism (Ile105Val) predicts cumulative neuropathy in patients receiving oxaliplatin-based chemotherapy. *Clin. Cancer Res.* 2006, 12, 3050-3056.

[94] Tulsyan, S.; Chaturvedi, P.; Agarwal, G.; Lal, P.; Agrawal, S.; Mittal, R. D.; Mittal, B. Pharmacogenetic influence of GST polymorphisms on anthracycline-based chemotherapy responses and toxicity in breast cancer patients: a multi-analytical approach. *Mol. Diagn. Ther.* 2013, 17, 371-379.

[95] Tecza, K.; Pamula-Pilat, J.; Lanuszewska, J.; Butkiewicz, D.; Grzybowska, E. Pharmacogenetics of toxicity of 5-fluorouracil, doxorubicin and cyclophosphamide chemotherapy in breast cancer patients. *Oncotarget.* 2018, 9, 9114-9136.

[96] Elhasid, R.; Krivoy, N.; Rowe, J. M.; Sprecher, E.; Adler, L.; Elkin, H.; Efrati, E. Influence of glutathione S-transferase A1, P1, M1, T1 polymorphisms on oral busulfan pharmacokinetics in children with congenital hemoglobinopathies undergoing hematopoietic stem cell transplantation. *Pediatr. Blood Cancer.* 2010, 55, 1172-1179.

[97] Ansari, M.; Huezo-Diaz, P.; Rezgui, M. A.; Marktel, S.; Duval, M.;< Bittencourt, H.; Cappelli, B.; Krajinovic, M. Influence of glutathione S-transferase gene polymorphisms on busulfan pharmacokinetics and outcome of hematopoietic stem-cell transplantation in thalassemia pediatric patients. *Bone Marrow Transplant.* 2016, 51, 377-383.

[98] Yin, J.; Xiao, Y.; Zheng, H.; Zhang, Y. C. Once-daily i.v. BU-based conditioning regimen before allogeneic hematopoietic SCT: a study of influence of GST gene polymorphisms on BU pharmacokinetics and clinical outcomes in Chinese patients. *Bone Marrow Transplant.* 2015, 50, 696-705.

[99] Vester, U.; Kranz, B.; Zimmermann, S.; Büscher, R.; Hoyer, P. F. The response to cyclophosphamide in steroid-sensitive nephrotic syndrome is influenced by polymorphic expression of glutathion-S-transferases-M1 and -P1. *Pediatr. Nephrol.* 2005, 20, 478-481.

[100] Sharda, S. V.; Gulati, S.; Tripathi, G.; Jafar, T.; Kumar, A.; Sharma, R. K.; Agrawal, S. Do glutathione-S-transferase polymorphisms influence response to intravenous cyclophosphamide therapy in idiopathic nephrotic syndrome? *Pediatr. Nephrol.* 2008, 23, 2001-2006.

[101] Zhong, S.; Huang, M.; Yang, X.; Liang, L.; Wang, Y.; Romkes, M.; Duan, W.; Chan, E.; Zhou, S. F. Relationship of glutathione S-transferase genotypes with side-effects of pulsed cyclophosphamide therapy in patients with systemic lupus erythematosus. *Br. J. Clin. Pharmacol.* 2006, 62, 457-472.

[102] Audemard-Verger, A.; Martin Silva, N.; Verstuyft, C.; Costedoat-Chalumeau, N.; Hummel, A.; Le Guern, V.; Sacré, K.; Meyer, O.; Daugas, E.; Goujard, C.; Sultan, A.; Lobbedez, T.; Galicier, L.; Pourrat, J.; Le Hello, C.; Godin, M.; Morello, R.; Lambert, M.; Hachulla, E.; Vanhille, P.; Queffeulou, G.; Potier, J.; Dion, J. J.; Bataille, P.; Chauveau, D.; Moulis, G.; Farge-Bancel, D.; Duhaut, P.; Saint-Marcoux, B.; Deroux, A.; Manuzak, J.; Francès, C.; Aumaitre, O.; Bezanahary, H.; Becquemont, L.; Bienvenu, B. Glutathione S transferases polymorphisms are independent prognostic factors in lupus nephritis treated with cyclophosphamide. *PLoS One*. 2016, 11, e0151696.

[103] Tumer, T. B.; Savranoglu, S.; Atmaca, P.; Terzioglu, G.; Sen, A.; Arslan, S. Modulatory role of GSTM1 null genotype on the frequency of micronuclei in pesticide-exposed agricultural workers. *Toxicol. Ind. Health*. 2016, 32, 1942-1951.

[104] Singh, S.; Kumar, V.; Singh, P.; Thakur, S.; Banerjee, B. D.; Rautela, R. S.; Grover, S. S.; Rawat, D. S.; Pasha, S. T.; Jain, S. K.; Rai, A. Genetic polymorphisms of GSTM1, GSTT1 and GSTP1 and susceptibility to DNA damage in workers occupationally exposed to organophosphate pesticides. *Mutat. Res*. 2011, 725, 36-42.

[105] Buzio, L.; De Palma, G.; Mozzoni, P.; Tondel, M.; Buzio, C.; Franchini, I.; Axelson, O.; Mutti, A. Glutathione S-transferases M1-1 and T1-1 as risk modifiers for renal cell cancer associated with occupational exposure to chemicals. *Occup. Environ. Med*. 2003, 60, 789-793.

[106] Karami, S.; Boffetta, P.; Rothman, N.; Hung, R. J.; Stewart, T.; Zaridze, D.; Navritalova, M.; Mates, D.; Janout, V.; Kollarova, H.; Bencko, V.; Szeszenia-Dabrowska, N.; Holcatova, I.; Mukeria, A.; Gromiec, J.; Chanock, S. J.; Brennan, P.; Chow, W. H.; Moore, L. E. Renal cell carcinoma, occupational pesticide exposure and modification by glutathione S-transferase polymorphisms. *Carcinogenesis.* 2008, 29, 1567–1571.

[107] Dekant, W. Chemical-induced nephrotoxicity mediated by glutathione S-conjugate formation. *Toxicol. Lett.* 2001, 124, 21-36.

[108] Cheng, X.; Zhang, T.; Zhao, J.; Zhou, J.; Shao, H.; Zhou, Z.; Kong, F.; Feng, N.; Sun, Y.; Shan, B.; Xia, Z. The association between genetic damage in peripheral blood lymphocytes and polymorphisms of three glutathione S-transferases in Chinese workers exposed to 1,3-butadiene. *Mutat. Res.* 2013, 750, 139-146.

[109] Fustinoni, S.; Soleo, L.; Warholm, M.; Begemann, P.; Rannug, A.; Neumann, H. G.; Swenberg, J. A.; Vimercati, L.; Colombi, A. Influence of metabolic genotypes on biomarkers of exposure to 1,3-butadiene in humans. *Cancer Epidemiol. Biomarkers Prev.* 2002, 11, 1082-1090.

[110] IARC Working Group on the Evaluation of Carcinogenic Risks to Humans. Chemical Agents and Related Occupations. *IARC Monogr. Eval. Carcinog. Risks. Hum.* 2012, No. 100F.

[111] Zarth, A. T.; Murphy, S. E.; Hecht, S. S. Benzene oxide is a substrate for glutathione S-transferases. *Chem. Biol. Interact.* 2015, 242, 390-395.

[112] Garte, S.; Taioli, E.; Popov, T.; Bolognesi, C.; Farmer, P.; Merlo, F. Genetic susceptibility to benzene toxicity in humans. *J. Toxicol. Environ. Health A.* 2008, 71, 1482-1489.

[113] Ye, L. L.; Zhang, G. H.; Huang, J. W.; Li, Y.; Zheng, G. Q.; Zhang, D. T.; Zhou, L. F.; Tao, X. D.; Zhang, J.; Ye, Y. J.; Sun, P.; Frank, A.; Xia, Z. L. Are polymorphisms in metabolism protective or a risk for reduced white blood cell counts in a Chinese population with low occupational benzene exposures? *Int. J. Occup. Environ. Health.* 2015, 21, 232-240.

[114] Angelini, S.; Kumar, R.; Bermejo, J. L.; Maffei, F.; Barbieri, A.; Graziosi, F.; Carbone, F.; Cantelli-Forti, G.; Violante, F. S.; Hemminki, K.; Hrelia, P. Exposure to low environmental levels of benzene: evaluation of micronucleus frequencies and S-phenylmercapturic acid excretion in relation to polymorphisms in genes encoding metabolic enzymes. *Mutat. Res.* 2011, 719, 7-13.

[115] Priya, K.; Yadav, A.; Kumar, N.; Gulati, S.; Aggarwal, N.; Gupta, R. Glutathione S-transferase gene polymorphisms: Modulator of genetic damage in gasoline pump workers. *Int. J. Toxicol.* 2015, 34, 500-504.

[116] Lin, L. C.; Chen, W. J.; Chiung, Y. M.; Shih, T. S.; Liao, P. C. Association between GST genetic polymorphism and dose-related production of urinary benzene metabolite markers, trans, trans-muconic acid and S-phenylmercapturic acid. Cancer Epidemiol. *Biomarkers Prev.* 2008, 17, 1460-1469.

[117] Carrieri, M.; Bartolucci, G. B.; Scapellato, M. L.; Spatari, G.; Sapienza, D.; Soleo, L.; Lovreglio, P.; Tranfo, G.; Manno, M.; Trevisan, A. Influence of glutathione S-transferases polymorphisms on biological monitoring of exposure to low doses of benzene. *Toxicol. Lett.* 2012, 213, 63-68.

[118] Cresci, M.; Foffa, I.; Ait-Ali, L.; Pulignani, S.; Gianicolo, E. A.; Botto, N.; Picano, E.; Andreassi, M. G. Maternal and paternal environmental risk factors, metabolizing GSTM1 and GSTT1 polymorphisms, and congenital heart disease. *Am. J. Cardiol.* 2011, 108, 1625-1631.

Biographical Sketches

Mariana Chantre Justino

Present Position

Postdoctoral fellow at State University of Rio de Janeiro (UERJ), Rio de Janeiro, RJ, Brazil

Affiliation

Laboratório de Marcadores Circulantes, Departamento de Patologia, Faculdade de Ciências Médicas, Universidade do Estado do Rio de Janeiro (UERJ), Brazil

Education

Biologist (2007) at Federal University of Rio de Janeiro (UFRJ); MSc (2009, UFRJ), PhD (2014, UFRJ)

Research and Professional Experience

Research projects conducted at the Federal University of Rio de Janeiro (UFRJ) and at the State University of Rio de Janeiro (UERJ), in partnership with public hospitals, such as the National Cancer Institute (INCa). She has experience in cancer research using molecular biology tools.

Short Biography

She has experience in the following topics: genomic instability, DNA repair, genetic polymorphisms, and molecular biology of urothelial tumors. At the present time, she investigates the molecular profile of human papillomavirus (HPV) in patients with recurrent respiratory papillomatosis.

Professional Appointments

Postdoctoral fellow at State University of Rio de Janeiro (UERJ).

Publications from the Last 3 Years

1. Chantre-Justino, M.; Alves, G.; Britto, C.; Cardoso, A.; Scherrer, L.; Moreira Ados, S.; Quirino, R.; Ornellas, A.; Leitão, A.; Lage C. Impact of reduced levels of APE1 transcripts on the survival of patients with urothelial carcinoma of the bladder. *Oncol Rep.* 2015, 34, 1667-1674.
2. Chantre Justino, M. Influence of genetic polymorphisms in benzene exposure. Vol. 1. pp. 53-68. In: *The health and environmental threats related to gas stations: Rio de Janeiro gas stations.;* Alves Brown, G., Ed.; Lambert Academic Publishing, Saarbrücken, 2015.

Gilda Alves

Present Position

Research Fellow at Universidade do Estado do Rio de Janeiro, Rio de Janeiro, RJ, Brazil

Affiliation

Laboratório de Marcadores Circulantes, Departamento de Patologia Faculdade de Ciências Médicas, Universidade do Estado do Rio de Janeiro (UERJ)

Education

Biologist (1984) by Universidade Federal do Rio de Janeiro (UFRJ); MSc (1988), UFRJ; PhD (1994), UFRJ.

Research and Professional Experience

Researcher at Instituto Nacional de Câncer (INCA) in Rio de Janeiro from 1986-2017; Fellowship in National Institutes of Health (1988), USA; in the Armed Forces Institute of Pathology (1992 – 1994), USA; and in the University of Jena, Germany (1999-2011, not in sequence).

Short Biography

Born on March 10th, 1963, Rio de Janeiro, Brazil, married, two children. She has publications on the molecular biology of Urological Cancers (Bladder, Kidney, Penis, Prostate), Breast Cancer, Leukemia, and Glioma. More recently, studying workers´ occupational risks related to cancer. She has large experience in cancer research using molecular biology, molecular cytogenetics and proteomics tools. She was fellow in the

USA (NIH/Frederick and AFIP, 1988 and 1991, respectively) and many times in Jena, Germany, 1999-2010.

Professional Appointments

Research Fellow at Universidade do Estado do Rio de Janeiro (UERJ) and supervision of students.

Publications from the Last 3 Years

1. Cypriano, A. S.; Alves, G.; Ornellas, A. A.; Scheinkman, J.; Almeida, R.; Scherrer, L.; Lage, C. Relationship between XPD, RAD51, and APEX1 DNA repair genotypes and prostate cancer risk in the male population of Rio de Janeiro, Brazil. *Genet. Mol. Biol.* 2017, 40, 751-758.
2. Leite, C.; Delmonico, L.; Alves, G.; Gomes, R. J.; Martino, M. R.; da Silva, A. R.; Moreira, A. D. S.; Maioli, M. C.; Scherrer, L. R.; Bastos, E. F.; Irineu, R.; Ornellas, M. H. Screening of mutations in the additional sex combs like 1, transcriptional regulator, tumor protein p53, and *KRAS* proto-oncogene, *GTPase/NRAS* proto-oncogene, *GTPase* genes of patients with myelodysplastic syndrome. *Biomed. Rep.* 2017, 7, 343-348.
3. Silvestre, R. T.; Delmonico, L.; Bravo, M.; Santiago, F.; Scherrer, L. R.; Moreira, A. D. S.; Tabalipa, M.; Otero, U.; Ornellas, M. H. F.; Alves, G. Health survey and assessment of the polymorphisms BRCA1/P871L, BRCA1/Q356R, and BRCA2/N372H in female gas station workers in Rio de Janeiro. *Environ. Mol. Mutagen.* 2017, 58, 730-734.
4. Ornellas, M. H. F.; Maioli, M. C. P.; Lucena, S. B. S. G.; Bastos, E. F.; Chaves, T. S.; Melo, K.V.; Ribeiro-Carvalho, M. M.; Liehr, T.; Alves, G. Complex karyotype including ring chromosome 11 in a patient with acute myeloid leukemia: case report. *Sao Paulo Med. J.* 2018, 136, 361-367.
5. Ornellas, P.; Ribeiro-Carvalho, M. M.; Ornellas, A. A.; Scherrer, L.; Koifman, L.; Magni, F.; Ornellas, M. H.; Alves, G. Experimental validation of the complement protein C3a down expression in the plasma of patients with squamous cell carcinoma of the penis. *Urol. Oncol.* 2017, 35, 545.e13-545.e18.
6. Afonso, L. A.; Carestiato, F. N.; Ornellas, A. A.; Ornellas, P.; Rocha. W. M.; Cordeiro, T. I.; Lisboa, D. C.; Alves, G. B.; Cavalcanti, S. M. B. Human papillomavirus, Epstein-Barr virus, and methylation status of $p16_{ink4a}$ in penile cancer. *J. Med. Virol.* 2017, 89,1837-1843.
7. Santiago, F.; Lima, S.; Pinheiro, T.; Silvestre, R.T.; Otero, U. B.; Tabalipa, M. M.; Kosyakova, N.; Ornellas, M. H.; Liehr, T.; Alves, G. Benzene poisoning, clinical and blood abnormalities in two Brazilian female gas station attendants: two case reports. *BMC Res. Notes.* 2017, 10, 52.

8. Delmonico, L.; Bravo, M.; Silvestre, R. T.; Ornellas, M. H.; De Azevedo, C. M.; Alves, G. Proteomic profile of saliva and plasma from women with impalpable breast lesions. *Oncol. Lett.* 2016, 12, 2145-2152.
9. Ornellas, M. H.; De França Silva, M.; Solza, C.; De Lucena Gonçalves, S.B.; Silva De Almeida, L.; De Paula Ayres-Silva, J.; Seixas, T. L.; Bastos, E. F.; Liehr, T.; Alves, G. Myelodysplastic syndrome without ring sideroblasts and with Janus kinase 2 gene mutation: An unusual case report. *Mol. Clin. Oncol.* 2016, 5, 227-230.
10. Camisasca, D. R.; da Rós Gonçalves, L.; Soares, M. R.; Sandim, V.; Nogueira, F. C.; Garcia, C. H.; Santana, R.; de Oliveira, S. P.; Buexm, L. A.; de Faria, P. A.; Dias, F. L.; Pereira, D. A.; Zingali, R. B.; Alves, G.; Lourenço, S. Q. A proteomic approach to compare saliva from individuals with and without oral leukoplakia. *J. Proteomics.* 2017, 151, 43-52.
11. da Silva, S. G.; Leon, L. A.; Alves, G.; Brito, S. M.; Sandes Vde, S.; Lima, M. M.; Nogueira, M. C.; Tavares Rde, C.; Dobbin, J.; Apa, A.; de Paula, V. S.; Oliveira, J. M.; Pinto, M. A.; Ferreir Oda, C. Jr.; Motta Ide, J. A. rare case of transfusion transmission of Hepatitis A virus to two patients with haematological disease. *Transfus. Med. Hemother.* 2016, 43,137-41.
12. Afonso, L. A.; Cordeiro, T. I.; Carestiato, F. N.; Ornellas, A. A.; Alves, G.; Cavalcanti, S. M. High risk human papillomavirus infection of the foreskin in asymptomatic men and patients with phimosis. *J. Urol.* 2016, 195, 1784-1789.
13. Sandim, V.; Pereira Dde, A.; Kalume, D. E.; Oliveira-Carvalho, A. L.; Ornellas, A. A.; Soares, M. R.; Alves, G.; Zingali, R. B. Proteomic analysis reveals differentially secreted proteins in the urine from patients with clear cell renal cell carcinoma. *Urol. Oncol.* 2016, 34, 5.e11-25.
14. Koifman, L.; Ornellas, P.; Ornellas, A. A.; Pereira Dde, A.; Zingali, B. R.; Cavalcanti, S. M.; Afonso, L. A.; Sandim, V.; Alves, G. Proteomics analysis of tissue samples from patients with squamous cell carcinoma of the penis and positive to human papillomavirus. *Int. Braz. J. Urol.* 2015, 41, 642-654.
15. Delmonico, L.; Moreira Ados, S.; Franco, M. F.; Esteves, E. B.; Scherrer, L.; Gallo, C. V.; do Nascimento, C. M.; Ornellas, M. H.; de Azevedo, C. M.; Alves, G. CDKN2A (p14(ARF)/p16(INK4a)) and ATM promoter methylation in patients with impalpable breast lesions. *Hum. Pathol.* 2015, 46, 1540-1547.
16. Chantre-Justino, M.; Alves, G.; Britto, C.; Cardoso, A.; Scherrer, L.; Moreira Ados, S.; Quirino, R.; Ornellas, A.; Leitão, A.; Lage, C. Impact of reduced levels of APE1 transcripts on the survival of patients with urothelial carcinoma of the bladder. *Oncol. Rep.* 2015, 34, 1667-1674.
17. Delmonico, L.; Areias, V. R.; Pinto, R. C.; Matos Cda, S.; Rosa, M. F.; De Azevedo, C.M.; Alves G. Protein identification from dried nipple aspirate fluid on Guthrie cards using mass spectrometry. *Mol. Med. Rep.* 2015, 12, 159-164.

18. Fonseca, R. F.; Rosas, S. L.; Oliveira, J. A.; Teixeira, A.; Alves, G.; Carvalho Mda, G. Frequency of Epstein-Barr virus DNA sequences in human gliomas. *Sao Paulo Med. J.* 2015, 133, 51-54.
19. Delmonico, L.; Alves, G.; do Amaral, L. F. P. A biologia do câncer de mama e testes moleculares de Prognóstico. *Revista HUPE, Rio de Janeiro*. 2015, 14(S1), 59-65.
20. Alves, G.; Delmonico, L. O presente e o futuro dos marcadores circulantes de câncer. *Revista HUPE, Rio de Janeiro*, 2015, 14(S1), 66-72.

In: Glutathione
Editor: Pál Perjési
ISBN: 978-1-53614-740-7
© 2019 Nova Science Publishers, Inc.

Chapter 8

THE ROLE OF GLUTATHIONE IN TOXICITY OF DRUGS AND XENOBIOTICS

Tomáš Roušar*
Department of Biological and Biochemical Sciences,
University of Pardubice, Pardubice, Czech Republic

ABSTRACT

The aim of this review is to describe the role of glutathione in biotransformation of xenobiotics. Glutathione can be conjugated with a number of xenobiotics increasing polarity of the parent molecules. The general overview of the glutathione-related biotransformation includes the function of glutathione-S-transferases, formation of mercapturates and action of β-lyases. In most cases, the conjugation leads to detoxification of the substrate, but in some cases the reactions can produce toxic derivatives. Thus, the review also aims to describe the role of glutathione in toxicity of selected halogenated compounds, quinones/hydrochinones and the aminophenol derivative acetaminophen (paracetamol).

Keywords: glutathione, glutathione conjugation, glutathione-S-transferases, redox cycling, glutathione toxicity

---

* Corresponding Author E-mail: Tomas.Rousar@upce.cz.

Glutathione-Dependent Biotransformation of Xenobiotics

Biotransformation of Xenobiotics

Liver is the most important organ participating in biotransformation and detoxification of endogenous compounds and exogenous substances (xenobiotics). These latter can be toxic to selected organs. The organism seeks to eliminate these substances from the body or at least to decrease their toxicity. Liver has got several mechanisms to modify chemical structure and biological effect of the compounds entering the organ.

Biotransformation reactions can be separated into two phases: Phase I and Phase II transformations (Figure 1). Both phases generally increase polarity of the parent compounds and facilitate their elimination from the organism through the urine or the feces. The biotransformation processes occurring in the liver cells, however, can generate reactive metabolites that can cause impairment of the organ. Thus, most of the pathological abnormalities caused by biotransformation of xenobiotics can be observed in the form of hepatotoxicity.

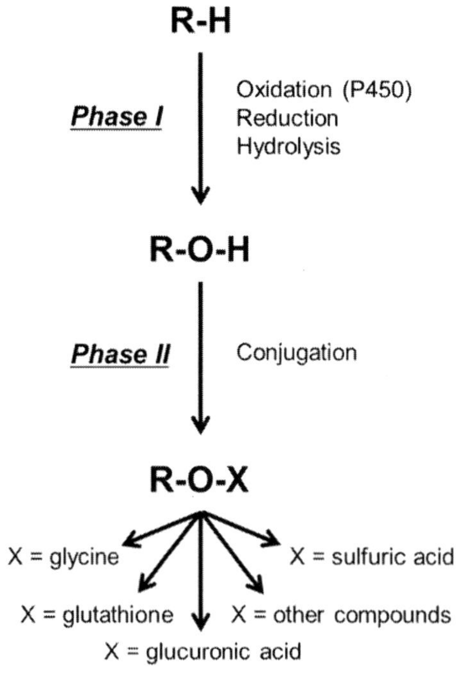

Figure 1. Biotransformation of xenobiotics (R-H, lipophilic substrate; R-OH, hydrophilic product; R-O-X, terminal conjugate of the parent substrate).

Phase I Transformations

The reactions belonging to the Phase I processes are: (a) oxidation, (b) reduction and (c) hydrolytic transformations. The parent molecule can be transformed by the cytochrome P450 (CYP450) and the flavin containing monooxygenase (FMO) systems, the monoamine oxidases (MAO) and the hydrolases. The result of these reactions is an increased polarity of the parent molecules. Specificity of the enzyme systems is not high and that is why they can catalyze the reactions with a wide range of structurally different compounds.

The cytochrome P450 (CYP) enzymes, the most important catalysts of the Phase I reactions, are heme-containing proteins localized in the smooth endoplasmic reticulum. The name of cytochrome P450 is derived from the spectrophotometric absorption band of the enzyme ($\lambda = 450$), which can be observed when its reduced state is bound to carbon monoxide (CO). The oxidation reactions are ensured through incorporation of an oxygen atom into the parent molecule during a redox cycle of the protoporphyrin IX moiety of the enzyme. During the catalytic cycle the iron(III) ion is reduced, molecular oxygen is bound to the iron(II) ion and then the substrate is oxidized [1].

Recently, more than a thousand of CYP subtypes have been described which can be distinguished regarding their structural similarities (Figure 2). There have been approximately sixty human CYP isoforms identified, which are expressed mostly in the liver, the kidney and the lungs [2, 3]. CYP is bound to the endoplasmic reticulum, but some isoforms have also been detected in mitochondria [4, 5]. In humans, 98% of the Phase I biotransformation reactions are catalyzed by six CYP isoenzymes: CYP1A2, CYP2C9, CYP2C19, CYP2D6, CYP2E1 and CYP3A4 [6].

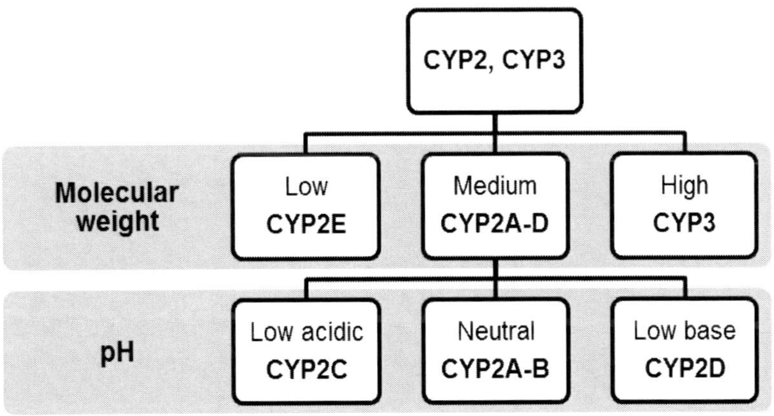

Figure 2. Substrate specificity of the CYP2 and CYP3 isoforms.

In humans, CYP3A4 has been found to be the most abundant cytochrome P450 isoform. Besides xenobiotics, CYP3A4 is also responsible for the metabolism of endogenous compounds, such as steroid hormones. The CYP3A subfamily accounts for biotransformation of more than 150 frequently used drugs currently available on the market

[7]. On the other hand, CYP2E1 is the only isoform of the CYP2E subfamily occurring in men. CYP2E1 accounts for biotransformation of several small, hydrophilic compounds (e.g., ethanol) and halogenated hydrocarbons. Substrate specificity of the CYP2 and CYP3 isoforms is shown in Figure 2.

Phase II Transformations

Most frequently, Phase I transformations are followed by Phase II reactions. In most of the Phase II processes the parent xenobiotics or their Phase I derivatives are linked to a polar endogenous molecule like glucuronic acid, sulfuric acid, glutathione, glycine, glutamine or taurine. Accordingly, Phase II reactions are generally conjugation reactions, which are most commonly catalyzed by transferases. The conjugates, or their further metabolites, are generally excreted through the bile and the urine

Role of Glutathione in Detoxification of Xenobiotics

Glutathione (GSH) is a tripeptide consisting of glutamate, cysteine and glycine, (L-γ-glutamyl-L-cysteinyl-glycine). Glutathione occurs in two free forms: the reduced (GSH) thiol, and the oxidized (GSSG) disulfide forms (Figure 3). In addition, it can be bound to proteins and to other thiols, which regulates their activity [8-11].

Glutathione is the most abundant non-protein thiol, which participates in a number of antioxidant reactions. Its intracellular levels reach millimolar range and the highest GSH concentration can be found in the liver, the kidney and the red blood cells. Glutathione occurs extracellularly at micromolar levels in blood plasma, but in contrast, its concentration is reaching 10 mM in the bile [12-15].

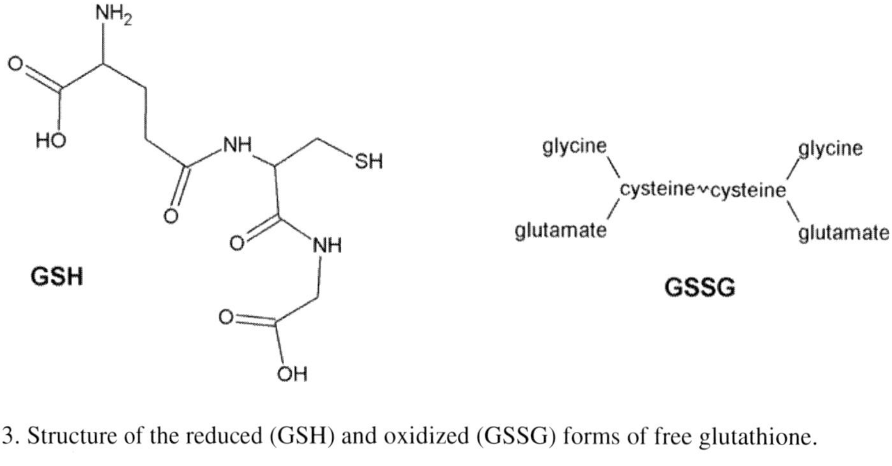

GSH

GSSG

Figure 3. Structure of the reduced (GSH) and oxidized (GSSG) forms of free glutathione.

In addition to the essential antioxidant role, glutathione also participates in Phase II conjugation reactions. These reactions are mostly catalyzed by glutathione-*S*-transferases

(GSTs). They utilize the thiol moiety of glutathione to bind GSH with endogenous or exogenous electrophilic substrates. The endogenous substrates for glutathione conjugation are leukotrienes, isoprostanes, catecholamines or electrophilic lipid peroxidation products. The exogenous substrates include a number of chemicals, drugs and toxins [16-18].

The synthesized glutathione conjugate can be excreted from hepatocytes through the canalicular pole into the bile or through the basolateral pole into the blood. At the canalicular membrane, the glutathione conjugates can be cleaved by the enzyme gamma-glutamyltransferase (GGT) to form the corresponding glycine-cysteine-conjugate, which can be further degraded by dipeptidase (DP) to form the respective cysteine conjugate (Figure 4). The cleavage process occurs in the bile and/or in the intestine where the excreted conjugates can be reabsorbed into the blood and transported to the liver. Most of the reabsorbed conjugates are transported through the hepatocytes into the bile again, but some portion of them can also be excreted through the urine [19]. Furthermore, the cysteine conjugates can be also acetylated in the presence of *N*-acetyltransferases (NATs) to form *N*-acetylcysteine conjugates (mercapturates), or can be directly cleaved through β-lyases.

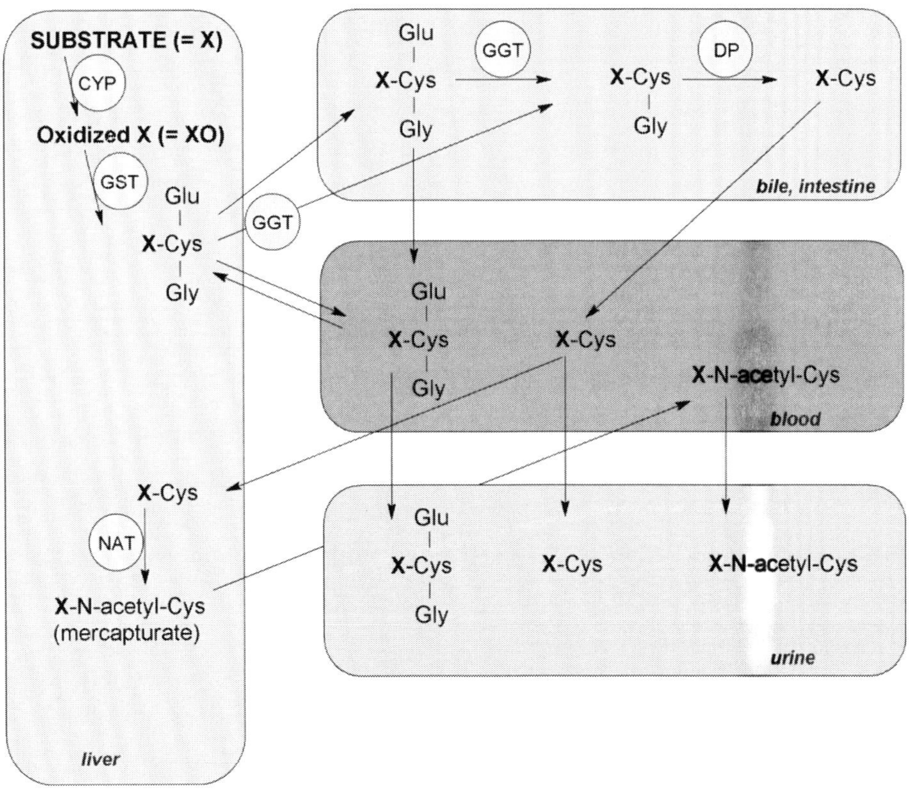

Figure 4. Biotransformation of glutathione conjugates (CYP, cytochrome P450; DP, dipeptidase; GGT, gamma-glutamyltransferase; GST, glutathione-S-transferase; NAT, N-acetyltransferase; X, substrate).

Glutathione-Related Enzymes Participating in Detoxification

Glutathione-S-Transferases

Glutathione-*S*-transferases (GSTs, EC 2.5.1.18) are the enzymes playing a crucial role in synthesis of glutathione (GSH) conjugates. The substrate of the GST enzymes can be any endogenous or exogenous compound that possesses an electrophilic center to react with the nucleophilic thiol (SH) group of glutathione. Accordingly, GSTs play an important role in detoxification of electrophilic xenobiotics/xenobiotic metabolites. In some occasions, however, GSH-conjugation results in bioactivation of chemicals leading to toxicity. The first report on GST was published in 1960, when conjugation of GSH with naphthalene was described [20].

Glutathione-*S*-transferases are ubiquitous intracellular enzymes present not only in mammals, but also in birds, insects, plants and bacteria. In eukaryotes, GSTs are divided according to their cellular localization into at least three major families of proteins, namely cytosolic GSTs, mitochondrial GSTs and microsomal GSTs.

Cytosolic GSTs are widely distributed and in turn divided into several major classes on the basis of their chemical, physical and structural properties. Mitochondrial GSTs are also known as kappa class GSTs and are soluble enzymes that bear structural similarities with cytosolic GSTs. On the contrary, microsomal GSTs, also known as MAPEG (membrane-associated proteins involved in eicosanoid and glutathione metabolism), are integral membrane proteins which are not evolutionary related to the other major classes [21].

Cytosolic GSTs are soluble enzymes present as dimers. The molecular weight of each subunit is approximately 25 kDa and each subunit contains a hydrophilic binding site for reduced glutathione (GSH). The hydrophobic binding site for a substrate is located in proximity to the glutathione binding site. Another superfamily of GSTs comprises of a group of microsomal enzymes, which are bound to a membrane. Due to their localization, MAPEGs are predominantly involved in metabolism of arachidonic acid and in detoxification of lipid peroxidation products of the cell membrane [22, 23].

Cytosolic glutathione-*S*-transferases can be divided into different classes. Now, seven GST classes have been described in humans. They can be classified as: alpha, mu, pi, theta, zeta, omega, and sigma. In addition to the seven cytosolic forms, kappa class of GST has been described in mitochondria [17]. All forms of GSTs are consisting of two subunits which can be present as homodimers or heterodimers. Usually dimers are made from identical chains but heterodimers made of two different chains from the same class are also found [24]. The standardized GST nomenclature identifies the species with a lower-case initial (hGSTP1-2, for human), the isoenzyme class with an upper-case letter (hGSTP1-2, for pi), followed by an Arabic numeral representing the class subfamily of both subunits (hGSTP1-2, for a heterodimer in the pi-class subfamily 1 and 2) [25].

Expression of most of the GSTs is tissue-specific. On the other hand, GST-pi expression is common in several tissues [26]. The liver contains a number of GSTs participating in biotransformation processes, e.g., alpha, mu, and pi. Another organ with a large activity of GSTs is the kidney, producing alpha, mu, and pi class GSTs. The highest activity of GSTs in the kidney has been found in the proximal and the distal tubular cells [27, 28].

Gama-Glutamyltransferase

Gamma-glutamyltransferase (GGT) is another essential enzyme participating in biotransformation of glutathione conjugates. It is a membrane-bound enzyme localized on the external part of the cell membrane of a number of cells involved in excretion processes. Therefore, the highest activity of GGT is found in the liver and the kidney. GGT is a unique enzyme because it is the only biomolecule possessing capability to cleave the glutathione molecule. Hence, GGT is capable to catalyze hydrolysis of the gamma-glutamyl bond in glutathione conjugates to form cysteinyl-glycine conjugates (Figure 4) [29].

GGT was firstly described in the kidney of rats but, at present, its activity has been found in a number of tissues. Gamma-glutamyltransferase consists of two different glycoprotein subunits. *N*-terminal part of the heavy subunit is composed of a hydrophobic chain anchoring GGT into the cell membrane. The glycosylated heavy subunit of GGT is non-covalently associated with the light subunit of the enzyme. The light subunit is catalytically active; it binds the gamma-glutamyl moiety of the substrate to the active site [30-32]. The specific inhibitor of GGT is acivicin ((2*S*)-amino[(5*S*)-3-chloro-4,5-dihydro-1,2-oxazol-5-yl]ethanoic acid).

In humans, it is the kidney that exhibits the largest GGT activity, especially in the proximal tubular cells of the apical membrane. Other localizations with large GGT activity are the canalicular membrane of the hepatocytes, the cholangiocytes, the endothelial cells in the nervous system, the pancreas, the testicular and the prostate cells [33, 34].

Mercapturic Acid Pathway and Cysteine-S-Conjugate B-Lyases

The cysteine conjugates formed by the action of GGT can be excreted, acetylated or cleaved. The cysteine conjugates can undergo acetylation by *N*-acetyltransferases (NATs) in the liver or the kidney, whereby *N*-acetylcysteine conjugates (mercapturates) are formed. The *N*-acetylcysteine conjugates are rather polar compounds; they can be excreted from the liver, and/or reabsorbed to the blood, and excreted through the urine (Figure 4). Acetylation of the cysteine conjugates is the final step of the mercapturate pathway of biotransfor-mation [35].

The cysteine conjugates can undergo another reaction as well, which is the cleavage of the molecule by cysteine-S-conjugate β-lyases (EC 4.4.1.139). These enzymes have large affinity to cysteine conjugates and therefore their contribution to biotransformation of cysteine conjugates is indispensable. The β-lyases have pyridoxal-5-phosphate in their structure, thus they also possess transaminase activity. They are capable to cleave the β-thiol bond of the cysteine conjugates to form ammonia, pyruvate and the respective thiol derivative of the parent compound (RSH) [35-37]. Accordingly, cysteine-S-conjugate β-lyases can react with cysteine conjugates formed by GGT-catalyzed cleavage of glutathione conjugates, or with deacetylated *N*-acetylcysteine conjugates. The potent inhibitor of cysteine-S-conjugate β-lyases activity is the aminooxyacetic acid.

β-Lyases occur both in eukaryotic and prokaryotic organisms. In humans, the largest β-lyase activity has been found in the kidney, the liver and the brain. In liver and kidney, the enzyme activity can be detected in the cytosol, the microsomes and the mitochondria [19]. β-Lyase activity has been characterized in at least eleven various enzymes, including glutamine transaminase-K, kynureninase, alanine aminotransferase (ALT), and aspartate aminotransferase (AST) [35]. β-Lyases can be divided into three groups, based on their molecular weight: low M_r (< 100 kDa), middle M_r and high M_r (> 300 kDa) [36].

Cellular Transport of Glutathione-Conjugates

Membrane Transporters in the Liver

Hepatocytes are playing a central role in biotransformation of endogenous compounds and xenobiotics. Because of their function is also linked with excretion of a variety of biotransformation products, cell membranes of hepatocytes consist of a number of transporter proteins, which belong to two major families of carrier proteins: multidrug resistance proteins (MRPs) and organic anion transporting polypeptides (OATPs).

The nine multidrug resistance proteins (MRPs) represent the major part of the 12 members of the MRP/CFTR subfamily belonging to the 48 human ATP-binding cassette (ABC) transporters. The MRP transporters belong to subfamily C (ABCC) of the ATP-binding cassette (ABC) transporter family. The ABCC family includes twelve proteins: nine human MRP genes (MRP1 to MRP9); the Cystic Fibrosis Transmembrane Conductance Regulator (CFTR/ABCC7), a channel gated by ATP binding and hydrolysis; and two ATP-dependent potassium channel regulators, sulfonylurea receptors 1 and 2 (SUR1/ABCC8 and SUR2/ABCC9) [38].

Multidrug Resistance Proteins are called in relation to their role in resistance to treatment of cancer cells, due to their non-specific transport of organic anions. On the other hand, the physiological roles of MRPs are associated with their capability to export a wide range of compounds, including cytotoxic substances. Another important physiological function of MRPs is associated with transport of glutathione and glutathione conjugates [39-41].

Table 1. MRP proteins with capability to transport glutathione-related compounds (ABCC, ATP-Binding Cassette; BCRP, Breast Cancer Resistance Protein; CFTR, Cystic Fibrosis Transmembrane Conductance Regulator; GSH and GS-, glutathione; MRP, Multidrug Resistance Protein) [38-41]

Type	ABCC Type	Glutathione Transport
MRP1	ABCC1	GSH + GSH-conjugates
MRP2	ABCC2	GSH + GSH-conjugates
MRP3	ABCC3	GSH-conjugates
MRP4	ABCC4	GSH + GSH-conjugates
MRP5	ABCC5	GSH + GSH-conjugates
CFTR	ABCC7	GSH
BCRP	ABCG2	GSH

The structure of each member of the MRPs is composed of two transmembrane domains, which pass six-times through the cell membrane. In addition, MRPs 1, 2, 3, 6 and 7 have a third transmembrane domain, which likely determines the specificity of a membrane pump. Cytosolic part of the MRPs is composed of nucleotide-binding domains, which are capable to bind ATP. Hydrolysis of ATP produces the energy that can be used for a change of protein conformation, whereby the substrate is transported through the MRP protein outside from the cell [38, 39].

In contrast to MRPs, the group of the OATP transporters consists of membrane pumps that ensure transport of compounds into the cells. The transport through the OATPs is not ATP-dependent, which is another difference to the MRPs. The OATP proteins are members of the superfamily of membrane proteins called solute carrier organic anion (SLCO). Therefore, the OATPs can be also called SLC21 [42].

MRP1 and MRP3 Transporters

The MRP1 protein (ABCC1) was the first MRP transporter to be identified. The molecular weight of MRP1 is 190 kDa and it is present in almost all mammalian cells, indicating its crucial role in cellular transports [38]. MRP1 expression is the most frequent cause of development of drug resistance in cancer cells.

The function of MRP1 in hepatocytes is also linked with glutathione transport. MRP1 is capable to transport both glutathione and glutathione conjugates into the extracellular space. A number of different mechanisms exert in glutathione transportation, including direct transport of GSH, co-transport of GSH with another substrate, or transport of a substrate in presence of GSH, without transport of GSH [40, 43].

The structure of the MRP3 transporter (ABCC3) is similar to that of MRP1. The similarity is approximately 58%, based on the amino acid sequences [39]. In hepatocytes, MRP3 occurs at the basolateral membrane together with MRP1. In addition to liver, the

MRP3 protein is also expressed in the lungs and the kidneys [44]. Interestingly, MRP3 is capable to transport only glutathione conjugates [40].

MRP2 Transporter

The MRP2 transporter (ABCC2) is also called Canalicular Multispecific Organic Anion Transporter (cMOAT). It is of 50% homology in amino acid sequence, in comparison with MRP1. Based on their structures, MRP2 and MRP1 have similar function since they can transport both glutathione and glutathione conjugates [40]. The topical difference between MRP2 and MRP1 is in their cellular localization. MRP2 is localized only at the apical membrane of polarized cells, at the canalicular membrane of hepatocytes. The higher protein expression of MRP2 can be also observed in the kidney, the brain and the enterocytes [44]. MRP2 has an essential role in metabolism of bilirubin; its secretion into the bile is ensured by MRP2. When secretion of glutathione and glutathione conjugates into the bile is impaired, the function of MRP2 can be replaced by MRP3 [39, 45, 46].

Other MRP Transporters

In comparison to MRPs 1, 2 and 3, importance of MRPs 4, 5, 6, 7, 8 and 9 seems to be rather low. The significance of MRP4 (ABCC4) and MRP5 (ABCC5) is higher in comparison to MRPs 6, 7, 8, and 9, because they are structurally similar to each other and their function in glutathione transport seems to be different from that of MRPs 6, 7, 8, and 9 [39]. Localization of MRP4 is similar to that of MRP3 [44]. In comparison to liver, where their presence was detected at the basolateral membrane of the hepatocytes, the highest MRP4 expression was found in the proximal tubular cells in the kidney [47]. In contrast, MRP5 expression seems to be strictly limited to cells of the gastrointestinal tract. The function of MRP4 and MRP5 is associated with the transport of glutathione and glutathione conjugates. In contrast to MRPs 1, 2 and 3, MRP4 and MRP5 are also enabled transport of cyclic nucleotides and their analogs [39].

OATP Transporters

OATPs are the family of organic anion transporters. They play a role in the transport of some classes of drugs across the cell membrane, particularly in the liver and the kidney. In the liver, OATPs are expressed on the basolateral membrane of hepatocytes, transporting compounds into the hepatocytes. The most well characterized human OATPs are OATP1A2, OATP1B1, OATP1B3 and OATP2B1. Their function is linked to the exchange of intracellular glutathione with an extracellular substrate. Therefore, the electrochemical energy of glutathione, which has intracellular levels more than 1000 times higher inside the cell in comparison to the extracellular space levels, is used for the transport of a substrate into the cell [39, 40]. The OATP transporters are present in most human tissues, including the liver, kidney and brain [48].

Membrane Transporters in the Kidney

In addition to liver, kidneys also play an essential role in biotransformation and excretion of the biotransformation products. Especially, the proximal tubular cells possess a number of membrane systems for intra- and extracellular transport of products of xenobiotic biotransformations. Because the glutathione conjugates are very polar anions, they can be easily transported through the basolateral membrane into the tubular cells from the blood, or through the apical membrane from the cells into the urine [29].

The basolateral membrane of the proximal tubular cells contains multi-specific organic anion transporters (OAT), which belong to the family of solute carrier proteins 22 (SLC22). The transport of substances through these carriers into the cells is generally dependent on the sodium gradient, but it can also be sodium-independent [49]. In humans, OAT1, OAT2 and OAT3 were detected in the proximal tubular cells. However, the transport of organic anions, including glutathione and glutathione conjugates, is ensured only by OATs 1 and 3 [50-53].

The efflux of organic anions from the proximal tubular cells has not been found to be associated with specific proteins. Based on the presence of transport proteins in the apical membrane, and their possible functions in the liver, it can be precluded, however, that OATP and MRP proteins account for such function. The OATP1 protein is expressed in the S3 segment of proximal tubule [49]. The MRP2 and MRP4 proteins are expressed in the apical membrane of the proximal tubular cells; therefore, they may have a role in extracellular transport of glutathione and glutathione conjugates from the cells into the urine [29, 54].

Glutathione-Related Toxicity of Drugs and Xenobiotics

Although glutathione has been considered as an antioxidant and detoxification agent, some studies reported that glutathione conjugation can induce or increase toxicity of the parent compound. It was demonstrated that glutathione conjugation can result in formation of compounds that are directly toxic or can be activated by subsequent reactions. Therefore, a number of enzymes participating in GSH metabolism, i.e., glutathione-*S*-transferases, gamma-glutamyltransferase, cysteine-S-conjugate β-lyases and *N*-acetyltransferases, can play important role also in glutathione-related toxicity.

The roles of the enzymes in glutathione-related toxicity are strongly dependent on the structure of the parent compounds. There are roughly three groups of compounds that can induce toxicity after GSH-conjugation: haloalkanes and haloalkenes; quinones and hydroquinones; and other compounds like epoxides and unsaturated ketones [19]. Glutathione-*S*-transferases play an important role in induction of toxicity in haloalkanes and haloalkenes, epoxides and unsaturated ketones [22]. In addition, activation of these molecules can also involve GGT and β-lyases. Toxicity of quinones and hydroquinones, for example, can be linked with acting of GSTs, GGT and β-lyases as well.

Because the GSH-related toxicity is frequently related to the activity of GGT, producing cysteine conjugates, it is obvious that cysteine conjugates and - in some cases *N*-acetyl-cysteine conjugates - can also have toxic effects. In such cases, dipeptidases can also participate in induction of toxicity. Toxicity of a number of compounds is related to activity of GGT, which was proved by experiments using acivicin, a specific inhibitor of the enzyme. For instance, its use caused a decrease or even disappearance of the toxic effect of 2-bromoquinone [55], or 1,2-dichlorovinyl-glutathione [56, 57]. The nephrotoxic effect of *S*-(1,2-dichlorovinyl)-L-cysteine and *S*-(2-benzothiazolyl)-L-cysteine was found to be dependent on the acting of β-lyases [37].

Localization of toxicity of glutathione-conjugates is directly linked to localization of glutathione-dependent biotransformations. Therefore, the most significant toxic events can be observed in the liver, and even more frequently in the kidney. Capability of the proximal tubular cells to transform and transport the biotransformation products, including glutathione conjugates, are the main reasons of their impairments. The biotransformation products can induce cell damage, death or carcinogenesis, and the toxic changes can be observed in the mitochondria, the cytosol and the nucleus [36]. Another biochemical basis of toxic action of GSH conjugates can be their ability to inhibit specific enzymes of glutathione metabolism, like glutathione reductase or glutathione-*S*-transferases.

Toxicity of Haloalkanes

Haloalkanes are the typical substrates to form glutathione conjugates. Conjugation of haloalkanes is mostly catalyzed by glutathione-*S*-transferases. 1,2-Dichloroethane, the potent mutagenic and carcinogenic compound, is converted to *S*-chloromethyl-glutathione and *S*-hydroxymethyl-glutathione. The formed glutathione conjugates can bind to DNA, which accounts for the mutagenic effects. Another pathway of bioactivation of 1,2-dichloroethane its cytochrome P450-catalyzed oxidation to chloroacetaldehyde, which can react with cellular proteins impairing their function. The same bioactivation pathway was described for 1,2-dibromoethane as well [19]. Their bioactivation causes nephrotoxicity.

1,2-Dibromo-3-chloropropane is a former pesticide, which can undergo similar bioactivation. Oxidation of 1,2-dibromo-3-chloropropane leads to formation of 2-bromopropane, which is conjugated with glutathione to form *S*-(2,3-hydroxypropyl)glutathione [58]. Due to its binding to DNA, it is a potent mutagen and likely accounts for nephrotoxicity of 1,2-dibromo-3-chloropropane and 2-bromopropane [59].

Another group of nephrotoxic compounds consists of polyhaloalkanes. They can form glutathione conjugates, which are cleaved by GGT to form cysteine conjugates. These latter derivatives can undergo additional cleavage by cysteine-S-conjugate β-lyases to form toxic compounds. Therefore, *S*-(1,2-dichlorovinyl)-L-cysteine, *S*-(1,1,2,2-tetrafluoroethyl)-L-cysteine, or similar aromatic compounds *S*-(2-benzothiazolyl)-L-cysteine and *S*-(1,2-dichlorovinyl)-L-cysteine are nephrotoxic [37, 58, 59].

Toxicity of Haloalkenes

Toxicity of haloalkenes is mostly detected as nephrotoxicity. In haloalkenes, cysteine-S-conjugate β-lyases play an essential role in toxic bioactivation of the parent cysteine-containing compounds [36]. The final products of haloalkene biotransformation can have carcinogenic effect [19]. For instance, hexachlorobutadiene and pentafluoropropene form glutathione conjugates, which are nontoxic. After additional biotransformation in the presence of GGT and β-lyases, however, they produce highly toxic thiol intermediates that can covalently bind to biomolecules [59, 60]. These products cause nephrotoxicity manifesting as necrosis in the proximal tubule, predominantly in the S3 segment [58].

Toxicity of Quinones and Hydroquinones

Quinones and hydroquinones are another groups of chemicals capable to induce toxicity after their conjugation with glutathione. When reacted with glutathione, the formed hydroquinone-glutathione conjugates can be involved in redox cycling [61]. Furthermore, redox cycling has also been described without participation of glutathione (Figure 5). Then, the redox cycling products can react with cellular macromolecules, including DNA and proteins, or can directly produce superoxide anion. The formed superoxide can serve as a substrate for reactions producing other, more toxic reactive oxygen species like hydroxyl radical or singlet oxygen, that can cause severe cell impairments [61]. In addition, hydroquinones can induce cell impairments as a result of their direct interaction with biomolecules.

Menadione (Figure 5) is an example of hydroquinone molecules. Its redox cycling has been used to produce superoxide in *in vitro* toxicological studies [62]. 2-bromohydroquinone is another example of toxic hydroquinones. It is a toxic metabolite of 2-bromobenzene, which can be transformed to the respective 2-bromoquinone. It is a reactive electrophile which reacts with glutathione to form mono- or bis-glutathione conjugates. The 2-bromoquinone-GSH conjugate accounts for the nephrotoxic effect of 2-bromobenzene, manifested as impairment of the proximal tubular cells [55, 59, 63, 64].

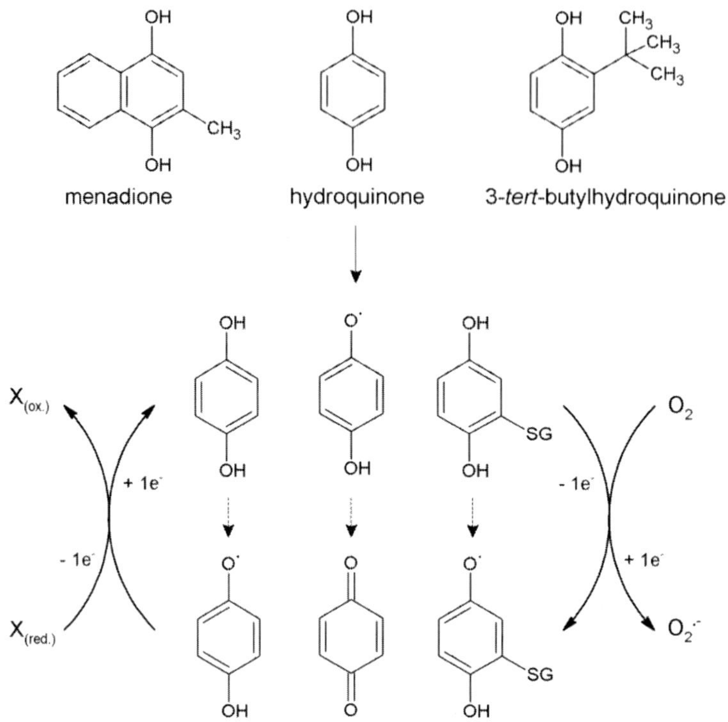

Figure 5. Redox cycling of hydroquinones.

3-*tert*-Butylhydroquinone is a food preservative. It has been considered as a nontoxic compound, but a number of recent studies showed that it can be nephrotoxic after conjugation with glutathione (Figure 6). The conjugates can induce carcinogenicity and cytotoxicity especially in the kidney [65-68].

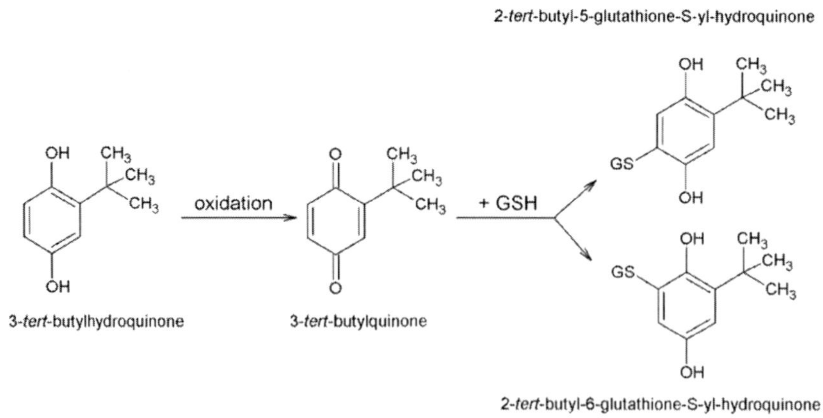

Figure 6. Biotransformation of 3-*tert*-butylhydroquinone.

Toxicity of Aminophenol

p-Aminophenol (PAP) is a former analgesic and antipyretic drug. It was withdrawn due to induction of acute proximal tubular necrosis in humans [69, 70]. *p*-Aminophenol can be oxidized by cytochrome P450 enzymes to form the highly reactive product, 1,4-benzoquinone imine, which can bind to a variety of biomolecules, including DNA and proteins. 1,4-Benzoquinone imine can also react with glutathione, which has been considered as a detoxication reaction. Thus, the mechanism of the toxic effect of PAP in the kidney was attributed to formation of 1,4-benzoquinone imine and glutathione depletion [71]. On the other hand, some studies reported that different mechanisms can also be involved in development of the observed nephrotoxicity.

A number of studies showed that PAP toxicity in kidneys could be induced by the PAP-SG conjugate. Formation of the PAP-SG conjugate followed by subsequent renal biotransformation leads to formation of PAP-SG metabolites that are likely responsible for the observed toxic effects, damaging especially the proximal tubular cells [72-76].

Toxicity of Acetaminophen

Acetaminophen (paracetamol; 4-hydroxyacetanilide; acetaminophenol, APAP) is a commonly used antipyretic and analgesic drug. It was developed as a nontoxic analog of *p*-aminophenol (Figure 7). Its pharmacological effect has been attributed to specific inhibition of cyclooxygenase in the hypothalamus. Acetaminophen is considered to be safe in therapeutic doses, but when overdosed, its use can lead to toxicity, which is the most often cause of acute liver failure (ALF) in the US and the UK [77, 78]. Therefore, study of acetaminophen toxicity is very actual and extensively considered. Despite the number of experiments, recent studies have been reporting that the entire mechanism of APAP toxicity is still unclear [79-84].

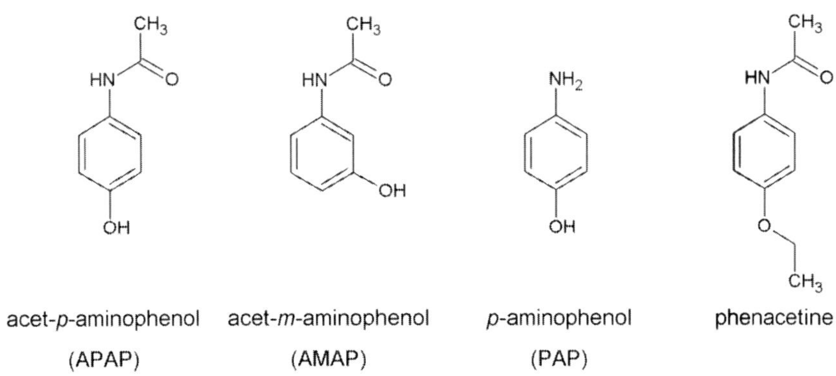

Figure 7. Derivatives of *p*-aminophenol.

Metabolism of Acetaminophen

Acetaminophen is commonly administered orally. It is absorbed in the intestine in 2 h and transported to the liver where is metabolized in three main metabolic pathways. The reactions result increased solubility and elimination of APAP from the organism.

The glucuronidation pathway is the most important in metabolic transformation of therapeutic doses of APAP. It accounts for 80% metabolism of the APAP dose and is catalyzed by glucuronyltransferases. Sulfate conjugation accounts for about 10% metabolic transformation of the therapeutic APAP dose. Approximately 10% of the drug is oxidized by cytochrome P450 enzymes, in humans especially by the CYP2E1, CYP1A2 and CYP3A4 isoenzymes [79, 85, 86]. In addition to the three major biotransformation pathways, other reactions, such as acetylation, hydroxylation or peroxidation can also participate in APAP metabolism [87]. However, those reactions are of minor importance in humans.

Mechanism of Acetaminophen Toxicity

In 1973, Mitchell and his coworkers reported four studies on acetaminophen hepatotoxicity [88-91]. They reported on the mechanism of APAP toxicity in the liver including typical centrilobular necrosis, glutathione depletion, and protective effect of *N*-acetylcysteine. Based on the reports, the fundamental principle of APAP toxicity related to APAP oxidation was described.

After APAP overdose, both glucuronidation and sulfation pathways are saturated and the remaining portion of APAP is metabolized through CYP oxidation (Figure 8). APAP oxidation leads to formation of a highly reactive substance, *N*-acetyl-*p*-benzoquinone imine (NAPQI), which can react with proteins and other biomolecules [79, 92]. Production of NAPQI was found to be linked with the extent of the liver toxicity, serving for development of two hypotheses of the molecular mechanism of APAP toxicity.

NAPQI is a very reactive quinoid-like compound. Its biological half-life was detected to be approximately 10 ms. Glutathione is an abundant thiol in the cells and capable to react with NAPQI. The reaction can be catalyzed by glutathione-S-transferases (GSTs) and the product of the reaction is 3-glutathione-*S*-yl-acetaminophen (APAP-SG) [93-95]. The APAP-SG conjugate has been generally considered as a detoxication product [79, 80, 81, 82], which can be exported from the hepatocytes into bile through the MRP-2 carrier protein [45]. The minor portion of APAP-SG can be transported into the blood through MRP-1 and MRP-3 [96].

The reaction of APAP with GSH can lead to exhaustion of GSH stores and depletion of the GSH level. This is the reason for the characteristic glutathione depletion observed in APAP overdose [88]. Then, NAPQI can react with other intracellular and intramitochondrial molecules forming typical APAP-protein adducts [80, 90]. By this time a number of APAP-protein adducts have been detected, including glutamate dehydrogenase, carbonic anhydrase and glutathione peroxidase [80, 97, 98].

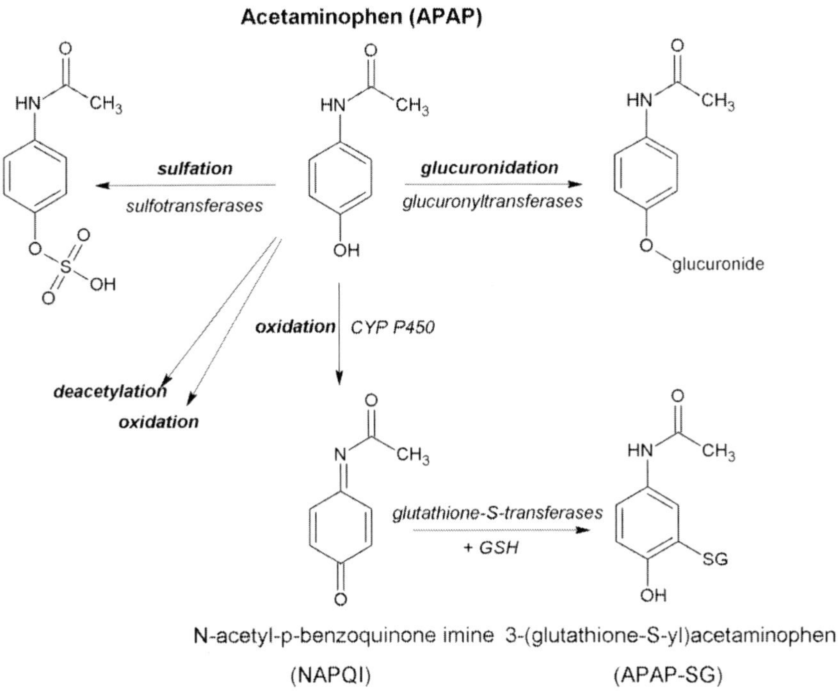

Figure 8. Biotransformation of acetaminophen.

Two hypotheses describing the mechanism of cell death have been formulated regarding the role of NAPQI. The first hypothesis has taken into consideration is likely impairment of protein function due to reaction with NAPQI [83]. However, this hypothesis has been withdrawn. Another hypothesis has been formulated according to the finding of increased oxidative stress in the cells. It hypothesized that increased oxidative stress linked with glutathione depletion, mitochondrial impairment, lipid peroxidation can lead to cell death [79]. That hypothesis seems to be more probable because it is capable to elucidate numerous findings of the experimental studies. However, neither the oxidative hypothesis can provide complete description and explanation of APAP toxicity.

Role of Acetaminophen-Glutathione Conjugate in Toxicity

As noted previously, reaction of NAPQI and GSH results in formation of an APAP-SG conjugate. The glutathione conjugate is predominantly transported from the hepatocytes into the bile. GGT and dipeptidases on the canalicular membrane of hepatocytes and cholangiocytes, as well as on the membrane of enterocytes can cleave the APAP-SG conjugate. The product the APAP-cysteine conjugate (APAP-Cys), which can be transported into the intestine or reabsorbed to the blood and metabolized in the kidney to form the respective mercapturate (Figure 9). Therefore, several APAP metabolites can occur in the blood in addition to APAP and APAP-SG [99].

Although hepatotoxicity is the major manifestation of APAP overdose, in some cases the toxic effect of acetaminophen can also be observed in the kidney. The molecular mechanism of the kidney injury, however, remains rather unclear. It is obvious that, in cases of APAP overdose, hepatotoxicity will appear in most of the cases but nephrotoxicity appears only in 1-2% of the APAP overdose cases. It follows that liver failure can appear without any evidence of nephrotoxicity, but occurrence of nephrotoxicity without any sign of hepatotoxicity is very rare [100-107].

The mechanism of kidney injury caused by APAP overdose has been studied extensively. The role of APAP oxidation in the kidney, inflammatory processes after an overdose and other mechanisms have been considered, but no general explanation of the surprising APAP-induced acute kidney injury has been stated. Therefore, the role of APAP-SG metabolites in nephrotoxicity has been also discussed in a number of studies [108-111]. A likely toxic role of APAP-S-conjugates in induction of kidney injury has been proven by means of use of GGT and cysteine-S-conjugate β-lyase inhibitors, which significantly decreased the APAP-induced nephrotoxicity. In addition, the authors investigated the origin of the APAP-SG conjugate that impairs renal functions. It was found the APAP-SG that impairs the kidney function is predominantly synthesized in the liver [108].

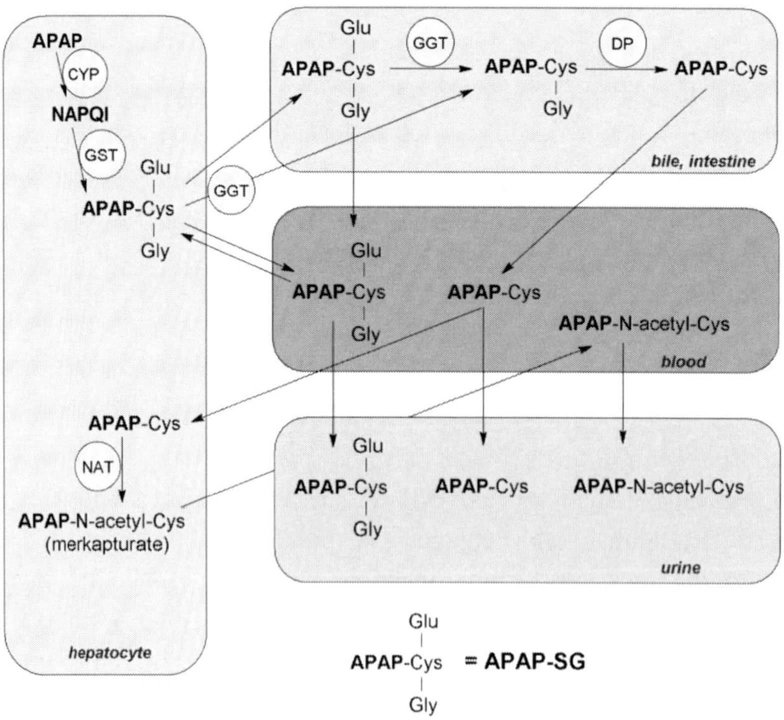

Figure 9. Biotransformation of the acetaminophen-glutathione conjugate (APAP, acetaminophen; CYP, cytochrome C; DP, dipeptidase; GGT, gamma-glutamyltransferase; GST, glutathione-*S*-transferase; NAPQI, *N*-acetyl-*p*-benzoquinone imine; NAT, *N*-acetyltransferase).

The possible toxic role of APAP-SG and other APAP-S-conjugates in APAP-induced cell death can be also supported by the outcomes of other papers. The results of a study showed that increased production of APAP-SG in the liver led to significantly increased hepatotoxicity and nephrotoxicity [109]. Therefore, increased concentration of glutathione in the liver can lead to enhanced APAP toxicity [108, 112, 113]. In addition, decreased production of APAP-SG in two animal studies was linked with decreased APAP toxicity both in the liver and the kidney [114-116]. The possible toxic roles of APAP-SG and APAP-Cys in kidney injury were also reported in other studies [110].

According to the results of the animal studies on the role of APAP-S-conjugates, the toxic role of those substances was also estimated in *in vitro* experiments. The results showed that APAP-SG is capable to induce production of ROS in mitochondria and it is a potent inhibitor of glutathione reductase [117-119]. Based on all the results, the role of the APAP-SG conjugate in APAP toxicity might be significant.

Inhibition of Enzymes Participating in Glutathione Metabolism

The toxicity of glutathione conjugates can also be associated with inhibition of enzymes participating in glutathione metabolism, i.e., inhibition of glutathione reductase and glutathione-*S*-transferases. Glutathione reductase (GR) is an essential antioxidant enzyme. It is the only enzyme in the cell that can reduce glutathione disulfide (GSSG) to glutathione (GSH). Reduction of GSSG is an essential biochemical process for maintenance of reduced glutathione (GSH) stores in the cell. Hence, inhibition of glutathione reductase has a large impact on the physiological reducing environment in the cells [120, 121].

Among others, ethacrynic acid, valproic acid, carmustine, heavy metals and glutathione conjugates are known inhibitors of glutathione reductase activity. Glutathione conjugates inhibiting glutathione reductase activity can be formed by conjugation of glutathione with ethacrynic acid, α,β-unsaturated aldehydes or aminophenol derivatives [122, 123].

Ethacrynic acid belongs to the group of α,β-unsaturated aldehydes. Those compounds can induce cytotoxicity and genotoxicity in treated cells due to the inhibition of glutathione reductase and glutathione-*S*-transferase activities [124, 125]. Ethacrynic is a highly electrophilic substance forming glutathione conjugates which are involved in induction of toxicity. Other α,β-unsaturated aldehydes, e.g., the lipid peroxidation product 4-hydroxynonenal, also form glutathione conjugates that might have toxic effect in the cells [126, 127].

1-Chloro-2,4-dinitrobenzene (CDNB), a substrate for GST, can form *S*-(2,4-dinitrophenyl)-glutathione conjugate. CDNB has been used as a marker of GST activity in cells but its glutathione conjugate has been proven to be a potent inhibitor of glutathione reductase, since the formed *S*-(2,4-dinitrophenyl)-glutathione conjugate can bind to the GSSG binding site of the enzyme [128, 129].

References

[1] Yasui, H.; Hayashi, S.; Sakurai, H. Possible involvement of singlet oxygen species as multiple oxidants in p450 catalytic reactions. *Drug. Metab. Pharmacokinet.* 2005, 20, 1-13.

[2] Dostálek, M. Enzymatický systém cytochromu P450. [Enzymatic system of cytochrome P450.] *Postgraduální medicína*. 2006, 8, 46-54.

[3] Guengerich, F. P. Cytochrome p450 and chemical toxicology. *Chem. Res. Toxicol.* 2008, 21, 70-83.

[4] Neve, E. P.; Ingelman-Sundberg, M. Identification and characterization of a mitochondrial targeting signal in rat cytochrome P450 2E1 (CYP2E1). *J. Biol. Chem.* 2001, 276, 11317-11322.

[5] Bai, J.; Cederbaum, A. I. Overexpression of CYP2E1 in mitochondria sensitizes HepG2 cells to the toxicity caused by depletion of glutathione. *J. Biol. Chem.* 2006, 281, 5128-5136.

[6] Guengerich, F. P. Role of cytochrome P450 enzymes in drug-drug interactions. *Adv. Pharmacol.* 1997, 43, 7-35.

[7] Tseng, E.; Walsky, R. L.; Luzietti, R. A. Jr.; Harris, J. J.; Kosa, R. E.; Goosen, T. C.; Zientek, M. A.; Obach, R. S. Relative contributions of cytochrome CYP3A4 versus CYP3A5 for CYP3A-cleared drugs assessed in vitro using a CYP3A4-selective inactivator (CYP3cide). *Drug. Metab. Dispos.* 2014, 42, 1163-1173.

[8] Deneke, S. M. Thiol-based antioxidants. *Curr. Top. Cell. Regul.* 2000, 36, 151-180.

[9] Sies, H. Glutathione and its role in cellular functions. *Free. Rad. Biol. Med. 1*999, 27, 916-921.

[10] Pastore, A.; Federici, G.; Bertini, E.; Piemonte, F. Analysis of glutathione: implication in redox and detoxification. *Clin. Chim. Act*a. 2003, 333, 19-39.

[11] Forman, H. J.; Torres, M.; Fukuto, *J. Redox signaling. Mol. Cell. Biochem.* 234-235, 49-62.

[12]. Wu, G.; Fang, Y. Z.; Yang, S.; Lupton, J. R.; Turner, N. D. Glutathione metabolism and its implications for health. *J. Nutr.* 2004, 134, 489-492.

[13] Meister, A.; Anderson, M. E. Glutathione. *Annu. Rev. Biochem.* 1983, 52, 711-760.

[14] Lu, S. C. Regulation of hepatic glutathione synthesis: current concepts and controversies. *FASEB J.* 1999, 13, 1169-1183.

[15] Forman, H. J.; Zhang, H.; Rinna, A. Glutathione: overview of its protective roles, measurement, and biosynthesis. *Mol. Aspects Med.* 2009, 30, 1-12.

[16] Blair, I. A. Endogenous glutathione adducts. *Curr. Drug. Metab.* 2006, 7, 853-872.

[17] Hayes, J. D.; Flanagan, J.; Jowsey, I. R. Glutathione transferases. *Annu. Rev. Pharmacol. Toxicol.* 2005, 45, 51-88.

[18] Tang, W.; Lu, A. Y. Metabolic bioactivation and drug-related adverse effects: current status and future directions from a pharmaceutical research perspective. *Drug. Metab. Rev.* 2010, 42, 225-249.

[19] Monks, T. J.; Anders, M. W.; Dekant, W.; Stevens, J. L.; Lau, S. S.; Van Bladeren, P. J. Glutathione conjugate mediated toxicities. *Toxicol Appl. Pharmacol.* 1990, 106, 1-19.

[20] Booth, J.; Boyland, E.; Sims, P. Metabolism of polycyclic compounds. 15. The conversion of naphthalene into a derivative of glutathione by rat-liver slices. *Biochem. J.* 1960, 74, 117-122.

[21] Allocati, N.; Masulli, M.; Di Ilio, C.; Federici, L. Glutathione transferases: substrates, inihibitors and pro-drugs in cancer and neurodegenerative diseases. *Oncogenesis.* 2018, 7, 8.

[22] Hayes, J. D.; Strange, R. C. Glutathione S-transferase polymorphisms and their biological consequences. *Pharmacology.* 2000, 61, 154-166.

[23] Oakley, A. Glutathione transferases: a structural perspective. *Drug. Metab. Rev.* 2011, 43, 138-151.

[24] Mannervik, B.; Board, P. G.; Hayes, J. D.; Listowsky, I.; Pearson, W. R. Nomenclature for mammalian soluble glutathione transferases. *Methods Enzymol.* 2005, 401, 1-8.

[25] Van Bladeren, P. J. Glutathione conjugation as a bioactivation reaction. *Chem. Biol. Interact.* 2000, 129, 61-76.

[26] Henderson, C. J.; Wolf, C. R. Disruption of the glutathione transferase pi class genes. *Methods Enzymol.* 2005, 401, 116-135.

[27] Cummings, B. S.; Lasker, J. M.; Lash, L. H. Expression of glutathione-dependent enzymes and cytochrome P450s in freshly isolated and primary cultures of proximal tubular cells from human kidney. *J. Pharmacol. Exp. Ther.* 2000, 293, 677-685.

[28]. Lash, L. H.; Putt, D. A.; Cai, H. Drug metabolism enzyme expression and activity in primary cultures of human proximal tubular cells. *Toxicology.* 2008, 244, 56-65.

[29] Commandeur, J. N. M.; Stijntjes, G. J.; Vermeulen, N. P. E. Enzymes and transport-systems involved in the formation and disposition of glutathione S-conjugates - Role in bioactivation and detoxication mechanisms of xenobiotics. *Pharmacol. Rev.* 1995, 47, 271-330.

[30] Tate, S. S.; Ross, M. E. Human kidney gamma-glutamyl transpeptidase. Catalytic properties, subunit structure, and localization of the gamma-glutamyl binding site on the light subunit. *J. Biol. Chem.* 1977, 252, 6042-6045.

[31] Meister, A.; Tate, S. S. Glutathione and related gamma-glutamyl compounds: biosynthesis and utilization. *Annu. Rev. Biochem.* 1976, 45, 559-604.

[32] Sakamuro, D.; Yamazoe, M.; Matsuda, Y.; Kangawa, K.; Taniguchi, N.; Matsuo, H.; Yoshikawa, H.; Ogasawara, N. The primary structure of human gamma-glutamyl transpeptidase. *Gene.* 1988, 73, 1-9.

[33] Hinchman, C. A.; Ballatori, N. Glutathione-degrading capacities of liver and kidney in different species. *Biochem. Pharmacol.* 1990, 40, 1131-1135.

[34] Endou, H.; Shimada, H.; Koseki, C.; Yokokura, Y.; Sakai, F. Distribution and possible functions of gamma-glutamyl-transpeptidase in the kidney. *Nihon Jinzo Gakkai Shi.* 1981, 23, 981-988.

[35] Cooper, A. J.; Krasnikov, B. F.; Niatsetskaya, Z. V.; Pinto, J. T.; Callery, P. S.; Villar, M. T.; Artigues, A.; Bruschi, S. A. Cysteine S-conjugate beta-lyases: important roles in the metabolism of naturally occurring sulfur and selenium-containing compounds, xenobiotics and anticancer agents. *Amino Acids.* 2011, 41, 7-27.

[36] Cooper, A. J.; Pinto, J. T. Cysteine S-conjugate beta-lyases. *Amino Acids.* 2006, 30, 1-15.

[37] Lohr, J. W.; Willsky, G. R.; Acara, M. A. Renal drug metabolism. *Pharmacol. Rev.*. 1998, 50, 107-141.

[38] Haimeur, A.; Conseil, G.; Deeley, R. G.; Cole, S. P. The MRP-related and BCRP/ABCG2 multidrug resistance proteins: biology, substrate specificity and regulation. *Curr. Drug Metab.* 2004, 5, 21-53.

[39] Ballatori, N.; Hammond, C. L.; Cunningham, J. B.; Krance, S. M.; Marchan, R. Molecular mechanisms of reduced glutathione transport: role of the MRP/CFTR/ABCC and OATP/SLC21A families of membrane proteins. *Toxicol. Appl. Pharmacol.* 2005, 204, 238-255.

[40] Bachhawat, A. K.; Thakur, A.; Kaur, J.; Zulkifli, M. Glutathione transporters. *Biochim. Biophys. Acta.* 2013, 1830, 3154-3164.

[41] Shugarts, S.; Benet, L. Z. The role of transporters in the pharmacokinetics of orally administered drugs. *Pharm. Res.* 2009, 26, 2039-2054.

[42] Franke, R. M.; Scherkenbach, L. A.; Sparreboom, A. Pharmacogenetics of the organic anion transporting polypeptide 1A2. *Pharmacogenomics.* 2009, 10, 339-344.

[43] Cole, S. P.; Deeley, R. G. Transport of glutathione and glutathione conjugates by MRP1. *Trends Pharmacol. Sci.* 2006, 27, 438-446.

[44] Kool, M.; De Haas, M.; Scheffer, G. L.; Scheper, R. J.; Van Eijk, M. J.; Juijn, J. A.; Baas, F.; Borst, P. Analysis of expression of cMOAT (MRP2), MRP3, MRP4, and MRP5, homologues of the multidrug resistance-associated protein gene (MRP1), in human cancer cell lines. *Cancer Res.* 1997, 57, 3537-3547.

[45] Chen, C.; Hennig, G. E.; Manautou, J. E. Hepatobiliary excretion of acetaminophen glutathione conjugate and its derivatives in transport-deficient (TR-) hyperbilirubinemic rats. *Drug Metab. Dispos.* 2003, 31, 798-804.

[46]. Silva, V. M.; Thibodeau, M. S.; Chen, C.; Manautou, J. E. Transport deficient (TR-) hyperbilirubinemic rats are resistant to acetaminophen hepatotoxicity. *Biochem. Pharmacol.* 2005, 70, 1832-1839.

[47] Van Aubel, R. A.; Smeets, P. H.; Peters, J. G.; Bindels, R. J.; Russel, F. G. The MRP4/ABCC4 gene encodes a novel apical organic anion transporter in human kidney proximal tubules: putative efflux pump for urinary cAMP and cGMP. *J. Am. Soc. Nephrol.* 2002, 13, 595-603.

[48] Van De Steeg, E.; Stranecky, V.; Hartmannova, H.; Noskova, L.; Hrebicek, M.; Wagenaar, E.; Van Esch, A.; De Waart, D.R.; Oude Elferink, R. P.; Kenworthy, K. E.; Sticova, E.; Al-Edreesi, M.; Knisely, A. S.; Kmoch, S.; Jirsa, M.; Schinkel, A. H. Complete OATP1B1 and OATP1B3 deficiency causes human Rotor syndrome by interrupting conjugated bilirubin reuptake into the liver. *J. Clin. Invest.* 2012, 122, 519-528.

[49] Van Aubel, R. A.; Masereeuw, R.; Russel, F. G. Molecular pharmacology of renal organic anion transporters. *Am. J. Physiol. Renal Physiol.* 2000, 279, F216-232.

[50] Hosoyamada, M.; Sekine, T.; Kanai, Y.; Endou, H. Molecular cloning and functional expression of a multispecific organic anion transporter from human kidney. *Am. J. Physiol.* 1999, 276(1 Pt 2), F122-128.

[51] Cha, S. H.; Sekine, T.; Fukushima, J. I.; Kanai, Y.; Kobayashi, Y.; Goya, T.; Endou, H. Identification and characterization of human organic anion transporter 3 expressing predominantly in the kidney. *Mol. Pharmacol.* 2001, 59, 1277-1286.

[52] Lash, L. H.; Jones, D. P. Renal glutathione transport - Characteristics of the sodium-dependent system in the basal-lateral membrane. *J. Biol. Chem.* 1984, 259, 4508-4514.

[53] Lash, L. H.; Putt, D. A. Renal cellular transport of exogenous glutathione: heterogeneity at physiological and pharmacological concentrations. *Biochem. Pharmacol.* 1999, 58, 897-907.

[54] Lash, L. H. Role of glutathione transport processes in kidney function. *Toxicol. Appl. Pharmacol.* 2005, 204, 329-342.

[55] Monks, T. J.; Lau, S. S.; Highet, R. J.; Gillette, J. R. Glutathione conjugates of 2-bromohydroquinone are nephrotoxic. *Drug Metabol. Disp.* 1985, 13, 553-559.

[56] Lash, L. H.; Jones, D. P. Uptake of the glutathione conjugate S-(1,2-dichlorovinyl)glutathione by renal basal-lateral membrane vesicles and isolated kidney cells. *Mol. Pharmacol.* 1985, 28, 278-282.

[57] Lash, L. H.; Anders, M. W. Cytotoxicity of S-(1,2-dichlorovinyl)glutathione and S-(1,2-dichlorovinyl)-L-cysteine in isolated rat kidney cells. *J. Biol. Chem.* 1986, 261, 13076-13081.

[58] Koob, M.; Dekant, W. Bioactivation of xenobiotics by formation of toxic glutathione conjugates. *Chem.-Biol. Interact.* 1991, 77, 107-136.

[59] Dekant, W. Chemical-induced nephrotoxicity mediated by glutathione S-conjugate formation. *Toxicol. Lett.* 2001, 124, 21-36.

[60] Anders, M. W. Glutathione-dependent bioactivation of haloalkanes and haloalkenes. *Drug Metabol. Rev.* 2004, 36, 583-594.

[61] Bolton, J. L.; Trush, M. A.; Penning, T. M.; Dryhurst, G.; Monks, T. J. Role of quinones in toxicology. *Chem. Res. Toxicol.* 2000, 13, 135-160.
[62] Singh, R.; Czaja, M. J. Regulation of hepatocyte apoptosis by oxidative stress. *J. Gastroenterol. Hepatol.* 2007, 22 Suppl 1, S45-48.
[63] Monks, T. J.; Lau, S. S. The pharmacology and toxicology of polyphenolic-glutathione conjugates. *Annu. Rev. Pharmacol. Toxicol.* 1998, 38, 229-255.
[64] Monks, T. J.; Lau, S. S. Glutathione, gamma-glutamyl transpeptidase, and the mercapturic acid pathway as modulators of 2-bromohydroquinone oxidation. *Toxicol. Appl. Pharmacol.* 1990, 103, 557-563.
[65] Nakamura, Y.; Kumagai, T.; Yoshida, C.; Naito, Y.; Miyamoto, M.; Ohigashi, H.; Osawa, T.; Uchida, K. Pivotal role of electrophilicity in glutathione S-transferase induction by tert-butylhydroquinone. *Biochemistry.* 2003, 42, 4300-4309.
[66] Haenen, H. E.; Bleijlevens, E.; Elzerman, H.; Temmink, J. H.; Koeman, J. H.; Van Bladeren, P. J. Cytotoxicity of 2-tert-butyl hydroquinone glutathione conjugates after apical and basolateral exposure of rat renal proximal tubular cell monolayers. *Toxicol. in Vitro.* 1996, 10, 141-148.
[67] Peters, M. M.; Rivera, M. I.; Jones, T. W.; Monks, T. J.; Lau, S. S. Glutathione conjugates of tert-butyl-hydroquinone, a metabolite of the urinary tract tumor promoter 3-tert-butyl-hydroxyanisole, are toxic to kidney and bladder. *Cancer Res.* 1996, 56, 1006-1011.
[68] Gharavi, N.; Haggarty, S.; El-Kadi, A. O. Chemoprotective and carcinogenic effects of tert-butylhydroquinone and its metabolites. *Curr. Drug Metab.* 2007, 8, 1-7.
[69] Gartland, K. P.; Bonner, F. W.; Timbrell, J. A.; Nicholson, J. K. Biochemical characterisation of para-aminophenol-induced nephrotoxic lesions in the F344 rat. *Arch. Toxicol.* 1989, 63, 97-106.
[70] Newton, J. F.; Kuo, C. H.; Gemborys, M. W.; Mudge, G. H.; Hook, J. B. Nephrotoxicity of p-aminophenol, a metabolite of acetaminophen, in the fischer 344 rat. Toxicol. *Appl. Pharmacol.* 1982, 65, 336-344.
[71]. Circu, M. L.; Aw, T. Y. Glutathione and apoptosis. *Free Rad. Res.* 2008, 42, 689-706.
[72]. Fowler, L. M.; Foster, J. R.; Lock, E. A. Nephrotoxicity of 4-amino-3-S-glutathionylphenol and its modulation by metabolism or transport inhibitors. *Arch. Toxicol.* 1994, 68, 15-23.
[73] Lock, E. A.; Cross, T. J.; Schnellmann, R. G. Studies on the mechanism of 4-aminophenol-induced toxicity to renal proximal tubules. *Human Exp. Toxicol.* 1993, 12, 383-388.
[74] Fowler, L. M.; Foster, J. R.; Lock, E. A. Effect of ascorbic-acid, acivicin and probenecid on the nephrotoxicity of 4-aminophenol in the Fischer-344 rat. *Arch. Toxicol.* 1993, 67, 613-621.

[75] Klos, C.; Koob, M.; Kramer, C.; Dekant, W. p-Aminophenol nephrotoxicity: biosynthesis of toxic glutathione conjugates. *Toxicol. Appl. Pharmacol.* 1992, 115, 98-106.

[76] Shao, R.; Tarloff, J. B. Lack of correlation between para-aminophenol toxicity in vivo and in vitro in female Sprague-Dawley rats. *Fundam. Appl. Toxico*l. 1996, 31, 268-278.

[77] Lee, W. M. Acetaminophen and the U.S. Acute Liver Failure Study Group: Lowering the risks of hepatic failure. *Hepatology.* 2004, 40, 6-9.

[78] Hinson, J. A. Mechanism of acetaminophen-induced liver disease.; In *Drug-induced liver disease,* 3rd ed.; Kaplowitz, N.; DeLeve, L. D. Eds.;. Academic Press: London, UK, 2013., p. 305-330.

[79] Jaeschke, H.; Bajt, M. L. Intracellular signaling mechanisms of acetaminophen-induced liver cell death. *Toxicol. Sci.* 2006, 89, 31-41.

[80] James, L. P.; Mayeux, P. R.; Hinson, J. A. Acetaminophen-induced hepatotoxicity. *Drug. Metab. Dispos.* 2003, 31, 1499-1506.

[81]. Mcgill, M. R.; Williams, C. D.; Xie, Y.; Ramachandran, A.; Jaeschke, H. Acetaminophen-induced liver injury in rats and mice: comparison of protein adducts, mitochondrial dysfunction, and oxidative stress in the mechanism of toxicity. *Toxicol. Appl. Pharmacol.* 2012, 264, 387-394.

[82] Jaeschke, H.; Mcgill, M. R.; Williams, C. D.; Ramachandran, A. Current issues with acetaminophen hepatotoxicity--a clinically relevant model to test the efficacy of natural products. *Life Sci.* 2011, 88, 737-745.

[83] Hinson, J. A.; Roberts, D. W.; James, L. P. Mechanisms of acetaminophen-induced liver necrosis. *Handb. Exp. Pharmacol.* 2010, 369-405.

[84] McGill, M. R.; Jaeschke, H. Metabolism and disposition of acetaminophen: Recent advances in relation to hepatotoxicity and diagnosis. *Pharm. Res.* 2013, 30, 2174-2187.

[85] Patten, C. J.; Thomas, P. E.; Guy, R. L.; Lee, M.; Gonzalez, F. J.; Guengerich, F. P.; Yang, C. S. Cytochrome P450 enzymes involved in acetaminophen activation by rat and human liver microsomes and their kinetics. *Chem. Res. Toxicol.* 1993, 6, 511-518.

[86] Thummel, K. E.; Lee, C. A.; Kunze, K. L.; Nelson, S. D.; Slattery, J. T. Oxidation of acetaminophen to N-acetyl-p-aminobenzoquinone imine by human CYP3A4. *Biochem. Pharmacol.* 1993, 45, 1563-1569.

[87] Bessems, J. G.; Vermeulen, N. P. Paracetamol (acetaminophen)-induced toxicity: molecular and biochemical mechanisms, analogs and protective approaches. *Crit. Rev. Toxicol.* 2001, 31, 55-138.

[88] Mitchell, J. R.; Jollow, D. J.; Potter, W. Z.; Davis, D. C.; Gillette, J. R.; Brodie, B. B. Acetaminophen-induced hepatic necrosis. I. Role of drug metabolism. *J. Pharmacol. Exp. Ther.* 1973, 187, 185-194.

[89] Mitchell, J. R.; Jollow, D. J.; Potter, W. Z.; Gillette, J. R.; Brodie, B. B. Acetaminophen-induced hepatic necrosis. IV. Protective role of glutathione. *J. Pharmacol. Exp. Ther.* 1973, 187, 211-217.

[90] Jollow, D. J.; Mitchell, J. R.; Potter, W. Z.; Davis, D. C.; Gillette, J. R.; Brodie, B. B. Acetaminophen-induced hepatic necrosis. II. Role of covalent binding in vivo. *J. Pharmacol. Exp. Ther.* 1973, 187, 195-202.

[91] Potter, W. Z.; Davis, D. C.; Mitchell, J. R.; Jollow, D. J.; Gillette, J. R.; Brodie, B. B. Acetaminophen-induced hepatic necrosis. 3. Cytochrome P-450-mediated covalent binding in vitro. *J. Pharmacol. Exp. Ther.* 1973, 187, 203-210.

[92] Potter, D. W.; Hinson, J. A. Mechanisms of acetaminophen oxidation to N-acetyl-p-benzoquinone imine by horseradish peroxidase and cytochrome P-450. *J. Biol. Chem.* 1987, 262, 966-973.

[93] Hinson, J. A.; Monks, T. J.; Hong, M.; Highet, R. J.; Pohl, L. R. 3-(glutathion-S-yl)acetaminophen: a biliary metabolite of acetaminophen. *Drug. Metab. Dispos.* 1982, 10, 47-50.

[94] Iida, S.; Mizuma, T.; Sakuma, N.; Hayashi, M.; Awazu, S. Transport of acetaminophen conjugates in isolated rat hepatocytes. *Drug. Metab. Dispos.* 1989, 17, 341-344.

[95] Dahlin, D. C.; Miwa, G. T.; Lu, A. Y.; Nelson, S. D. N-acetyl-p-benzoquinone imine: a cytochrome P-450-mediated oxidation product of acetaminophen. *Proc. Natl. Acad. Sci. USA.* 1984, 81, 1327-1331.

[96] Manautou, J. E.; De Waart, D. R.; Kunne, C.; Zelcer, N.; Goedken, M.; Borst, P.; Elferink, R. O. Altered disposition of acetaminophen in mice with a disruption of the Mrp3 gene. *Hepatology.* 2005, 42, 1091-1098.

[97] Qiu, Y.; Benet, L.Z.; Burlingame, A.L. Identification of the hepatic protein targets of reactive metabolites of acetaminophen in vivo in mice using two-dimensional gel electrophoresis and mass spectrometry. *J. Biol. Chem.* 1998, 273, 17940-17953.

[98]. Halmes, N. C.; Hinson, J. A.; Martin, B. M.; Pumford, N. R. Glutamate dehydrogenase covalently binds to a reactive metabolite of acetaminophen. *Chem. Res. Toxicol.* 1996, 9, 541-546.

[99] Mcgill, M. R.; Lebofsky, M.; Norris, H. R.; Slawson, M. H.; Bajt, M. L.; Xie, Y.; Williams, C. D.; Wilkins, D. G.; Rollins, D. E.; Jaeschke, H. Plasma and liver acetaminophen-protein adduct levels in mice after acetaminophen treatment: dose-response, mechanisms, and clinical implications. *Toxicol. Appl. Pharmacol.* 2013, 269, 240-249.

[100] Mour, G.; Feinfeld, D.A.; Caraccio, T.; Mcguigan, M. Acute renal dysfunction in acetaminophen poisoning. *Ren. Fail.* 2005, 27, 381-383.

[101] Mazer, M.; Perrone, J. Acetaminophen-induced nephrotoxicity: pathophysiology, clinical manifestations, and management. *J. Med. Toxicol.* 2008, 4, 2-6.

[102] Loh, C. S.; Ponampalam, R. Nephrotoxicity associated with acute paracetamol overdose: a case report and review of the literature. *Hong Kong J. Emerg. Med.* 2006, 13, 105-110.

[103] Kleinman, J. G.; Breitenfield, R. V.; Roth, D. A. Acute renal failure associated with acetaminophen ingestion: report of a case and review of the literature. *Clin. Nephrol.* 1980, 14, 201-205.

[104] Blakely, P.; Mcdonald, B. R. Acute-renal-failure due to acetaminophen ingestion - a case-report and review of the literature. *J. Am. Soc. Nephrol.* 1995, 6, 48-53.

[105] Campbell, N. R. C.; Baylis, B. Renal impairment associated with an acute paracetamol overdose in the absence of hepatotoxicity. *Postgrad. Med. J.* 1992, 68, 116-118.

[106] Prescott, L. F.; Proudfoot, A. T.; Cregeen, R. J. Paracetamol-induced acute-renal-failure in the absence of fulminant liver-damage. *British Med. J.* 1982, 284, 421-422.

[107] Cobden, I.; Record, C. O.; Ward, M. K.; Kerr, D. N. S. Paracetamol-induced acute-renal-failure in the absence of fulminant liver-damage. *British Med. J.* 1982, 284, 21-22.

[108] Hart, S. G. E.; Wyand, D. S.; Khairallah, E. A.; Cohen, S. D. Acetaminophen nephrotoxicity in the CD-1 mouse .2. Protection by probenecid and AT-125 without diminution of renal covalent binding. *Toxicol. Appl. Pharmacol.* 1996, 136, 161-169.

[109] Rzucidlo, S. J.; Bounous, D. I.; Jones, D. P.; Brackett, B. G. Acute acetaminophen toxicity in transgenic mice with elevated hepatic glutathione. *Vet. Hum. Toxicol.* 2000, 42, 146-150.

[110].Stern, S. T.; Bruno, M. K.; Hennig, G. E.; Horton, R. A.; Roberts, J. C.; Cohen, S. D. Contribution of acetaminophen-cysteine to acetaminophen nephrotoxicity in CD-1 mice: I. Enhancement of acetaminophen nephrotoxicity by acetaminophen-cysteine. *Toxicol. Appl. Pharmacol.* 2005, 202, 151-159.

[111] Stern, S.T.; Bruno, M.K.; Horton, R.A.; Hill, D.W.; Roberts, J.C.; Cohen, S.D. Contribution of acetaminophen-cysteine to acetaminophen nephrotoxicity II. Possible involvement of the gamma-glutamyl cycle. *Toxicol. Appl. Pharmacol.* 2005, 202, 160-171.

[112] Trumper, L.; Monasterolo, L. A.; Elias, M. M. Nephrotoxicity of acetaminophen in male Wistar rats: role of hepatically derived metabolites. *J. Pharmacol. Exp. Ther.* 1996, 279, 548-554.

[113] Trumper, L.; Monasterolo, L. A.; Elias, M. M. Probenecid protects against In vivo acetaminophen-induced nephrotoxicity in male Wistar rats. *J. Pharmacol. Exp. Ther.* 1998, 284, 606-610.

[114] Henderson, C. J.; Wolf, C. R.; Kitteringham, N.; Powell, H.; Otto, D.; Park, B. K. Increased resistance to acetaminophen hepatotoxicity in mice lacking glutathione S-transferase Pi. *Proc. Natl. Acad. Sci. USA.* 2000, 97, 12741-12745.

[115] Kitteringham, N. R.; Powell, H.; Jenkins, R. E.; Hamlett, J.; Lovatt, C.; Elsby, R.; Henderson, C. J.; Wolf, C. R.; Pennington, S. R.; Park, B. K. Protein expression profiling of glutathione S-transferase pi null mice as a strategy to identify potential markers of resistance to paracetamol-induced toxicity in the liver. *Proteomics.* 2003, 3, 191-207.

[116] Arakawa, S.; Maejima, T.; Fujimoto, K.; Yamaguchi, T.; Yagi, M.; Sugiura, T.; Atsumi, R.; Yamazoe, Y. Resistance to acetaminophen-induced hepatotoxicity in glutathione S-transferase Mu 1-null mice. *J. Toxicol. Sci.* 2012, 37, 595-605.

[117] Rousar, T.; Nydlova, E.; Cesla, P.; Stankova, P.; Kucera, O.; Parik, P.; Cervinkova, Z. Purified acetaminophen-glutathione conjugate is able to induce oxidative stress in rat liver mitochondria. *Physiol. Res.* 2012, 61 Suppl 2, S103-109.

[118] Rousar, T.; Parik, P.; Kucera, O.; Bartos, M.; Cervinkova, Z. Glutathione reductase is inhibited by acetaminophen-glutathione conjugate in vitro. *Physiol. Res.* 2010, 59, 225-232.

[119] Nydlova, E.; Vrbova, M.; Cesla, P.; Jankovicova, B.; Ventura, K.; Rousar, T. Comparison of inhibtory effects between acetaminophen-glutathione conjugate and reduced glutathione in human glutathione reductase. *J. Appl. Toxicol.* 2014, 34, 968-973.

[120] Ho, H. Y.; Cheng, M. L.; Chiu, D. T. Y. Glucose-6-phosphate dehydrogenase - from oxidative stress to cellular functions and degenerative diseases. *Redox Report.* 2007, 12, 109-118.

[121] Kamerbeek, N. M.; Van Zwieten, R.; De Boer, M.; Morren, G.; Vuil, H.; Bannink, N.; Lincke, C.; Dolman, K. M.; Becker, K.; Schirmer, R. H.; Gromer, S.; Roos, D. Molecular basis of glutathione reductase deficiency in human blood cells. *Blood.* 2007, 109, 3560-3566.

[122] Karplus, P. A.; Krauthsiegel, R. L.; Schirmer, R. H.; Schulz, G. E. Inhibition of human glutathione-reductase by the nitrosourea drugs 1,3-bis(2-chloroethyl)-1-nitrosourea and 1-(2-chloroethyl)-3-(2-hydroxyethyl)-1-nitrosourea - a crystallographic analysis. *Eur. J. Biochem.* 1988, 171, 193-198.

[123] Jopperi-Davis, K. S.; Park, M. S.; Rogers, L. K.; Backes, C. H.; Lim, I. K.; Smith, C. V. Compartmental inhibition of hepatic glutathione reductase activities by 1,3-bis(2-chloroethyl)-N-nitrosourea (BCNU) in Sprague-Dawley and Fischer-344 rats. *Toxicol. Lett.* 2004, 147, 219-228.

[124] Seefeldt, T.; Zhao, Y.; Chen, W.; Raza, A. S.; Carlson, L.; Herman, J.; Stoebner, A.; Hanson, S.; Foll, R.; Guan, X. Characterization of a novel dithiocarbamate glutathione reductase inhibitor and its use as a tool to modulate intracellular glutathione. *J. Biol. Chem.* 2009, 284, 2729-2737.

[125] Tandogan, B.; Ulusu, N. N. The inhibition kinetics of yeast glutathione reductase by some metal ions. *J. Enzyme Inhib. Med. Chem.* 2007, 22, 489-495.

[126] Ritter, D.; Knebel, J. W.; Aufderheide, M.; Mohr, U. Development of a cell culture model system for routine testing of substances inducing oxidative stress. *Toxicol. in Vitro.* 1999, 13, 745-751.

[127] Vander Jagt, D. L.; Hunsaker, L. A.; Vander Jagt, T. J.; Gomez, M. S.; Gonzales, D. M.; Deck, L. M.; Royer, R. E. Inactivation of glutathione reductase by 4-hydroxynonenal and other endogenous aldehydes. *Biochem. Pharmacol.* 1997, 53, 1133-1140.

[128] Guengerich, F. P. Cytochrome P450 and chemical toxicology. *Chem. Res. Toxicol.* 2008, 21, 70-83.

[129] Bilzer, M.; Krauth-Siegel, R. L.; Schirmer, R. H.; Akerboom, T. P.; Sies, H.; Schulz, G. E. Interaction of a glutathione S-conjugate with glutathione reductase. Kinetic and X-ray crystallographic studies. *Eur. J. Biochem.* 1984, 138, 373-378.

BIOGRAPHICAL SKETCH

Tomáš Roušar

Present Position

Head of cell toxicology, Associated Professor of Physiology

Affiliation

Department of Biological and Biochemical Sciences, Faculty of Chemical Technology, University of Pardubice, University of Pardubice, Czech Republic

Education

Physiology

Research and Professional Experience

Cell toxicology, Drug toxicity, Oxidative stress, Bioenergetics, Hepatotoxicity, Nephrotoxicity

Short Biography

Dr. Roušar has been experienced in studying cell toxicity *in vitro*. The topical attention has been given especially to study the kidney and liver toxicity with focus on oxidative processes participating in cell impairment after treatment with drugs and toxins. In our laboratory, we have developed some assays that can be used for monitoring cellular oxidative status. In addition, we described some new mechanisms causing cell impairment regarding glutathione-related toxicity.

Professional Appointments

Associate Professor of Physiology at University Pardubice (till present),
Associate Professor of Physiology at Charles University in Prague (until 2016)

Publications from the Last 3 Years

1. Majtnerová, P.; Roušar, T. An overview of apoptosis assays detecting DNA fragmentation. *Mol. Biol. Rep.* 2018, 45, 1469-1478.
2. Čapek, J.; Hauschke, M.; Brůčková, L.; Roušar, T. Comparison of glutathione levels measured using optimized monochlorobimane assay with those from ortho-phthalaldehyde assay in intact cells. *J. Pharmacol. Toxicol. Methods.* 2017, 88, 40-45.
3. Hauschke, M.; Roušarová, E.; Flídr, P.; Čapek, J.; Libra, A.; Roušar, T. Neutrophil gelatinase-associated lipocalin production negatively correlates with HK-2 cell impairment: Evaluation of NGAL as a marker of toxicity in HK-2 cells. *Toxicol. In Vitro.* 2017, 39, 52-57.
4. Vrbová, M.; Roušarová, E.; Brůčková, L.; Česla, P.; Roušar T. Characterization of acetaminophen toxicity in human kidney HK-2 cells. *Physiol. Res.* 2016, 65, 627-635.
5. Mezera, V.; Kucera, O.; Moravcova, A.; Peterova, E.; Rousar, T.; Rychtrmoc, D.; Sobotka, O.; Cervinkova, Z. Comparison of acetaminophen toxicity in primary hepatocytes isolated from transgenic mice with different appolipoprotein E alleles. *J. Physiol. Pharmacol.* 2015, 66, 863-873.
6. Hromadkova, L.; Kupcik, R.; Jankovicova, B.; Rousar, T.; Ripova, D.; Bilkova, Z. Difficulties associated with the structural analysis of proteins susceptible to form aggregates: The case of Tau protein as a biomarker of Alzheimer's disease. *J. Sep. Sci.* 2016, 39, 799-807.

SESSION III. IMPLICATIONS-APPLICATIONS

In: Glutathione
Editor: Pál Perjési
ISBN: 978-1-53614-740-7
© 2019 Nova Science Publishers, Inc.

Chapter 9

DETECTION AND QUANTITATION OF GSH AND GSSG

Hui Wang*, Xiaojing Yang* and Jiang Zheng*,†
Wuya College of Innovation, Shenyang Pharmaceutical University, Shenyang, Liaoning, P. R. China;
State Key Laboratory of Functions and Applications of Medicinal Plants, Key Laboratory of Pharmaceutics of Guizhou Province, Guizhou Medical University, Guiyang, P. R. China

ABSTRACT

Among small-molecular bio-thiols, glutathione (GSH) is the most abundant intracellular thiol as well as a major cellular antioxidant. GSH can be readily oxidized to its disulfide (GSSG). The GSH/GSSG system plays a fundamental role in cell homeostasis, and the concentrations of GSH and GSSG and their molar ratio in biological samples are often used as indicators of oxidative stress and injury in toxicological studies. GSH monitoring has been catching public concerns because it serves a vital role in human pathologies, such as acting both as a nucleophile and a reductant to prevent the interactions between critical cellular constituents, especially nucleic acids as well as proteins, and electrophilic or oxidizing species. In this chapter, an overview of various methodologies developed for the detection and/or quantification of GSH and GSSG in a biological system is presented. It starts with a brief description of GSH-mediated biological functions in redox homeostasis, biosynthesis, and pharmacological events. The chapter covers two parts: (1) general preparation methods for assessment of GSH and GSSG in biological samples, including (a) sample collection and storage, which describes the basic pretreatment of various biological samples, such as urine, blood, tissue homogenates and

---

* All authors contributed equally to this work.
† Corresponding Author E-mail: zhengneu@yahoo.com.

cell lysates; (b) a brief review of reduction of disulfides in biological system, along with a collection of varieties of reducing reagents, such as thiol-containing reductants, phosphines, and sodium borohydride; and (c) the significance and challenge of derivatization of GSH and GSSG are provided, and the most commonly used derivatizing reagents and their individual chemical properties are reviewed in terms of their practical advantages and drawbacks; (2) detection techniques without separation, including ultraviolet (UV) spectrophotometry, fluorescence detection (FLD), electrochemiluminescence (ECL), electrochemical detection (ECD), nuclear magnetic resonance (NMR), and surface-enhanced Raman scattering (SERS); and (3) detection techniques with a separation step including high-performance liquid chromatography (HPLC), gas chromatography (GC), and capillary electrophoresis (CE) that coupled with UV, FLD, ECD, mass spectrometry (MS), or MS/MS.

Keywords: GSH, GSSG, sample preparation, sample storage, detection, quantitation, HPLC, CE, MS

INTRODUCTION

Glutathione (γ-glutamylcysteinylglycine, GSH), (Figure 1), a well-known thiol-containing tripeptide of glutamic acid, cysteine, and glycine, is the most abundant intracellular thiol species at millimolar concentrations [1-3]. GSH is synthesized by γ-glutamylcysteine synthetase [4] primarily in the liver which is responsible for supplying 90% of circulating GSH [5]. GSH and its oxidized form glutathione disulfide (GSSG), (Figure 1), play a vital and distinct role in various biological processes such as regulating redox homeostasis and pharmacological events [6]. Due to the high nucleophilicity of the sulfhydryl (-SH) group, GSH also plays an important role in detoxification of a variety of electrophilic xenobiotics and/or free radicals [7, 8].

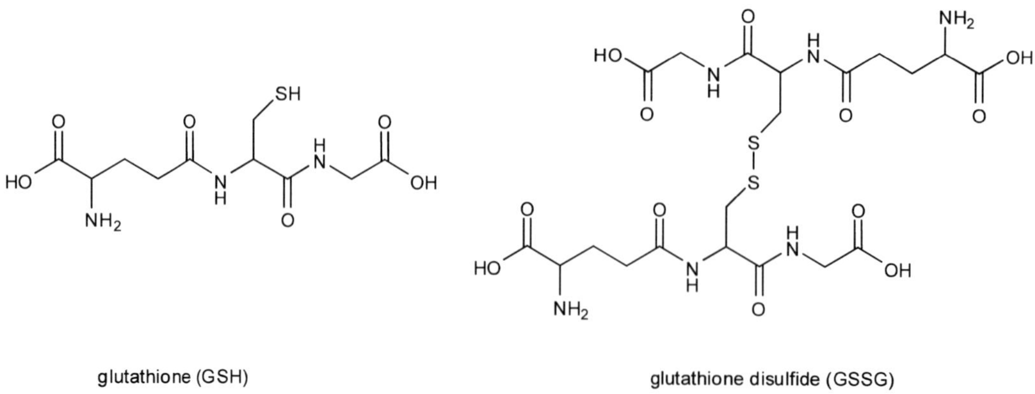

Figure 1. Chemical structures of GSH and GSSG.

In cells, total glutathione refers to GSH and GSSG content (free glutathione forms), along with protein-bound fraction. Under physiological conditions, the GSH/GSSG ratio in the most cells is ~100:1, or higher [9]. Under oxidative conditions, the sulfur atoms of two GSH molecules could donate one electron each to produce glutathione disulfide (GSSG). GSSG would also be reduced by the action of glutathione reductase (GR). In addition, GSH can be bound to proteins, leading to the formation of glutathionyl proteins. The intracellular level of GSH in mammalian cells is in the millimolar range (~1-2 mM in most cells), which varies depending on not only the type of cells, but also during cell proliferation [10, 11]. The concentration of GSH in hepatocytes is ~10 mM, which is the result of the balance between its synthesis and efflux from the cells. In human blood, most of GSH located in the red blood cells (~3 mM per RBC). GSH in mitochondria (10-15% of the total cellular GSH) is similar to that found in the cytosol [12]. Concentration of GSH is much lower in the extracellular matrices, e.g., human plasma, where a substantial amount of -SH is present as free cysteine, whereas GSH concentration is in the range of 2–20 μM [13, 14].

A decrease in cellular GSH concentration and/or in the GSH/GSSG ratio are often considered as evidence of redox imbalance and have been implicated in many human diseases such as diabetes mellitus, malignancy, renal failure, and Alzheimer's disease [15, 16]. Assessment of GSH and GSSG levels and their manipulation in cells or living systems could provide solid evidence to investigate the crucial roles of GSH and GSSG in physiological activities. This fact prompts great efforts to be paid to the detection and quantitation of GSH and/or GSSG in biological systems.

The main challenge of the GSH and GSSG assays lies in the unfavorable physicochemical properties of the analytes. 1) They are very polar, water soluble substances found in low concentrations, which makes their extraction from the biological matrices difficult. 2) Measurement of GSH needs great care, because of its tendency to be oxidized to GSSG. 3) They have no signal generating physico-chemical properties (e.g., strong absorption in the UV–Vis regions or native fluorescence) needed for high detection sensitivity.

The GSH/GSSG ratio is a promising biomarker of oxidative stress. It is very sensitive in response to slight oxidative perturbations [17]. To accuratelely measure the GSH/GSSG ratio, however, is very difficult. Based on previous experience, once the sample is collected, the GSH/GSSG ratio will alter in both directions, i.e., decrease in GSH or GSSG depending on several variables. On one hand, cellular reductases may continue working after sample collection, thus artificially reducing the GSSG concentration. Additionally, when GSH in the sample is depleted, glutathione reductase (GR) can reduce GSSG into GSH quickly, as a consequence of the shift toward the right of the following reversible reaction: GSSG + NADPH$\leftrightarrow$2 GSH + $NADP^+$. On the other hand, auto-oxidation of GSH may increase the GSSG level in the media [18-20]. Thiols can be easily oxidized by molecular oxygen. It is reported that the oxidation could be stimulated by traces of metal ion contaminants of the reagent solutions. Cellular GSH levels are 300- to 800-fold higher than those of GSSG. Therefore, oxidation of a small amount of GSH can induce a dramatic increase in the GSSG levels leading to an inaccurate measure of the GSH/GSSG ratios.

For these reasons, sample preparation and sample handling of GSH containing samples need special attention during thiol determination. Several factors and approaches have been explored in order to obtained accurate estimation of GSH, GSSG and the GSH/GSSG ratio in biological samples.

In recent decades, a variety of detection approaches have been developed for GSH and/or GSSG monitoring, including ultraviolet (UV) spectrophotometry, fluorescence detection (FLD), electrochemiluminescence (ECL), electrochemical detection (ECD), nuclear magnetic resonance (NMR), and surface-enhanced Raman scattering (SERS). In view of the complex components of the biological systems, high-performance liquid chromatography (HPLC), gas chromatography (GC), and capillary electrophoresis (CE) have been widely utilized to sample separation before a UV, FLD, ECD, mass spectrometric (MS), or MS/MS detection. Taking into consideration their unfavorable physicochemical properties, a great many of derivative strategies have been developed, which allow us to yield clean, stable and detectable samples. This article covers these methodologies currently in use for GSH and/or GSSG detection and quantitation.

PRE-ANALYTICAL TREATMENT

Sample Collection and Storage

One of the major problems in GSH and GSSG determination is poor stability of GSH and GSSG in aqueous solutions. Samples must be carefully treated for the accurate quantification of GSH and GSSG, since non-enzymatic oxidation of GSH, thiol-disulfide exchange (Figure 2), other side reactions, and hemolysis can occur during the pretreatment. These reactions may consume GSH or GSSG in the samples. Therefore, samples should be pretreated as soon as possible after their collection. Refrigeration is needed during sample processing and protein precipitation [18, 21, 22]. There must be a short time between sampling and analysis. Bayle and co-workers determined the mean concentrations of GSH in whole blood cooled on ice with delayed centrifugation (0, 1 or 4 h after blood collection). Significantly decreased GSH level was observed after 4 h even in the presence of EDTA [23]. Prolonged isolation time of cells from blood, separation of subcellular organelles, and preparation of tissues are not allowed for an accurate analysis [24].

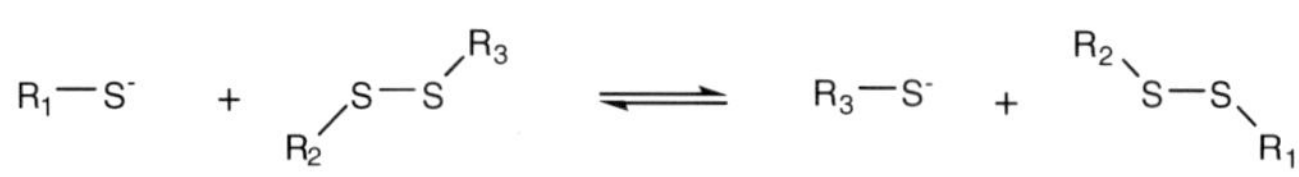

Figure 2. Mechanism of thiol-disulfide exchange reactions.

GSH could be oxidized by molecular oxygen and the reaction could be stimulated by traces of metal ion contaminants in the reagent solutions or on the surface of the vessels, including Fe^{2+}, Cu^{2+}, Fe^{3+}, Ni^{2+}, Co^{2+}, especially copper and iron ions [25]. Therefore, oxygen in sample solutions is suggested to be eliminated by nitrogen or argon bubbling, and metal chelating agents should be added. EDTA is the most commonly used metal chelating agent effectively attenuating thiol oxidation [23, 26, 27]. Other chelating agents, such as 1,10-phenanthroline [28] and diethylenetriaminopentaacetic acid (DETAPAC) [29] have also been suggested.

Oxidation and thiol-disulfide exchange reactions are pH and temperature dependent, therefore lower the pH and the temperature is recommended during sample pretreatment and storage. GSH can easily be oxidized non-enzymatically at pH values greater than 7, and γ-GT, which is responsible for the enzymatic proteolysis of GSH, has an optimum of activity around neutral pH [30, 31]. Therefore, it is recommended to keep the pH below 7 [24]. GSSG is also unstable under physiological conditions. When released into the blood, GSSG is degraded by proteolysis and has a half-life of less than 2 min [32]. This may lead the assay artifacts or data misinterpretation. Acidification could suppress thiol–disulfide exchange, but low pH is often not compatible with other pre-treatments, such as reduction and derivatization. Acid quenching is reversible, undesired oxidation could occur, however, as the pH is raised (e.g., to reduce disulfides or to alkylate thiols).

After collection of whole blood, plasma should be immediately separated from blood cells and frozen. In clinical studies, to minimalize invasive damage to the organism, whole blood or separated red blood cell are the most frequently used samples to measure the ratio of GSH/GSSG to provide valuable information on the redox balance [33].

The most commonly used anticoagulants are EDTA, heparin or sodium citrate. But the influence of different anticoagulants on GSH and GSSG determination vary, depending on the different analytical methods used. The most widely used anticoagulant is EDTA. It serves as an anticoagulant and chelating agent entrapping several transition metals to attenuate thiol oxidation [23, 26, 27].

The hemolytic effect can have significant influence on the GSH and GSSG assays. Balye and co-workers found that hemolysis significantly increases the concentration of GSH in plasma or serum [23]. Hemolysis could cause overestimation of GSH in plasma since erythrocytes contain 500-fold higher GSH levels than plasma [34].

Sample Preparation

The most common methodologies for GSH and GSSG quantification involve several steps described below. 1. Protein precipitation to obtain clear and protein-free supernatants after centrifugation. 2. Quenching the samples with proper agents to trap cellular thiol-disulfide status. This step is to prevent artificial oxidation of GSH and reduction of GSSG during cell lysis and sample pretreatment. 3. Analysis of quenched GSH. 4. Reduction of GSSG with proper agents and analysis of the additionally formed GSH. The details of these steps are described below.

Protein Precipitation

Proteins are the most abundant components in biological samples, and their presence in injected samples may influence accuracy of the method and shorten the lifetime of the instrumentation. Just a few methods (e.g., NMR) are suitable to perform measurements in intact cells or non-deproteinized samples. In most cases, separation of proteins must be performed before analysis. Generally, three methods can be used: 1) acidic method, i.e., 5-sulfosalicylic acid (5-SSA), trifluoroacetic acid (TCA), perchloric acid (PCA), or metaphosphoric acid (MPA) [35, 36]; 2) organic solvent methods, e.g., acetonitrile (ACN) methanol (MeOH) or acetone [27, 37, 38]; and 3) ultrafiltration.

On acid precipitation, GSH and GSSG remain soluble and can be easily separated from protein thiols by centrifugation. The acids could help to suppress oxidation of GSH by keeping low pH and remove proteins from the media. In some early studies, the acidic agents used for this purpose were compared. Oxidation of the SH groups have been observed in PCA- and SSA-treated samples [20, 39-41], while MPA was found to leave plenty of proteins in the treated samples [42]. Therefore, TCA proved to be the most proper acidic agent for this purpose.

Organic solvents are preferable over acidic agents when mass spectrometry (MS) is used for detection [37, 38]. On the other hand, neutralization step and sample-processing time could be saved by dissolving fluorogenic reagent in the organic solvents and treat the samples with the solution.

Quenching Samples

Probably the most critical step when working with redox biology is quenching of samples to trap the cellular thiol–disulfide status. The purpose of this step includes 1) avoiding autooxidation during the sample manipulation; 2) avoiding any cross-reaction with the "new GSH" released from GSSG during analysis of the latter; and 3) avoiding perturbation of the thiol–disulfide equilibrium of the biological samples. Thiol blocking must precede any sample treatment if the aim is GSSG analysis, otherwise autooxidation of GSH is not avoided. There are some strategies to suppress thiol–disulfide exchange: addition of a chelating agent, acidification of the samples, and alkylation of reduced thiol groups. Acid quenching is reversible. When the pH is raised (e.g., to reduce disulfides or to alkylate thiols), however, undesired oxidation can occur. On the contrary, free thiols are blocked irreversibly when alkylating agent is used to quench the thiol-disulfide exchange of the system.

Blocking Agents of Free Thiols

An ideal reagent must react rapidly and specifically with thiols at the lowest possible temperature and slightly acidic pH to prevent oxidation of the analytes leading to artefactual results of the analysis. In this section, a collection of varieties of thiol alkylating agents is provided, and their practical merits and drawbacks are presented.

NEM

N-Ethylmaleimide (NEM) is a widely used reagent for alkylation of thiols. The reaction is a Michael-type addition of the thiols onto the polar double bond of NEM to form thioether derivatives (Figure 3A). The absorption spectrum of NEM has a maximum at 305 nm [43] which is abolished when the reagent decomposes or reacts with a thiol compound. The notable advantages of using NEM are as follows: i) it rapidly reacts with thiols at neutral or slightly acidic pH values with rate constants that are some 3 to 4 orders of magnitude faster than that of iodoacetamide under comparable conditions [44-46]; ii) cells are permeable to NEM and this has encouraged its use for quenching thiols in intact cells; iii) NEM is able to irreversibly inhibit GSSG reductase to suppress the enzymatic-reduction of GSSG. This means that, when NEM is added to any biological sample, it could quickly enter the cells and react with free thiols inside the cells. Furthermore, inhibition of the reactivity of GSSG reductase by NEM could avoid perturbation of the cellular GSH/GSSG ratio. It is suggested that addition of NEM to the biological samples, at the right concentration (e.g., about 40 mM in the whole blood) as soon as possible after the sample collection, can derivatize most of the free thiols. Combining addition of NEM with acidification is not a method of choice; reactivity of NEM would decrease under acidic conditions.

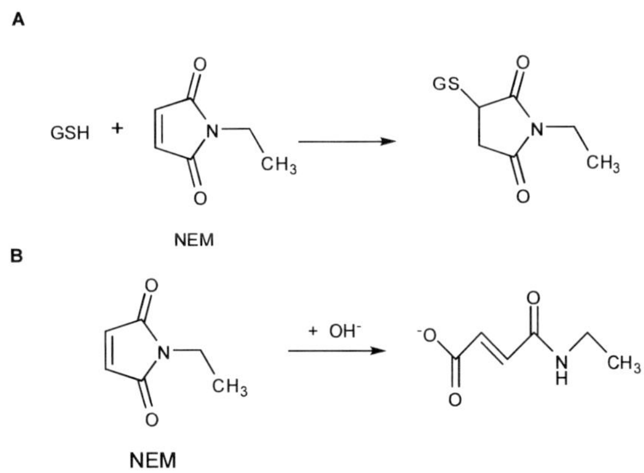

Figure 3. (A) Reaction of GSH with NEM. (B) Hydrolysis of NEM with formation of N-ethylmaleamic acid.

Several chemical issues should be taken into consideration when using NEM. It is reported that NEM is stable at or below pH 6 [47]. Its hydrolytic half-life at pH 7 is approximately 45 h, whereas at pH 9 it is less than 1 h [48]. Its instability in alkaline media is caused by hydrolytic ring cleavage leading to the formation of *N*-ethylmaleamic acid (Figure 3B), which results in loss of thiol reactivity [48, 49]. Furthermore, the GSH-NEM adduct is subject to release the free thiols at or above pH 7 [50]. In addition, under basic conditions NEM can react with other nucleophilic groups, such as the amino group (-NH_2) of lysine, as well.

Another attractive feature of NEM derivatization is that the GSH-NEM adducts can be detected by HPLC-UV/Vis or by LC-MS/MS without any further derivatization [51, 52]. In fact, the maximal absorbance of the GSH-NEM adducts is revealed at 265 nm, which makes it possible to analyze the GSH-NEM conjugates by HPLC-UV.

Iodoacetic Acid and Iodoacetamide

Iodoacetic acid (IAA) and iodoacetamide (IAM) (Figure 4) irreversibly react with thiols in a nucleophilic substitution reaction, leading to formation of the corresponding carboxymethyl or carboxamidomethyl derivatives. IAA and IAM are water-soluble and can be used as 1 M and 0.5 M solutions of pH 8, respectively [53]. They are light sensitive and must be protected from light during both storage and use.

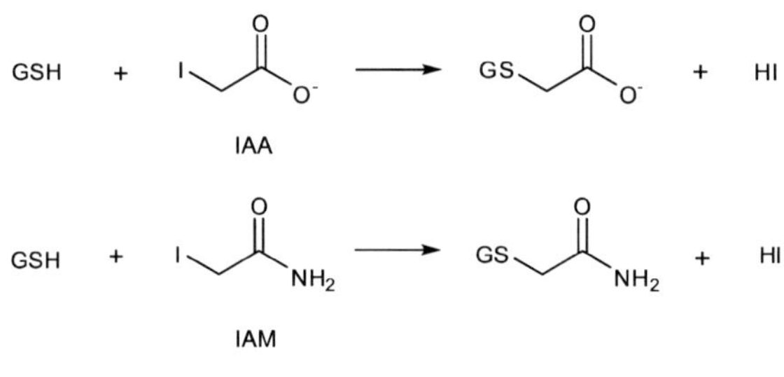

Figure 4. Mechanism of reaction of GSH with IAA or IAM.

There are also some considerations that should be taken into account when using IAA and IAM. When the reaction time is too short, derivatization of some of the sulfhydryl groups could be incomplete. Longer reaction times, however, may result in undesired reactions, i.e., adduction with other nucleophilic amino acid residues such as lysine, histidine, methionine, and tyrosine [54].

The reactions could not be specific for thiols at or above pH 7. The reagents may react slowly with the hydroxyl group of tyrosine to form *O*-carboxamidomethyl derivatives, the imidazole group of histidine to be converted to 3- or 1-carbamidomethyl derivatives [56-58], the ε–amino group of lysine [59, 60], and the *N*-terminus of proteins [57], which bear peripheral polar functional groups. The extent and reaction rate of the side reactions (e.g., reaction with amines) depend on the nucleophilicity of the targeted amino group.

2-VP

Thiols can react with the double bond of 2-vinylpyridine (2-VP) *via* Michael addition, as shown in Figure 5. The reactions are typically performed in neutral or slightly acidic (pH 6-7) solutions using 170 mM of 2-VP [61]. Unlike NEM, 2-VP is not able to inhibit GR. This is the main reason why 2-VP is used to mask thiols for detection of GSSG by the GSH recycling method [62]. However, other cellular oxidoreductases may work during the manipulation which may lead to oxidation of the free GSH.

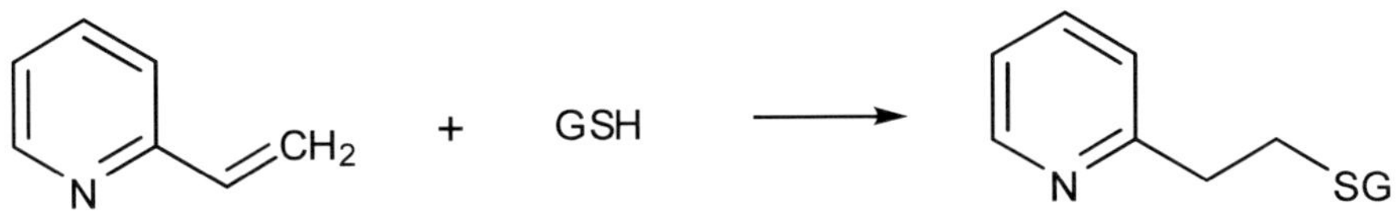

Figure 5. Reaction of 2-VP with GSH.

The main disadvantage of use of 2-VP is that complete masking of GSH takes approximately 1 h at room temperature (the reaction is about 500-fold slower than that of NEM) and it could not permeate the cell membranes. During the longer incubation time GSH can be oxidized yielding reports of erroneously high GSSG/GSH values. Furthermore, specificity of 2-VP to thiols would decrease under alkaline condition [64]. The toxic properties of 2-VP also make it difficult to use.

Practical Consideration

An effective quenching is considered that the alkylating agent could not only completely block the free thiols but inhibit the redox enzymes as well. When GSH is depleted by blocking agents, GR will reduce GSSG into GSH possibly because of perturbation of the redox equilibrium [41]. The reaction conditions should be carefully chosen since high pH and temperature would accelerate autooxidation of GSH. Furthermore, any excess of alkylating agents must be removed before reduction of disulfides to avoid the newly released GSH to be alkylated. The disulfide reducing agents could be considered to consume excess of the alkylating agents if their reactivity is higher than reduction of the disulfide bonds.

Reduction of Disulfides

As previously described, specific measurement of the disulfide bonds generally involves three steps: blocking the free thiol groups, reducing the disulfides, and detecting the newly formed thiols. Reduction of disulfides can be achieved by the following ways: 1. reducing by enzymes, such as GR [64]; 2. electrolysis [36]; and 3. reducing through chemical reactions. The third method could involve using a variety of reducing agents, such as thiol containing reagents (dithiothreitol, dithioerythritol, 2-mercaptoethanol, etc.), several phosphines and sodium borohydride [65]. The reagents and the reaction conditions are discussed in detail.

Reducing Agents

Thiol Containing Reagents

Thiol containing reagents used for reduction of disulfides include dithiothreitol (DTT), dithioerythritol (DTE), dimercaptopropane sulfonic acid (DMPS), 2-mercaptoethane sulfonic acid (MESNA), and 2-mercaptoethanol (ME).

DTT is a superior reducing reagent compared with others, owing to its high conformational propensity to form a six-membered ring with an internal disulfide bond (Figure 6A) [62]. Reduction of a typical disulfide bond proceeds by two sequential thiol–disulfide exchange reactions [66]. The intermediate mixed-disulfide state is unstable and the second thiol of DTT would attack it to close the ring to form oxidized DTT and produce a reduced disulfide bond [45, 67]. The reaction can be performed in phosphate buffer of approximately neutral pH [68]. However, being thiol compound itself, DTT could negatively affect quantitation of GSH through competitively reacting with the thiol derivatizing agents. Therefore, selection of a proper concentration of DTT which is enough to obtain the efficacious disulfide reduction and to avoid the side reaction and interfering peak, is important [69].

Dithionite (DT, sodium hydrosulfite), (Figure 6B) is also used to reduce disulfides. It is worth noting that stoichiometry of GSSG reduction by DTT is different from that of other reducing reagents. The number of moles of released GSH is equal to that of GSSG in the DTT reduction, instead of releasing two GSH molecules from one GSSG when using other reducing reagents.

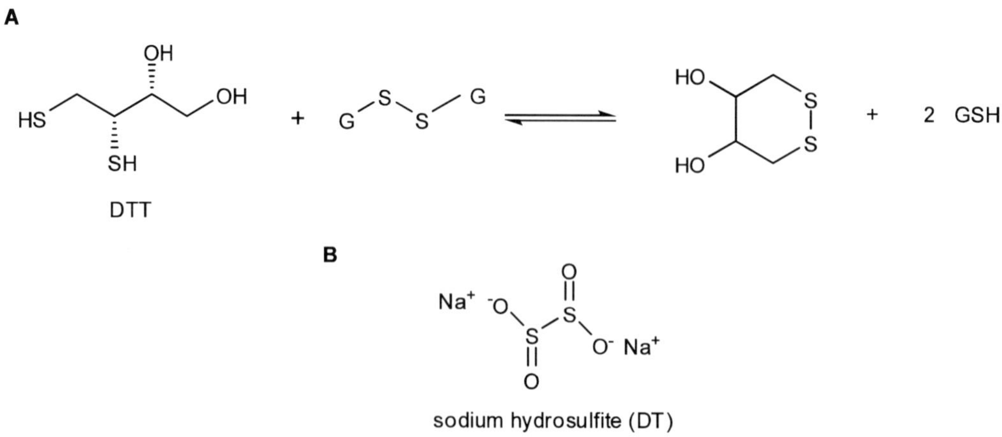

Figure 6. (A) Reaction of DTT with GSSG. (B) Chemical structure of DT.

Phosphines

The commonly used phosphines to reduce disulfides includes tri-*n*-butylphosphine (TBP), triphenylphosphine (TPP), and *tris*-(2-carboxyethyl)phosphine. Their chemical structures and proposed mechanism of the reductive reaction are shown in Figure 7.

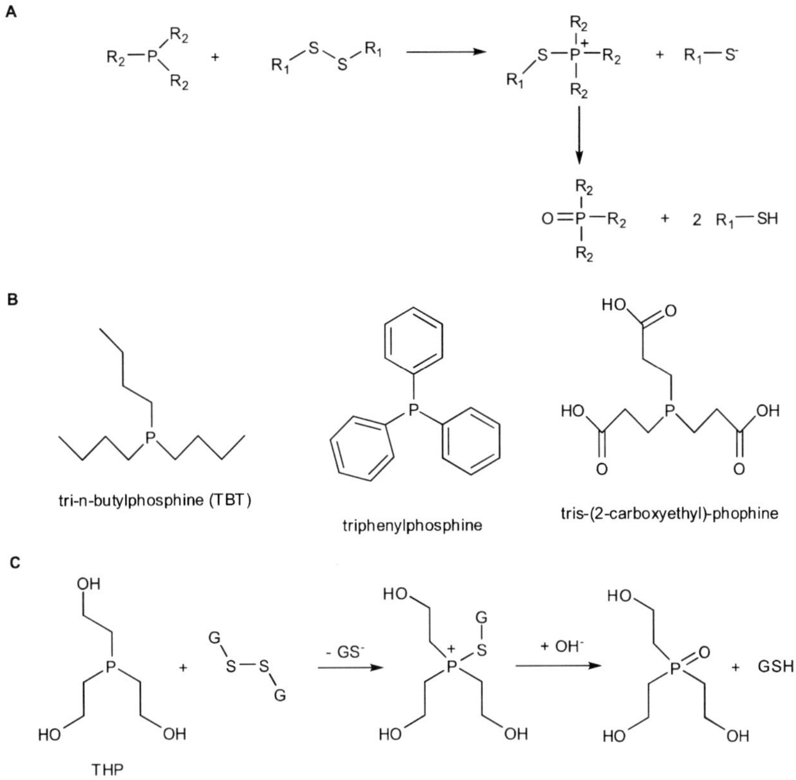

Figure 7. (A) Mechanism of reaction of phosphines with GSSG. (B) Chemical structure of commonly used phosphines. (C) Reaction of THP with GSSG.

The first and rate-limiting step of the reductions is the attack of the disulfide by the phosphine nucleophile to form a thiophosphonium salt (Figure 7A). Next, rapid hydrolysis releases the second thiol fragment and phosphine oxide. TBP and TPP are powerful reductants that need only a slight molar excess to completely reduce the disulfide. Most of the agents do not react with thiol derivatizing agents and do not give rise to gas formation during the reactions [70]. TBP is not used very frequently, because of its explosive property at relatively high concentration [36]. Another phosphine is *tris*-(2-carboxyethyl)phosphine (TCEP), which is highly soluble in water, nonvolatile and relatively stable while exposed to humidity and air [71]. The main disadvantage of TCEP is that it can react with certain disulfide containing derivatizing agents, such as DTNB [72]. Mono-, di-, and trimethyl ester analogs of TCEP are still reactive even at low pH. Esterification of TCEP could improve their lipid solubility and allow them to pass phospholipid bilayers 30 times faster than DTT. Besides, there is another water-soluble member of this family, *tris*(hydroxypropyl)phosphine (THP) (Figure 7C). It is more reactive toward low molecular weight disulfides and surface-exposed protein disulfides than TCEP [72]. However, some researchers reported that phosphines could react quickly with some thiol alkylating agents, such as IAA, IAM and NEM, even under acidic conditions [73-76].

Sodium or Potassium Borohydride

Sodium or potassium borohydride (BH) are kinds of powerful reducing agents, due to the high S-nucleophilicity of the borohydride ion [77]. They are highly reactive and used for the reduction of protein disulfides [78-80]. The main problems that they are unstable in aqueous solution and the resulting hydrogen gas forms foam in the samples during the reaction. Adding phosphoric acid, octanol or hexanol to the borohydride solution could prevent foaming [34]. The main benefit of BH is that excess of the reagent can be easily removed by the addition of acid or acetone.

Practical Consideration

It is notable that the reagents listed above are susceptible to oxidation. Therefore, the storage of the agents must be careful, and the working solution should be prepared freshly on a daily basis. If stock solution is used, it should be kept at low temperature and the concentration needs to be confirmed before use.

For many analytical purposes, it is necessary to completely remove the rest of the reducing agent prior to the further derivatization.

Thiol Detection

Derivatization with a suitable reagent could improve stabilization of the free SH and increase sensitivity of spectrophotometric or spectrofluorimetric determinations. Furthermore, derivatization could increase hydrophobicity of thiols increasing separation efficiency in RP-HPLC measurements. GSH has three functional groups susceptible to derivatization, i.e., the carboxylic, amino and thiol functionalities. The thiol moiety is preferred to be modified for its specificity and for its function as a protective group. Derivatization can be applied before and after the chromatographic separations (i.e., pre-column and post-column labeling). The major advantage of the pre-column derivatization is that the reagents could block the thiol group as soon as possible after sample collection. In general, the reagents that can instantaneously, selectively and quantitatively derivatize thiols under mild conditions are more suitable for thiol labeling. Selection of thiol reagents should address the following aspects: 1. they must form GSH adducts to increase the sensitivity of detection; 2. specificity and chemical reactivity must be high, so that they can react rapidly and specifically with thiols to form stable products in several minutes; 3. stability of the reagents and the resulted derivatives must be adequate; and 4. the reagents and the derivatives should be soluble in aqueous solutions. A high number of reagents for UV, FLD and MS detection have been employed for determination of GSH and GSSG, and their use will be discussed in the following section.

Thiol Detection Agents

Thiol Reagents that Introduce a Chromophore

The reagents used to cover SH protection and improve UV detection are NEM and monoiodoacetic acid (MIAA) (Figure 8). Both reagents can react with SH within a few minutes to yield the respective thioethers [37, 81-83]. *N*-(2,4-dinitrophenylaminoethyl)-maleimide, eosin-5-maleimide (EMA), and pyrenyl-maleimide are NEM analogs (Figure 8) designed to introduce a chromophore into GSH and related ATs [84-86].

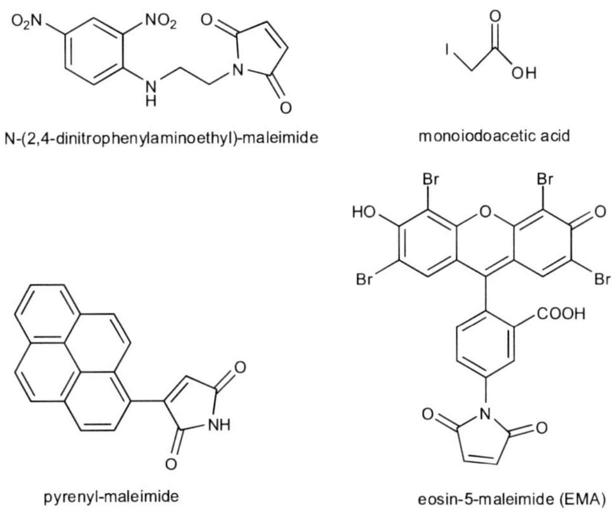

Figure 8. Chemical structures of monoiodoacetic acid and NEM analogs.

5,5'-Dithiobis(2-nitrobenzoic acid) (DTNB), also known as Ellman's reagent, invented by Ellman in 1959 (Figure 9), is the most commonly used thiol detection agent [87]. DTNB can react with thiols in biological samples in a thiol–disulfide exchange reaction with formation of 2-nitro-5-mercaptobenzoic acid (TNB), which possesses an absorption maximum at 412 nm ($\varepsilon = 14{,}150\ M^{-1}\ cm^{-1}$ at pH 7.3) [88] and oxidized glutathione-TNB adduct. TNB is a good leaving group because of the electron withdrawing character of the nitro group. Using excess of DTNB (commonly 1 mM), derivatization reactions could complete rapidly, in 1 min at pH 7.3. The reactions have the great benefits of being fast, irreversible, inexpensive, precise and accurate. DTNB is an especially proper reagent for quantitation of low molecular weight thiols, and DTNB-based colorimetric methods that enable accurate measurement of GSH in most tissues.

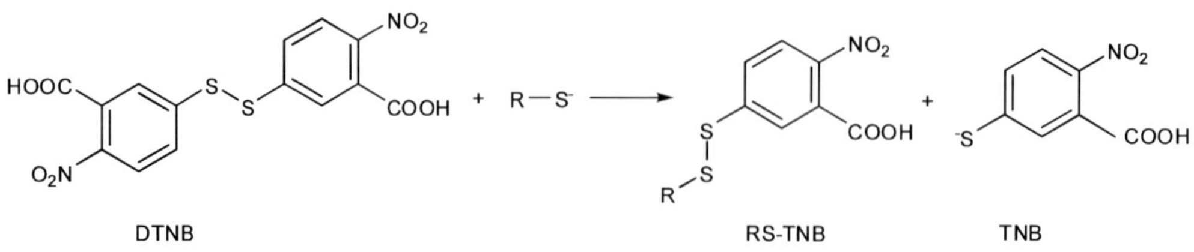

Figure 9. The reaction of DTNB with thiols.

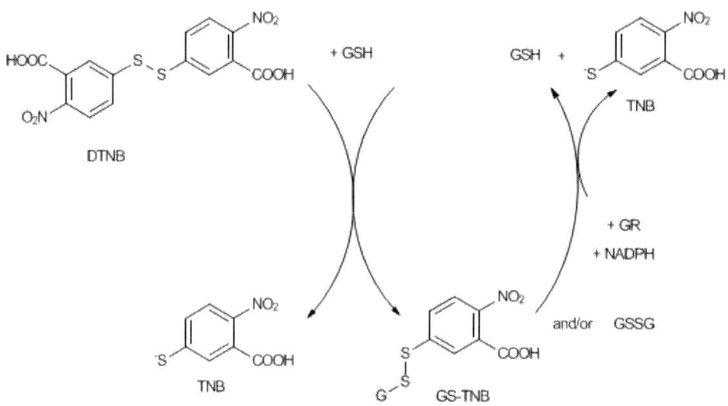

Figure 10. The enzymatic recycling of GSH from GSSG by glutathione reductase (GR) in the presence of NADPH.

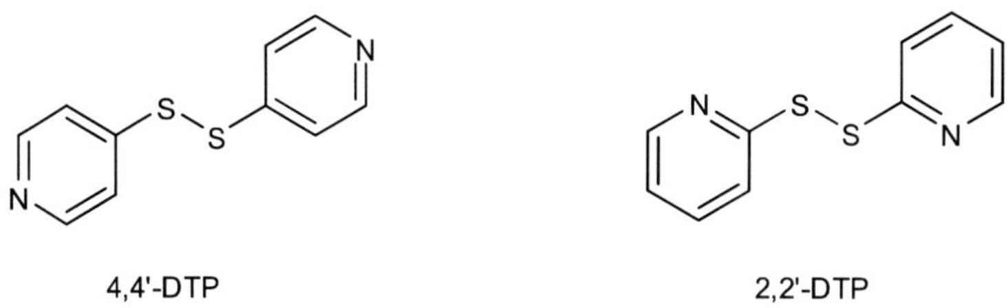

Figure 11. Chemical structure of 4,4'-DTP and 2,2'-DTP.

The method using DTNB as detecting agent is named GSH recycling assay. It is the most popular method to measure GSH by spectrophotometry, because of its high specificity and sensitivity [21]. Figure 10 shows the outline of the process. DTNB reacts with GSH to form GS-TNB and TNB. The rate of formation of TNB is proportional to the concentration of GSH in the sample. Then, the GS–TNB is reduced by GR in the presence of NADPH, recycling GSH back into the reaction. The amount of GSH measured represents the sum of reduced and oxidized GSH in the sample ($[GSH]_{total} = [GSH] + 2 \times [GSSG]$). The rate of change in absorbance ($\Delta A_{412\ nm}\ min^{-1}$) is linearly proportional to the total concentration of GSH. The difference between the total and initial GSH values is used to determine the GSSG concentration in the sample.

As an alternative, 4,4'-dithiopyridine (4,4'-DTP) and 2,2'-ditiopyridine (2,2'-DTP) (Figure 11) are used in place of Ellman's reagent [89]. They react with GSH *via* thiol-disulfide exchange with forming mixed disulfides. Their reactivity towards SH is much higher than that of DTNB, thus pH of the reaction can be set lower (pH>4.5) than pH 8 [90].

Thiol Reagents That Introduce a Fluorophore

Precolumn derivatization of thiols with fluorogenic reagents followed by chromatographic or electrophoretic separation has become popular, because of its high selectivity and sensitivity [91]. There are some frequently used fluorogenic reagents and the selection should address the following aspects. 1. It must form GSH (thiol) adducts with high enough fluorescence yield to evaluate GSH concentrations at the picomolar or lower range. 2. It must be nonfluorescent to avoid interfering measurement of GSH (thiol) conjugates. There are chemically different kinds of reagents available, including maleimides, halides, bimanes, halogeno-benzofurazans, *ortho*-phthalaldehyde, etc.

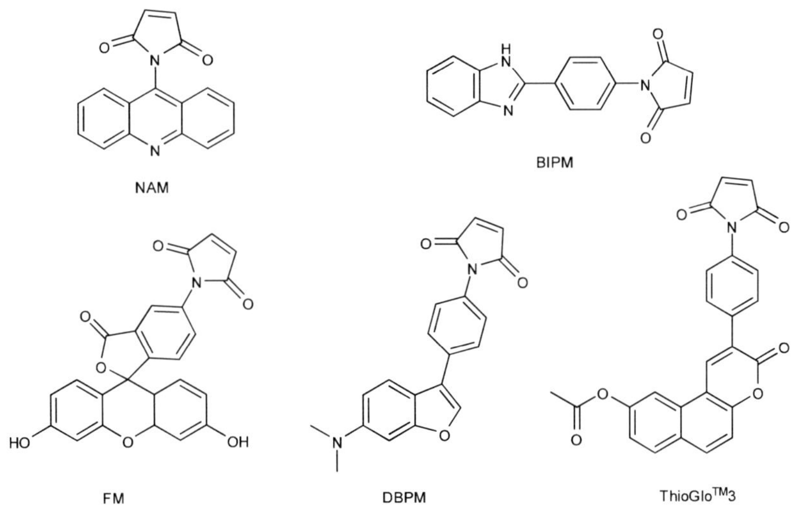

Figure 12. Chemical structure of typical maleimide-type reagents.

Maleimides

Maleimide-type reagents, e.g., *N*-(9-acridinyl)-maleimide (NAM), *N*-(*p*-(2-(6-dimethylamino)benzofuranyl)phenyl)-maleimide (DBPM), *N*-[*p*-(2-benzimidazolyl) phenyl]-maleimide (BIPM) and fluorescein-5-maleimide (FM), are frequently used reagents for derivatization of thiols. Their chemical structures are shown in Figure 12. Their reactivities towards thiols are based on the maleimide moiety. Therefore, the reaction conditions and the derivatization efficiencies are comparable. The fluorescence properties significantly depend on the structure of fluorophors connected to the maleimide moiety.

The maleimide derivative ThioGlo^{TM}3 (9-acetoxy 2-(4-(2,5-dihydro-2,5-dioxo-1H-pyrrol-1-yl)phenyl)-3-oxo-3H-naphtho[2,1-b]pyran) (Figure 12) has also been employed for fluorescent labeling. It shows high affinity towards SH groups, low baseline fluorescence, and high selectivity along with high yield of fluorescent products [64, 92].

Halogeno-Benzofurazans

Halogeno-benzofurazans (Figure 13), such as ammonium 7-fluoro-2,1,3-benzoxadiazole-4-sulfonate (SBD-F) [93-95], 4-(aminosulfonyl)-7-fluoro-2,1,3-benzoxadiazole (ABD-F) [96, 97] and 4-(N,N-dimethylaminosulfonyl)-7-fluoro-2,1,3-benzoxadiazole (DBDF) [98] are reactive fluorescent halides used as thiol labeling agents. These reagents have similar chemical properties and their thiol reactive sites are the same. Excitation and emission maxima of the derivatives obtained from SBD-F, ABD-F and DBD-F are very similar (appr. ex. 380 nm, em. 540 nm).

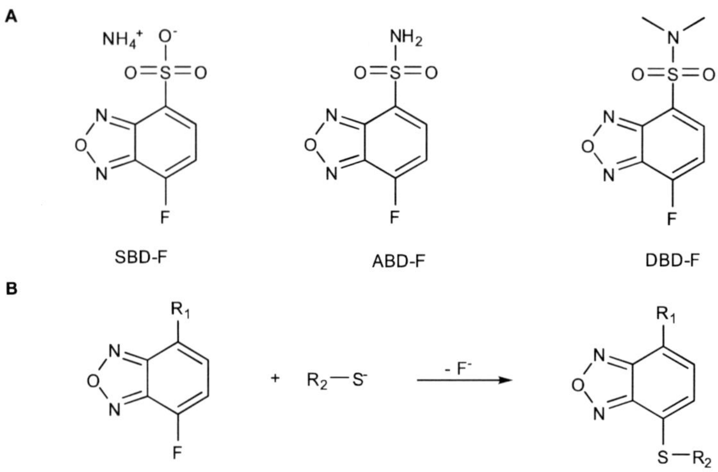

Figure 13. (A) Chemical structures of typical halogeno-benzofurazans. (B) Mechanism of reaction of halogeno-benzofurazans with thiols.

Furthermore, these reagents can be used for labeling thiols in the presence of some types of reducing agents, since they could not cross-react with phosphine-type reductant agents, such as TBP and TCEP [99-101]. They are nonfluorescent, water soluble, and the thiol adducts are stable [102, 103]. More importantly, the resulted derivatives show strong fluorescence in rather long wavelength regions. Since they are water soluble, they can be used for determination of thiols in biological samples as well. The main problem of the reagents is their low-reactivity. Use of the reagents needs high pH and temperature (pH 9.5 and 60°C) and long (1 h) incubation times [94]. The harsh reaction conditions may lead to reoxidation of the reduced GSH. Compared to SBD-F, ABD-F could react relatively fast and quantitatively under mild conditions [104, 105].

Bimanes

Monobromobimane (mBBr) (Figure 14) is a weakly fluorescent reagent which reacts with thiols to form highly fluorescent thioethers [106]. Using HPLC-FLD after derivatization with mBBr GSH can be determined at pmol levels [107]. The reaction proceeds under mild conditions, such as room temperature and neutral pH. The resulted derivatives emit light in the range of 470-480 nm. One drawback of this reagent is its selfluorescence and fluorescence of its photodegradation products [108]. These products give rise to several reagent peaks, which may interfere with GSH determination. Therefore, the reaction must be carried out in the dark. Another disadvantage of mBBr is its possible cross-reactivity with thiol-based reductants and phosphines [109]. Moreover, mBBr is not specific to GSH and it could react with other thiols of the samples [36, 110].

Monochlorobimane (mBCl) (Figure 15), a chlorine analog of mBBr, is also used for labeling GSH in biosamples [111]. Its chemical reactivity is slightly lower than that of mBBr, because bromide (Br^-) is a better leaving group. The fluorescent products derived from mBCl are the same as the corresponding derivatives of mBBr.

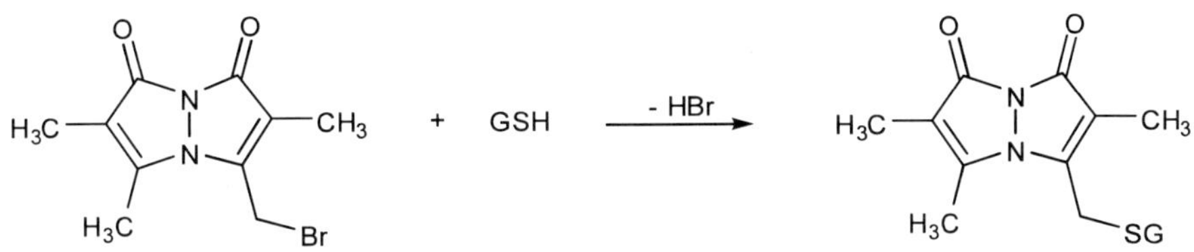

Figure 14. Reaction of mBBr with GSH.

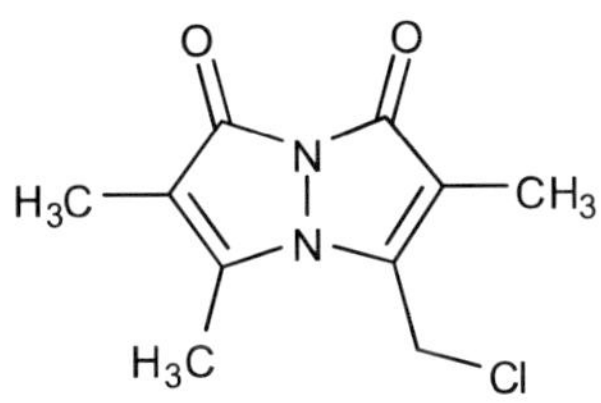

Figure 15. Chemical structure of mBCl.

ortho-Phthalaldehyde

ortho-Phthalaldehyde (OPA) (Figure 16) is a well-known thiol labeling agent. The major advantages of OPA derivatization include mild reaction conditions, widespread experience with its use, and the possible determination of other thiols and amino acids [112, 113]. It is a heterobifunctional reagent that could react with a primary amino group in the presence of a thiol or *vice versa*, with formation of a highly fluorescent isoindole derivative. A co-reagent such as 2-aminoethanol, glycine (Gly), 2-mercaptoethanol, *N*-acetylcysteine (NAC), 3-mercaptopropionic acid or ethanethiol is commonly used to participate in the heterobifunctional reaction to direct the OPA derivatization selectively towards thiol or amino functions [114]. GSH and other aminothiols could react with OPA to produce a tricyclic isoindole derivative without the need for a thiol co-reagent [115, 116]. GSH and GSSG in biological samples (plasma, whole blood, rat hepatocytes) can be determined by OPA and NEM derivatization [117]. Careser and co-workers achieved quantification of total GSH at femtomole level by OPA derivatization without co-reagent [118]. However, Michaelsen et al. found that derivatization of GSH with OPA in the cytosol of red blood cells (RBC) yielded 2 derivatives; addition of *N*-acetylcysteine ethyl ester, however, resulted in formation only a single peak [119]. There are also reports that OPA reacts very rapidly (within 1 min) with *S*-carboxymethylglutathione (formed in the reaction of IAA and GSH) at room temperatures (e.g., 20–25°C) in an autosampler and the derivatives can be immediately injected into HPLC column without needing extraction [120-122].

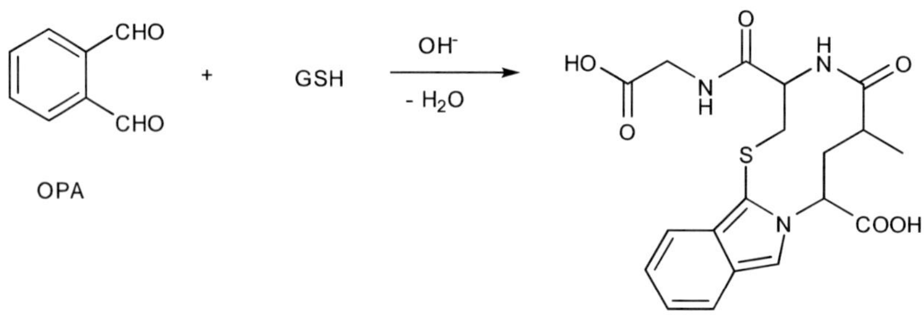

Figure 16. The reaction of OPA with GSH.

Derivatization reaction of GSH with OPA is pH dependent; the optimal pH to obtain the maximal fluorescent yield is between 9.5 and 12 [114]. Although GSSG would be degraded at pH 12.0 to produce GSH, there are methods using post-column derivatization with OPA to analyze fluorescent derivatives of the separated GSH and GSSG [123]. Another attractive feature of the OPA derivatization is stability the produced fluorescent derivatives under various storage conditions.

Naphthalene-2,3-dicarboxaldehyde (NDA) (Figure 17), a reagent with similar structure to OPA, was also used for GSH quantification [124].

Another approach to introduce a fluorescent group is the use of reagents specific for the amino functional group. Among them, *N*-hydroxyuccinimidyl-α-(9-acridine)-acetate (HSAA) and *N*-hydroxysuccimidyl-α-(9-phenanthrene)-acetate (HSMPA) showed specificity and selectivity towards the primary amino group [125].

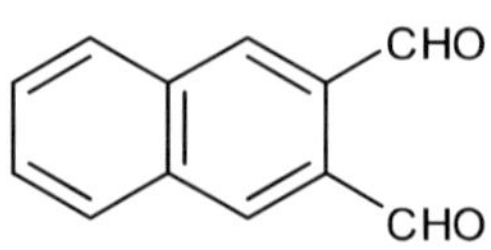

Figure 17. Chemical structure of naphthalene-2,3-dicarboxaldehyde (NDA).

Practical Consideration

The choice of derivatizing reagents is important not only to increase sensitivity of the detection but also for the stabilization of thiols, improvement of chromatographic separation, increase MS ionization, or introduction a charge into the molecule (CE).

DETECTION TECHNIQUES FOR THE DETERMINATION OF GSH AND/OR GSSG

Detection Techniques without Separation

Ultraviolet (UV) Spectrophotometry

The classical spectrophotometric method is based on GSH recycling assay first reported by Tietze [21]. This method takes advantage of the specificity of GR to reduce GSSG to GSH in the presence of NADPH. GSH, in turn reacts with 5,5′-dithiobis-(2-nitrobenzoic acid) (DTNB) to form disulfide GSH-TNB and the yellow TNB, which possesses a maximum of absorbance at 412 nm. The resulting GSH-TNB can be further reduced to GSH and TNB by GR and NADPH or turned to GSSG by direct reaction with the excess GSH existing in the assay. The slope of the absorbance changes of TNB is proportional to the total glutathione (GSH plus GSSG) concentration. The levels of whole blood GSH and GSSG reported here are 1.161 ± 151 and 3.51 ± 2.09 μM, respectively. GSSG is determined following preliminary blocking of GSH with excess of NEM in a separate sample [126-128]. The GSH recycling assay is still one of the most popular techniques utilized to detect GSH and GSSG for its satisfactory sensitivity and relatively low cost. However, the main disadvantages are the unsatisfactory selectivity and the need for the removal of the excess NEM due to its potent GR-inhibiting property [21], which makes the technique quite complex and time-consuming. Later, Adams et al. modified this method by using solid phase extraction (SPE) instead of solvent extraction to remove the excess of NEM [129]. Application of the SPE method has been proved to be more accurate for assessment of GSSG [130]. Over the years, NEM has been substituted with other GSH-masking agents. Griffith [62] introduced 2-VP as a new thiol-masking agent for analysis of plasma GSSG. In contrast to NEM, 2-VP shows no inhibition to GR and does not need to be eliminated from the samples before analysis. This method has become the standard for quantification of GSSG in most biological samples, such as human whole blood [41, 131], and rat bile and liver samples [132].

Other spectrophotometric techniques include the glutathione *S*-transferase (GST)-1-chloro-2,4-dinitrobenzene (CDNB) endpoint method, measuring only GSH [133], and a GSSG-endpoint method which is based on determination of NADPH consumption by GR

[127]. Even though with more precision, these methods offer much less sensitivity than that of the GSH recycling assay.

Fluorescence Detection (FLD)

Recent years, FL detection of bio-thiols has been becoming a popular approach in consequence of its operational simplicity, high sensitivity and selectivity, and real time monitoring. Based on special chemical reactions, a host of thiol-specific fluorescent probes have been developed and employed for assessment of GSH in various biological media including cells, mitochondria, human blood, plasma, and serum. These probes are designed mainly to exploit the strong nucleophilicity of the sulfhydryl group, which enables reactions such as Michael addition, cleavage of S-S bonds, Se-N bonds, or cleavage of sulfonamides and sulfonate esters. Though highly sensitive, the probes have some disadvantages such as tedious synthesis, instability to light, or poor water solubility. Various organic small-molecular probes used over the past decade are summarized in Table 1.

Table 1. Summary of fluorescent probes for the detection of GSH and/or GSSG

Probe	Biological system	Analytes	Label	LOD	Ref.
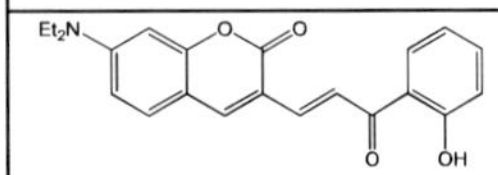 Ex: 466 nmEm: 553 nm.	HeLa cells	GSH	LPA or NEM	N.R.	[134]
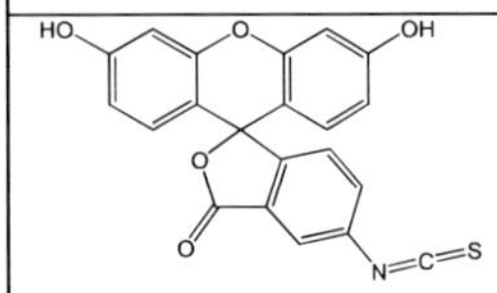 Ex: 473 nm/Em: 525 nm.	Mouse hepatocytes	GSH	2,3-Naphthalenedicarboxaldehyde (NDA)	44.7 nM	[135]
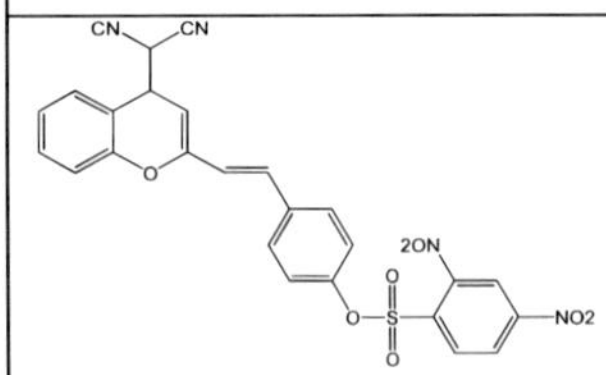 Ex: 486 nm/Em: 690 nm.	Hela cells	GSH	None	18 pM	[136]
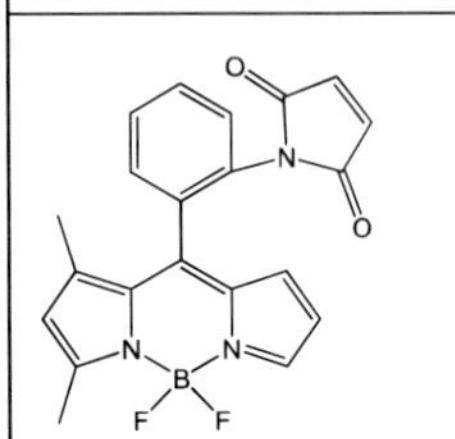 Ex: 475 nm/Em: 530 nm	Macrophage RAW264.7 cells	GSH	NEM	3 nM	[137]

Probe	Biological system	Analytes	Label	LOD	Ref.
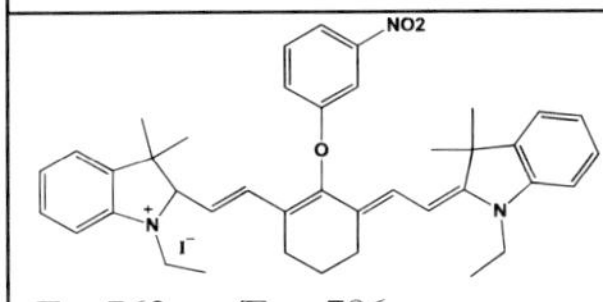 Ex: 763 nm/Em: 786 nm	Hela cells	GSH	NEM	0.25 □M	[138]
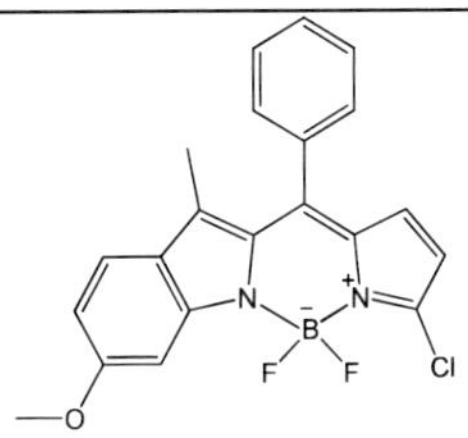 Ex: 579 nm/Em: 602 nm	HeLa cells and living MKN-45 cells.	GSH	*N*-Methylmaleimide (NMM)	85 pM	[139]
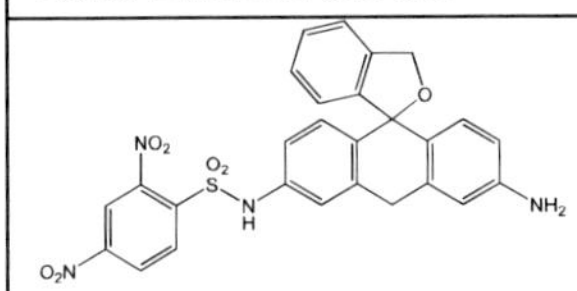 Ex: 495 nm/Em: 520 nm	SHIN3 cells and HUVEC cells	GSH	NEM	N.R.	[140]
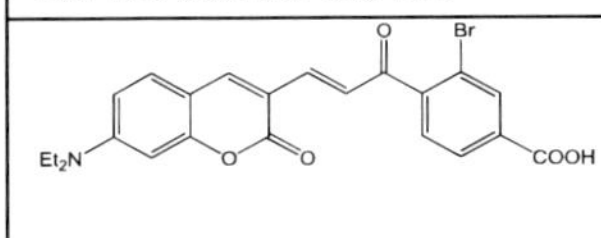 Ex: 488 nm/Em: 590 nm	3T3-L1, HepG2, HeLa, PANC-1, and PANC-28 cells	GSH	Diethyl maleate (DEM)	N.R.	[141]
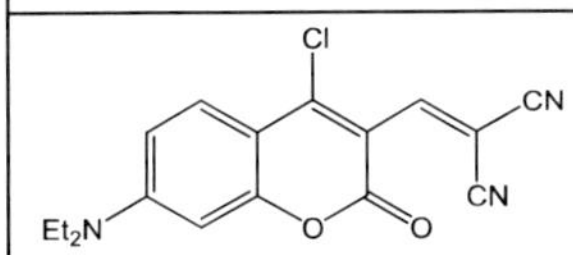 Ex: 430 nm/Em: 537 nm	MCF-7 cells	GSH	NEM	87 nM	[142]
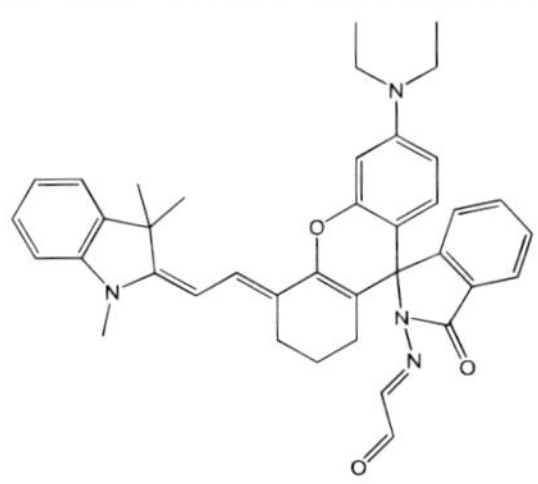 Ex: 698 nm/Em: 735 nm	HeLa cells and liver tissues	GSH	NEM	0.15 μm	[143]
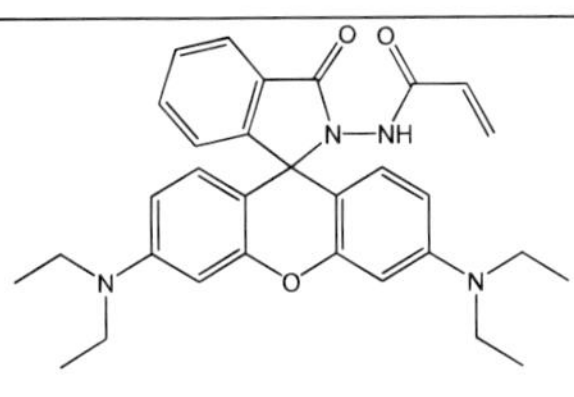 Ex: 520 nm/Em: 582 nm	HeLa cells	GSH	NEM	190 nM	[144]

Table 1. (Continued)

Probe	Biological system	Analytes	Label	LOD	Ref.
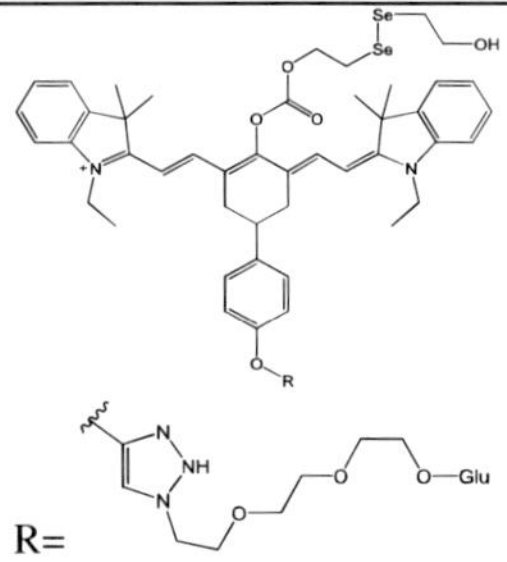 R= Ex: 615 nm/Em: 785 nm	HepG2 and HL-7702 cells	GSH	NEM	N.R.	[145]
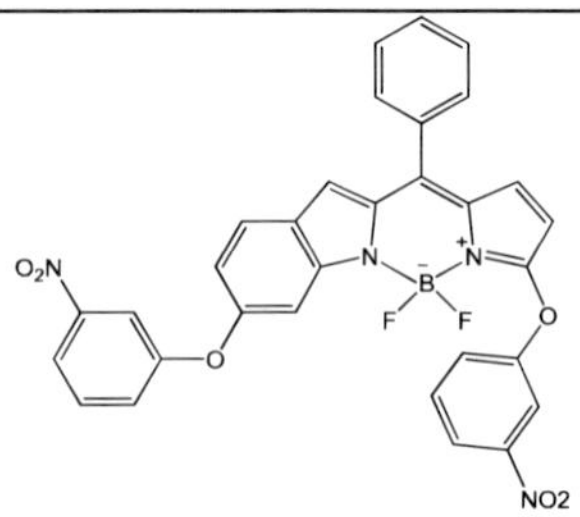 Ex: 579 nm/Em: 596 nm	BHK cells	GSH	NEM	7.6 nM and 76 nM	[146]
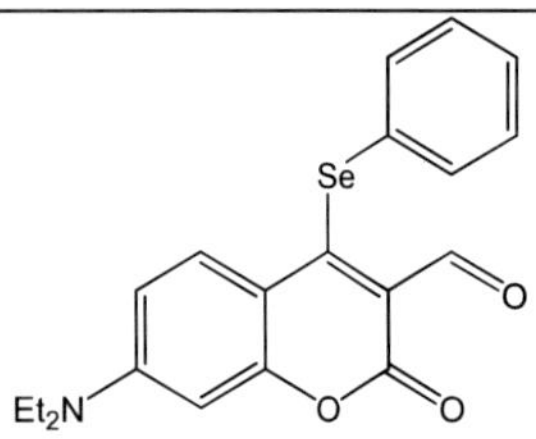 Ex: 518 nm/Em: 550 nm	Hep3B cells	GSH	NEM	27 nM	[147]
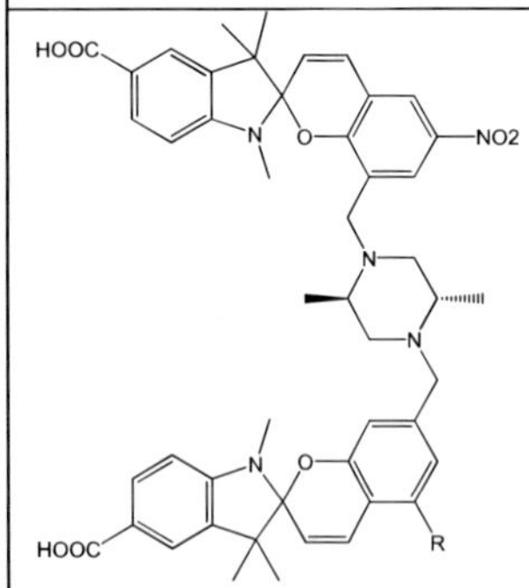 Ex: 532 nm/Em: 650 nm	HEK293 cells	GSH	None	N.R.	[148]
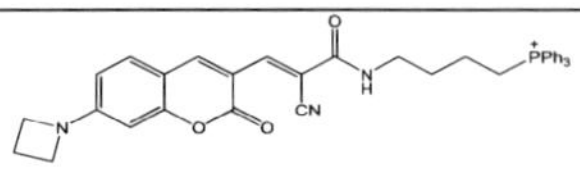 Ex: 405 nm/Em: 488 nm	Mitochondria in HeLa cells	GSH	None	N.R.	[152]

Probe	Biological system	Analytes	Label	LOD	Ref.
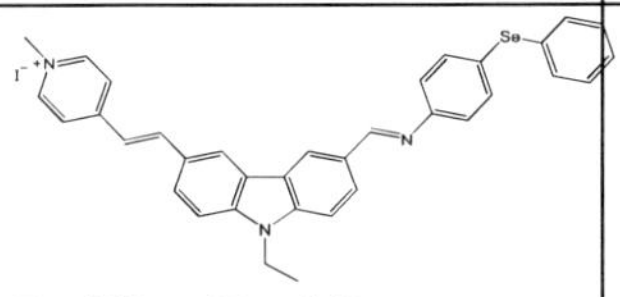 Ex: 445 nm/Em: 565 nm	Mitochondria in human hepatocellular liver, carcinoma cell line RAW264.7	GSH	None	31 nM	[153]
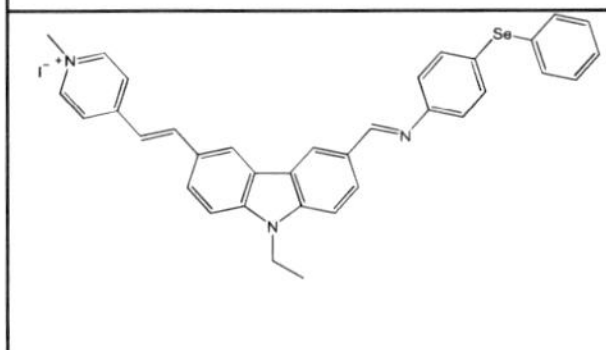 Ex: 600 nm/Em: 695 nm	Fetal bovine serum	GSH, GSSG	GR	N.R.	[154]
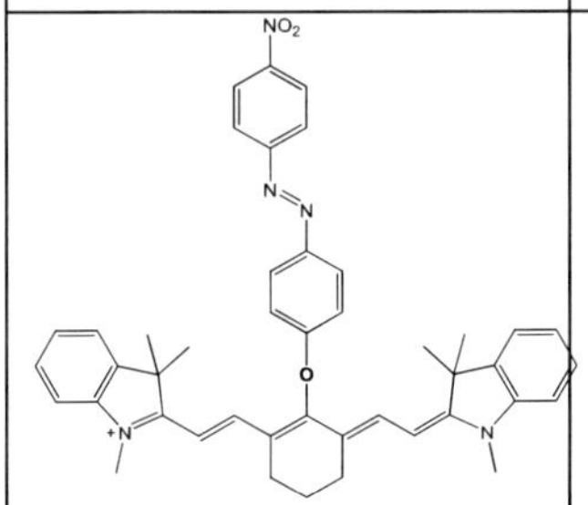 Ex: 804 nm/Em: 832 nm	Human blood samples	GSH, GSSG	GR	87.2 nM	[155]

Recently, another kind of fluorescent sensor, fluorescent nanomaterials, have attracted increasing attentions for GSH detection such as gold nanoparticles (Au NCs) [156-158], silver nanoparticles [159], upconversion nanoparticles (UCNPs) [160], quantum dots (QDs) [161, 162], graphene quantum dots [163, 164], carbon quantum dots (CQDs) [165-167], UiO-type metalorganic frameworks (MOFs) [168, 169], or DNA based sensor [170, 171]. Among which, CQDs have shown great potential as biosensors, due to their exceptional advantages including highly stable fluorescence intensity, low toxicity, good resistance to light and chemical degradation, and excellent biocompatibility. By applying these strategies, GSH sensing in urine and saliva samples as well as in food samples such as grape juice and wine have been realized.

Electrochemiluminescence (ECL) Method

Electrochemiluminescence (ECL) is also called electrogenerated chemiluminescence, which involves a light emission that arises from the electron-transfer reaction at electrode surfaces. ECL has important advantages over conventional chemiluminescence (CL), in which luminescence is initiated *in situ* and controlled by switching an electrode voltage. ECL is an attractive means of detection for biological substance analysis, due to its low detection limit (nanomolar levels), wide linear ranges, and excellent sensitivity [172].

Chen et al. [173] found that GSH would enhance the ECL of $Ru(bpy)_3^{2+}$ at a hemin glassy carbon electrode, and the enhanced ECL intensity was linear with the concentration of GSH in a wide range from 0.01- 0.1 μM. The detection limit of GSH was 20 pM. Wang et al. [174] reported a graphene oxide amplified ECL of quantum dots platform for the selective detection of GSH from thiol-containing compounds. The proposed method showed a detection limit of 8.3 μM for GSH, and it was successively utilized for GSH-containing drug detection without pretreatment. Zhang et al. [175] designed a photoluminescent (PL) and ECL dual-signaling probe ($[Ru(bpy)_2(DNBSO\text{-}bpy)](PF_6)_2$) for the detection of bio-thiols (GSH and Cys) in aqueous media and HeLa cells. The dose-dependent ECL response exhibited a good linear relationship against the GSH concentration with the detection limit of 19.8 nM.

Recently, a novel solid-state ECL sensor for ultrasensitive and selective sensing of GSH based on $Fe(CN)_6^{3-/4-}$ induced ECL amplification of carbon dots was constructed [176]. The GSH sensor showed a detection limit of 54.3 nM and a wide linear range from 0.1-1.0 μM. Gao et al. developed a new ECL method for GSH detection utilizing lucigenin as the ECL luminophore in the presence of MnO_2 nanosheets. GSH could reduce MnO_2 to Mn^{2+} and consequently, resulted in the ECL inhibition of lucigenin. The efficiencies of ECL inhibition increased linearly with the concentrations of GSH (10-2,000 nm), and the detection limit was 3.7 nM. Meanwhile, GSH detection in human serum samples was conducted using this method. The obtained satisfactory recoveries indicate its promising application in biological systems [177]. Yu et al. successfully measured the GSH/GSSG ratio in whole blood and serum samples, using a mixture of ZnAl-LDH nanosheets and quantum dots to reach a high ECL resonance energy transfer efficiency [178]. Peng et al. [179] reported a highly sensitive ECL nanosensor for GSH sensing, based on a novel high-quantum-yield AuNC probe. This AuNC-based ECL sensor presented an exceptionally wide dynamic range (1.0 pM to 10 nM) and low detection limit (32 pM), together with high sensitivity and selectivity, making the method successfully applied for detection of GSH in human urine and blood samples.

Electrochemical Detection (ECD)

Electrochemical detection (ECD) of GSH has become a popular technique owing to its admirable advantages like easy preparation, convenient operation, low cost, high sensitivity and specificity. Routine electrode materials on the signal amplification mainly consist of nanomaterial and enzyme-based transducer elements.

Currently, numerous nanomaterials have been applied to GSH sensing by ECD, such as carbon paste/Nafion® modified with gold nanoparticles [180], TiO_2-Pt nanowhiskers [181], NiHCF/CTAB/Au nanoparticles [182], 2-chlorobenzoyl ferrocene/Ag-ZnO nanoplates modified carbon paste electrode [183], rGO/Fe_3O_4/PANI (rGFP) nanocomposites [184], cobalt nanoparticle modified indium tin oxide (CoNP/ITO) electrodes [185], ZnO/CNTs nanocomposite [186], nanoscale $Cu(OH)_2$ carbon ionic liquid electrode [187], and nanocomposite molecular material based on CoPc/MWCNTf/GC [188]. The last two ECD methods are used for the simultaneous determination of GSH and GSSG in blood serum or rat urine samples. In addition to nanomaterials, Zhang et al.

developed a pair of MIP-modified electrochemical sensors for the detection of GSH and GSSG in arsenic trioxide-treated HL-60 cells [189].

The strategies based on enzyme-mediated amplification have also been employed for GSH detection. Ru et al. [191] constructed a GSH biosensor using coenzyme Q10 (CoQ10) for the first time as a sensing element and transducer. The CoQ10-based biosensor exhibited high sensitivity, fast response and good stability. The peak current of CoQ10 increased with the increasing of GSH concentrations (0.004-0.2 nM) with a low detection limit (32 pM). Another biosensor (GSH-Px(immob)/Pt-NP/GCPE) was developed by Anik et al., using glutathione peroxidase (GSH-Px) as the enzyme transducer. The biosensor response of GSH was monitored by the electrochemical oxidation of GSH in the presence of H_2O_2 (GSH/H_2O_2 ratios) [191]. Recently, Barman et al. [192] reported a novel approach for detection of GSH by means of glutathione *S*-transferase (GST)-ZnO composite nanoparticles. The electrochemical response proved to be linear over the 100 nM - 10 mM, GSH concentration range and the detection limit was determined to be 41.9 nM.

Moreover, DNA-based electrochemical sensors for detection of GSH have attracted much interest. The design of this sensor is based on the inhibition of Hg^{2+} -ion mediated strand displacement reaction, owing to the ability of GSH to chelate Hg^{2+}-ions. This method can easily distinguish GSH in complex samples such as saliva [193], serum samples [194] and cell extractions [195].

Fenton-type reaction between H_2O_2 and Cu^{2+}-ion was utilized in development of electrochemical sensor for GSH. This indirect and sensitive method is based on the inhibition of electroreduction of the oxidized dye polymer (PTH) by GSH [196].

Nuclear Magnetic Resonance (NMR) Method

Nuclear magnetic resonance (NMR) is a technique that allows detection and mapping of GSH and GSSG in intact cells and in the brain *in vivo*. Currently, different pulse sequences such as stimulated echo acquisition mode (STEAM), J-difference editing, point-resolved spectroscopy (PRESS), and Mescher-Garwood point-resolved spectroscopy (MEGA-PRESS) were developed to detect GSH.

Terpstra et al. used ^{1}H NMR in combination with STEAM spectroscopy at 4 T to detect the difference of GSH contents in the brain of healthy volunteers and schizophrenic patients. No differences in the GSH concentrations were observed in neurochemical measurements (1.6 ± 0.4 μmol/g) and those in schizophrenia versus control (1.4 ± 0.4 μmol/g) [197]. Reproducibility study was performed on a very short echo time (TE) phase rotation stimulated echo acquisition mode (STEAM) sequence at 3T with a focus on detection of GSH in ten healthy subjects. Excellent reproducibility was obtained for GSH detection using PR-STEAM at a TE of 6.5 ms [198]. Deelchand et al. also demonstrated the sensitivity and specificity of using short TE ^{1}H magnetic resonance spectroscopy (MRS) to measure changes and differences in human brain GSH concentration [199].

Spectral editing using J-difference is also a reliable method for GSH measurement. An et al. developed a ^{1}H NMR method at 3 T using J-difference spectroscopy for the measurement of GSH in healthy volunteers and stroke patients. The GSH levels were found to be 1.20 ± 0.14 and 1.19 ± 0.24 mM, respectively [200]. The simultaneous determination of GSH and γ-aminobutyric acid (GABA) was achieved with ^{1}H NMR by using J-difference editing at 7 T [201]. TE optimization for J-difference editing of GSH *in vivo* was conducted at 3 T by Chan et al.; the optimal TE was estimated to be 120 ms due to the improved editing efficiency [202].

Another pulse sequence applied in NMR measurement is PRESS. Trabesinger et al. detected GSH in the human brain by means of ^{1}H NMR with double quantum coherence (DQC) filtering in combination with PRESS localization sequence. The mean GSH concentration estimated in the frontal lobe of 12 healthy volunteers was 2-5 mM [203]. Zhao et al. introduced a new DQC filtering combined with PRESS for localized detection of GSH *in vivo*. Compared to the conventional DQC filter, the new filter presented frequency selective pulses, making it possible to detect GSH concentrations in the left (0.91 ± 0.16 mM) and right (0.89 ± 0.16 mM) parietal lobes of 5 healthy volunteers [204]. Reliable detection of glutamate (Glu), glutamine (Gln), and GSH was achieved by minimizing the *N*-acetylaspartate (NAA) multiplet signals at 2.49 ppm by means of PRESS coupled with a novel J-suppression pulse. This new method has been utilized to measure Glu, Gln, and GSH in gray matter and white matter of healthy human volunteers [205]. Similar work was done by Lally et al. using an adapted TE optimized PRESS pulse sequence at 7 T to detect Glu, Gin, and GSH in the healthy human brain [206].

Due to the overlap of GSH resonance with other brain metabolites in ^{1}H NMR, unambiguous measurement of GSH cannot be achieved by PRESS pulse sequences. Therefore, MEGA-PRESS pulse sequence has been developed, and its application to detect GSH in brain region has been demonstrated by many researchers. Emir et al. reported simultaneous assessment of ascorbate and GSH in the elderly human brain using double editing with (DEW) MEGA-PRESS [207]. Mandal et al. investigated GSH content in different brain regions of healthy male/female, mild cognitive impairment (MCI) and Alzheimer's disease (AD) patients by means of MEGA-PRESS pulse sequence [208]. Raschke et al. measured cerebral and cerebellar GSH in children using ^{1}H MEGA-PRESS MRS at 3 T [209]. Nezhad et al. compared accuracy and reliability of GSH determinations by PRESS and MEGA-PRESS. They concluded that physiological concentrations of GSH cannot be reliably measured using PRESS (TE = 35 ms) at 3 T, and recommended using MEGA-PRESS GSH editing sequence (TE = 130 ms) instead [210].

Surface-Enhanced Raman Scattering (SERS) Method

Surface-enhanced Raman scattering (SERS) is a vibrational-spectroscopy technique based on amplification of Raman scattering by appropriate nanostructures [211]. It is generally believed that electromagnetic (EM) and charge transfer (CT) mechanisms are accounted for the origin of SERS, EM being awarded the main contribution [212]. Apart from high sensitivity and specificity, SERS provides specific molecular information on bio-thiols in aqueous solutions or in complex matrices such as whole blood and living cells.

Direct detection of GSH in aqueous solutions utilizing SERS has been proposed by Huang et al. This group used rhodamine B, a reversed reporting agent, as a probe for the detection of GSH in aqueous solutions and achieved a detection limit of 1 μM [214]. A heat-induced SERS sensing method has been developed for rapid determination of GSH within 5 min and made a great achievement by decreasing the detection limit to 50 nM [214].

Recently, a few attentions have been focused on direct determination of cellular GSH. Saha et al. reported a "turn-off" SERS approach for reliable detection of cellular GSH and GSSG [215]. This approach exploited GSH-induced replacement of a Raman probe from a magnetic-plasmonic nanoparticle, followed by generation of electromagnetic hot spots. To determine GSSG, NEM is used to block GSH. Wang et al. created a novel SERS nanoprobe, 4-mercaptophenol (4-MP)-fabricated gold flowers (AuF/MP), for real-time monitoring ClO^- and GSH in living cells [216]. The single SERS probe for GSH detection exhibited high selectivity against other amino acids and ROS which may exist in biological samples.

SERS has also been employed for assessment of GSH in blood samples. Kuligowski et al. carried out SERS detection of GSH in umbilical cord whole blood. GSH concentrations ranging from 13 to 2200 μM were calculated by employing a Partial Least Squares (PLS) model. This approach achieved the detection of GSH in small blood volumes of only 50 μL, presenting its potential application in clinical investigations [217]. Ouyang et al. proposed a sensitive and selective method for detection of trace amount of GSH based on Fe_3O_4/Ag nanoparticles and crystal violet probe [218]. Quantitative determination of GSH in the range from 50 to 700 nM, with a detection limit of 40 nM was obtained using this approach.

Despite the various advantages, SERS has its limitations, such as possible side reactions due to the catalysis of the metal nanoparticles, and susceptibility of intensity of the SERS signals to the volume of the nanoparticles [219].

Detection Techniques with a Separation Step

High-Performance Liquid Chromatography (HPLC)

Several high-performance liquid chromatography (HPLC) and ultra-performance liquid chromatography (UPLC) methods for the determination of GSH and GSSG have recently been developed. Most assays are based on reversed-phase (RP) chromatography. To maintain GSH in non-ionized form, the mobile phase is usually buffered at acidic pH to facilitate interaction with the stationary phase. Trifluoroacetic acid (TFA), an ion-pairing reagent, is usually added to the mobile phase to avoid peak tailing of GSSG. HPLC in combination with several detection techniques, such as UV, FLD, ECD, MS and MS/MS is the most reported methodology.

UV Detection

For UV detection, derivatization of GSH is required to obtain higher sensitivity, due to the lack of chromophores or fluorophores. 1-Benzyl-2-chloropyridinium bromide (BCPB) and 2-chloro-1-methylquinolinium tetrafluoroborate (CMQT) are the two derivatizing reagents recently utilized for GSH determination.

Kusmierek et al. used BCPB as a derivatizing reagent which rapidly reacts with thiols (Cys, Hcy, GSH, and Cys-Gly) to form stable *S*-pyridinium derivatives with intense UV absorption at 315 nm. The resulting derivatives were separated and quantitated by HPLC with a detection limit of 0.2 μM GSH [220]. Głowacki et al. measured total albumin, Cys, Hcy, GSH, and Cys-Gly in plasma using automated precolumn derivatization with CMQT, followed by on-line dialysis, HPLC separation, and UV detection at 355 nm [221, 222].

Fluorescence Detection (FLD)

FLD-based methods are the most widely used detection techniques in combination with HPLC, owing to its high selectivity and sensitivity. Different types of derivatizing reagents have been used, including OPA, mBBr, ammonium 7-fluorobenzo-2-oxa-1,3-diazole-4-sulphonate (SBD-F), ammonium 5-bromo-7-fluorobenzo-2-oxa-1,3-diazole-4-sulphonate (SBD-BF), 2,3-naphthalenedialdehyde (NDA), 7-flouro-4-nitrobenzo-2-oxa-1,3-diazole (NBD-F), and 4-(N,N-dimethylaminosulfonyl)-7-fluoro-2,1,3-benzoxadiazole (DBD-F).

Kand'ar et al. developed an HPLC method with FL detection for the detection of both GSH and GSSG in various biological samples. OPA reacts with GSH and GSSG (without reduction) at pH 8 and 12, respectively, to produce stable fluorescent derivatives. For measurement of GSSG, free thiols were blocked with NEM [223]. In another study, Michaelsen et al. reported that unlike in aqueous solution, derivatization of GSH in red blood cell cytosol yielded two closely related derivatives in the absence of *N*-acetylcysteine ethyl ester (NACET) and only the GSH-OPA derivative in the presence of NACET. The resulting GSH-OPA derivative was detected with FLD (excitation/emission 338/458 nm) [119]. GSH and GSSG were simultaneously determined in as small as 50 μL plasma samples using dual mode FLD and column switching HPLC method [224]. By means of this method, GSH was detected by a fluorescence detector after derivatization with mBBr, and GSSG was determined by another fluorescence detector after online post-column derivatization with OPA. Ju et al. established an HPLC method for simultaneous determination of Cys, GSH and phytochelatins by fluorescence detection (excitation/ emission 380/470 nm) after derivatization with mBBr [225].

SBD-F, a benzofurazan reagent, is the most often used fluorescence derivatizing reagent, since it selectively reacts with thiols and has no fluorescence itself. Most previous studies have employed SBD-F as a labeling reagent for thiols detection in combination with HPLC [226-228]. However, simultaneous detection of various thiols has been failed, owing to the difficulty in retaining the high polarity of SBD-thiols. Hence, Tsunoda et al. developed an HPLC-FLD method for detection of Cys, GSH, CysGly, and γ-GluCys in human serum, using HILIC column [228, 229]. Cevasco et al. synthesized SBD-BF, as a new fluorescent labeling reagent, whose reaction rate with bio-thiols at room temperature was nearly three times faster than that of SBD-F at 60°C [230].

Recently, naphthalene-2,3-dicarboxaldehyde (NDA) has been employed as a fluorophore to derivatize GSH in wines [231] or dried blood spot samples [232] for HPLC determination. Other fluorescent derivatizing reagents include NBD-F [233] and DBD-F [234]. These latter methods have been applied to determine GSH in plasma samples from healthy individuals.

Electrochemical Detection (ECD)

HPLC with EC detection is a sensitive and economical approach that can directly measure thiols and disulfides without the problems of sample dilution and reaction efficiency associated with derivatization processes. Recently, several studies have described simultaneous measurement of thiols such as GSH, GSSG, Cys, HCys, and NAC by HPLC-ECD.

Bailey et al. described an HPLC-ECD method with a stable boron-doped diamond (BDD) electrode for direct measurement of aminothiols, disulfides and thioethers [235]. In this method, plasma and tissue levels of GSH and GSSG were well determined. BDD has become a useful electrode in electrochemical detection. Park et al. proposed an HPLC-ECD method using BDD electrode to assess hepatic GSH redox status [236]. Childs et al. measured the GSH and GSSG ratio in prostate cancer cells by HPLC-ECD with BDD electrode. The ratios of GSH and GSSG in LNCaP and DU145 cells were determined to be 8:1 and 32:1, respectively [237]. Other applications of HPLC-ECD include the measurement of GSH, GSSG, and GSNO in biological samples [238], analysis of the common monothiols and disulfides in human plasma and erythrocytes [239], and rapid detection of GSH in HepG2 cells [240].

Mass Spectrometric Detection (MSD)

Recently, HPLC combined with MS or MS/MS has been widely used for the assessment of GSH, GSSG, and related compounds. The advantages of LC-MS/MS detection include high-throughput analysis, small amounts of analyte and excellent specificity and sensitivity.

Iwasaki et al. developed a column switching hydrophilic interaction chromatography coupled with mass spectrometry (CS-HILIC–MS) method for the quantification of GSH and GSSG in mouse serum [241]. The blood samples undergo two derivatization steps using *N*-benzylmaleimide (NBenzM) and *N*-cyclohexylmaleimide (NCycloM), sequentially. After deproteinization using 20% sulfosalicylic acid, the samples were analyzed by HPLC-MS using selected ion monitoring (SIM). Another study utilized HPLC-MS using silver nitrate as a post-column reagent to determine GSH and GSSG in human dermis [242].

Numerous studies have reported application of HPLC-MS/MS in detection of biothiols after derivatization of free thiols to prevent auto-oxidation. To prevent thiol auto-oxidation and improve chromatographic resolution, sequential derivatization with iodoacetamide (IAM) and isopropylchloroformate (IPCF) was conducted for sulfur amino acids in blood samples [243]. Huang et al. synthesized ω-bromoacetonyl-quinolinium bromide (BQB) and d^7-BQB for thiol derivatization and applied them for measurement of GSH and GSSG in urine [244]. GSSG was derivatized after reducing it by TCEP. Urine samples were analyzed using a HILIC column by ESI-MS/MS detection. Moore et al. used stable-isotope labeled GSH-^{13}C and GSSG-$^{13}C,^{15}N$ as internal standards, and NEM as derivatizing reagent for precise and selective analysis of blood GSH and GSSG [245]. The GSH and GSSG content of erythrocytes was determined by LC-MS/MS using GSH-$^{13}C_3,^{15}N$ and GSSG-$^{13}C_4,^{15}N_2$ as internal standards after derivatization of GSH with iodoacetic acid [246]. Our group developed an mBBr derivatization and mass spectrometry-based method for measurement of GSH [247]. The derivatization reaction was conducted in the dark at room temperature for 20 min to produce a GSH-mBBr conjugate, followed by deproteinization with 10% 5-sulfosalicylic acid for biological samples. This method is the most sensitive approach compared to the previous MS-based GSH assays, with the detection limit as low as 500 amol GSH. The method has been successfully applied to simultaneous assessment of GSH, Cys, NAC, Hcy, and CysGly, within 10 min [248].

Several simple and direct HPLC-MS/MS methods without derivatization have been developed for assessment of GSH and GSSG. An HPLC-ESI-MS/MS method was established to simultaneously quantitate eight plasma aminothiols [249]. By using hydrophilic column, optimized pH value of the mobile phase and multiple reaction monitoring (MRM) transitions, the method offered unique advantages in determination of compounds with high polarity and instability. Squellerio et al. presented a simple LC-MS/MS method to accurately and selectively measure whole blood GSH and GSSG levels without derivatization [250]. GSH decrease is prevented by rapid treatment of blood samples with 10% TCA containing 1 mM EDTA. Good separation of hydrophilic GSH and GSSG is achieved using a pentafluorophenylpropyl bonded silica column for LC-MS/MS. Norris et al. introduced an HPLC-MS/MS method for GSH assay, including precipitation with methanol, addition of stable isotope labeled GSH as an internal standard, combined with storage at -70°C, which showed superior storage stability up to 99 days [251]. Simultaneous quantification of GSH and GSSG in wine was achieved by means of UPLC-MS/MS methods [252, 253]. Kritzinger et al. reported a detection limit of 0.002

mg/L for GSH and 0.001 mg/L for GSSG [252]. A LOD as low as 0.0014 mg/L in white wine and 0.0018 mg/L in rosé wine was obtained by the UPLC-MRM-MS method [253].

Capillary Electrophoresis (CE) Method

Capillary electrophoresis (CE) is another widely used separation technique for determination of bio-thiols, in which the analytes are separated in a narrow capillary filled with a background electrolyte. CE methods are characterized by small sample consumption, short analysis time, high separation efficiency, good reproducibility and simple operation. CE approaches coupled with various types of detection including UV-Vis, FLD, laser-induced fluorescence (LIF), MS and ECD have been elaborated.

In most CE-UV absorbance methods, GSH and GSSG are measured in their underivatized forms. Recently, Kazarjan et al. developed a CE method to separate GSH and its four analogs [254]. Baseline separation was achieved in the pH range of 7-9 and the analytes were monitored by DAD at 195 nm. Hempe et al. [255] described modifications of a capillary zone electrophoresis (CZE) method with UV detection developed by Serru et al. [256]. In this method, 5% rather than 10% metaphosphoric acid (MPA) was used to precipitate blood proteins, and the pH of running buffer was lowered from 8.4 to 7.8 to improve reproducibility of both peak area and migration times. Simultaneous analysis of GSH and GSSG in mouse and human red blood cells has been achieved by using the modified CZE-UV method. Kowalska et al. compared blood GSH and GSSG in healthy adult women with women taking oral contraceptives, pregnant women, and female smokers [257]. In that paper, GSH was measured in two methods. One of them was based on the reaction of GSH with alloxan using UV detection at 305 nm. The other was based on a CE method with simultaneous determination of GSSG.

Omitting derivatization procedure, however, can increase autooxidation of GSH. Tsardaka et al. utilized methyl propiolate, for the first time, as a thiol derivatizing reagent for CZE detection of GSH in baker's yeast [258]. The best separation efficiency was achieved at pH 9.5, 75 mmol/L of borate concentration and separation voltage of 20 kV. The LOD of GSH was determined to be 1.5 mol/L.

CE with FL detection is frequently used in the assessment of GSH and GSSG. Recently, Tu et al. measured intracellular GSH in RAW264.7 cells, based on microchip electrophoresis (MCE) with FL detection using a highly sensitive fluorogenic probe, 1,3,5,7-tetramethyl-8-phenyl-(2-maleimide)-difluoroboradiaza-s-indacene (TMPAB-o-M), as the derivatizing reagent [259]. FLD offers better sensitivity than UV detection and application of LIF leads to further improvements. Zhang et al. developed a CZE with LIF detection method for analysis of GSH [260]. A new BODIPY-based probe named 1,7-dimethyl-3,5-distyryl-8-phenyl-(4-iodoacetamide)difluoroboradiaza-*s*-indacene (DMDSPAB-I) was used as labeling reagent. Excitation wavelength of DMDSPAB-I is 620 nm, which is close to that of the wavelength of the laser line (635 nm). Similar CZE-LIF method with DMDSPAB-I as labeling reagent was applied for analysis of GSH in cell homogenates and in single cells [261] as well as in human urine [262]. Eosin-5-maleimide (EMA), a fluorescent tag for GSH, was used for the first time in CE-LIF studies of GSH in biological samples (exhaled breath condensate and saliva) [263]. EMA has an excitation maximum at 515 nm, which matches the wavelengths of the laser module, yielding a LOD in the subnanomolar range. To provide more insight into the anticancer mechanisms of xylitol selenite, CE-LIF was employed to measure the changes in GSH content in SMMC-7221 cells treated with xylitol selenite [264]. In this method, 5-iodoacetamidofluorescein (5-IAF) was utilized to label GSH to form a negatively charged derivative.

Recently, a few methods have coupled CE with MS Yuan et al. reported that GSH and GSSG were quantified, using CE-orbitrap MS with isotope coded affinity tags (ICAT) [265]. The application of ICAT, by which GSH is isolated, enhanced selectivity of MS detection by quantitation of $^{12}C/^{13}C$ isotope tags. D'Agostino et al. introduced an improved CE-ESI-MS method for estimation of plasma GSH and other low molecular weight thiols [266]. In this method, *N*-(2-aminoethyl)maleimide trifluoroacetate salt (NAEM) or *N*-[2-(trimethylammonium)-ethyl]maleimide chloride (NTAM) was used as derivatizing reagent to enhance ionization efficiency of GSH, and two steps of quenching procedure were needed with sodium 2-mercaptoethane-sulfonate (MESNA) and *N*-methylmaleiminde (NMM) sequentially.

ECD, typically operated with multiple detection electrodes, can be coupled with CE to achieve sensitive and selective determination of thiols. A method based on CE-ECD was employed for the separation and measurement of GSH, Hcy, Cys, ascorbic acid, and uric acid [170]. Recently, Wang et al. determined GSH in single HepG2 cells by CE-ECD with reduced graphene oxide modified microelectrode [255]. The detection limit was 15 fmol and the average GSH content in single HepG2 cells was determined to be 7.13±1.11 fmol. This method has been successfully applied to quantify GSH in human liver cancer cells.

Gas Chromatography (GC)

The poor thermal stability and lack of volatility of GSH represent a great challenge for its GC analysis. GC-MS was successfully applied to GSH analysis after its derivatization to the *N,S*-ethoxycarbonyl methyl esters [171, 267]. Tea et al. reported derivatization of GSH with ethyl chloroformate (ECF), followed by gas chromatography/combustion/isotope ratio mass spectrometric (GC/C/IRMS) analysis [171]. The method was applied to determine low ^{13}C-enrichments in GSH in erythrocyte extracts. Determination of GSH in microvolumes of neonatal blood samples was performed on GC-MS after ECF derivatization by the same research group [267]. Most recently, a simple "one-pot" derivatization combined with liquid-liquid extraction was developed for GC-MS assessment of GSH in erythrocytes [268]. Blood samples were treated with ECF after deproteinization with 5% TCA. A good linearity in the range of 25.4-813.4 μM was observed. GC-MS analysis of the above *N,S*-ethoxycarbonyl methyl esters requires an electron ionization source which lacks sensitivity. Tsikas et al. proposed a highly sensitive

GC-MS method for GSH assay involving acylation of GSH dimethyl ester and ^{2}H-labeled GSH dimethyl ester (IS) with pentafluoropropionic anhydride (PFPA) and application of electron-capture negative-ion chemical ionization (ECNICI) detection [269].

CONCLUSION

Development of appropriate analytical methods to assess GSH and GSSG in physiological fluids determines the success in characterization of oxidative stress s and diagnosis of several human diseases. This stimulates investigations of measurement of these two important sulfur-containing compounds, and intensive research on development of GSH-GSSG assessment is ongoing nowadays. Up to now, numerous methods for quantification/detection of GSH and GSSG have been established. The advances of these techniques and their combinations have facilitated enhancement of detection sensitivity and accuracy of the methods. However, the concentration of GSH reported in the literature varies over a 10-fold range, especially in those sample containing a trace amount of GSH. Because of its instability resulting from autoxidation, accurate detection of GSH/GSSG ratio is challenging. To date, no golden standard method has been acknowledged for assessment of these important biomolecules, and each methodology exhibits benefits and potential limitations. This chapter listed the basic sample handling processes, reagents and various detecting techniques, coupled with some practical considerations in usage of them. The authors hope this chapter would provide readers with helpful information for further experimental design and would benefit advanced research of this area.

REFERENCES

[1] Lu, S. C. Regulation of glutathione synthesis. *Mol. Aspects Med.* 2009, 30, 42-59.

[2] Forman, H. J.; Zhang, H.; Rinna, A. Glutathione: overview of its protective roles, measurement, and biosynthesis. *Mol. Aspects Med.* 2009, 30, 1-12.

[3] Lu, S. C. Glutathione synthesis. *Biochim. Biophys. Acta.* 2013, 1830, 3143-53.

[4] Meister, A.; Tate, S. S. Glutathione and related gamma-glutamyl compounds: Biosynthesis and utilization. *Ann. Rev. Biochem.* 1976, 45, 559-604.

[5] Valencia, E.; Marin, A.; Hardy, G. Glutathione-nutritional and pharmacological viewpoints: Part III. *Nutrition.* 2001, 17, 696-697.

[6] Dalle-Donne, I.; Colombo, G.; Gagliano, N.; Colombo, R.; Giustarini, D.; Rossi, R.; Milzani, A. S-glutathiolation in life and death decisions of the cell. *Free Rad. Res.* 2011, 45, 3-15.

[7] Anderson, M. E.; Luo, J. L. Glutathione therapy: Fom prodrugs to genes. *Semin. Liver. Dis.* 1998, 18, 415-424.

[8] Briviba, K.; Klotz, L. O.; Sies, H. Defenses against peroxynitrite. *Methods Enzymol.* 1999, 301, 301-311.

[9] Dalle-Donne, I.; Milzani, A.; Gagliano, N.; Colombo. R.; Giustarini, D.; Rossi, R. Molecular mechanisms and potential clinical significance of S-glutathionylation. *Antioxid. Redox. Signal.* 2008, 10, 445-473.

[10] Kirlin, W. G.; Cai, J.; Thompson, S. A.; Diaz, D.; Kavanagh, T. J.; Jones, D. P. Glutathione redox potential in response to differentiation and enzyme inducers. *Free Rad. Biol. Med.* 1999, 27, 1208-1218.

[11] Ortega, A. L.; Mena, S.; Estrela, J. M. Glutathione in cancer cell death. *Cancers (Basel).* 2011, 3, 1285-1310.

[12] Griffith, O. W.; Meister, A. Origin and turnover of mitochondrial glutathione. *Proc. Natl. Acad. Sci. USA.* 1985, 82, 4668-4672.

[13] Jones, D. P.; Mody, V. C; Carlson, J. L; Lynn, M. J.; Sternberg, P. Redox analysis of human plasma allows separation of pro-oxidant events of aging from decline in antioxidant defenses. *Free Rad. Biol. Med.* 2002, 33, 1290-1300.

[14] Giustarini, D.; Dalle-Donne, I.; Lorenzini, S.; Milzani, A.; Rossi, R. Age-related influence on thiol, disulfide, and protein-mixed disulfide levels in human plasma. *J. Gerontol. A.* 2006, 61, 1030-1038.

[15] Giustarini, D.; Dalle-Donne, I.; Tsikas, D.; Rossi, R. Oxidative stress and human diseases: Origin, link, measurement, mechanisms, and biomarkers. *Crit. Rev. Clin. Lab. Sci.* 2009, 46, 241-281.

[16] Ballatori, N.; Krance, S. M.; Notenboom, S.; Shi, S.; Tieu, K.; Hammond, C. L. Glutathione dysregulation and the etiology and progression of human diseases. *Biol. Chem.* 2009, 390, 191-214.

[17] Rossi, R.; Dalle-Donne, I.; Milzani, A.; Giustarini, D. Oxidized forms of glutathione in peripheral blood as biomarkers of oxidative stress. *Clin. Chem.* 2006, 52, 1406-1414.

[18] Srivastava, S,K,; Beutler, E. Oxidized glutathione levels in erythrocytes of glucose-6-phosphate-dehydrogenase-deficient subjects. *Lancet.* 1968, 2, 23-24.

[19] Hansen, R E.; Winther, J. R. An introduction to methods for analyzing thiols and disulfides: Reactions, reagents, and practical considerations. *Anal. Biochem.* 2009, 394, 147-158.

[20] Stempak, D; Dallas, S.; Klein, J.; Bendayan, R.; Koren, G.; Baruchel, S. Glutathione stability in whole blood: effects of various deproteinizing acids. *Ther. Drug. Monit.* 2001, 23, 542-549.

[21] Tietze, F. Enzymic method for quantitative determination of nanogram amounts of total and oxidized glutathione: applications to mammalian blood and other tissues. *Anal. Biochem.* 1969, 27, 502-522.

[22] Giustarini, D.; Tsikas, D.; Colombo, G.; Milzani, A.; Dalle-Donne, I.; Fanti, P.; Rossi, R. Pitfalls in the analysis of the physiological antioxidant glutathione (GSH) and its disulfide (GSSG) in biological samples: An elephant in the room. *J. Chromatogr. B.* 2016, 1019, 21-28.

[23] Bayle, C.; Issac, C.; Salvayre, R.; Couderc, F.; Causse, E. Assay of total homocysteine and other thiols by capillary electrophoresis and laser-induced fluorescence detection. II. Pre-analytical and analytical conditions. *J. Chromatogr. A*. 2002, 979, 255-260.

[24] Lenton, K.J.; Therriault, H.; Wagner, J. R. Analysis of glutathione and glutathione disulfide in whole cells and mitochondria by postcolumn derivatization high-performance liquid chromatography with ortho-phthalaldehyde. *Anal. Biochem.* 1999, 274, 125-130.

[25] Winther, J. R.; Thorpe, C. Quantification of thiols and disulfides. *Biochim. Biophys. Acta.* 2014, 1840, 838-846.

[26] Camera, E.; Picardo, M. Analytical methods to investigate glutathione and related compounds in biological and pathological processes. *J. Chromatogr. B.* 2002, 781, 181-206.

[27] Caussé, E.; Issac, C.; Malatray, P.; Bayle, C.; Valdiguié, P.; Salvayre, R.; Couderc, F. Assays for total homocysteine and other thiols by capillary electrophoresis-laser-induced fluorescence detection. I. Preanalytical condition studies. *J. Chromatogr. A.* 2000, 895, 173-178.

[28] Yoshida, T. Determination of reduced and oxidized glutathione in erythrocytes by high-performance liquid chromatography with ultraviolet absorbance detection. *J. Chromatogr. B.* 1996, 678, 157-164.

[29] Truong, D. H.; Eghbal, M. A.; Hindmarsh, W.; Roth, S. H.; O'Brien, P. J. Molecular mechanisms of hydrogen sulfide toxicity. *Drug. Metab. Rev.* 2006, 38, 733-744.

[30] Luo, J. L.; Hammarqvist, F.; Cotgreave, I. A.; Lind, C.; Andersson, K.; Wernerman, J. Determination of intracellular glutathione in human skeletal muscle by reversed-phase high-performance liquid chromatography. *J Chromatogr B Biomed. Appl.* 1995, 670, 29-36.

[31] Anderson, D. J.; Guo, B.; Xu, Y.; Ng L. M.; Kricka, L. J.; Skogerboe, K. J.; Hage, D. S.; Schoeff, L.; Wang, J.; Sokoll, L.J.; Chan, D. W.; Ward, K. M.; Davis, K. A. *Clinical Chemistry. Anal. Chem.* 1997, 69, 165R-229R.

[32] Senft, A. P.; Dalton, T. P., Shertzer, H. G. Determining glutathione and glutathione disulfide using the fluorescence probe o-phthalaldehyde. *Anal. Biochem.* 2000, 280, 80-86.

[33] Margaritelis, N. V.; Veskoukis, A. S.; Paschalis, V.; Vrabas, I. S.; Dipla, K.; Zafeiridis, A.; Kyparos, A.; Nikolaidis, M. G. Blood reflects tissue oxidative stress: A systematic review. *Biomarkers.* 2015, 20, 97-108.

[34] Kleinman, W. A.; Richie, J. P. Jr. Status of glutathione and other thiols and disulfides in human plasma. *Biochem. Pharmacol.* 2000, 60, 19-29.

[35] Ivanov, A. R.; Nazimov, I. V.; Baratova, L. Determination of biologically active low-molecular-mass thiols in human blood. I. Fast qualitative and quantitative, gradient and isocratic reversed-phase high-performance liquid chromatography with photometric and fluorescence detection. *J. Chromatogr. A.* 2000, 895,157-66.

[36] Ivanov, A. R.; Nazimov, I. V.; Baratova, L. A. Qualitative and quantitative determination of biologically active low-molecular-mass thiols in human blood by reversed-phase high-performance liquid chromatography with photometry and fluorescence detection. *J. Chromatogr. A.* 2000, 870, 433-442.

[37] Camera, E.; Rinaldi, M.; Briganti, S.; Picardo, M.; Fanali, S. Simultaneous determination of reduced and oxidized glutathione in peripheral blood mononuclear cells by liquid chromatography-electrospray mass spectrometry. *J. Chromatogr. B.* 2001, 757, 69-78.

[38] Loughlin, A.F.; Skiles, G. L.; Alberts, D. W.; Schaefer, W. H. An ion exchange liquid chromatography/mass spectrometry method for the determination of reduced and oxidized glutathione and glutathione conjugates in hepatocytes. *J. Pharm. Biomed. Anal.* 2001, 26, 131-142.

[39] Asensi, M.; Sastre, J.; Pallardo, F. V.; Garcia de la Asuncion J, Estrela, J. M.; Vina, J. A high-performance liquid chromatography method for measurement of oxidized glutathione in biological samples. *Anal. Biochem.* 1994, 217, 323-328.

[40] Cortese, M. S.; Baird, J. P.; Uversky, V. N.; Dunker, A. K. Uncovering the unfoldome: enriching cell extracts for unstructured proteins by acid treatment. *J. Proteome Res.* 2005, 4, 1610-1618.

[41] Rossi, R.; Milzani, A.; Dalle-Donne, I.; Giustarini, D.; Lusini, L.; Colombo, R.; Di Simplicio, P. Blood glutathione disulfide: in vivo factor or in vitro artifact? *Clinical Chem.* 2002, 48, 742-753.

[42] Yang, Q.; Krautmacher, C.; Schilling, D.; Pittelkow, M. R.; Naylor, S. Simultaneous analysis of oxidized and reduced glutathione in cell extracts by capillary zone electrophoresis. *Biomed. Chromatogr.* 2002, 16, 224-228.

[43] Riordan, J. F.; Vallee, B. L. Reactions with N-ethylmaleimide and p-mercuribenzoate. *Methods Enzymol.* 1972, 25, 449-456.

[44] Bednar, R. A. Reactivity and pH dependence of thiol conjugation to N-ethylmaleimide: Detection of a conformational change in chalcone isomerase. *Biochemistry.* 1990, 29, 3684-3690.

[45] Gilbert, H. F. Thiol/disulfide exchange equilibria and disulfide bond stability. *Methods Enzymol.* 1995, 251, 8-28.

[46] MacQuarrie, R. A.; Bernhard, S. A. Mechanism of alkylation of rabbit muscle glyceraldehyde 3-phosphate dehydrogenase. *Biochemistry.* 1971, 10, 2456-2466.

[47] Titani, Y.; Tsuruta, Y. Some chemical and biological characteristics of showdomycin. *J. Antibiot.* 1974, 27, 956-962.

[48] Gregory, J. D. The stability of N-ethylmaleimide and its reaction with sulfhydryl groups. *J. Am. Chem. Soc.* 1955, 77, 3922-3923.

[49] Knight, P. Hydrolysis of p-NN'-phenylenebismaleimide and its adducts with cysteine. Implications for cross-linking of proteins. *Biochem. J.* 1979, 179, 191-197.

[50] Beutler, E.; Srivastava, S. K.; West, C. The reversibility of N-ethylmaleimide (NEM) alkylation of red cell glutathione. *Biochem. Biophys. Res. Commun.* 1970, 38, 341-347.

[51] Giustarini, D.; Dalle-Donne, I.; Milzani, A.; Rossi, R. Detection of glutathione in whole blood after stabilization with N-ethylmaleimide. *Anal. Biochem.* 2011, 415, 81-83.

[52] Steghens, J. P.; Flourie, F.; Arab, K.; Collombel, C. Fast liquid chromatography-mass spectrometry glutathione measurement in whole blood: micromolar GSSG is a sample preparation artifact. *J. Chromatogr. B.* 2003, 798, 343-349.

[53] Zander, T.; Phadke, N. D.; Bardwell, J. C. Disulfide bond catalysts in *Escherichia coli*. *Methods Enzymol.* 1998, 290, 59-74.

[54] Lundell, N.; Schreitmuller, T. Sample preparation for peptide mapping--A pharmaceutical quality-control perspective. *Anal. Biochem.* 1999, 266, 31-47.

[55] Cotner, R. C, Clagett, C. O. O-carboxamidomethyl tyrosine as a reaction product of the alkylation of proteins with iodoacetamide. *Anal. Biochem.* 1973, 54, 170-177.

[56] Crestfield, A. M.; Stein, W. H.; Moore, S. Alkylation and identification of the histidine residues at the active site of ribonuclease. *J. Biol. Chem.* 1963, 238, 2413-2419.

[57] Fruchter, R. G.; Crestfield, A. M. The specific alkylation by iodoacetamide of histidine-12 in the active site of ribonuclease. *J. Biol. Chem.* 1967, 242, 5807-5812.

[58] Yang, Z.; Attygalle, A. B. LC/MS characterization of undesired products formed during iodoacetamide derivatization of sulfhydryl groups of peptides. *J. Mass Spectr.* 2007, 42, 233-243.

[59] Boja, E. S.; Fales, H. M. Overalkylation of a protein digest with iodoacetamide. *Anal. Chem.* 2001, 73, 3576-3582.

[60] Gundlach, H. G.; Stein, W. H.; Moore, S. The nature of the amino acid residues involved in the inactivation of ribonuclease by iodoacetate. *J. Biol. Chem.* 1959, 234, 1754-1760.

[61] Anderson, M. E. Determination of glutathione and glutathione disulfide in biological samples. *Methods Enzymol.* 1985, 113, 548-555.

[62] Griffith, O. W. Determination of glutathione and glutathione disulfide using glutathione reductase and 2-vinylpyridine. *Anal. Biochem.* 1980, 106, 207-212.

[63] Friedman, M.; Krull, L. H.; Cavins, J. F. The chromatographic determination of cystine and cysteine residues in proteins as s-beta-(4-pyridylethyl)cysteine. *J. Biol. Chem.* 1970, 245, 3868-3871.

[64] Ercal, N.; Yang, P.; Aykin, N. Determination of biological thiols by high-performance liquid chromatography following derivatization by ThioGlo maleimide reagents. *J. Chromatogr. B.* 2001, 753, 287-292.

[65] Lenton, K.; Therriault, H.; Wagner, J. R. Analysis of glutathione and glutathione disulfide in whole cells and mitochondria by postcolumn derivatization high-performance liquid chromatography with ortho-phthalaldehyde. *Anal. Biochem.* 1999, 274, 125-130.

[66] Gilbert, H. F. Molecular and cellular aspects of thiol-disulfide exchange. *Adv. Enzymol. Relat. Areas Mol. Biol.* 1990, 63, 69-172.

[67] Cleland, W. W. Dithiothreitol, a new protective reagent for SH groups. *Biochemistry.* 1964, 3, 480-482.

[68] Yang, C. S.; Chou, S. T.; Liu, L.; Tsai, P. J.; Kuo, J. S. Effect of ageing on human plasma glutathione concentrations as determined by high-performance liquid chromatography with fluorimetric detection. *J. Chromatogr. B.* 1995, 674, 23-30.

[69] Chassaing, C.; Gonin, J.; Wilcox, C. S.; Wainer, I. W. Determination of reduced and oxidized homocysteine and related thiols in plasma by thiol-specific pre-column derivatization and capillary electrophoresis with laser-induced fluorescence detection. *J. Chromatogr. B.* 1999, 735, 219-227.

[70] Pastore, A.; Federici, G.; Bertini, E.; Piemonte, F. Analysis of glutathione: implication in redox and detoxification. *Clin. Chim. Acta; Intl. J. Clin. Chem.* 2003, 333, 19-39.

[71] Krijt, J.; Vackova, M.; Kozich, V. Measurement of homocysteine and other aminothiols in plasma: advantages of using tris(2-carboxyethyl)phosphine as reductant compared with tri-n-butylphosphine. *Clin. Chem.* 2001, 47, 1821-1828.

[72] Cline, D.J.; Redding, S. E.; Brohawn, S. G.; Psathas, J. N.; Schneider, J. P.; Thorpe, C. New water-soluble phosphines as reductants of peptide and protein disulfide bonds: reactivity and membrane permeability. *Biochemistry*. 2004, 43, 15195-15203.

[73] Getz, E. B.; Xiao, M.; Chakrabarty, T.; Cooke, R.; Selvin, P. R. A comparison between the sulfhydryl reductants tris(2-carboxyethyl)phosphine and dithiothreitol for use in protein biochemistry. *Anal. Biochem.* 1999, 273, 73-80.

[74] Pretzer, E.; Wiktorowicz, J. E. Saturation fluorescence labeling of proteins for proteomic analyses. *Anal. Biochem.* 2008, 374, 250-262.

[75] Shafer, D. E.; Inman, J. K.; Lees A. Reaction of Tris(2-carboxyethyl)phosphine (TCEP) with maleimide and alpha-haloacyl groups: anomalous elution of TCEP by gel filtration. *Anal. Biochem.* 2000, 282, 161-164.

[76] Tyagarajan, K.; Pretzer, E.; Wiktorowicz, J. E. Thiol-reactive dyes for fluorescence labeling of proteomic samples. *Electrophoresis.* 2003, 24, 2348-2358.

[77] Krull, L. H.; Friedman, M. Reduction of protein disulfide bonds by sodium hydride in dimethyl sulfoxide. *Biochem. Biophys. Res. Commun.* 1967, 29, 373-377.

[78] Duane Brown W. Reduction of protein disulfide bonds by sodium borohydride. *Biochim. Biophys. Acta.* 1960, 44, 365-367.

[79] Cavallini, D.; Graziani, M. T.; Dupre, S. Determination of disulphide groups in proteins. *Nature.* 1966, 212, 294-295.

[80] Habeeb, A. F. A sensitive method for localization of disulfide containing peptides in column effluents. *Anal. Biochem.* 1973, 56, 60-65.

[81] Jones, D. P.; Carlson, J. L.; Samiec, P. S.; Sternberg, P.; Mody, V. C.; Reed, R. L.; Brown, L. A. Glutathione measurement in human plasma. Evaluation of sample collection, storage and derivatization conditions for analysis of dansyl derivatives by HPLC. *Clin. Chim. Acta; Intl. J. Clin. Chem.* 1998, 275, 175-184.

[82] Flatman, S.; Hurst, J. S.; McDonald-Gibson, R. G.; Jonas, G. E.; Slater, T. F. Biochemical studies on a 12-lipoxygenase in human uterine cervix. *Biochim. Biophys. Acta.* 1986, 883, 7-14.

[83] Piccoli, G.; Fiorani, M.; Biagiarelli, B.; Palma, F.; Potenza, L.; Amicucci, A.; Stocchi, V. Simultaneous high-performance capillary electrophoretic determination of reduced and oxidized glutathione in red blood cells in the femtomole range. *J. Chromatogr. A.* 1994, 676, 239-246.

[84] Ohmori, S.; Kawase, T.; Higashiura, M.; Chisaka, Y.; Nakata, K.; Yamasaki, Y. High-performance liquid chromatographic method to analyze picomole levels of glutathione, cysteine and cysteinylglycine and its application to pre-cancerous rat livers. *J. Chromatogr. B.* 2001, 762, 25-32.

[85] Majima, E.; Goto, S.; Hori, H.; Shinohara, Y.; Hong, Y. M.; Terada, H. Stabilities of the fluorescent SH-reagent eosin-5-maleimide and its adducts with sulfhydryl compounds. *Biochim. Biophys. Acta.* 1995, 1243, 336-342.

[86] Ercal, N.; Le, K.; Treeratphan, P.; Matthews, R. Analysis of thiol-containing compounds in biological samples by capillary zone electrophoresis. *Biomed. Chromatogr.* 1996, 10, 15-18.

[87] Ellman, G. L. Tissue sulfhydryl groups. *Arch. Biochem. Biophys.* 1959, 82, 70-77.

[88] Riddles, P.W.; Blakeley, R. L.; Zerner, B. Reassessment of Ellman's reagent. *Methods Enzymol.* 1983, 91, 49-60.

[89] Hansen, R. E.; Ostergaard, H.; Norgaard, P.; Winther, J. R. Quantification of protein thiols and dithiols in the picomolar range using sodium borohydride and 4,4'-dithiodipyridine. *Anal. Biochem.* 2007, 363, 77-82.

[90] Egwim, I. O.; Gruber, H. J. Spectrophotometric measurement of mercaptans with 4,4'-dithiodipyridine. *Anal. Biochem.* 2001, 288, 188-194.

[91] Imai, K.; Uzu, S.; Toyo'oka, T. Fluorogenic reagents, having benzofurazan structure, in liquid chromatography. *J. Pharm. Biomed. Anal.* 1989, 7, 1395-1403.

[92] Benkova, B.; Lozanov, V.; Ivanov, I. P.; Todorova, A.; Milanov, I.; Mitev, V. Determination of plasma aminothiols by high performance liquid chromatography after precolumn derivatization with N-(2-acridonyl)maleimide. *J. Chromatogr. B.* 2008, 870, 103-108.
[93] Imai, K.; Toyo'oka, T.; Watanabe, Y. A novel fluorogenic reagent for thiols: ammonium 7-fluorobenzo-2-oxa-1,3-diazole-4-sulfonate. *Anal. Biochem.* 1983, 128, 471-473.
[94] Toyo'oka, T.; Imai, K. High-performance liquid chromatography and fluorometric detection of biologically important thiols, derivatized with ammonium 7-fluorobenzo-2-oxa-1,3-diazole-4-sulphonate (SBD-F). *J. Chromatogr.* 1983, 282, 495-500.
[95] Nolin, T. D.; McMenamin, M. E.; Himmelfarb, J. Simultaneous determination of total homocysteine, cysteine, cysteinylglycine, and glutathione in human plasma by high-performance liquid chromatography: Application to studies of oxidative stress. *J. Chromatogr. B.* 2007, 852, 554-561.
[96] Toyooka, T.; Imai, K. New fluorogenic reagent having halogenobenzofurazan structure for thiols: 4-(aminosulfonyl)-7-fluoro-2,1,3-benzoxadiazole. *Anal. Chem.* 1984, 56, 2461-2464.
[97] Toyo'oka, T.; Imai, K. Isolation and characterization of cysteine-containing regions of proteins using 4-(aminosulfonyl)-7-fluoro-2,1,3-benzoxadiazole and high-performance liquid chromatography. *Anal. Chem.* 1985, 57, 1931-1937.
[98] Toyooka, T.; Suzuki, T.; Saito, Y.; Uzu, S.; Imai, K. Fluorogenic reagent for thiols: 4-(N,N-dimethylaminosulphonyl)-7-fluoro-2,1,3-benzoxadiazole. *Analyst.* 1989, 114, 413-419.
[99] Satoh, S.; Shindoh, M.; Min, J. Z.; Toyo'oka, T.; Fukushima, T.; Inagaki, S. Selective and sensitive determination of lipoyllysine (protein-bound alpha-lipoic acid) in biological specimens by high-performance liquid chromatography with fluorescence detection. *Analytica chimica acta.* 2008, 618, 210-217.
[100] Kirley, T. L. Reduction and fluorescent labeling of cyst(e)ine-containing proteins for subsequent structural analyses. *Anal. Biochem.* 1989, 180, 231-236.
[101] Chin, C. C.; Wold, F. The use of tributylphosphine and 4-(aminosulfonyl)-7-fluoro-2,1,3-benzoxadiazole in the study of protein sulfhydryls and disulfides. *Anal. Biochem.* 1993, 214, 128-134.
[102] Abukhalaf, I. K.; Silvestrov, N. A.; Menter, J. M.; von Deutsch, D. A.; Bayorh, M. A.; Socci, R. R.; Ganafa, A. A. High performance liquid chromatographic assay for the quantitation of total glutathione in plasma. *J. Pharm. Biomed. Anal.* 2002, 28, 637-643.
[103] Oe, T.; Ohyagi, T.; Naganuma, A. Determination of gamma-glutamylglutathione and other low-molecular-mass biological thiol compounds by isocratic high-performance

liquid chromatography with fluorimetric detection. *J. Chromatogr. B.*. 1998, 708, 285-289.

[104] Kang, S.H.; Wie, W.; Yeung, E. S. On-column derivatization for the analysis of homocysteine and other thiols by capillary electrophoresis with laser-induced fluorescence detection. *J. Chromatogr. B.* 2000, 744, 149-156.

[105] Weber, S.U.; Thiele, J. J.; Cross, C. E.; Packer L. Vitamin C, uric acid, and glutathione gradients in murine stratum corneum and their susceptibility to ozone exposure. *J. Investig. Dermatol.* 1999, 113, 1128-1132.

[106] Kosower, E.M.; Kosower, N. S. Bromobimane probes for thiols. *Methods Enzymol.* 1995, 251, 133-148.

[107] Chou, S.-T.; Ko, L.-E.; Yang, C.-S. High performance liquid chromatography with fluorimetric detection for the determination of total homocysteine in human plasma: method and clinical applications. *Anal. Chim. Acta.* 2001, 429, 331-336.

[108] Baeyens, W.; Van der Weken, G.; Ling, B.L.; De Moerloose, P. Hplc determination of N-acetylcysteine in pharmaceutical preparations after pre-column derivatization with thiolyteR MB using fluorescence detection. *Anal. Letters.* 1988, 21, 741-757.

[109] Graham, D.E.; Harich, K. C.; White, R. H. Reductive dehalogenation of monobromobimane by tris(2-carboxyethyl)phosphine. *Anal. Biochem.* 2003, 318, 325-328.

[110] van der Ven, A. J.; Mier, P.; Peters, W. H.; Dolstra, H.; van Erp, P. E.; Koopmans, P. P.; van der Meer, J. W. Monochlorobimane does not selectively label glutathione in peripheral blood mononuclear cells. *Anal. Biochem.* 1994, 217, 41-47.

[111] Kamencic, H.; Lyon, A.; Paterson, P. G.; Juurlink, B. H. Monochlorobimane fluorometric method to measure tissue glutathione. *Anal. Biochem.* 2000, 286, 35-37.

[112] Paroni, R.; De Vecchi, E.; Cighetti, G.; Arcelloni, C.; Fermo, I.; Grossi, A.; Bonini, P. HPLC with o-phthalaldehyde precolumn derivatization to measure total, oxidized, and protein-bound glutathione in blood, plasma, and tissue. *Clin. Chem.* 1995, 41, 448-454.

[113] Cereser, C.; Guichard, J.; Drai, J.; Bannier, E.; Garcia, I.; Boget, S.; Parvaz, P.; Revol, A. Quantitation of reduced and total glutathione at the femtomole level by high-performance liquid chromatography with fluorescence detection: application to red blood cells and cultured fibroblasts. *J. Chromatogr. B.* 2001, 752, 123-132.

[114] Parmentier, C.; Leroy, P.; Wellman, M.; Nicolas, A. Determination of cellular thiols and glutathione-related enzyme activities: versatility of high-performance liquid chromatography-spectrofluorimetric detection. *J. Chromatogr. B.* 1998, 719, 37-46.

[115] Neuschwander-Tetri, B. A.; Roll, F. J. Glutathione measurement by high-performance liquid chromatography separation and fluorometric detection of the glutathione-orthophthalaldehyde adduct. *Anal. Biochem.* 1989, 179, 236-241.

[116] Cohn, V. H.; Lyle, J. A fluorometric assay for glutathione. *Anal. Biochem.* 1966, 14, 434-440.

[117] Kand'ar, R.; Zakova, P.; Lotkova, H.; Kucera, O.; Cervinkova, Z. Determination of reduced and oxidized glutathione in biological samples using liquid chromatography with fluorimetric detection. *J. Pharm. Biomed. Anal.* 2007, 43, 1382-1387.

[118] Cereser, C.; Guichard, J.; Drai, J.; Bannier, E.; Garcia, I.; Boget, S.; Parvaz, P.; Revol, A. Quantitation of reduced and total glutathione at the femtomole level by high-performance liquid chromatography with fluorescence detection: application to red blood cells and cultured fibroblasts. *J. Chromatogr. B.* 2001, 752, 123-132.

[119] Michaelsen, J. T.; Dehnert, S.; Giustarini, D.; Beckmann, B.; Tsikas, D. HPLC analysis of human erythrocytic glutathione forms using OPA and N-acetyl-cysteine ethyl ester: Evidence for nitrite-induced GSH oxidation to GSSG. *J. Chromatogr.* B. 2009, 877, 3405-3417.

[120] Yi, D.; Hou, Y.; Wang, L.; Long, M.; Hu, S.; Mei, H.; Yan, L.; Hu, C.-A. A.; Wu, G. N. N-acetylcysteine stimulates protein synthesis in enterocytes independently of glutathione synthesis. *Amino acids.* 2016, 48, 523-533.

[121] Wang, W.; Dai, Z.; Wu, Z.; Lin, G.; Jia, S.; Hu, S.; Dahanayaka, S.; Wu, G. Glycine is a nutritionally essential amino acid for maximal growth of milk-fed young pigs. *Amino acids.* 2014, 46, 2037-2045.

[122] Tekwe, C. D.; Lei, J.; Yao, K.; Rezaei, R.; Li, X.; Dahanayaka, S.; Carroll, R. J.; Meininger, C. J.; Bazer, F. W.; Wu, G. Oral administration of interferon tau enhances oxidation of energy substrates and reduces adiposity in Zucker diabetic fatty rats. *Biofactors.* 2013, 39, 552-563.

[123] Yan, C. C.; Huxtable, R. J. Fluorimetric determination of monobromobimane and o-phthalaldehyde adducts of gamma-glutamylcysteine and glutathione: application to assay of gamma-glutamylcysteinyl synthetase activity and glutathione concentration in liver. *J. Chromatogr. B.* 1995, 672, 217-224.

[124] Parmentier, C.; Wellman, M.; Nicolas, A.; Siest, G.; Leroy, P. Simultaneous measurement of reactive oxygen species and reduced glutathione using capillary electrophoresis and laser-induced fluorescence detection in cultured cell lines. *Electrophoresis.* 1999, 20, 2938-2944.

[125] Sian, J.; Dexter, D. T.; Cohen, G.; Jenner, P. G.; Marsden, C. D. Comparison of HPLC and enzymatic recycling assays for the measurement of oxidized glutathione in rat brain. *J. Pharm. Pharmacol.* 1997, 49, 332-335.

[126] Colman, R. F.; Black, S. On the role of flavin adenine dinucleotide and thiol grups in the catalytic mechanism of yeast glutathione reductase. *J. Biol. Chem.* 1965, 240, 1796-1803.

[127] Guntherberg, H.; Rost, J. The true oxidized glutathione content of red blood cells obtained by new enzymic and paper chromatographic methods. *Anal. Biochem.* 1966, 15, 205-210.

[128] Giustarini, D.; Dalle-Donne, I.; Milzani, A.; Fanti, P.; Rossi, R. Analysis of GSH and GSSG after derivatization with N-ethylmaleimide. *Nature Protocols.* 2013, 8, 1660-1669.

[129] Adams, J. D. Jr.; Lauterburg, B. H.; Mitchell, J. R. Plasma glutathione and glutathione disulfide in the rat: regulation and response to oxidative stress. *J. Pharmacol. Exp. Ther.* 1983, 227, 749-754.

[130] McGill, M. R.; Jaeschke, H. A direct comparison of methods used to measure oxidized glutathione in biological samples: 2-vinylpyridine and N-ethylmaleimide. *Toxicol. Mechanisms Methods.* 2015, 25, 589-595.

[131] Araujo, A. R.; Saraiva, M. L.; Lima, J. L. Determination of total and oxidized glutathione in human whole blood with a sequential injection analysis system. *Talanta.* 2008, 74, 1511-1519.

[132] Shaik, I. H.; Mehvar, R. Rapid determination of reduced and oxidized glutathione levels using a new thiol-masking reagent and the enzymatic recycling method: application to the rat liver and bile samples. *Anal. Bioanal. Chem.* 2006, 385, 105-113.

[133] Brigelius, R.; Muckel, C.; Akerboom, T. P.; Sies, H. Identification and quantitation of glutathione in hepatic protein mixed disulfides and its relationship to glutathione disulfide. *Biochem. Pharmacol.* 1983, 32, 2529-2534.

[134] Kim, G. J.; Lee, K.; Kwon, H.; Kim, H. J. Ratiometric fluorescence imaging of cellular glutathione. *Org. Letters.* 2011,13, 2799-2801.

[135] Hao, M.; Liu, R.; Zhang, H.; Li, Y.; Jing, M. Detection of glutathione within single mice hepatocytes using microfluidic chips coupled with a laser-induced fluorescence system. *Spectrochim. Acta. A.* 2014, 125, 7-11.

[136] Li, M.; Wu, X.; Wang, Y.; Li, Y.; Zhu, W.; James, T. D. A near-infrared colorimetric fluorescent chemodosimeter for the detection of glutathione in living cells. *Chem. Commun. (Camb).* 2014, 50, 1751-1753.

[137] Tu, F. Q.; Zhang, L. Y.; Guo, X. F.; Zhang, Z. X.; Wang, H.; Zhang, H. S. Dual labeling for simultaneous determination of nitric oxide, glutathione and cysteine in macrophage RAW264.7 cells by microchip electrophoresis with fluorescence detection. *J. Chromatogr. A.* 2014, 1359, 309-316.

[138] Wang, X.; Lv, J.; Yao, X.; Li, Y.; Huang, F.; Li, M.; Yang, J.; Ruan, X.; Tang, B. Screening and investigation of a cyanine fluorescent probe for simultaneous sensing of glutathione and cysteine under single excitation. *Chem. Commun. (Camb).* 2014, 50, 15439-15442.

[139] Wang, F.; Guo, Z.; Li, X.; Li, X.; Zhao, C. Development of a small molecule probe capable of discriminating cysteine, homocysteine, and glutathione with three distinct turn-on fluorescent outputs. *Chemistry. A European J.* 2014, 20, 11471-11478.

[140] Yoshida, M.; Kamiya, M.; Yamasoba, T.; Urano, Y. A highly sensitive, cell-membrane-permeable fluorescent probe for glutathione. *Bioorg. Med. Chem. Letters.* 2014, 24, 4363-4366.

[141] Jiang, X.; Yu, Y.; Chen, J.; Zhao, M.; Chen, H.; Song, X.; Matzuk, A. J.; Carroll, S. L.; Tan, X.; Sizovs, A.; Cheng, N.; Wang, M. C.; Wang, J. Quantitative imaging of glutathione in live cells using a reversible reaction-based ratiometric fluorescent probe. *ACS Chem. Biol.* 2015,10, 864-874.

[142] Li, H.; Peng, W.; Feng, W.; Wang, Y.; Chen, G.; Wang, S.; Li, S.; Li, H.; Wang, K.; Zhang, J. A. A novel dual-emission fluorescent probe for the simultaneous detection of H_2S and GSH. *Chem Commun (Camb).* 2016, 52, 4628-4631.

[143] Xie, J.-Y.; Li, C.-Y.; Li, Y.-F.; Fei, J.; Xu, F.; Ou-Yang, J.; Liu, J. Near-infrared fluorescent probe with high quantum yield and its application in the selective detection of glutathione in living cells and tissues. *Anal. Chem.* 2016, 88, 9746-9752.

[144] Huang, R.; Wang, B.-B.; Si-Tu, X.-M.; Gao, T.; Wang, F.-F.; He, H.; Fan, X.-Y.; Jiang, F.-L.; Liu, Y. A lysosome-targeted fluorescent sensor for the detection of glutathione in cells with an extremely fast response. *Chem Commun (Camb).* 2016, 52, 11579-11582.

[145] Han, X.; Song, X.; Yu, F.; Chen, L. A ratiometric fluorescent probe for imaging and quantifying anti-apoptotic effects of GSH under temperature stress. *Chem Sci.* 2017, 8, 6991-7002.

[146] Gong, D.; Han, S.-C.; Iqbal, A.; Qian, J.; Cao, T.; Liu, W.; Liu, W.; Qin, W.; Guo, H.. Fast and selective two-stage ratiometric fluorescent probes for imaging of glutathione in living cells. *Anal. Chem.* 2017, 89, 13112-13119.

[147] Kim, Y.; Mulay, S. V.; Choi, M.; Yu, S. B.; Jon, S.; Churchill, D. G. Exceptional time response, stability and selectivity in doubly-activated phenyl selenium-based glutathione-selective platform. *Chem. Sci.* 2015, 6, 5435-5439.

[148] Heng, S.; Zhang, X.; Pei, J.; Abell, A. D. A rationally designed reversible 'turn-off' sensor for glutathione. *Biosensors.* (Basel). 2017, 7, 36-49.

[149] He, L.; Yang, X.; Xu, K.; Kong, X.; Lin, W. A multi-signal fluorescent probe for simultaneously distinguishing and sequentially sensing cysteine/homocysteine, glutathione, and hydrogen sulfide in living cells. *Chem. Sci.* 2017, 8, 6257-6265.

[150] Xue, Z.; Xiao, L.; Chen, H.; Zhou, T.; Qian, Y.; Suo, J.; Hua, Q.; Zhou, B.; Ye, R.; Bao, X.; Zhu, J. Synthesis and evaluation of a novel 'off-on' chemical sensor based on rhodamine B and the 2,5-pyrrolidinedione moiety for selective discrimination of glutathione and its bioimaging in living cells. *Bioorg. Med. Chem.* 2018, 26, 1823-1831.

[151] Lim, S. Y.; Hong, K. H.; Kim, D. I.; Kwon, H.; Kim, H. J. Tunable heptamethine-azo dye conjugate as an NIR fluorescent probe for the selective detection of mitochondrial glutathione over cysteine and homocysteine. *J. Am. Chem. Soc.* 2014,136, 7018-7025.

[152] Chen, J.; Jiang, X.; Zhang, C.; MacKenzie, K. R.; Stossi, F.; Palzkill, T.; Wang, M. C.; Wang, J. Reversible reaction-based fluorescent probe for real-time imaging of glutathione dynamics in mitochondria. *ACS Sens.* 2017, 2, 1257-1261.

[153] Sun, C.; Du, W.; Wang, P.; Wu, Y.; Wang, B.; Wang, J.; Xie, W. A novel mitochondria-targeted two-photon fluorescent probe for dynamic and reversible detection of the redox cycles between peroxynitrite and glutathione. *Biochem. Biophys. Res. Commun.* 2017, 494, 518-525.

[154] Maity, D.; Govindaraju, T. A turn-on NIR fluorescence and colourimetric cyanine probe for monitoring the thiol content in serum and the glutathione reductase assisted glutathione redox process. *Org Biomol. Chem.* 2013, 11, 2098-2104.

[155] Liu, C.-H.; Qi, F.-P.; Wen, F.-B.; Long, L.-P.; Liu, A.-J, Yang, R.-H. Fluorescence detection of glutathione and oxidized glutathione in blood with a NIR-excitable cyanine probe. *Meth. Appl. Fluoresc.* 2018, 6.

[156] Park, K. S.; Kim, M. I.; Woo, M. A.; Park, H. G. A label-free method for detecting biological thiols based on blocking of Hg^{2+}-quenching of fluorescent gold nanoclusters. *Biosens. Bioelectron.* 2013,45, 65-69.

[157] Guan, Z.; Li, S.; Cheng, P. B.; Zhou, N.; Gao, N.; Xu, Q. H. Band-selective coupling-induced enhancement of two-photon photoluminescence in gold nanocubes and its application as turn-on fluorescent probes for cysteine and glutathione. *ACS Appl. Mater. Interfaces.* 2012, 4, 5711-5716.

[158] Chen, T. H.; Tseng, W. L. (Lysozyme type VI)-stabilized Au8 clusters: Synthesis mechanism and application for sensing of glutathione in a single drop of blood. *Small.* 2012, 8, 1912-1919.

[159] Han, B.; Wang, E. Oligonucleotide-stabilized fluorescent silver nanoclusters for sensitive detection of biothiols in biological fluids. *Biosens. Bioelectron.* 2011, 26, 2585-2589.

[160] Deng, R.; Xie, X.; Vendrell, M.; Chang, Y. T.; Liu, X. Intracellular glutathione detection using MnO(2)-nanosheet-modified upconversion nanoparticles. *J. Am. Chem. Soc.* 2011, 133, 20168-20171.

[161] Liu, J.; Bao, C.; Zhong, X.; Zhao, C.; Zhu, L. Highly selective detection of glutathione using a quantum-dot-based OFF-ON fluorescent probe. *Chem. Commun. (Camb).* 2010, 46, 2971-2973.

[162] Sheng, Z.; Chen, L. Switch-on fluorescent strategy based on crystal violet-functionalized CdTe quantum dots for detecting L-cysteine and glutathione in water and urine. *Anal. Bioanal. Chem.* 2017, 409, 6081-6090.

[163] Wu, Z.; Li, W.; Chen, J.; Yu, C. A graphene quantum dot-based method for the highly sensitive and selective fluorescence turn on detection of biothiols. *Talanta.* 2014, 119, 538-543.

[164] Xu, Y.; Niu, X.; Zhang, H.; Xu, L.; Zhao, S.; Chen, H.; Chen, X. Switch-on fluorescence sensing of glutathione in food samples based on a graphitic carbon

nitride quantum dot (g-CNQD)-Hg(2)(+) chemosensor. *J. Agric. Food Chem.* 2015, 63, 1747-1755.

[165] Pan, J.; Zheng, Z.; Yang, J.; Wu, Y.; Lu, F.; Chen, Y.; Gao, W. A novel and sensitive fluorescence sensor for glutathione detection by controlling the surface passivation degree of carbon quantum dots. *Talanta.* 2017, 166, 1-7.

[166] Jahanbakhshi, M.; Habibi, B. A novel and facile synthesis of carbon quantum dots via salep hydrothermal treatment as the silver nanoparticles support: Application to electroanalytical determination of H_2O_2 in fetal bovine serum. *Biosens. Bioelectron.* 2016, 81, 143-150.

[167] Cai, Q.-Y.; Li, J.; Ge, J.; Zhang, L.; Hu, Y.-L.; Li, Z.-H.; Qu, L.-B. A rapid fluorescence "switch-on" assay for glutathione detection by using carbon dots-MnO2 nanocomposites. *Biosens. Bioelectron.* 2015, 72, 31-36.

[168] Li, L.; Wang, X.; Li, Q.; Liu, P.; Xu, K.; Chen, H.; Tang, B. An accurate mass spectrometric approach for the simultaneous comparison of GSH, Cys, and Hcy in L02 cells and HepG2 cells using new NPSP isotope probes. *Chem. Commun. (Camb).* 2015, 51, 11317-11320.

[169] Sun, Z. J.; Jiang, J. Z.; Li, Y. F. A sensitive and selective sensor for biothiols based on the turn-on fluorescence of the Fe-MIL-88 metal-organic frameworks-hydrogen peroxide system. *Analyst.* 2015, 140, 8201-8208.

[170] Yao, X.; Wang, Y.; Chen, G. Simultaneous determination of aminothiols, ascorbic acid and uric acid in biological samples by capillary electrophoresis with electrochemical detection. *Biomed. Chromatogr.* 2007, 21, 520-526.

[171] Tea, I.; Ferchaud-Roucher, V.; Küster, A.; Darmaun, D.; Robins, R. J. Determination of 13C isotopic enrichment of glutathione and glycine by gas chromatography/combustion/isotope ratio mass spectrometry after formation of the N- or N,S-ethoxycarbonyl methyl ester derivatives. *Rapid. Commun. Mass Spectrom.* 2007, 21, 3245-3252.

[172] Ma, Y.; Kind, T.; Vaniya, A.; Gennity, I.; Fahrmann, J. F.; Fiehn, O. An in silico MS/MS library for automatic annotation of novel FAHFA lipids. *J. Cheminform.* 2015, 7, 53.

[173] Nana, C. G.; Jian, W.; Xi, C.; Pinga, D. J.; Feng, Z. Z.; Qing, C. H. The enhanced electrogenerated chemiluminescence of Ru(bpy)3(2+) by glutathione on a glassy carbon electrode modified with some porphine compounds. *Analyst.* 2000, 125, 2294-2298.

[174] Wang, Y.; Lu, J.; Tang, L.; Chang, H.; Li, J. Graphene oxide amplified electrogenerated chemiluminescence of quantum dots and its selective sensing for glutathione from thiol-containing compounds. *Anal. Chem.* 2009, 81, 9710-9715.

[175] Zhang, W.; Zhang, R.; Zhang, J.; Ye, Z.; Jin, D.; Yuan, J. Photoluminescent and electrochemiluminescent dual-signaling probe for bio-thiols based on a ruthenium(II) complex. *Anal. Chim. Acta.* 2012, 740, 80-87.

[176] Niu, W.-J.; Zhu, R.-H.; Cosnier, S.; Zhang, X.-J.; Shan, D. Ferrocyanide-ferricyanide redox couple induced electrochemiluminescence amplification of carbon dots for ultrasensitive sensing of glutathione. *Anal. Chem.* 2015, 87, 11150-11156.

[177] Gao, W.; Liu, Z.; Qi, L.; Lai, J.; Kitte, S. A.; Xu, G. Ultrasensitive glutathione detection based on lucigenin cathodic electrochemiluminescence in the presence of MnO_2 Nanosheets. *Anal. Chem.* 2016, 88, 7654-7659.

[178] Yu, Y.; Shi, J.; Zhao, X.; Yuan, Z.; Lu, C.; Lu, J. Electrochemiluminescence detection of reduced and oxidized glutathione ratio by quantum dot-layered double hydroxide film. *Analyst.* 2016, 141, 3305-3312.

[179] Peng, H.-P.; Jian, M.-L.; Huang, Z.-N.; Wang, W.-J.; Deng, H.-H.; Wu, W.-H.; Liu, A.-L.; Xia, X.-H.; Chen, W. Facile electrochemiluminescence sensing platform based on high-quantum-yield gold nanocluster probe for ultrasensitive glutathione detection. *Biosens. Bioelectron.* 2018, 105, 71-76.

[180] Atta, N. F.; Galal, A.; Azab, S. M. Novel sensor based on carbon paste/Nafion(R) modified with gold nanoparticles for the determination of glutathione. *Anal. Bioanal. Chem.* 2012, 404, 1661-1672.

[181] Chen, G.; Wang, J.; Wu, C.; Li, C.; Jiang, H.; Wang, X. Photoelectrocatalytic oxidation of glutathione based on porous TiO_2-Pt nanowhiskers. *Langmuir.* 2012, 28, 12393-12399.

[182] He, H.; Du, J.; Hu, Y.; Ru, J.; Lu, X. Detection of glutathione based on nickel hexacyanoferrate film modified Pt ultramicroelectrode by introducing cetyltrimethylammonium bromide and Au nanoparticles. *Talanta.* 2013, 115, 381-385.

[183] Beitollahi, H.; Gholami, A.; Ganjali, M. R. Preparation, characterization and electrochemical application of Ag-ZnO nanoplates for voltammetric determination of glutathione and tryptophan using modified carbon paste electrode. *Mater. Sci. Eng. C.* 2015, 57, 107-112.

[184] Zhu, W.; Jiang, G.; Xu, L.; Li, B.; Cai, Q.; Jiang, H.; Zhou, X. Facile and controllable one-step fabrication of molecularly imprinted polymer membrane by magnetic field directed self-assembly for electrochemical sensing of glutathione. *Anal. Chim. Acta.* 2015, 886, 37-47.

[185] Wang, T.; Su, W.; Xiao, Z.; Hao, S.; Li, Y.; Hu, J. Highly sensitive determination of reduced glutathione based on a cobalt nanoparticle implanted-modified indium tin oxide electrode. *Analyst.* 2015, 140, 5176-5183.

[186] Raoof, J. B.; Teymoori, N.; Khalilzadeh, M. A.; Ojani, R. A high sensitive electrochemical nanosensor for simultaneous determination of glutathione, NADH and folic acid. *Mater. Sci. Eng. C.* 2015, 47, 77-84.

[187] Safavi, A.; Maleki, N.; Farjami, E.; Mahyari, F. A. Simultaneous electrochemical determination of glutathione and glutathione disulfide at a nanoscale copper

hydroxide composite carbon ionic liquid electrode. *Anal. Chem.* 2009, 81, 7538-7543.

[188] Olmos Moya, P. M.; Martínez Alfaro, M.; Kazemi, R.; Alpuche-Avilés, M. A.; Griveau, S.; Bedioui, F.; Gutiérrez Granados, S. Simultaneous electrochemical speciation of oxidized and reduced glutathione. Redox profiling of oxidative stress in biological fluids with a modified carbon electrode. *Anal. Chem.* 2017, 89, 10726-10733.

[189] Zhang, B.; Liu, J.; Ma, X.; Zuo, P.; Ye, B.-C.; Li, Y. Ultrasensitive and selective assay of glutathione species in arsenic trioxide-treated leukemia HL-60 cell line by molecularly imprinted polymer decorated electrochemical sensors. *Biosens. Bioelectron.* 2016, 80, 491-496.

[190] Ru, J.; Du, J.; Qin, D.-D.; Huang, B.-M.; Xue, Z.-H.; Zhou, X.-B.; Lu, X.-Q. An electrochemical glutathione biosensor: ubiquinone as a transducer. *Talanta.* 2013, 110, 15-20.

[191]Anik, Ü.; Çubukçu, M.; Ertaş, F. N. An effective electrochemical biosensing platform for the detection of reduced glutathione. *Artif. Cells Nanomed. Biotechnol.* 2016, 44, 971-977.

[192] Barman, U.; Mukhopadhyay, G.; Goswami, N.; Ghosh, S. S.; Paily, R. P. Detection of glutathione by glutathione-S-transferase-nanoconjugate ensemble electrochemical device. *IEEE Trans. Nanobiosci.* 2017, 16, 271-279.

[193] Lotfi Zadeh Zhad, H. R.; Lai, R. Y. A Hg(II)-mediated "signal-on" electrochemical glutathione sensor. *Chem. Commun. (Camb).* 2014, 50, 8385-8387.

[194] Wang, Y.; Jiang, L.; Leng, Q.; Wu, Y.; He, X.; Wang, K. Electrochemical sensor for glutathione detection based on mercury ion triggered hybridization chain reaction signal amplification. *Biosens. Bioelectron.* 2016, 77, 914-920.

[195] Lv, Y.; Yang, L.; Mao, X.; Lu, M.; Zhao, J.; Yin, Y. Electrochemical detection of glutathione based on Hg(2+)-mediated strand displacement reaction strategy. *Biosens. Bioelectron.* 2016, 85, 664-668.

[196] Liu, B.; Ma, C.; Li, Y.; Kou, Y.; Lu, J.; Jiang, X.; Tan, L. Voltammetric determination of reduced glutathione using poly(thionine) as a mediator in the presence of Fenton-type reaction. *Talanta.* 2017, 170, 399-405.

[197] Terpstra, M.; Vaughan, T. J.; Ugurbil, K.; Lim, K. O.; Schulz, S. C.; Gruetter, R. Validation of glutathione quantitation from STEAM spectra against edited ^{1}H NMR spectroscopy at 4T: application to schizophrenia. *MAGMA.* 2005, 18, 276-282.

[198] Wijtenburg, S. A.; Gaston, F. E.; Spieker, E. A.; Korenic, S. A.; Kochunov, P.; Hong, L. E.; Rowland, L. M. Reproducibility of phase rotation STEAM at 3T: focus on glutathione. *Magn. Reson. Med.* 2014, 72, 603-609.

[199] Deelchand, D. K.; Marjańska, M.; Hodges, J. S.; Terpstra, M. Sensitivity and specificity of human brain glutathione concentrations measured using short-TE (1)H MRS at 7 T. *NMR Biomed.* 2016, 29, 600-606.

[200] An, L.; Zhang, Y.; Thomasson, D. M.; Latour, L. L.; Baker, E. H.; Shen, J.; Warach, S. Measurement of glutathione in normal volunteers and stroke patients at 3T using J-difference spectroscopy with minimized subtraction errors. *J. Magn. Reson. Imaging.* 2009, 30, 263-270.

[201] Prinsen, H.; de Graaf, R. A.; Mason, G. F.; Pelletier, D.; Juchem, C. Reproducibility measurement of glutathione, GABA, and glutamate: Towards *in vivo* neurochemical profiling of multiple sclerosis with MR spectroscopy at 7T. *J. Magn. Reson. Imaging.* 2017, 45, 187-198.

[202] Chan, K. L.; Puts, N. A. J.; Snoussi, K.; Harris, A. D.; Barker, P. B.; Edden, R. A. E. Echo time optimization for J-difference editing of glutathione at 3T. *Magn. Reson. Med.* 2017, 77, 498-504.

[203] Trabesinger, A. H.; Weber, O. M.; Duc, C. O.; Boesiger, P. Detection of glutathione in the human brain *in vivo* by means of double quantum coherence filtering. *Magn. Reson. Med.* 1999, 42, 283-289.

[204] Zhao, T.; Heberlein, K.; Jonas, C.; Jones, D. P.; Hu, X. New double quantum coherence filter for localized detection of glutathione in vivo. *Magn. Reson. Med.* 2006, 55, 676-680.

[205] An, L.; Li, S.; Murdoch, J. B.; Araneta, M. F.; Johnson, C.; Shen, J. Detection of glutamate, glutamine, and glutathione by radiofrequency suppression and echo time optimization at 7 tesla. *Magn. Reson. Med.* 2015, 73, 451-458.

[206] Lally, N.; An, L.; Banerjee, D.; Niciu, M. J.; Luckenbaugh, D. A.; Richards, E. M.; Roiser, J. P.; Shen, J.; Zarate, C. A.; Nugent, A. C. Reliability of 7T (1) H-MRS measured human prefrontal cortex glutamate, glutamine, and glutathione signals using an adapted echo time optimized PRESS sequence: A between- and within-sessions investigation. *J. Magn. Reson. Imaging.* 2016, 43, 88-98.

[207] Emir, U. E.; Raatz, S.; McPherson, S.; Hodges, J. S.; Torkelson, C.; Tawfik, P.; White, T.; Terpstra, M. Noninvasive quantification of ascorbate and glutathione concentration in the elderly human brain. *NMR in Biomedicine.* 2011, 24, 888-894.

[208] Mandal, P. K.; Tripathi, M.; Sugunan, S. Brain oxidative stress: detection and mapping of anti-oxidant marker 'Glutathione' in different brain regions of healthy male/female, MCI and Alzheimer patients using non-invasive magnetic resonance spectroscopy. *Biochem. Biophys. Res. Commun.* 2012, 417, 43-48.

[209] Raschke, F.; Noeske, R.; Dineen, R. A.; Auer, D. P. Measuring cerebral and cerebellar glutathione in children using (1)H MEGA-PRESS MRS. *Am. J. Neuroradiol.* 2018, 39, 375-379.

[210] Sanaei Nezhad, F.; Anton, A.; Parkes, L. M.; Deakin, B.; Williams, S. R. Quantification of glutathione in the human brain by MR spectroscopy at 3 Tesla: Comparison of PRESS and MEGA-PRESS. *Magn. Reson. Med.* 2017, 78, 1257-1266.

[211] Schlucker, S. Surface-enhanced Raman spectroscopy: concepts and chemical applications. *Angew. Chem. Int. Ed. Engl.* 2014, 53, 4756-4795.

[212] Cialla, D.; März, A.; Böhme, R.; Theil, F.; Weber, K.; Schmitt, M.; Popp, J. Surface-enhanced Raman spectroscopy (SERS): progress and trends. *Anal. Bioanal. Chem.* 2012, 403, 27-54.

[213] Huang, G. G.; Hossain, M. K.; Han, X. X.; Ozaki, Y. A novel reversed reporting agent method for surface-enhanced Raman scattering; highly sensitive detection of glutathione in aqueous solutions. *Analyst.* 2009, 134, 2468-2474.

[214] Huang, G. G.; Han, X. X.; Hossain, M. K.; Ozaki, Y. Development of a heat-induced surface-enhanced Raman scattering sensing method for rapid detection of glutathione in aqueous solutions. *Anal. Chem.* 2009, 81, 5881-5888.

[215] Saha, A.; Jana, N. R. Detection of cellular glutathione and oxidized glutathione using magnetic-plasmonic nanocomposite-based "turn-off" surface enhanced Raman scattering. *Anal. Chem.* 2013, 85, 9221-9228.

[216] Wang, W.; Zhang, L.; Li, L.; Tian, Y. A single nanoprobe for ratiometric imaging and biosensing of hypochlorite and glutathione in live cells using surface-enhanced Raman scattering. *Anal. Chem.* 2016, 88, 9518-9523.

[217] Kuligowski, J.; EL-Zahry, M. R.; Sánchez-Illana, Á.; Quintás, G.; Vento, M.; Lendl, B. Surface enhanced Raman spectroscopic direct determination of low molecular weight biothiols in umbilical cord whole blood. *Analyst.* 2016, 141, 2165-2174.

[218] Ouyang, L.; Zhu, L.; Jiang, J.; Tang, H. A surface-enhanced Raman scattering method for detection of trace glutathione on the basis of immobilized silver nanoparticles and crystal violet probe. *Anal. Chim. Acta.* 2014, 816, 41-49.

[219] Abalde-Cela, S.; Aldeanueva-Potel, P.; Mateo-Mateo, C.; Rodríguez-Lorenzo, L.; Alvarez-Puebla, R. A.; Liz-Marzán, L. M. Surface-enhanced Raman scattering biomedical applications of plasmonic colloidal particles. *J. Royal Soc. Interface.* 2010, 7, S435-S450.

[220] Kuśmierek, K.; Bald, E. Reversed-phase liquid chromatography method for the determination of total plasma thiols after derivatization with 1-benzyl-2-chloropyridinium bromide. *Biomed. Chromatogr.* 2009, 23, 770-775.

[221] Głowacki, R.; Bald, E. Fully automated method for simultaneous determination of total cysteine, cysteinylglycine, glutathione and homocysteine in plasma by HPLC with UV absorbance detection. *J. Chromatogr. B.* 2009, 877, 3400-3404.

[222] Borowczyk, K.; Wyszczelska-Rokiel, M.; Kubalczyk, P.; Głowacki, R. Simultaneous determination of albumin and low-molecular-mass thiols in plasma by HPLC with UV detection. *J. Chromatogr. B.* 2015, 981-982, 57-64.

[223] Kand'ár, R.; Žáková, P.; Lotková, H.; Kučera, O.; Červinková, Z. Determination of reduced and oxidized glutathione in biological samples using liquid chromatography with fluorimetric detection. *J. Pharm. Biomed. Anal.* 2007, 43, 1382-1387.

[224] Sakhi, A. K.; Blomhoff, R.; Gundersen, T. E. Simultaneous and trace determination of reduced and oxidized glutathione in minute plasma samples using dual mode fluorescence detection and column switching high performance liquid chromatography. *J. Chromatogr. A.* 2007, 1142, 178-184.

[225] Ju, X. H.; Tang, S.; Jia, Y.; Guo, J.; Ding, Y.; Song, Z.; Zhao, Y. Determination and characterization of cysteine, glutathione and phytochelatins (PC(2)(-)(6)) in *Lolium perenne* L. exposed to Cd stress under ambient and elevated carbon dioxide using HPLC with fluorescence detection. *J. Chromatogr. B.* 2011, 879, 1717-1724.

[226] Ichinose, S.; Nakamura, M.; Maeda, M.; Ikeda, R.; Wada, M.; Nakazato, M.; Ohba, Y.; Takamura, N.; Maeda, T.; Aoyagi, K.; Nakashima, K. A validated HPLC-fluorescence method with a semi-micro column for routine determination of homocysteine, cysteine and cysteamine, and the relation between the thiol derivatives in normal human plasma. *Biomed. Chromatogr.* 2009, 23, 935-939.

[227] Ferin, R.; Pavao, M. L.; Baptista, J. Methodology for a rapid and simultaneous determination of total cysteine, homocysteine, cysteinylglycine and glutathione in plasma by isocratic RP-HPLC. *J. Chromatogr. B.* 2012, 911, 15-20.

[228] Isokawa, M.; Funatsu, T.; Tsunoda, M. Fast and simultaneous analysis of biothiols by high-performance liquid chromatography with fluorescence detection under hydrophilic interaction chromatography conditions. *Analyst.* 2013, 138, 3802-3208.

[229] Kono, Y.; Iizuka, H.; Isokawa, M.; Tsunoda, M.; Ichiba, H.; Sadamoto, K.; Fukushima, T. HPLC-fluorescence determination of thiol compounds in the serum of human male and female subjects using HILIC-mode column. *Biomed. Chromatogr.* 2014, 28, 589-593.

[230] Cevasco, G.; Piatek, A. M.; Scapolla, C.; Thea, S. An improved method for simultaneous analysis of aminothiols in human plasma by high-performance liquid chromatography with fluorescence detection. *J. Chromatogr. A.* 2010, 1217, 2158-2162.

[231] Marchand, S.; de Revel, G. A HPLC fluorescence-based method for glutathione derivatives quantification in must and wine. *Anal. Chim. Acta.* 2010, 660, 158-163.

[232] Kanďár, R.; Štramová, X.; Drábková, P.; Brandtnerová, M. Determination of total glutathione in dried blood spot samples using a high-performance liquid chromatography. *J. Chromatogr. Sci.* 2015, 53, 879-885.

[233] Wang, X.; Chi, D.; Song, D.; Su, G.; Li, L.; Shao, L. Quantification of glutathione in plasma samples by HPLC using 4-fluoro-7-nitrobenzofurazan as a fluorescent labeling reagent. *J. Chromatogr. Sci.* 2012, 50, 119-122.

[234] Wada, M.; Hirose, M.; Kuroki, M.; Ikeda, R.; Sekitani, Y.; Takamura, N.; Kuroda, N.; Nakashima, K. Simultaneous determination of homocysteine, methionine and cysteine in maternal plasma after delivery by HPLC-fluorescence detection with DBD-F as a label. *Biomed. Chromatogr.* 2013, 27, 708-713.

[235] Bailey, B.; Waraska, J.; Acworth, I. Direct determination of tissue aminothiol, disulfide, and thioether levels using HPLC-ECD with a novel stable boron-doped diamond working electrode. *Methods Mol. Biol.* 2010, 594, 327-339.
[236] Park, H. J.; Mah, E.; Bruno, R. S. Validation of high-performance liquid chromatography-boron-doped diamond detection for assessing hepatic glutathione redox status. *Anal. Biochem.* 2010, 407, 151-159.
[237] Childs, S.; Haroune, N.; Williams, L.; Gronow, M. Determination of cellular glutathione:glutathione disulfide ratio in prostate cancer cells by high performance liquid chromatography with electrochemical detection. *J. Chromatogr. A.* 2016, 1437, 67-73.
[238] Yap, L.-P.; Sancheti, H.; Ybanez, M. D.; Garcia, J.; Cadenas, E.; Han, D. Determination of GSH, GSSG, and GSNO Using HPLC with electrochemical detection. Thiol redox transitions in cell signaling, Part A: Chemistry and biochemistry of low molecular weight and protein thiols. *Methods in Enzymol.* 2010, 137-47.
[239] Khan, A.; Khan, M. I.; Iqbal, Z.; Shah, Y.; Ahmad, L.; Nazir, S.; Watson, D. G.; Khan, J. A.; Nasir, F.; Khan, A.; Ismail. A new HPLC method for the simultaneous determination of ascorbic acid and aminothiols in human plasma and erythrocytes using electrochemical detection. *Talanta.* 2011, 84, 789-801.
[240] Bayram, B.; Rimbach, G.; Frank, J.; Esatbeyoglu, T. Rapid method for glutathione quantitation using high-performance liquid chromatography with coulometric electrochemical detection. *J. Agricult. Food Chem.* 2013, 62, 402-408.
[241] Iwasaki, Y.; Nakano, Y.; Mochizuki, K.; Ogawa, T.; Oda, M.; Ito, R.; Saito, K.; Nakazawa, H. Quantification of reduced and oxidized thiols in mouse serum by column-switching hydrophilic interaction chromatography coupled with mass spectrometry. *J. Pharm. Biomed. Anal.* 2011, 56, 103-113.
[242] Robin, S.; Leveque, N.; Courderot-Masuyer, C.; Humbert, P. LC-MS determination of oxidized and reduced glutathione in human dermis: a microdialysis study. *J. Chromatogr. B.* 2011, 879, 3599-3606.
[243] Suh, J. H.; Kim, R.; Yavuz, B.; Lee, D.; Lal, A.; Ames, B. N.; Shigenaga, M. K.l. Clinical assay of four thiol amino acid redox couples by LC-MS/MS: utility in thalassemia. *J. Chromatogr. B.* 2009, 877, 3418-3427.
[244] Huang, Y.-Q.; Ruan, G.-D.; Liu, J.-Q.; Gao, Q.; Feng, Y.-Q. Use of isotope differential derivatization for simultaneous determination of thiols and oxidized thiols by liquid chromatography tandem mass spectrometry. *Anal. Biochem.* 2011, 416, 159-166.
[245] Moore, T.; Le, A.; Niemi, A.-K.; Kwan, T.; Cusmano-Ozog, K.; Enns, G. M.; Cowan, T. M. A new LC–MS/MS method for the clinical determination of reduced and oxidized glutathione from whole blood. *J. Chromatogr B.* 2013, 929, 51-55.

[246] Reinbold, J.; Koehler, P.; Rychlik, M. Quantitation of glutathione and its oxidation products in erythrocytes by multiple-label stable-isotope dilution. *Anal. Biochem.* 2014, 445, 41-48.

[247] Yu, J.; Li, C.; Shen, S.; Liu, X.; Peng, Y.; Zheng, J. Mass spectrometry based detection of glutathione with sensitivity for single-cell analysis. *Rapid Commun. Mass Spectrom.* 2015, 29, 681-689.

[248] Sun, Y.; Yao, T.; Guo, X.; Peng, Y.; Zheng, J. Simultaneous assessment of endogenous thiol compounds by LC-MS/MS. *J. Chromatogr. B.* 2016, 1029-1030, 213-221.

[249] Jiang, Z.; Liang, Q.; Luo, G.; Hu, P.; Li, P.; Wang, Y. HPLC-electrospray tandem mass spectrometry for simultaneous quantitation of eight plasma aminothiols: application to studies of diabetic nephropathy. *Talanta.* 2009, 77, 1279-1284.

[250] Squellerio, I.; Caruso, D.; Porro, B.; Veglia, F.; Tremoli, E.; Cavalca, V. Direct glutathione quantification in human blood by LC-MS/MS: comparison with HPLC with electrochemical detection. *J. Pharm. Biomed. Anal.* 2012, 71, 111-118.

[251] Norris, R. L.; Paul, M.; George, R.; Moore, A.; Pinkerton, R.; Haywood, A.; Charles, B. A stable-isotope HPLC-MS/MS method to simplify storage of human whole blood samples for glutathione assay. *J. Chromatogr. B.* 2012, 898, 136-140.

[252] Kritzinger, E. C.; Stander, M. A.; Du Toit, W. J. Assessment of glutathione levels in model solution and grape ferments supplemented with glutathione-enriched inactive dry yeast preparations using a novel UPLC-MS/MS method. *Food Addit. Contam. Part A. Chem. Anal. Control. Expo. Risk Assess.* 2013, 30, 80-92.

[253] Vallverdú-Queralt, A.; Verbaere, A.; Meudec, E.; Cheynier, V.; Sommerer, N. Straightforward method to quantify GSH, GSSG, GRP, and hydroxycinnamic acids in wines by UPLC-MRM-MS. *J. Agric. Food Chem.* 2015, 63, 142-149.

[254] Kazarjan, J.; Vaher, M.; Mahlapuu, R.; Hansen, M.; Soomets, U.; Kaljurand, M. Separation of glutathione and its novel analogs and determination of their dissociation constants by capillary electrophoresis. *Electrophoresis.* 2013, 34, 1820-1827.

[255] Hempe, J. M.; Ory-Ascani, J. Simultaneous analysis of reduced glutathione and glutathione disulfide by capillary zone electrophoresis. *Electrophoresis.* 2014, 35, 967-971.

[256] Serru, V.; Baudin, B.; Ziegler, F.; David, J. P.; Cals, M. J.; Vaubourdolle, M.; Mario, N. Quantification of reduced and oxidized glutathione in whole blood samples by capillary electrophoresis. *Clinical Chem.* 2001, 47, 1321-1324.

[257] Kowalska, K.; Zalewska, M.; Milnerowicz, H. The application of capillary electrophoresis in the determination of glutathione in healthy women's blood. *J. Chromatogr. Sci.* 2015, 53, 353-359.

[258] Tsardaka, E.-C.; Zacharis, C. K.; Tzanavaras, P. D.; Zotou, A. Determination of glutathione in baker's yeast by capillary electrophoresis using methyl propiolate as derivatizing reagent. *J. Chromatogr. A.* 2013, 1300, 204-208.

[259] Tu, F.-Q.; Zhang, L.-Y.; Guo, X.-F.; Wang, H.; Zhang, H.-S. Development of a potential method based on microchip electrophoresis with fluorescence detection for the sensitive determination of intracellular thiols in RAW264.7 cells. *Electrophoresis.* 2014, 35, 1188-1195.

[260] Zhang, L.-Y.; Tu, F.-Q.; Guo, X.-F.; Wang, H.; Wang, P.; Zhang, H.-S. Rapid and sensitive determination of free thiols by capillary zone electrophoresis with near-infrared laser-induced fluorescence detection using a new BODIPY-based probe as labeling reagent. *Electrophoresis.* 2014, 35, 2951-2958.

[261] Guo, X.-F.; Arceo, J.; Huge, B. J.; Ludwig, K. R.; Dovichi, N. J. Chemical cytometry of thiols using capillary zone electrophoresis-laser induced fluorescence and TMPAB-o-M, an improved fluorogenic reagent. *Analyst.* 2016, 141, 1325-1330.

[262] Duan, Y.-J.; Zhang, L.-Y.; Guo, X.-F.; Wang, H. A CE-LIF method based on long wavelength fluorescence labeling for the analysis of thiols in human urine. *Electrophoresis.* 2016, 37, 2097-2105.

[263] Hodáková, J.; Preisler, J.; Foret, F.; Kubán, P. Sensitive determination of glutathione in biological samples by capillary electrophoresis with green (515 nm) laser-induced fluorescence detection. *J. Chromatogr. A.* 2015, 1391, 102-108.

[264] Wu, X.; Cao, Y.; Zhang, J.; Lei, M.; Deng, X.; Zahid, K. R.; Liu, Y.; Liu, K.; Yang, J.; Xiong, G,; et al. Determination of glutathione in apoptotic SMMC-7221 cells induced by xylitol selenite using capillary electrophoresis. *Biotechnol. Lett.* 2016, 38, 761-766.

[265] Yuan, W.; Edwards, J. L. Thiol metabolomics of endothelial cells using capillary liquid chromatography mass spectrometry with isotope coded affinity tags. *J. Chromatogr. A.* 2011, 1218, 2561-2568.

[266] D'Agostino, L. A.; Lam, K. P.; Lee, R.; Britz-McKibbin, P. Comprehensive plasma thiol redox status determination for metabolomics. *J. Proteome Res.* 2011, 10, 592-603.

[267] Küster, A.; Tea, I.; Sweeten, S.; Rozé, J.-C.; Robins, R. J.; Darmaun, D. Simultaneous determination of glutathione and cysteine concentrations and 2H enrichments in microvolumes of neonatal blood using gas chromatography-mass spectrometry. *Anal. Bioanal. Chem.* 2008, 390, 1403-1412.

[268] Manca, A.; Alladio, E.; Massarenti, P.; Puccinelli, M. P.; De Francesco, A.; Del Grosso, E.; Mengozzi, G.; Pazzi, M.; Vincenti, M. "One-pot" ethyl chloroformate derivatization and liquid-liquid extraction of reduced glutathione in erythrocyte and its quantitative GC-MS analysis. *J. Chromatogr. B.* 2017, 1070, 15-22.

[269] Tsikas, D.; Hanff, E.; Kayacelebi, A. A.; Böhmer, A. Gas chromatographic-mass spectrometric analysis of the tripeptide glutathione in the electron-capture negative-ion chemical ionization mode. *Amino Acids.* 2016, 48, 593-598.

BIOGRAPHICAL SKETCH

Jiang Zheng

Present Position

Professor

Affiliation

Shenyang Pharmaceutical University and Guizhou Medical University, Shenyang, Liaoning, China

Education

1978 – 1982:	B.S. in Pharmacy, Zhejiang Medical University, China.
1983 – 1985:	M.S. in Pharmacology, Zhejiang Medical University, China.
1988 – 1992:	M.S. and Ph.D. in Medicinal Chemistry, School of Pharmacy, University of Kansas, USA.

Research and Professional Experience

1993 – 1995	Post-doctoral Fellow, Department of Environmental Toxicology, University of California at Davis.
1993 – 1995	Pharmacologist, Department of Internal Medicine, School of Medicine, University of California at Davis.
1999 – 2006	Assistant Professor, Department of Pharmaceutical Sciences, School of Pharmacy, Northeastern University.
2006 – 2015	Associate Professor and Principal Investigator, Division of Gastroenterology and Hepatology, Department of Pediatrics, University of Washington School of Medicine.
2016 – present	Professor, Shenyang Pharmaceutical University and Guizhou Medical University , China.

Short Biography

Professor Jiang Zheng received his Ph.D. in Medicinal Chemistry at University of Kansas. Then he relocated to University of California at Davis to obtain his postdoctoral training. Dr. Zheng worked at Northeastern University School of Pharmacy and then University of Washington School of Medicine as Assistant Professor, Associate Professor, and Principal Investigator. Currently, he is employed as Professor at Guizhou Medical University as well as Shenyang Pharmaceutical University. His research interests cover three major areas, including drug metabolism, drug toxicity, and drug design. Dr. Zheng has published over 100 peer-reviewed scientific papers. Professor Zheng was a recipient of "Highly prolific author (most published articles in the past 5 years) in Chemical Research in Toxicology", presented by American Chemical Society in 2017. Dr. Zheng was awarded as "American Cancer Society Research Scholar" presented by American Cancer Society. He has been serving on Editorial Advisory Board of Chemical Research in Toxicology and Journal of Biochemical and Molecular Toxicology.

Publications from the Last 3 Years:

1. Zhang, T.; Shen, S.; Zhu, Z.; Lu, S. Yin, X.; Zheng, J.; Jin, J. Homoharringtonine binds to and increases myosin-9 in myeloid leukemia. *Brit. J. Pharmacol.* 2016, 173, 212-221.
2. Chi, M.; Peng, Y.; Zheng, J. Characterization of glutathione conjugates derived from reactive metabolites of bakuchiol. *Chem. Biol. Interact.* 2016, 244, 178-186.
3. Lu, D.; Zheng, L.; Cao, J.; Ji, L.; Peng, Y.; Zheng, J. Mechanism-based inactivation of CYP2B6 by isopsoralen. *Xenobiotica.* 2016, 46, 335-341.
4. Li, W.; Lin, D.; Gao, H.; Xu, Y.; Smith, C. V.; Peng, Y.; Zheng, J. Metabolic activation of the furan moiety makes diosbulbin B hepatotoxic. *Arch. Toxicol.* 2016, 90, 863-872.
5. Chen, M.; Li, L.; Zhong, D.; Zheng, J.; Chen, X. Metabolic activation of retronecine-type pyrrolizidine alkaloids in rats and in human liver microsomes, and glutathione conjugation of their reactive metabolites. *Chem. Res. Toxicol.* 2016, 29, 180-189.
6. Guo, X.; Lin, D.; Li, W.; Wang, K.; Peng, Y.; Zheng, J. Electrophilicities and protein covalent binding of demethylation metabolites of colchicines. *Chem. Res. Toxicol.* 2016, 29, 296-302.
7. Li, W.; Lin, G.; Peng, Y.; Zheng, J. Lysine adduction by reactive metabolite(s) of monocrotaline. *Chem. Res. Toxicol.* 2016, 29, 333-341.
8. Lin, D.; Li, W.; Peng, Y.; Jiang, C.; Xu, Y.; Gao, H.; Zheng, J. Role of metabolic activation in 8-epidiosbulbin E acetate-induced liver injury: mechanism of action of the hepatotoxic furanoid. *Chem. Res. Toxicol.* 2016, 29, 359-366.
9. Wang, K.; Wang, H.; Peng, Y.; Zheng, J. Identification of epoxide-derived metabolite(s) of benzbromarone. *Drug Metab. Dispos.* 2016, 44, 607-615.
10. Xie, H.; Liu, Y.; Peng, Y.; Zhao, D.; Zheng, J. Detection of protein modification derived from quinone methide metabolite of dauricine. *Anal. Bioanal. Chem.* 2016, 408, 4111-4119.
11. Yao, H.; Peng, Y; Zheng, J. Identification of glutathione and cysteine conjugates derived from reactive metabolites of methyleugenol in rats. *Chem. Biol. Interact.* 2016, 253, 143-152.

12. Tian, Y.; Shen, S.; Jiang, Y.; Shen, Q.; Zeng, S.; Zheng, J. CYP3A5 mediates bioactivation and cytotoxicity of tetrandrine. *Arch. Toxicol.* 2016, 90, 1737-1748.
13. Sun, Y.; Yao, T.; Guo, X.; Peng, Y.; Zheng, J. Simultaneous assessment of endogenous thiol compounds by LC-MS/MS. *J. Chromat. B.* 2016, 1029-1030, 213-221.
14. Chen, J.; Peng, Y.; Zheng, J. Cytochromes P450 mediated metabolic activation of saracatinib. *Chem. Res. Toxicol.* 2016, 29, 1835-1842.
15. Qin, B.; Xu, Y.; Chen, J.; Peng, Y.; Zheng, J. Chemical reactivity of emodin and its oxidative metabolites to thiols. *Chem. Res. Toxicol.* 2016, 29, 2114-2124.
16. Wang, H.; Peng, Y.; Zheng, J. Cysteine-based protein adduction by epoxide(s) metabolite of bromobenzmarone. *Chem. Res. Toxicol.* 2016, 29, 2145-2152.
17. Lin, D.; Wang, K.; Gao, H.; Peng, Y.; Zheng, J. Lysine- and cysteine-based protein adductions derived from toxic metabolites of 8-epidiosbulbin E acetate. *Toxicol. Lett.* 2016, 264, 20-28.
18. Li, H.; Lin, D.; Peng, Y.; Zheng, J. Oxidative bioactivation of nitrofurantoin in rat liver microsomes. *Xenobiotica.* 2017, 47, 103-111.
19. Yang, X.; Li, W.; Sun, Y.; Guo, X.; Peng, Y.; Zheng, J. Comparative study of hepatotoxicity of retrorsine and monocrotaline. *Chem. Res. Toxicol.* 2017, 30, 532-539.
20. Feng, Y. Peng, Y.; Zheng, J. Chemical interactions of cysteine residues with reactive metabolites of methyleugenol. *Chem. Res. Toxicol.* 2017, 30, 564-573.
21. Sun, Y.; Peng, Y.; Zheng, J. *In vitro* and *in vivo* metabolic activation of berbamine to quinone methide intermediate. *J. Biochem. Mol. Toxicol.* 2017, 31, 1-8.
22. Zhang, Z; Lin, D.; Li, W.; Peng, Y.; Gao, H.; Zheng, J. Sensitive bromine-based screening of potential toxic furanoids in *Dioscorea Bulbifera*. *J. Chromatog. B Anal. Technol. Biomed. Life Sci.* 2017, 1057, 1-14.
23. Li, N.; Zhang, F.; Lian, W.; Wang, H.; Li, W.; Zheng, J.; Lin, G. Immunoassay approach for diagnosis of exposure to pyrrolizidine alkaloids. *J. Environ. Sci. Health C. Environ. Carcinog. Ecotoxicol. Rev.* 2017, 35, 127-139.
24. Wang, K.; Lin, D.; Guo, X.; Peng, Y.; Zheng, J. Chemical identity of interaction of protein with reactive metabolite of diosbulbin B *in vitro* and *in vivo. Toxins.* 2017, 9, pii: E249.
25. Wang, H.; Peng, Y.; Zhang, T.; Lan, Q.; Zhao, H.; Wang, W.; Zhao, Y.; Wang, X.; Pang, J.; Wang, S.; Zheng, J. Metabolic epoxidation is a critical step for the development of benzbromarone-induced hepatotoxicity. *Drug Metab. Dispos.* 2017, 45, 1354-1363.
26. Xu, Y.; Mao, X.; Qin, B.; Peng, Y.; Zheng, J. *In vitro* and *in vivo* metabolic activation of rhein. *Chem. Biol. Interact.* 2018, 283, 1-9.
27. Feng, Y.; Wang, S.; Peng, Y.; Zheng, J. Urinary methyleugenol-deoxyadenosine adduct as potential biomarker of methyleugenol exposure. *J. Agric. Food Chem.* 2018, 66, 1258-1263.

28. Li, X.; Yang, X.; Xiang, E.; Luo, J.; Qiu, S.; Fang, Y.; Zhang, L.; Guo, Y.; Zheng, J.; Wang, H. Metabolic activation of pyrolizidine alkaloids in placenta and fetal liver caused developmental toxicity in different gender of fetal rats. *Drug Metab. Dispos.* 2018, 46, 422-428.
29. Zhao, H.; Li, S.; Yang, Z.; Peng, Y.; Zheng, J. Identification of ketene reactive intermediate of erlotinib possibly responsible for inactivation of P450 enzymes. *Drug Metab. Dispos.* 2018, 46, 442-450.
30. Hu, Z.; Li, W.; Lin, D.; Zhang, N.; Zhou, S.; Peng, Y.; Zheng, J. Development of polyclonal antibodies for detection of diosbulbin B derived *cis*-enedial protein adducts. *Chem. Res. Toxicol.* 2018, 31, 231-237.
31. Qin, X.; Peng, Y.; Zheng, J. *In vitro* and *in vivo* studies of the electrophilicity of physcion and its oxidative metabolites. *Chem. Res. Toxicol.* 2018, 31, 340-349.
32. Sun, Y.; Xin, X.; Zhang, K.; Cui, T.; Peng, Y.; Zheng, J. Cytochrome P450 mediated metabolic activation of chrysophanol. *Chem. Biol. Interact.* 2018, 289, 57-67.
33. Yu, J.; Zheng, Q.; Peng, Y.; Zheng, J. Metabolite profile analysis and pharmacokinetic study of emodin, baicalin and geniposide in rats. *Xenobiotica*. 2018, 48, 927-937.
34. Mao, X.; Hu, Z.; Wang, Q.; Zhang, N.; Zhou, S.; Peng, Y.; Zheng, J. Nitidine chloride is a mechanism-based inactivator of CYP2D6. *Drug Metab. Dispos.* 2018, 46, 57-67.
35. Liu, Y.; Cui, T.; Peng, Y.; Ji, M.; Zheng, J. Mechanism-based inactivation of cytochrome P450 2D6 by chelidonine. *J. Biochem. Mol. Toxicol.* doi: 10.1002/jbt.22251.
36. Yang, X.; Li, W.; Wang, X.; Chen, Y.; Guo, X.; Peng, Y.; Zheng, J. Difference in internal exposure made infant mice more susceptible to hepatotoxicity of retrorsine than adult mice. *Chem. Res. Toxicol.* (Accepted)

In: Glutathione
Editor: Pál Perjési
ISBN: 978-1-53614-740-7
© 2019 Nova Science Publishers, Inc.

Chapter 10

A GLUTATHIONE SENSOR BASED ON CdO-DECORATED CNT NANOCOMPOSITES

Mohammed M. Rahman[1,2,*], Mohammad M. Hussain[3] and Abdullah M. Asiri[1,2]
[1]Chemistry Department, Faculty of Science, King Abdulaziz University, Jeddah, Saudi Arabia
[2]Center of Excellence for Advanced Material Research (CEAMR), King Abdulaziz University, Jeddah, Saudi Arabia
[3]Department of Pharmacy, Faculty of Life and Earth Sciences, Jagannath University, Dhaka, Bangladesh

ABSTRACT

Glutathione (GSH) is an important antioxidant in plants, animals, fungi, and some bacteria and archaea, which is capable of preventing damage to important cellular components caused by reactive oxygen species such as free radicals, peroxides, lipid peroxides, and heavy metals. In this approach, CdO decorated CNT nanocomposites (NCs) were synthesized using a wet-chemical technique. The optical, morphological, and structural characteristics of CdO.CNT NCs were determined using FT-IR, UV-Vis; FESEM combined with XEDS, XPS, and XRD methods. An L-glutathione (GSH; L-glutamyl-L-cysteinylglycine) selective glassy carbon electrode (GCE) was constructed with a thin-layer of CdO.CNT NCs. Electrochemical responses including sensitivity, linear dynamic range and long-term stability towards GSH were recorded using the fabricated sensor. The calibration curve was found linear over a wide range of GSH concentration. Based on the slope of the calibration curve, limit of detection (LOD) and sensitivity were calculated. The prepared CdO.CNT NCs is an organized way for development of sensors based on nano-material (NM) embedded GCE using electrochemical oxidation

[*] Corresponding Author E-mail: mmrahman@kau.edu.sa.

phenomena. To best of our knowledge, this is the first report on detection of GSH by CdO.CNT NCs using electrochemical method. The pioneering development of GSH sensitive sensor using CdO.CNT NCs with a facile and reliable electrochemical method coluld be a versatile application of biomolecules in the near future.

Keywords: glutathione, CdO-CNT nanocomposites, sensitivity, nanoparticles, electrochemical method, glassy carbon electrode, healthcare safety

BACKGROUND STUDY

A sensor is an analytical tool that performs two functions (i) capturing biological targets, and (ii) transducing target binding at measurable signals. There have been huge efforts over the past decade to develop new biomolecular sensors for fundamental studies in biology and biotechnology such as determination of analytes, enzymatic activities, and identification of biomolecules (BM) (oligonucleotides, proteins, and sugars). Different traditional techniques have been developed for detection of BM like fluorescence, electrochemical signal transducing, surface-enhanced Raman spectroscopy, and surface plasmon resonance (SPR). Among them, the electrochemical signal transducing method is advantageous over other signal transducing methods in terms of cost, accuracy, sensitivity, and selectivity [1]. The unique electronic, mechanical, and optical properties of carbon materials have fueled extensive, ground breaking research in areas such as composite materials, electronics, energy conversion devices, and sensors [2]. GSH is synthesized in the body from the amino acids L-cysteine, L-glutamic acid, and glycine. L-Cysteine concentration is the rate limiting factor in glutathione biosynthesis because L-cysteine is relatively rare in food supplements and is usually synthesized in the human body [3]. Glutathione exists in reduced (GSH) and oxidized forms (GSSG) in the body. In the reduced form, the thiol (-SH) group of GSH can donate protons and electrons to other molecules such as reactive oxygen species. Therefore, GSH acts as an important antioxidant and prevents damage to important cellular components caused by reactive oxygen species such as free radicals, peroxides, and heavy metals [6]. After donating protons and electrons, two reactive glutathione molecules combine and form GSSG. GSH can be regenerated from GSSG by the enzyme, glutathione reductase [4]. In healthy cells and tissues, more than 90% of total glutathione is found in the reduced (GSH) form and less than 10% exists as GSSG [5]. An increased GSSG/GSH ratio is considered to be indicative of oxidative stress. GSH is used in metabolic and biochemical reactions such as DNA synthesis and repair, protein synthesis, amino acid transport, and enzyme activation [7]. GSH also plays an important role in iron metabolism. Hence, every system in the body can be affected by the glutathione state especially the immune system, the nervous system, and the lungs. GSH is also an essential antidote to the overdose of paracetamol (analgesic and antipyretic drug). The abnormal concentration of GSH in humans is directly related to various diseases such as Alzheimer's disease, Parkinson's disease, diabetes, macular degeneration, cancer and HIV/AIDS [8, 9].

Cadmium oxide (CdO) nanostructures exhibit remarkable optical and electrical transparency in the visible region [10] and high *n*-type electrical conductivity, which made it as a transparent conductive oxide layer in solar cells [11]. To avail the remarkable properties of cadmium oxides in nanotechnology, this material has been recently grown and studied in the form of nanowires, nanowalls, nanotubes, nano-needles, porous nanobelts, nanopowder, etc. [12]. One of the advantages of using CdO nanoparticles (NPs) over bulk CdO is the possibility to tune the optical band gap (3.57~3.62 eV) and emission energy at a lower wavelength (398–385 nm). Various techniques like vapor transport, chemical-bath deposition, thermal evaporation, wet-chemical, hydrothermal, co-precipitation methods, solvothermal process, microwave combustion, and successive ionic layer adsorption have been used to grow CdO nanostructures (CdO NSs) [13]. Most of the CdO NSs have been manufactured employing a seed layer or a sacrificial template to offer nucleation points for the NSs growth or involving processes with post-annealing treatments [14]. Several studies reported on CdO thin films and NSs, to the best of our knowledge, however, there are only a few studies dealing with the growth of isolated CdO NPs on flat-substrates [15, 16]. Transition-metal doped semiconductor nanostructure materials offer particular patterns for the growth of doped-materials with promising optical, electrochemical, opto-electronic, and magnetic properties compared to undoped nano-materials (NMs). Semiconductors display enormously high absorption behaviors based on their band-gap energy. The excellent nanostructure of doped morphology allows them to absorb solar-light and promote photo-catalysis with higher efficiency [17, 18]. Considering the solar-light absorption of the NSs, it is intended simulated block-structures to act as photo-catalysts as well as solid phase adsorbents with high capacity and sensitivity. Carbon nanotubes (CNT) have been attractive candidates for fundamental research studies owing to their potential applications such as miniature biological electronic devices and sensors for carcinogenic metal ion determination. The small dimensions, strength and the remarkable physical properties of the functionalized structures make them unique materials with a wide range of promising applications. Since the discovery of CNT by Iijima in 1991 using transmission electron microscopy [19], it has been introduced the subject of numerous investigations in chemical, physical, sensors, adsorption, and material areas owing to their novel structural, electrical, mechanical, electronic and chemical properties [20]. Depending on their atomic structure, CNTs behave electrically as metals or semiconductors [21]. Functionalization of CNT has attracted great attention in the past few years. These works are of great importance to explore possible applications of functionalized nanotubes. Recent developments in sensitive and efficient methods have become of concern for monitoring toxic metal ions as well as organic compounds in the environment.

L-GSH is a non-protein, low molecular weight thiol present in its reduced (GSH) and oxidized (GSSG) form in mammalian tissues and plays a significant role in numerous biological functions such as antioxidant defense, signal transduction, and cell proliferation. GSH is a small peptide of great relevance in many biological processes, and its molecular speciation is a valuable aid to understand the cellular health status. Important information can be obtained, for example, from quantification of its reduced and oxidized forms, from both the ratio and the absolute amounts of the two species. Due to its multiple functions, reduction of its antioxidant capacity (ratio of GSH/GSSG) has an important role in the molecular basis of different illnesses, metabolism, and aging [22]. GSH is widely present in all forms of life and has an important role in the homeostasis of organisms, especially the aerobic ones [23, 24]. Qualitative and quantitative alterations of the GSH/GSSG ratio is an indication of oxidative damage, and GSH level can be used as an indicator of some diseases (e.g., Alzheimer disease, Parkinson's disease and diabetes) [25]. Therefore, quantitative determination of GSH has become more important for examination of cellular functions, including clinical diagnosis of related diseases. Due to its biological importance, a wide variety of techniques such as chemiluminescence, colorimetric assays, electrochemical methods, enzymatic techniques, high performance liquid chromatography, mass spectrometry, optical spectroscopy, and surface enhanced Raman scattering have been developed to monitor its concentration in biological samples [26, 27]. Although these techniques provide enough selectivity and sensitivity, some problem remains to solve such as complexity of the biological matrix, activity of enzymes of the enzymatic methods, pretreatment of the compound in the samples. So, development of novel analytical methods for GSH assay is still an important task for biological, clinical, and medicinal reasons. The objective of this study was to develop a novel method based on constructed CdO.CNT NCs modified electrode using electrochemical technique.

EXPERIMENTAL SECTION

Methodology

Analytical grade sodium hydroxide, nafion (5% ethanolic solution), ethanol, creatine, D-fructose, D-glucose, glycine, L-cystine, L-glutamic acid, L-glutathione, L-tyrosine, L-leuine, Penicillin G potassium salt, thiourea, and uric acid were purchased from Sigma Aldrich and used without further purification. FT-IR and UV-Vis spectra of the dried CdO.CNT NCs were performed on a Thermo scientific NICOLET iS50 FTIR spectrometer (Madision, WI, USA) and a 300 UV-Vis spectrophotometer (Thermo Scientific), respectively. XPS examinations for calculation of binding energies (eV) among C, Cd and O were performed on a K-α spectrometer (Thermo Scientific, K-α 1066) with an excitation radiation source (A1 Kα, Beam spot size = 300.0 μm, pass energy = 200.0 eV, pressure 10⁻

^{8}torr). Arrangement, elemental analysis, morphology and particle size of the CdO.CNT NC samples were investigated using Field Emission Scanning Electron Microscopy (FESEM, JEOL, JSM-7600F, Japan) coupled with X-ray Electron Dispersive Spectroscopy (XEDS). X-ray Diffraction (XRD) experiments (to determine the crystalline nature of CdO.CNT NCs) were performed under ambient conditions. Electrochemical experiments using the constructed CdO.CNT NCs were carried out by a Kethley 6517A electrometer at room conditions.

Nanocomposites from CdO NPs and CNT

The wet-chemical process [28, 29] is a traditional solid-state method and widely used in the preparation of doped or undoped NM resulting in formation of smaller grains with shorter duration of phase formation. According to this technique, CNT (1.0 wt%, 0.25 μg) was added with dissolved [$Cd(NO_3)_2.4H_2O$] (0.1 M, 3.10 g) in distilled water (100.0 mL) in a conical flask (250.0 mL) under continuous stirring. The pH of the resultant mixture was adjusted to pH 10.25 by adding NaOH and then continuously stirred at 80.0°C. After 6 h continuous stirring, the flask was washed thoroughly with water and acetone subsequently, and dried for 12 h in open air at room temperature. The obtained ash grown products (CdO.CNT NCs) were dried in oven at 60.0°C for 24 h, grinded into powders, and dried at 60.0°C in oven for 24 h. The cadmium oxide nanoparticles (CdO NPs) without CNT were also synthesized using the same method. The possible reactions of formation of CdO.CNT NCs as follows [(**i**)-(**iv**)].

$$NaOH_{(s)} \rightarrow Na^{+}_{(aq)} + OH^{-}_{(aq)} \quad \text{(i)}$$

$$Cd\,(NO_3)_2 \rightarrow Cd^{2+}_{(aq)} + 2\,NO_3^{-}{}_{(aq)} \quad \text{(ii)}$$

$$Na^{+}_{(aq)} + 2\,OH^{-}_{(aq)} + Cd^{2+}_{(aq)} + NO_3^{-}{}_{(aq)} \rightarrow Cd(OH)_{2(aq)} + NaNO_{3(aq)} \quad \text{(iii)}$$

$$2\,Cd(OH)_{2(aq)} + 2\,CNT\text{ (dispersed)} \rightarrow 2\,CdO.CNT_{(s)}\downarrow + 2\,H_2O_{(aq)} \quad \text{(iv)}$$

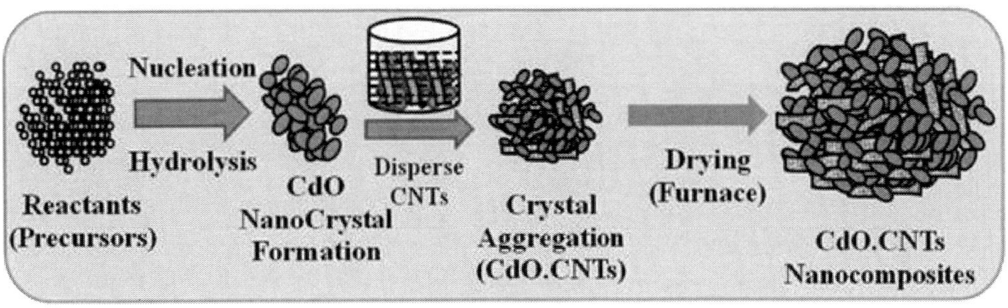

Scheme 1. Schematic representation of growth mechanism of CdO.CNT NCs.

In the CdO.CNT NCs growth method, initially formation of CdO nuclei takes place. The formed nuclei grow in size and the aggregates form CdO nanocrystals according to the Ostwald-ripening method. Nanocrystals re-aggregate with each other's counter parts in presence of disperse CNT through Van der-Waals forces and reforms the CdO.CNT porous morphology (Scheme 1).

Construction of GCE with CdO.CNT NCs

Phosphate buffer saline (PBS, 0.1 M, pH 7.0) was prepared using Na_2HPO_4 (0.2 M), NaH_2PO_4 (0.2 M) and distilled water (200.0 mL). Ethanol and conducting binder, nafion (Nf) were used to construct GCE with CdO.CNT NCs. The fabricated electrode was kept for 6 h at room temperature for complete drying with uniform film formation. The constructed GCE and platinum (Pt) wire were used as a working and counter electrode, respectively, to determine the electrochemical responses. Different concentrations of GSH were used as standard solutions and the volume of PBS was kept constant (10.0 mL) throughout the entire experiment. During examination of the selected BM, GSH solution (0.25 μL) was added from the lower to the higher concentration (0.1 nM ~ 0.1 M) with the constructed NCs electrode. Based on the slope and the ratio of noise to the slope (Slope, S and Noise, N = 3), sensitivity and limit of detection (LOD) of the sensor were calculated from the linear dynamic range (LDR) of the calibration curve, respectively. An electrometer was used as a constant current source for the working and the counter electrode and GSH was detected as the target BM using the constructed CdO.CNT NCs electrode.

RESULTS AND DISCUSSION

Optical and Structural Properties

Optical property is one of the significant characteristics of the ash grown CdO.CNT NCs for assessment of photo-catalytic activity. Based on the UV-Vis principle, the outermost electrons of an atom may be shifted to the higher energy level after adsorption of radiant energy. The spectrum along with band-gap energy of the metal oxide could be achieved due to the adsorption of radiant energy. The UV-Vis spectra of the CdO NPs and CdO.CNT NCs have been recorded in the range of 200 - 800 nm and a broad absorption band at around 309.5 nm and 304.7 nm was found, respectively (Figure 1a-c). Based on the wavelength of the absorption maximum, the band-gap energy (E_{bg}) of the CdO NPs and CdO.CNT NCs was calculated using equation (**v**). According to the Tauc law (direct band-gap rule) the absorption coefficient (α) due to interband transition near the band-gap is well described (**vi**). Thus, $(\alpha h\nu)^2$ *vs* $h\nu$ was plotted and then extrapolated to the x-axis. From the extrapolated curve, the band-gap energy for CdO NPs and CdO.CNT NCs were found ~2.5 and 2.3 eV, respectively (Figure 1b-d) [30-32].

$$Ebg= 1240/\lambda max\ (eV) \quad (v)$$

$$(\alpha hv)1/r = A\ (hv\text{-}Ebg) \quad (vi)$$

where, E_{bg} = Band gap energy, λ_{max}= Absorption maximum, α = Absorption coefficient, *A* = Constant related to the effective mass of the electrons, r = 0.5 (Direct transition), *h* = Plank's constant, and *v* = frequency.

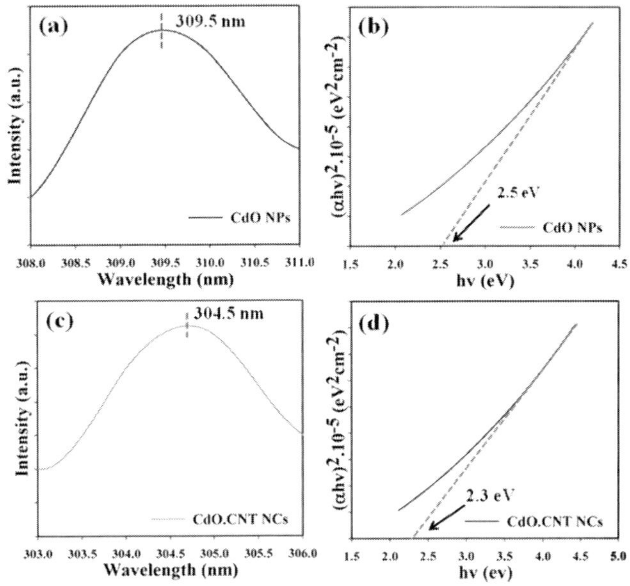

Figure 1. Optical properties. (a-c) UV spectra and (b-d) Band-gap energy plot of **CdO** NPs and CdO.CNT NCs, respectively.

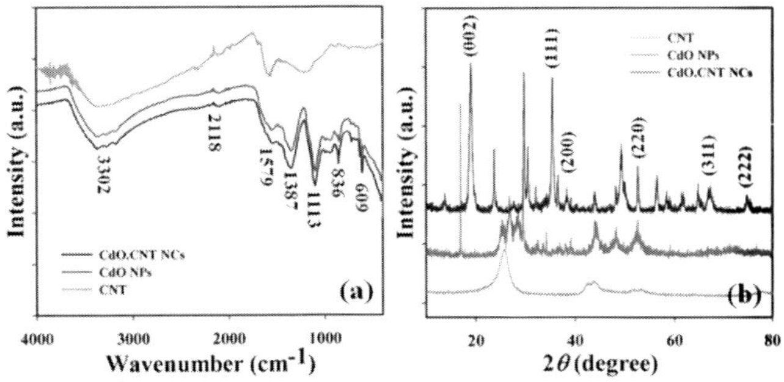

Figure 2. Structural properties. (a) FT-IR spectrum and (b) XRD pattern of CNT, CdO NPs and CdO.CNT NCs.

To identify the functional nature of the NCs, the ash grown CdO.CNT NCs were investigated regarding atomic and molecular vibrations. To that end FT-IR spectra were recorded in the region of 4000-400 cm^{-1} under normal conditions. FT-IR spectra of the NPs and NCs showed peaks at 3302 (br), 2118 (w), 1579 (m), 1387 (m), 1113 (s), 836 (w), and 609 (m) cm^{-1} which indicated the presence of O-H (stretching), −C≡C− (stretching), −C=C− (stretching), C−H (rocking), Cd−O−Cd (stretching), C−H and Cd=O (stretching) in the CdO NPs and CdO.CNT NCs, respectively (Figure 2a). The observed peak at 609 cm^{-1} indicated formation of metal-oxide (Cd=O) bond, which recognized the configuration of the CdO.CNT NCs [33].

The crystalline pattern is the indication of metal-oxygen framework. X-ray powder diffraction (XRD) analysis was performed in the range (2θ) of 20-80° to identify the crystalline nature of the synthesized CdO.CNT NCs. The potential peaks with indication for 2θ values were observed at 24.0 (002), 36.0 (111), 38.0 (200), 53.0 (220), 68.0 (311), and 75.0 (222) degrees (Figure 2b). The peak at 24.0 (002) is assigned for carbon of CNT. All the observed peaks in the spectra were assigned by using the JCPDS file (05-0640 and 78-0653). The strong peaks indicated the crystalline pattern and purity of the NCs. According to the XRD analysis it can be suggested that a big amount of crystalline CdO was present in the prepared NCs [34,35].

Morphological and Elemental Characteristics

Field-emission scanning electron microscopy (FESEM) is an established method to examine morphology of NCs and nanomaterials. Morphology and elemental pattern of the synthesized ash grown CdO NPs and CdO.CNT NCs were studied using FESEM and energy-dispersive X-ray spectroscopy (XEDS), respectively. The typical shapes of the ash grown CdO NPs and CdO.CNT NCs were recorded using FESEM from low to high magnified range. The magnified images indicated that CdO NPs was well dispersed on the surface of CNT (Figure 3c-d). The conductance of CNT may be increased by adding of CdO, which correlates with calculation of the band-gap energy of the two materials.

According to the XEDS analysis, only oxygen (31.11 m/m%) and cadmium (68.89 m/m%) were present in the prepared CdO NPs (Figure 4a-b). According to the expected structure, carbon (52.48 m/m%), oxygen (22.05 m/m %) and cadmium (25.47 m/m%) were present in the prepared ash grown CdO.CNT NCs. (Figure 4c-d). According to the elemental analysis, CdO NPs were well dispersed with bright contrast by CNT. No other peaks regarding impurities were found on FESEM coupled with XEDS spectra, which indicated that the NCs were composed of only carbon, oxygen, and cadmium.

Analysis of Binding Energy

The surface-sensitive quantitative spectroscopic technique, X-ray photoelectron spectroscopy (XPS) can be used to measure the chemical state and electronic state of the elements that exist within a material. XPS spectra are recorded by irradiating a material with an X-ray beam while simultaneously measuring the kinetic energy and the number of electrons that escape the surface of the material being analyzed. Based on XPS spectra, carbon, cadmium, and oxygen were present in the prepared CdO.CNT NCs (Figure 5a). The C1s, spin orbit Cd $3d_{3/2}$ and Cd $3d_{5/2}$, and O1s spectra give the main peak at 293.5, 409.6-416.3, and 535.8 eV, respectively, which indicate that carbon (C), (Cd^{2+}), and oxygen (O^{2-}) were present in the NCs (Figure 5b-d) [36-40]. A comparison of binding energies of CNT, CdO NPs and CdO.CNT NCs are presented in Table 1.

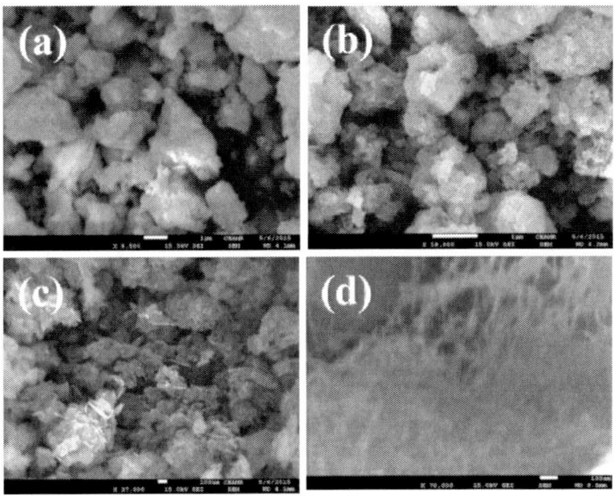

Figure 3. Morphological evaluation. Low to high magnified FESEM images of CdO NPs and CdO.CNT NCs.

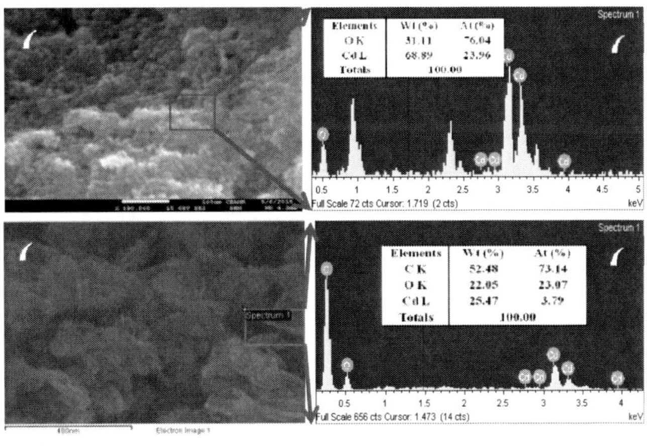

Figure 4. Elemental analysis. (a-b) CdO NPs and (c-d) CdO.CNT NCs.

Table 1. Binding energies of CNT, CdO NPs and CdO.CNT NCs

Materials	Binding energies (eV)		
	C1s	Cd3d	O1s
CNT	285.0	-	
CdO NPs	-	407, 414	536.0
CdO.CNT NCs	288.0	410, 417	536.0

Detection of L-Glutathione by CdO.CNT NCs

Enhancement of the constructed electrode with NCs is the initial stage of application it as a sensor. Application of CdO.CNT NCs assembled as a biological sensor onto a GCE has been evaluated for detection and determination of GSH in a buffer medium. The CdO.CNT NCs/GCE sensors have various advantages such as chemically inert, easy to fabricate, non-toxic, simple to assemble, stable in air and safe. Based on the electrochemical technique principle, the current (CR) of CdO.CNT NCs/GCE considerably changed during adsorption of GSH by electrochemical oxidation. The surface of CdO.CNT NCs/GCE can be made uniformed with the conducting binder, Nf.

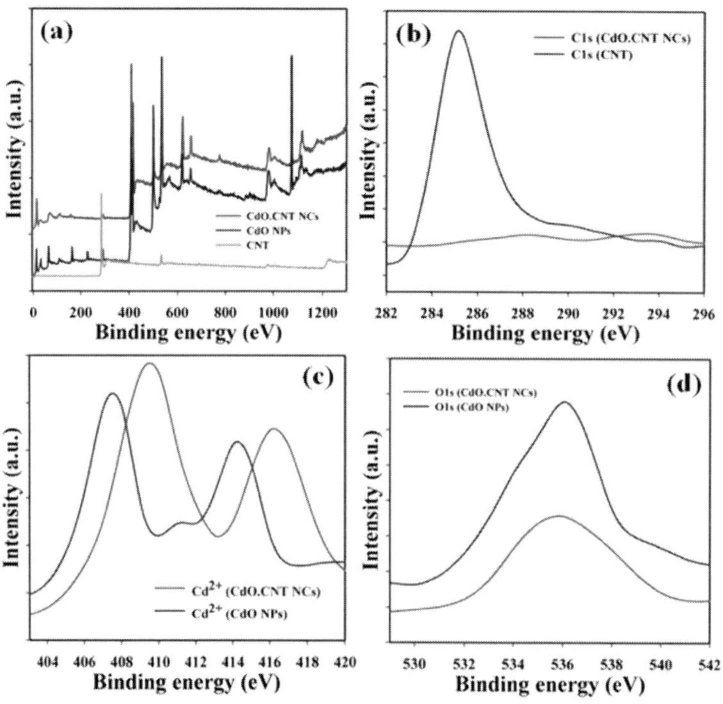

Figure 5. X-ray photoelectron spectroscopy (XPS) for CdO.CNT NCs, (a) Full spectrum of CdO.CNT NCs, (b) C1s, (c) Cd $3d_{3/2}$ and Cd $3d_{5/2}$, and (d) O1s level achieved with MgKα1 radiations. X-ray beam. Spot size is 400.0 μm; Pass-energy is 200.0 eV; Pressures less than 10^{-8} Torr.

Electrochemical responses of GSH were anticipated having CdO.CNT NCs/GCE on thin-film as a function of current-potential. The resultant electrical signals of the target GSH were executed by simple and reliable electrochemical technique using CdO.CNT NCs/GCE. A significant amplification in the current response with applied potential was noticeably confirmed having the holding time of the electrometer 1.0 s. The constructed-surface of the CdO.CNT NCs sensor was prepared by means of the conducting binder Nf on the GCE surface, which is presented in Scheme 2a. The probable electrochemical curve is presented in Scheme 2b. The simple and possible reaction mechanism in presence of GSH on the CdO.CNT NCs/GCE sensor surfaces by electrochemical method is generalized in Scheme 2c. The resultant practical electrical response of GSH was investigated by simple and reliable electrochemical technique using CdO.CNT NCs/GCE sensor, which is presented in Scheme 2d. A significant enhancement in the current response with applied potential was noticeably recorded.

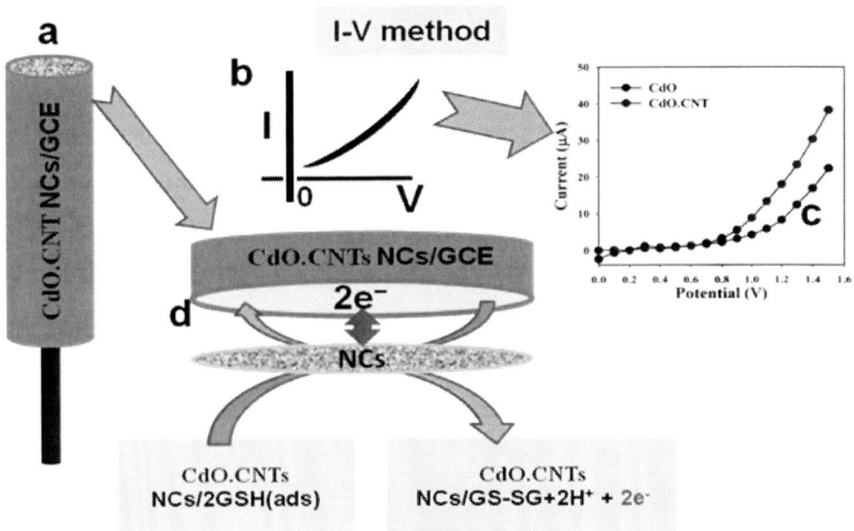

Scheme 2. Schematic view of (a) CdO.CNT NCs coated rod-shape round disc-GCE electrode with Nf, (b) Expected electrochemical curve, (c) Observed electrochemical response by CdO.CNT NCs/GCE, (d) proposed detection mechanism of glutathione detection onto CdO.CNT NCs/GCE electrodes in presence of target BM.

The CdO.CNT NCs assembled onto a glassy carbon electrode as a sensor has been used for detection of target compounds found in biological systems. The CR (potential range: 0.0 ~ + 1.5 V) for the bare GCE and the CdO.CNT NCs coated GCE are presented in Figure 6a. The differences of the CR between the bare and the coated GCE occurred because the current signals have been increased by coating (Figure 6a). Creatine, D-fructose, D-glucose, glycine, L-cystine, L-glutamic acid, L-glutathione, L-tyrosine, L-leucine, penicillin G potassium salt, thiourea, and uric acid were examined to find out the maximum CR to the CdO.CNT NCs coated electrode. Our results clearly demonstrated that the sensor was much more selective towards GSH than to any other investigated chemical (Figure 6b). The current signal without GSH (green-dotted) and with GSH (red dotted) was also performed (Figure 6c). An increment of recorded CR regarding the modified CdO.CNT electrode can be explained as a result of the large surface area with better coverage in absorption and adsorption potentiality onto the porous NCs surfaces of the target molecule (GSH).

Resistance of the CdO.CNT NCs modified GCE sensors can be declined with increment of the active surface area, which is an important criterion of the NCs materials [41, 42]. These reactions can occur in bulk-system/air-liquid interface/neighboring atmosphere owing to the small carrier concentration, which increases the resistance. Sensitivity of GSH toward CdO.CNT NCs may be the result of a higher-oxygen lacking conducts and increased oxygen adsorption. Activity of GSH would have been extremely higher in comparison to the other biological agents with the surface under indistinguishable conditions [43, 44]. Based on the modified two-electrode system, electrochemical responses of the CdO.CNT NCs coated GCE was activated as a function of GSH concentration and an increased CR was recorded. The CR of the CdO.CNT NCs/GCE film became higher with increasing GSH concentrations; similar phenomena for hazardous chemical determination have also been reported earlier [45-47].

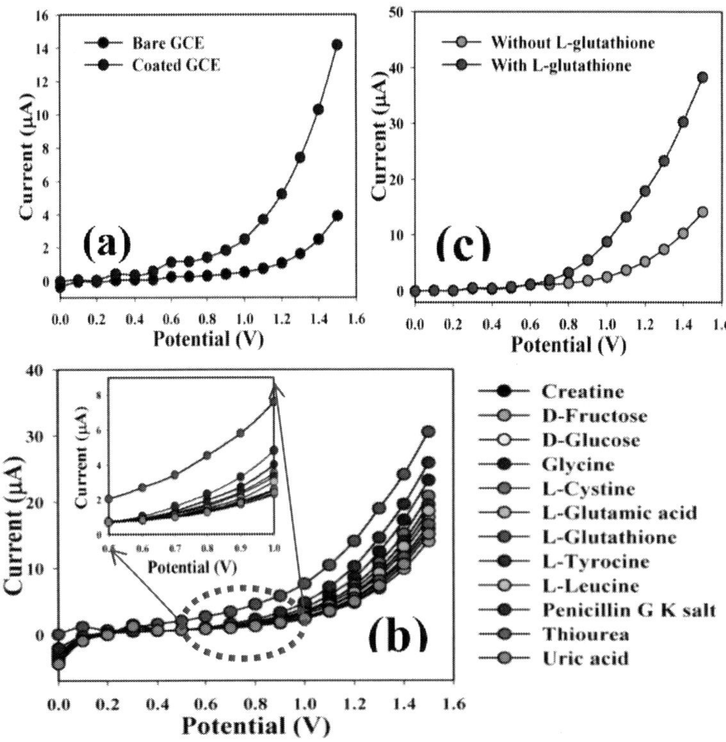

Figure 6. (a) Bare and coated electrode, (b) Selectivity (inset: magnified view of response in 0.5 ~ 1.0 V range), and (c) Absence and presence of GSH with CdO.CNT NCs electrode.

The electrochemical responses of GSH with different concentration (0.1 nM ~ 0.1 M) towards the CdO.CNT NCs modified electrode were conducted with indication of the changes of current of the constructed electrode. It was found that the CR increased gradually from the lower to the higher concentration of the target molecule (Figure 7a). A wide range of GSH concentrations (0.1 nM ~ 0.1 M) was examined from the low to the high potentials (0.0 ~ + 1.5 V) to find the possible analytical limits. The linear calibration curve (current *vs* GSH concentration) at + 0.5 V was plotted and shown in Figure 7b. Regression coefficient (R^2 = 0.99), sensitivity (949.37 $nAnM^{-1}cm^{-2}$), and LOD (10.0 ± 0.1 pM) at signal to noise ratio of ~ 3 were calculated from the calibration curve. LDR (1.0 nM ~ 0.1 M) of the electrode were also calculated from the graph.

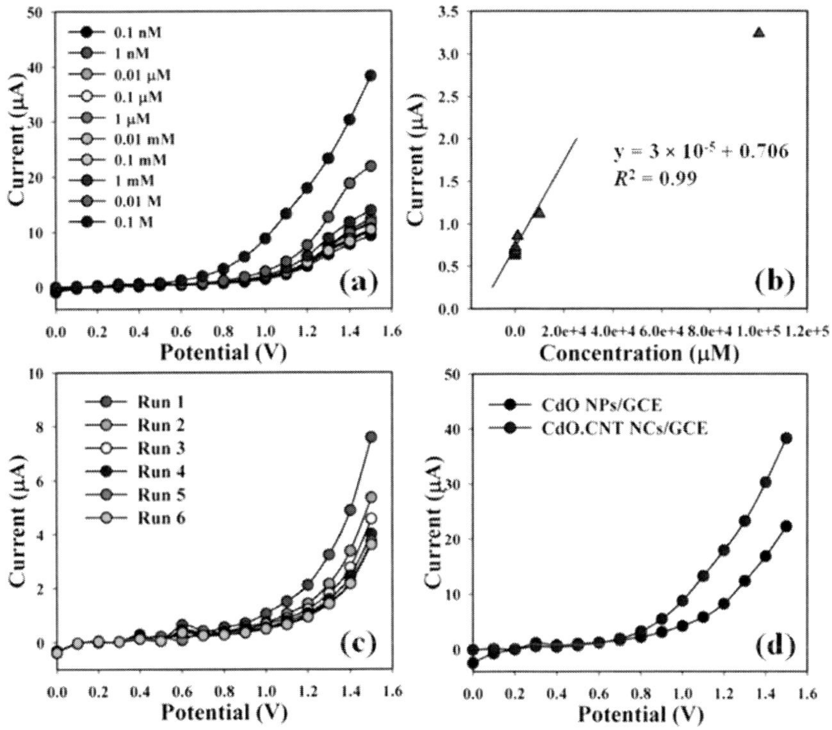

Figure 7. CdO.CNT NCs coated electrode for GSH sensing electrochemical responses: (a) concentration variation, (b) calibration curve, (c) reproducibility study, and (d) control experiment.

Sensitivity of the CdO.CNT NCs coated electrode was tested to check reproducibility and possible storage period. In this regard, a series of six successive measurements of GSH solution (0.1 μM) was investigated. It was found that the electrochemical responses practically did not change after washing the CdO.CNT NCs electrode after each experiment (Figure 7c). A control experiment was also performed at the same (0.1 μM) GSH concentration using the two different electrodes. A remarkable increase of CR was found of the CdO.CNT NCs electrode compared to that of the CdO NPs one (Figure 7d).

Table 2. Detection of GSH using various composites or materials by electrochemical approaches

Electrode Assembly	Methods	LOD* (μM)	LDR** (μM)	Sensitivity (μAμM^{-1})	Ref.
FePt/CNT nanocomposite CPE	CV	1.0 nM	0.004–340	---	[51]
2,7-BFEF/Carbon paste electrode	Voltamogram	0.50	0.92-11	1.997	[52]
ZnO/CNT nanocomposite CPE	SWV	0.0008	0.002–720	---	[53]
Ferrocene/Carbon paste electrode	CV	2.10	2.2-3000	0.017	[54]
NiO/MWCNT nanocomposite CPE	CV	0.006	0.01–200	1.056	[55]
Aligned/CNT/GCE	CV	0.20	0.4-16.4	0.01	[56]
MWCNT paste electrode	CV	0.1	0.5–400	---	[57]
Horseradish peroxidase/GCE	CV	0.03	0.04-90	0.024	[58]
FePt/CNT nanocomposite electrode	SWV	0.05	0.08–220	---	[59]
MWCNT/CPE modified with dinitrobenzamide	SWV	0.05	0.09–300	---	[60]
p-Aminophenol MWCNT paste electrode	SWV	0.9 nM	20 μM-100 mM	---	[61]
Ferrocene dicarboxylic acid modified CNT paste electrode	DPV	0.2	0.5–24	----	[62]
MWCNT paste with chloropromazine	DPV	0.16	0.3–18.3	0.541	[63]
DHB-modified CNT/paste electrode	DPV	70 nM	0.4–700	0.2077	[64]
Poly-*m*-aminophenol modified GCE	CA	0.095	0.1–5.0	---	[65]
CdO.CNT NCs/GCE	I-V	10.0 pM	0.1 nM ~ 0.1 M	949.37 nAnM^{-1}cm^{-2}	[66]

*LOD: Limit of detection; **LDR: Linear dynamic range.

Responses of the NCs sensor were measured as a function of storage time to determine the effect of long term storage stability. Sensitivity remained almost the same as the initial response up to two weeks; after that period the electrode response gradually declined. It was revealed that the constructed sensor can be used without any significant loss of sensitivity up to two weeks.

The enhanced sensitivity of the fabricated electrode can be the result of the good absorption and adsorption ability as well as the high catalytic-decomposition activity of the CdO.CNT NCs. The estimated sensitivity of the fabricated sensor is relatively higher and its LOD is comparatively lower than the previously reported chemical sensors based on other nanocomposite or nanomaterial modified electrodes measured by electrochemical approaches [48-50]. Having more specific surface area, CdO.CNT NCs provides a favorable nano-environment for detection of GSH. The higher sensitivity of CdO.CNT NCs/GCE sensor provides more effective electron communication features, which accelerate the direct electron transfer between the active sites of CdO.CNT NCs and the coated GCE. A comparison of GSH recognition based on different modified electrodes is presented in Table 2.

Real Sample Analysis

To confirm validity of the electrochemical method, the CdO.CNT NC/GCE sensor was used to determine GSH in various real samples (rabbit serum). In these studies, the standard addition method [67-74] was used to estimate accuracy of the GSH determination. A fixed amount (25.0 μL) of each sample were mixed and analyzed in phosphate buffer (10.0 mL, 0.1 M) by the fabricated CdO.CNT NC/GCE electrodes. LOD of GSH in rabbit serum samples was found to be 8.47 μM, which apparently confirmed that the electrochemical method was satisfactory, reliable, and suitable for analyzing real sample with the CdO.CNT NC/GCE sensor assembly.

CONCLUSION

Here, CdO.CNT NCs were prepared using a wet-chemical method that was cost effective and efficient in alkaline medium. The electrochemical characteristics of the NCs were examined by UV/Vis, FT-IR, FESEM, XEDS, XPS and XRD techniques. A simple fabrication method has been used to construct CdO.CNT NCs electrode onto flat GCE using conducting coating binder. A selective and sensitive of GSH sensor was successfully prepared based on the electrode embedded with CdO.CNT NCs by electrochemical method. The electrochemical nature of the fabricated GSH sensor was excellent in term of LOD, LDR, sensitivity, and repeatability. The CdO.CNT NCs electrode exhibited higher sensitivity and lower LOD compared to previously reported articles. A well-organized technique can be introduced in this novel approach for development of efficient biological sensor in the health care and biomedical fields on a broad scale.

REFERENCES

[1] Kim, S.; Kim, J.; Park, H. R.; Kim, M. K.; Chong, Y.; Yeo, W. S. Electrochemical detection of glutathione on SAMs on gold using an electroactive quininoid-type molecule. *Bull. Korean. Chem. Soc.* 2009, 30, 2574-2576.

[2] Kotchey, G. P.; Gaugler, J. A.; Kapralov, A. A.; Kagan, V. E.; Star, A. Effect of antioxidants on enzyme-catalysed biodegradation of carbon nanotubes. *J. Mater. Chem. B.* 2013, 1, 302-309.

[3] Pompella, A.; Visvikis, A.; Paolicchi, A.; de Tata, V.; Casini, A. F. The changing faces of glutathione, a cellular protagonist. *Biochem. Pharmacol.* 2003, 66, 1499–1503.

[4] Harfield, J. C.; Batchelor-McAuley, C.; Compton, R. G. Electrochemical determination of glutathione: a review. *Analyst.* 2012, 137, 2285–2296.

[5] Couto, N.; Malys, N.; Gaskell, S. J.; Barber, J. Partition and turnover of glutathione reductase from Saccharomyces cerevisiae: a proteomic approach. *J. Proteome Res.* 2013, 12, 2885–2894.

[6] Pastore, A.; Piemonte, F.; Locatelli, M.; Lo, R. A.; Gaeta, L. M.; Tozzi, G.; Federici, G. Determination of blood total, reduced, and oxidized glutathione in pediatric subjects. *Clin. Chem.* 2001, 47, 1467–1469.

[7] Huber, W. W.; Parzefall, W. Thiols and the chemoprevention of cancer. *Curr. Opin. Pharmacol.* 2007, 7, 404–409.

[8] Naisbitt, D. J.; Vilar, F. J.; Stalford, A. C.; Wilkins, E. G.; Pirmohamed, M.; Park, B. K. Plasma cysteine deficiency and decreased reduction of nitrososulfamethoxazol with HIV infection. *AIDS Res. Hum. Retroviruses.* 2000, 16, 1929–1938.

[9] El Sayed, S.; Gimenez, C.; Aznar, E.; Martinez-Manez, R.; Sancenon, F.; Licchelli, M. Highly selective and sensitive detection of glutathione using mesoporous silica nanoparticles capped with disulfide-containing oligo(ethylene glycol) chains. *Org. Biomol. Chem.* 2015, 13, 1017–1021.

[10] Farahani, S. K. V.; Munoz-Sanjose, V.; Zuniga-Perez, J.; McConville, C. F.; Veal, T. D. Temperature dependence of the direct bandgap and transport properties of CdO. *Appl. Phys. Lett.* 2013, 102, 022102.

[11] Santos-Cruz, J.; Torres-Delgado, G.; Castanedo-Perez, R.; Jimenez-Sandoval, S.; Marquez-Marin, J.; Zelaya-Angel, O. Au–Cu/p–CdTe/n–CdO/glass-type solar cells. *Sol. Energy Mater. Sol. Cells.* 2006, 90, 2272-2279.

[12] Kuo, T. J.; Huang, M. H. Gold-catalyzed low-temperature growth of cadmium oxide nanowires by vapor transport. *J. Phys. Chem. B.* 2006, 110, 13717-13721.

[13] Jia, Z.; Tang, Y.; Luo, L.; Li, B. Shape-controlled synthesis of single-crystalline $CdCO_3$ and corresponding porous CdO nanostructures. *Cryst. Growth Des.* 2008, 8, 2116-2120.

[14] Yang, Z. X.; Zhong, W.; Yin, Y. X.; Du, X.; Deng, Y.; Au, C.; Du, Y. W. Controllable synthesis of single-crystalline CdO and $Cd(OH)_2$ nanowires by a simple hydrothermal approach. *Nanoscale Res. Lett.* 2010, 5, 961-965.

[15] Ebihara, M.; Suemune, I.; Kumano1, H.; Nakashita, T.; Machida, H. Formation of CdO dots on atomically flat ZnO surfaces. *Phys. Status Solidi C.* 2006, 3, 933-937.

[16] Rahman, M. M.; Jamal, A.; Khan, S. B.; Faisal, M.; Asiri, A. M. Highly sensitive methanol chemical sensor based on undoped silver oxide nanoparticles prepared by a solution method. *Microchim. Acta.* 2012, 178, 99-106.

[17] Wang, F.; Zhao, D.; Guo, Z.; Liu, L.; Zhang, Z.; Shen, D. Artificial leaf structures as a UV detector formed by the self-assembly of ZnO nanoparticles. *Nanoscale*. 2013, 5, 2864-2869.

[18] Zheng, Y.; Chen, C.; Zhan, Y.; Lin, X.; Zheng, Q.; Wei, K.; Zhu, J. Photocatalytic activity of Ag/ZnO heterostructure nanocatalyst: Correlation between structure and property. *J. Phys. Chem. C.* 2008, 112, 10773-10777.

[19] Iijima, S. Helical microtubules of graphitic carbon. *Nature*. 1991, 354, 56-58.

[20] Baughrman, R. H.; Zakhidov, A. A.; Heer, W. A. Carbon nanotubes–The route towards applications. *Science*. 2002, 297, 787.

[21] Odom, T. W.; Huang, J. L.; Kim, P.; Lieber, C. Electronic density of states of atomically resolved single-walled carbon nanotubes: Van Hove singularities and end states. *Nature*. 1998, 391, 62.

[22] Lee, P. T.; Goncalves, L. M.; Compton, R. G. Electrochemical determination of free and total glutathione in human saliva samples. *Sens. Actuators B*. 2015, 221, 962–968.

[23] Zhao, X.; Zhou, S.; Shen, Q.; Jiang, L. P.; Zhu, J. J. Fabrication of glutathione photoelectrochemical biosensor using graphene-CdS nanocomposites. *Analyst*. 2012, 137, 3697-3703.

[24] Timur, S.; Odaci, D.; Dincer, A.; Zihnnioglu, F.; Telefoncu, A. Biosensing approach for glutathione detection using glutathione reductase and sulfhydryl oxidase bioenzymatic system. *Talanta*. 2008, 74, 1492-1497.

[25] Valle, M. A. D.; Chen, Y. Y.; Alarcon, A.; Ramos, A.; Hernandez, L. A.; Canales, C.; Gonzalez, B. Inclusion of glutathione peroxidase on polypyrrole modified electrode, to be assayed as glutathione sensor. *J. Chil. Chem. Soc.* 2013, 58, 1991-1993.

[26] Smith, Z. M.; Terry, J. M.; Barnett, N. W.; Gray, L. J.; Wright, D. J.; Francis, P. S. Enhancing permanganate chemiluminescence detection for the determination of glutathione and glutathione disulfide in biological matrices. *Analyst*. 2014, 139, 2416-2422.

[27] Zhao, J.; Chen, C.; Zhang, L.; Jiang, J.; Shen, G.; Yu, R. A Hg^{2+}-mediated label-free fluorescent sensing strategy based on G-quadruplex formation for selective detection of glutathione and cysteine. *Analyst*. 2013, 138, 1713-1718.

[28] Hussain, M. M.; Rahman, M. M.; Asiri, A. M. Ultrasensitive and selective 4-aminophenol chemical sensor development based on nickel oxide nanoparticles decorated carbon nanotube nanocomposites for green environment. *J. Environ. Sci.* 2017, 53, 27-38.

[29] Hussain, M. M.; Rahman, M. M.; Asiri, A. M.; Awual, M. R. Non-enzymatic simultaneous detection of L-glutamic acid and uric acid using mesoporous Co_3O_4nanosheets. *RSC Adv.* 2016, 6, 80511-80521.

[30] Rahman, M. M.; Hussain, M. M.; Asiri, A. M. Ultrasensitive and label-free detection of creatine based on CdO nanoparticles: A real sample approach. *New J. Chem.* 2017, 41, 6667-6677.

[31] Rahman, M. M.; Hussain, M. M.; Asiri, A. M. Bilirubin sensor based on CuO-CdO composites deposited in a nafion/glassy carbon electrode matrices. *Progress Nat. Sci: Mater. Intl.* 2017, 27, 566-573.

[32] Rahman, M. M.; Hussain, M. M.; Asiri, A. M. Fabrication of 3-methoxyphenol sensor based on Fe_3O_4 decorated carbon nanotube nanocomposites for environmental safety. *PLoS One*. 2017, 12, e0177817.

[33] Jain, U.; Narang, J.; Rani, K.; Burna, S.; Chauhan, N. Synthesis of cadmium oxide and carbon nanotube based nanocomposites and their use as a sensing interface for xanthine detection. *RSC Adv.* 2015, 5, 29675-29683.

[34] Feng, J.; Xiong, S.; Qian, Y.; Yin, L. Synthesis of nanosized cadmium oxide (CdO) as a novel high capacity anode material for lithium-ion batteries: influence of carbon nanotubes decoration and binder choice. *Electrochim. Acta*. 2014, 129, 107-112.

[35] Uplane, M. D.; Kshirsagar, P. N.; Lokhande, B. J.; Bhosale, C. H. Characteristic analysis of spray deposited cadmium oxide thin films. *Mater. Chem. Phys.* 2000, 64, 75-78.

[36] Chen, S.; Zhang, X.; Zhang, Q.; Tan, W. Trioctylphosphine as both solvent and stabilizer to synthesize CdS nanorods. *Nanoscale Res. Lett.* 2009, 4, 1159-1165.

[37] Patel, J.; Mighri, F.; Ajji, A. Simple non-aqueous fabrication route for oleic acid capped luminescent cadmium sulphide quantum dots at relatively low temperature. *Soft. Nanosci. Lett.* 2011, 1, 61-65.

[38] Hussain, M. M.; Rahman, M. M.; Asiri, A. M. Efficient 2-nitrophenol chemical sensor development based on Ce_2O_3 nanoparticles decorated CNT nanocomposites for environmental safety. *PLoS One*. 2016, 11, e0166265.

[39] Hussain, M. M.; Rahman, M. M.; Asiri, A. M. Sensitive L-leucine sensor based on a glassy carbon electrode modifed with SrO nanorods. *Microchim. Acta.* 2016, 183, 3265-3273.

[40] Rahman, M. M.; Hussain, M. M.; Asiri, A. M. A novel approach towards the hydrazine sensor development by SrO.CNT nanocomposites. *RSC Adv.* 2016, 6, 65338-65348.

[41] Zhou, W.; Zhou, Y.; Liang, Y.; Feng, X.; Zhou, H. Silver nanoparticles on carboxyl-functionalized $Fe_3O_4$with high catalytic activity for 4-nitrophenol reduction. *RSC Adv*. 2015, 5, 50505-50511.

[42] Rahman, M. M.; Khan, S. B.; Gruner, G.; Al-Ghamdi, M. S.; Daous, M. A.; Asiri, A. M. Chloride ion sensors based on low-dimensional α-MnO_2-Co_3O_4 nanoparticles fabricated glassy carbon electrodes by simple I-V technique. *Electrochim. Acta*. 2013, 103, 143-150.

[43] Zhang, Y.; Cui, Z.; Li, L.; Guo, L.; Yang, S. Two-dimensional structure Au nanosheets are super active for the catalytic reduction of 4-nitrophenol. *Phys. Chem. Chem. Phys.* 2015, 17, 14656-14661.

[44] Rahman, M.; Khan, S. B.; Asiri, A. M. Chemical sensor development based on polycrystalline gold electrode embedded low-dimensional Ag_2O nanoparticles. *Electrochim. Acta*. 2013, 112, 422–430.

[45] Zhao, X. L.; Huo, L. H.; Gao, S.; Zhao, J. G. ZnO nanoparticulate thin film: preparation, characterization and gas-sensing property. *Sens. Actuators B*. 2004, 102, 248-252.

[46] Srivastava, J. K.; Pandey, P.; Mishra, V. N.; Dwivedi, R. Structural and micro structural studies of PbO-doped SnO_2 sensor for detection of methanol, propanol and acetone. *J. Natur. Gas Chem.* 2011, 20, 179-183.

[47] Rahman, M. M.; Khan, S. B;, Asiri, A. M. A microchip based fluoride sensor based on the use of CdO doped ferric oxide nanocubes. *Microchim. Acta*. 2015, 182, 487-494.

[48] Rahman, M. M.; Khan, S. B.; Asiri, A. M.; Alamry, K. A.; Khan, A. A. P.; Khan, A.; Rub, M. A.; Azum, N. Acetone sensor based on solvothermally prepared ZnO doped with Co_3O_4 nanorods. *Microchim. Acta*. 2013, 180, 675-685.

[49] Jordan, E.; Bell, R. G.; Wilmer, D.; Koller, H. Anion-promoted cation motion and conduction in zeolites. *J. Am. Chem. Soc.* 2006, 128, 558–567.

[50] Rahman, M. M.; Jamal, A.; Khan, S. B.; Faisal, M. Highly sensitive ethanol chemical sensor based on Ni-doped SnO_2 zeolite materials. *Biosens. Bioelectron.* 2011, 28, 127-134.

[51] Karimi-Maleh, H.; Tahernejad-Javazmi, F.; Ensafi, A. A.; Moradi, R.; Mallakpour, S.; Beitollahi, H. A high sensitive biosensor based on FePt/CNT nanocomposite/N-(4-hydroxyphenyl)-3,5-dinitrobenzamide modified carbon paste electrode for simultaneous determination of glutathione and piroxicam, *Biosens. Bioelectron.* 2014, 60, 1–7.

[52] Raoof, J. B.; Ojani, R.; Karimi-Maleh, H. Electrocatalytic oxidation of glutathione at carbon paste electrode modified with 2,7-bis(ferrocenyl ethyl)fluoren-9-one: application as a voltammetric sensor. *J. Appl. Electrochem.* 2009, 39, 1169–1175.

[53] Karimi-Maleh, H.; Tahernejad-Javazmi, F.; Gupta, V. K.; Ahmar, H.; Asadi, M. H. A novel biosensor for liquid phase determination of glutathione and amoxicillin in

biological and pharmaceutical samples using a ZnO/CNT nanocomposite/catechol derivative modified electrode. *J. Mol. Liq.* 2014, 196, 258–263.

[54] Raoof, J. B.; Ojani, R.; Kolbadinezhad, M. Voltammetric sensor for glutathione determination based on ferrocene-modified carbon paste electrode. *J. Solid State Electrochem.* 2009, 13, 1411–1416.

[55] Shahmiri, M. R.; Bahari, A.; Karimi-Maleh, H.; Hosseinzadeh, R.; Mirnia, N. Ethynylferrocene–NiO/MWCNT nanocomposite modified carbon paste electrode as a novel voltammetric sensor for simultaneous determination of glutathione and acetaminophen. *Sens. Actuators B.* 2013, 177, 70–77.

[56] Tang, H.; Chen, J.; Nie, L.; Yao, S.; Kuang, Y. Electrochemical oxidation of glutathione at well-aligned carbon nanotube array electrode. *Electrochim. Acta.* 2006, 51, 3046–3051.

[57] Keyvanfard, M.; Karimi-Maleh, H.; Alizad, K. Multiwall carbon nanotube paste electrode with 3,4-dihydroxy-cinnamic acid as mediator for the determination of glutathione in pharmaceutical and urine samples. *Chin. J. Catal.* 2013, 34, 1883–1889.

[58] Ruiz-Díaz, J. J. J.; Torriero, A. A. J.; Salinas, E.; Marchevsky, E. J.; Sanz, M. I.; Raba, J. Enzymatic rotating biosensor for cysteine and glutathione determination in a FIA system. *Talanta.* 2006, 68, 1343–1352.

[59] Moradi, R.; Sebt, S. A.; Karimi-Maleh, H.; Sadeghi, R.; Karimi, F.; Bahari, A.; Arabi, H. Synthesis and application of FePt/CNT nanocomposite as a sensor and novel amide ligand as a mediator for simultaneous determination of glutathione, nicotinamide adenine dinucleotide and tryptophan. *Phys. Chem. Chem. Phys.* 2013, 15, 5888–5897.

[60] Ensafi, A. A.; Karimi-Maleh, H.; Mallakpour, S. A new strategy for the selective determination of glutathione in the presence of nicotinamide adenine dinucleotide (NADH) using a novel modified carbon nanotube paste electrode. *Colloids Surf. B: Biointerfaces.* 2013, 104, 186–193.

[61] Ensafi, A. A.; Dadkhah-Tehrani, S.; Karimi-Maleh, H. Voltammetric determination of glutathione in haemolysed erythrocyte and tablet samples using modified-multiwall carbon nanotubes paste electrode. *Drug Test Anal.* 2012, 4, 978–985.

[62] Raoof, J. B.; Ojani, R.; Karimi-Maleh, H.; Hajmohamadi, M. R.; Biparva, P. Multi-wall carbon nanotubes as a sensor and ferrocene dicarboxylic acid as a mediator for voltammetric determination of glutathione in hemolyzed erythrocyte. *Anal. Method.* 2011, 3, 2637-2643.

[63] Ensafi, A. A.; Taei, M.; Khayamian, T.; Karimi-Maleh, H.; Hasanpour, F. Voltammetric measurement of trace amount of glutathione using multiwall carbon nanotubes as a sensor and chlorpromazine as a mediator. *J. Solid State Electrochem.* 2010, 14, 1415–1423.

[64] Mazloum-Ardakani, M.; Sheikh-Mohseni, M. A.; Mirjalili, B. F. Selective and simulataneous voltammetric determination of glutathione, uric acid and penicillamine by a modified carbon nanotube paste electrode. *Electroanal.* 2013, 25, 2021–2029.
[65] Oztekin, Y.; Ramanaviciene, A.; Ramanavicius, A. Electrochemical glutathione sensor based on electrochemically deposited poly-*m*-aminophenol. *Electroanal.* 2011, 23, 701–709.
[66] Rahman, M. M.; Hussain, M. M.; Asiri, A. M. A glutathione biosensor based on a glassy carbon electrode modified with CdO nanoparticle-decorated carbon nanotube in a nafion matrix. *Microchim. Acta.* 2016, 183, 3255-3263.
[67] Hussain, M. M.; Rahman, M. M.; Arshad, M. N.; Asiri, A. M. Hg^{2+} sensor development based on (*E*)-*N′*-Nitrobenzylidine benzene sulfonohydrazide (NBBSH) derivatives fabricated on a glassy carbon electrode with a nafion matrix. *ACS Omega.* 2017, 2, 420-431.
[68] Hussain, M. M.; Rahman, M. M.; Arshad, M. N.; Asiri, A. M. Electrochemical detection of Ni^{2+} ions using synthesized (*E*)-*N′*-chlorobenzylidene-4-methylbenzenesulfonohydrazide derivatives modified with a nafion matrix. *Chem. Select.* 2017, 2, 7455-7464.
[69] Hussain, M. M.; Rahman, M.M.; Arshad, M. N.; Asiri, A. M. Trivalent Y^{3+} ionic sensor development based on (*E*)-Methyl-*N′*-nitrobenzylidene-benzene-sulphonohydrazide (MNBBSH) derivatives modified with nafion matrix. *Sci. Rep.* 2017, 7: 5832.
[70] Asiri, A. M.; Hussain, M. M.; Arshad, M. N.; Rahman, M. M. Sensitive and selective heavy metal ion, Mn^{2+} sensor development based on the synthesized (*E*)-N′-chlorobenzylidene-benzenesulfonohydrazide (CBBSH) molecules modified with nafion matrix. *J. Ind. Eng. Chem.* 2018, 63, 312-321.
[71] Rahman, M. M.; Alam, M.M.; Hussain, M. M.; Asiri, A. M.; Zayed, M. E. M. Hydrothermally prepared Ag_2O/CuO nanomaterial for an efficient chemical sensor development for environmental remediation. *Environ. Nanotech. Monitor. Manage.* 2018, 10, 1-9.
[72] Hussain, M. M.; Arshad, M. N.; Asiri, A. M.; Rahman, M .M. Development of selective Co^{2+} ionic sensor based on variuos derivatives of benzenesulfonohydrazide (BSH) compounds: An electrochemical approach. *Chem. Eng. J.* 2018, 339, 133-143.
[73] Asiri, A. M.; Hussain, M. M.; Arshad, M. N.; Rahman, M. M. A Ce^{2+} sensor based on napthalen-1-yl-methylene-benzenesulfonohydrazide (NMBSH) molecules: ecological sample analysis. *New J. Chem.* 2018, 42, 4465-4473.
[74] Hussain, M. M.; Rahman, M. M.; Arshad, M. N.; Asiri, A. M. Fabrication of Ga^{3+} sensor probe based on methoxybenzylidene benzenesulfonohydrazide (MBBSH) by an electrochemical approach. *New J. Chem.* 2018, 42, 1169-1180.

BIOGRAPHICAL SKETCHES

Mohammed Muzibur Rahman

Present Position

Professor

Affiliation

King Abdulaziz University, Jeddah, Saudi Arabia.

Education

PhD (Chemistry)

Research and Professional Experience

I worked as postdoctoral fellows and assistant professor in research centers and universities in South Korea (Pusan National University), Japan (Toyohashi University of Technology), and Saudi Arabia (Najran University) between 2007 and 2011. Since 2011 I have been working as an Associate Professor in Center of Excellence for Advanced Materials Research (CEAMR) and Chemistry Department at King Abdulaziz University, Saudi Arabia. My research work has been largely on the fields of biosensors, nanoparticles, carbon nanotubes, nanotechnology, ionic liquids, surface chemistry, electrochemistry, instrumental science, nanomaterials, self-assembled monolayers, photochemistry, and μ-chips, and devices.

Short Biography

Mohammed Muzibur Rahman received his B.Sc. and M.Sc. from Shahjalal University of Science & Technology, Sylhet, Bangladesh in 1999 and 2001, respectively. He received his Ph.D. from the Chonbuk National University, South Korea, in 2007. After his Ph.D., he worked as postdoctoral fellow and assistant professor in research centers and universities in South Korea, Japan, and Saudi Arabia (from 2007 to 2011). Presently, he is working as an associate professor in Center of Excellence for Advanced Materials Research (CEAMR) and Chemistry Department at King Abdulaziz University, Saudi Arabia since 2011. He has published more than 250 research articles; 4 patents, 13 book chapters, 12 books, and attended over 50 international conferences.

Professional Appointments

Professor at King Abdulaziz University, Saudi Arabia.

Publications from the Last 3 Years: (2016-2018)

1. Subhan, M. A.; Saha, P. C.; Sumon, S. A.; Ahmed, J.; Asiri, A. M.; Rahman, M. M. Enhanced photocatalytic activity and ultra-sensitive benzaldehyde sensing performances of $SnO_2 \cdot ZnO \cdot TiO_2$ nanomaterials. *RSC Adv*. 2018, 8, 33048-33058.
2. Salam, M. A.; Rahman, M. M. Synthesis, spectroscopic characterization, crystal structure and anti-bacterial activity of diorganotin(IV) complexes with 5-bromo-2-hydroxybenzaldehyde-N(4)-ethylthiosemicarbazon. *J. Coord. Chem*. 2018, 71, 1593-1605.
3. Rahman, M. M.; Alam, M. M.; Asiri, A. M. Carbon black co-adsorbed ZnO nanocomposites for selective benzaldehyde sensor development by electrochemical approach for environmental safety. *J. Industrial Eng. Chem*. 2018, 65, 300–308.
4. Sheikh, T. A.; Rahman, M. M.; Asiri, A. M.; Marwani, H. M.; Awual, M. R. 4-Hexylresorcinol sensor development based on wet-chemically prepared $Co_3O_4@Er_2O_3$ nanorods: A practical approach. *J. Indust. Eng. Chem*. 2018, 66, 446-455.
5. Wahid, A.; Asiri, A. M.; Rahman, M. M. Fabrication of isopropyl alcohol sensor development based on $Co_3O_4@Nd_2O_3$ nanocomposites. *Environ. Nanotechnol, Monitor. Manag.*, 2018, 10, 314–321.
6. Sheikh, T. A.; Arshad, M. N.; Asiri, A. M.; Rahman, M. M. Development of selective and sensitive Ga^{3+} sensor for environmental safety: A comparative study between cyclohexane and aromatic bis-sulphonamides fabricated glassy carbon electrodes. *New J. Chem*. 2018, 42, 13589-13601.
7. Rahman, M. M.; Alam, M. M.; Asiri, A. M. Selective hydrazine sensor fabrication with facile low-dimensional Fe_2O_3/CeO_2nanocubes. *New J. Chem*. 2018, 42, 10263-10270.
8. El-Shishtawy, R. M.; Al-Ghamdi, H. A.; Alam, M. M.; Al-Amshany, Z. M.; Asiri, A. M.; Rahman, M. M. Development of Cd^{2+} sensor based on BZNA/Nafion/Glassy carbon electrode by electrochemical approach. *Chem. Eng. J*. 2018, 352, 225–231.
9. Rahman, M. M.; Wahid, A.; Alam, M. M.; Asiri, A. M. Efficient 4-Nitrophenol sensor development based on facile $Ag@Nd_2O_3$ nanoparticles. *Mater. Today Commun*. 2018, 16, 307–313.
10. Rahman, M. M.; Alenazi, N. A.; Hussein, M. A.; Alam, M. M.; Alamry, K. A.; Asiri, A. M. Hybride ZnCdCrO embedded aminated polyethersulfone nanocomposites for the development of H^{g2+} ionic sensor. *Mater. Res. Express*, 2018, 5, 065019.
11. Rahman, M. M.; Hussein, M. A.; Aly, K. I.; Asiri, A. M. Thermally stable hybrid polyarylidene(azomethine-ether)s polymers (PAAP): An ultrasensitive arsenic(III) sensor approach. *Des. Monomers Polym.*, 2018, 21, 82-89.
12. Rahman, M. M.; Sheikh, T. A.; El-Shishtawy, R. M.; Arshad, M. N.; Al-Zahrani, F. A. M.; Asiri, A. M. Fabrication of S^{b3+} sensor based on 1,1'-(-(naphthalene-2,3-diylbis(azanylylidene))bis (methanylylidene))bis(naphthalen-2-ol)/Nafion/glassy

carbon electrode assembly by electrochemical approach. *RSC Adv.* 2018, 8, 19754-19764.

13. Asiri, A. M.; Hussain, M. M.; Arshad, M. N.; Rahman, M. M. Sensitive and selective heavy metal ion, Mn^{2+} sensor development based on the synthesized (*E*)-N'-chlorobenzylidene-benzenesulfonohydrazide (CBBSH) molecules modified with nafion matrix. *J. Industrial Eng. Chem.* 2018, 63, 312-321.
14. Hossain, M. A.; Elias, M.; Sarker, D. R.; Diba, Z. R.; Mithun, J. M.; Azad, M. A. K.; Siddiquey, I. A.; Rahman, M. M.; Uddin, J.; Uddin, M. N. Synthesis of Fe- or Ag-doped TiO_2–MWCNT nanocomposite thin films and their visible-light-induced catalysis of dye degradation and antibacterial activity. *Res. Chem. Intermed.*, 2018, 44, 2667-2683.
15. Rahman, M. M.; Alam, M. M.; Hussain, M. M.; Asiri, A. M.; Zayed, M. E. M. Hydrothermally prepared Ag_2O/CuO nanomaterial for an efficient chemical sensor development for environmental remediation. *Environ. Nanotechnol. Monitor. Manag.*, 2018, 10, 1-9.
16. Rahman, M. M.; Alam, M. M.; Asiri, A. M. Sensitive 1,2-dichlorobenzene chemi-sensor development based on solvothermally prepared FeO/CdO nanocubes for environmental safety. *J. Indust. Eng. Chem.* 2018, 62, 392-400.
17. Rahman, M. M.; Hussein, M. A.; Alamry, K. A.; Al-Shehry, F. M.; Asiri, A. M. Polyaniline/graphene/carbon nanotubes nanocomposites for sensing environmentally hazardous 4-aminophenol. *Nano-struct. Nano-Objects.* 2018, 15, 63-74.
18. Alam, M. M.; Asiri, A. M.; Uddin, M. T.; Islam, M. A.; Rahman, M. M. Wet-chemically prepared low-dimensional $ZnO/Al_2O_3/Cr_2O_3$ nanoparticles for the xanthine sensor development by an electrochemical method. *RSC Adv.* 2018, 8, 12562-12572.
19. Rahman, M. M.; Ahmed, J.; Asiri, A. M. Thiourea sensor development based on hydrothermally prepared CMO nanoparticles for environmental safety. *Biosens. Bioelectron.* 2018, 99, 586-592.
20. Hussain, M. M.; Asiri, A. M.; Arshad, M. N.; Rahman, M. M. Development of selective Co^{2+} ionic sensor based on various derivatives of benzenesulfonohydrazide (BSH) compounds: An electrochemical approach. *Chem. Eng. J.* 2018, 339, 133–143.
21. Ahmed, J.; Rahman, M. M.; Siddiquey, I. A.; Asiri, A. M.; Hasnat, M. A. Efficient hydroquinone sensor based on zinc, strontium and nickel based ternary metal oxide (TMO) composites by differential pulse voltammetry. *Sens. Actuators B.* 2018, 256, 383-392.
22. Awual, M. R.; Khraisheh, M.; Alharthi, N. H.; Luqman, M.; Islam, A.; Karim, M. R.; Rahman, M. M.; Khaleque, M. A. Efficient detection and adsorption of cadmium(II) ions using innovative nano-composite materials. *Chem. Eng. J.* 2018, 343, 118-127.

23. Alam, M. M.; Rashed, M. A.; Rahman, M. M.; Rahman, M. M.; Nagao, Y.; Hasnat, M. A. Electrochemical oxidation of As(III) on Pd immobilized Pt surface: kinetics and sensing performance. *RSC Adv.* 2018, 8, 8071-8079.
24. Hussain, M. M.; Asiri, A. M.; Arshad, M. N.; Rahman, M. M. Fabrication of a Ga^{3+} sensor probe based on methoxybenzylidene benzenesulfonohydrazine (MBBSH) by an electrochemical approach. *New J. Chem.* 2018, 42, 1169-1180.
25. Rahman, M. M.; Alam, M. M.; Asiri, A. M.; Islam, M. A. 3,4-Diaminotoluene sensor development based on hydrothermally prepared $MnCo_xO_y$ nanoparticles. *Talanta.* 2018, 176, 17–25.
26. Abdel-Latif, I. A.; Rahman, M. M.; Khan, S. B. Neodymium cobalt oxide as a chemical sensor. *Res. Phys.* 2018, 8, 578-583.
27. Sheikh, T. A.; Rahman, M. M.; Asiri, A. M.; Marwani, H. M. Sensitive 3-chlorophenol sensor development based on facile Er_2O_3/CuO nanomaterials for environmental safety. *New J. Chem.* 2018, 42, 3936-3946.
28. Rahman, M. M.; Ahmed, J. Cd-doped Sb_2O_4 nanostructures modified glassy carbon electrode for efficient detection of melamine by electrochemical approach. *Biosens. Bioelectron.* 2018, 102, 631-636.
29. Rahman, M. M.; Alam, M. M.; Asiri, A. M. 2-Nitrophenol sensor-based wet-chemically prepared binary doped Co_3O_4/Al_2O_3 nanosheets by an electrochemical approach. *RSC Adv.* 2018, 8, 960-970.
30. Subhan, M. A.; Saha, P. C.; Rahman, M. M.; Ahmed, J.; Asiri, A. M.; Al-Mamun, M. Fabrication of a 2,4-dinitrophenol sensor based on Fe_3O_4@Ag@Ni nanomaterials and studies of their antibacterial properties. *New J. Chem.* 2018, 42, 872-881.
31. Subhan, M. A.; Saha, P. C.; Alam, M. M.; Asiri, A. M.; Al-Mamun, M.; Rahman, M. M. Development of Bis-Phenol a sensor based on $Fe_2MoO_4 \cdot Fe_3O_4 \cdot ZnO$ nanoparticles for sustainable environment. *J. Environ. Chem. Eng.* 2018, 6, 1396-1403.
32. Asiri, A. M.; Hussain, M. M.; Arshad, M. N.; Rahman, M. M. A $Ce2^{+}$ sensor based on napthalen-1-yl-methylene-benzenesulfonohydrazine (NMBSH) molecules: ecological sample analysis. *New J. Chem.* 2018, 42, 4465-4473.
33. Khan, A.; Khan, A. A.P.; Rahman, M. M.; Asiri, A. M.; Alfaifi, S. Y. M.; Taib, L. A. Towards facile preparation and design of mulberry shaped poly(N–methylaniline)$Ce_2(WO_4)_3$@CNT nanocomposite and its application for electrochemical Cd^{2+} ions detection for environment remediation. *Polymer-Plastics Technol. Eng.* 2018, 57, 335-345.
34. Hussein, M. A.; Rahman, M. M.; Asiri, A. M. A novel facial conducting polyamides based dithiophenylidenecyclyhexanone moiety utilized for selective Cu^{2+} sensing. *Polymer Plastics Technol. Eng.* 2018, 57, 812-825.
35. Algethami, F. K.; Marwani, H. M.; Asiri, A. M.; Rahman, M. M. Comparative performances of phenolic sensors based on various CeO_2-carbon material nanocomposites for environmental safety. *Sensor Rev.* 2018, 1-12.

36. Rahman, M. M. A label-free kanamycin sensor development based on CuO-NiO hollow-spheres: Food samples analyses. *Sens. Actuators B*. 2018, 264, 84-91.
37. Rahman, M. M.; Abu-Zied, B. M.; Asiri, A. M. Cu-loaded ZSM-5 zeolites: An ultra-sensitive phenolic sensor development for environmental safety. *J. Indust. Eng. Chem.* 2018, 61, 304–313.
38. Khan, A. A. P.; Khan, A.; Rahman, M. M.; Asiri, A. M.; Oves, M. Chemical sensor development and Antibacterial activities based on polyaniline/gemini surfactants for environmental safety. *J. Polymers Environ*. 2018, 26, 1673-1684.
39. Khan, A.; Khan, A. A. P.; Rahman, M. M.; Asiri, A. M.; Inamuddin; Alamry, K. A.; hamed, S. A. Preparation and characterization of PANI@G/CWO nanocomposite for enhanced 2-nitrophenol sensing. *Appl. Surface Sci.* 2017, 433, 696-704.
40. Rahman, M. M.; Hussain, M. M.; Asiri, A. M. Bilirubin sensor based on CuO-CdO composites deposited in a nafion/glassy carbon electrode matrixes. *Prog. Nat. Sci. Mater. Int.* 2017, 27, 566-573.
41. Rahman, M. M.; Alam, M. M.; Asiri, A. M.; Awual, M. R. Fabrication of 4-aminophenol sensor based on hydrothermally prepared ZnO/Yb_2O_3 nanosheets. *New J. Chem.* 2017, 41, 9159-9169.
42. Rahman, M. M.; Hussein, M. A.; Salam, M. A.; Asiri, A. M. Fabrication of L-glutathione sensor based on PEG conjugated functionalized CNT nanocomposites: A real sample analysis. *New J. Chem.* 2017, 41, 10761-10772.
43. Rahman, M. M.; Hussain, M. M.; Asiri, A. M. Fabrication of 3-methoxyphenol sensor based on Fe_3O_4 decorated carbon nanotube nanocomposites for environmental safety: Real sample analyses. *PLOS ONE*. 2017, 12, e0177817.
44. Rahman, M. M.; Alam, M. M.; Asiri, A. M. Fabrication of acetone sensor based on facile ternary $MnO_2/Gd_2O_3/SnO_2$nanosheets for environmental safety. *New J. Chem.* 2018, 42, 9938-9946.
45. Ahmad, I.; Arshad, M. N.; Rahman, M. M.; Asiri, A. M.; Sheikh, T. A.; Aqlan, F. M. Crystal Structure of N'-[(*E*)-(3-hydroxynaphthalen-2-yl) methylidene]-4-benzenesulfonohydrazide (HNMBSH) and its application as Pb^{+2} ion sensor by its fabrication onto glassy carbon electrode. *Inorg. Chim. Acta*, 2017, 467, 297-306.
46. Hussain, M. M.; Rahman, M. M.; Arshad, M. N.; Asiri, A. M. Electrochemical detection of Ni^{2+} ions using synthesized (*E*)-N′-chlorobenzylidene-4-methylbenzenesulfonohydrazide derivatives modified with a nafion matrix. *Chemistry Select*. 2017, 2, 7455-7464.
47. Subhan, M. A.; Saha, P. C.; Rahman, M. M.; Akand, M. A. R.; Asiri, A. M.; Al-Mamun, M. Enhanced photocatalytic and chemical sensor development based on ternary $B_2O_3.Zn_6Al_2O_9.ZnO$ nanomaterials for environmental safety. *New J. Chem.* 2017, 41, 7220-7231.

48. Hussain, M. M.; Rahman, M. M.; Asiri, A. M. Trivalent Y^{3+} ionic sensor development based on (E)-Methyl-N′-nitrobenzylidene-benzenesulfonohydrazide (MNBBSH) derivatives modified with nafion matrix. *Sci. Rep.* 2017, 7, 5832.
49. Rahman, M. M.; Marwani, H. M.; Algethami, F. K.; Asiri, A. M. Xanthine sensor development based on ZnO-CNT, ZnO-CB, ZnO-GO and ZnO nanoparticles: An electrochemical approach. *New J. Chem.* 2017, 41, 6262-6271.
50. Rahman, M. M.; Hussain, M. M.; Asiri, A. M. Ultrasensitive and label-free detection of creatine based on CdO nanoparticles: A real sample approach. *New J. Chem.* 2018, 41, 6667-6677.
51. Ahmed, J.; Rahman, M. M.; Siddiquey, I. A.; Asiri, A. M.; Hasnat, M. A. Efficient bisphenol-A detection based on the ternary metal oxide (TMO) composites by electrochemical approaches. *Electrochim. Acta.* 2017, 246, 597–605.
52. Rahman, M. M.; Marwani, H. M.; Algethami, F. K.; Asiri, A. M. Comparative performance of hydrazine sensor development on various Mn_3O_4/Carbon-nanotubes, Mn_3O_4/Graphene-oxides, Mn_3O_4/Carbon-black nanocomposites. *Mater. Express.* 2017, 7, 169-179.
53. Rahman, M. M.; Alfonso, V. G.; Fabregat-Santiago, F.; Bisquert, J.; Asiri, A. M.; Alshehri, A. A.; Albar, H. A. Hydrazine sensors development based on a glassy carbon electrode modified with a nanostructured TiO_2 films by electrochemical approach. *Microchim. Acta.* 2017, 184, 2123-2129.
54. Subhan, M. A.; Fahim, A. M. M.; Saha, P. C.; Rahman, M. M.; Begum, K.; Azad, A. K. Structural study, photoluminescence and photocatalytic properties of $La_2O.Fe_3O_4{\cdot}ZnO$, $AgO{\cdot}NiO{\cdot}ZnO$ and $La_2O_3{\cdot}AgO{\cdot}ZnO$ nanocomposites. *Nano Struct. Nano Objects*, 2017, 10, 30-41.
55. Rahman, M. M.; Marwani, H. M.; Algethami, F. K.; Asiri, A. M.; Hameed, S. A.; Alhogbi, B. Ultra-sensitive p-nitrophenol sensing performances based on various Ag_2O conjugated carbon material composites. *Envn. Nanotechnol. Monitor. Manag.* 2017, 8, 73-82.
56. Sheikh, T. A.; Arshad, M. N.; Rahman, M. M.; Asiri, A. M.; Marwani, H. M.; Awual, M. R.; Bawazir, W. A. Trace electrochemical detection of Ni^{2+} ions with bidentate N,N'-(ethane-1,2-diyl)bis(3,4-dimethoxybenzenesulfonamide) [EDBDMBS] as a chelating agent. *Inorg. Chim. Acta.* 2017, 464157-166.
57. Rahman, M. M.; Balkhoyor, H. B.; Asiri, A. M. Ultra-sensitive xanthine sensor development based on wet-chemically prepared Co/ZnO nanoparticles. *Mater. Express.* 2017, 7, 93-103.
58. Awual, M. R.; Alharthi, N. H.; Hasan, M. M.; Karim, M. R.; Islam,A.; Znad, H.; Hossain, M. A.; Halim, M.; Rahman, M. M.; Khaleque, M. A. Inorganic-organic based novel nano-conjugate material for effective cobalt(II) ions capturing from wastewater. *Chem. Eng. J.* 2017, 324, 130-139.

59. Alam, M. K.; Rahman, M. M.; Elzwawy, A.; Torati, S. R.; Islam, M. S.; Todo, M.; Asiri, A. M.; Kim, D., Kim, C. G. Highly sensitive and selective detection of bisphenol A based on reduced graphene oxide decorated hydroxyapatite nanocomposites. *Electrochim. Acta*. 2017, 214, 353-361.
60. Rahman, M. M.; Algethami, F. K.; Asiri, A. M.; Marwani, H. M.; Alhogbi, B. A comparative study on 4-aminophenol sensor development with various CdO nanocomposites. *Nano Struct. Nano Objects*, 2017, 10, 141–150.
61. Rahman, M. M.; Alam, M. M.; Asiri, A. M.; Islam, M. A. Ethanol sensor development based on ternary-doped metal oxides ($CdO/ZnO/Yb_2O_3$) nanosheets for environmental safety. *RSC Adv*. 2017, 7, 22627-22639.
62. Rahman, M. M.; Alam, M. M.; Asiri, A. M.; Islam, M. A. Fabrication of selective chemical sensor with ternary $ZnO/SnO_2/Yb_2O_3$ nanoparticles. *Talanta*, 2017, 170, 215-223.
63. Rahman, M. M.; Abu-Zied, B. M.; Asiri, A. M. Ultrasensitive hydrazine sensor fabrication based on Co-doped ZSM-5 zeolites for environmental safety. *RSC Adv*. 2017, 7, 21164–21174.
64. Hussain, M. M.; Rahman, M. M.; Asiri, A. M. Ultrasensitive and selective 4-aminophenol chemical sensor development based on nickel oxide nanoparticles decorated carbon nanotube nanocomposites for green environment. *J. Environ. Sci*. 2017, 53, 27–38.
65. Awual, M. R.; Alharthi, N. H.; Okamoto, Y.; Karim, M. R.; Halim, M. E.; Hasan, M. M.; Rahman, M. M.; Islam, M. M. Ligand field effect for dysprosium(III) and lutetium(III) adsorption and EXAFS coordination with novel composite nanomaterials. *Chem. Eng. J*. 2017, 320, 427-435.
66. Rahman, M. M.; Ahmed, J.; Asiri, A. M. A glassy carbon electrode modified with γ-Ce_2S_3-decorated CNT nanocomposites for uric acid sensor development: a real sample analysis. *RSC Adv*. 2017, 7, 14649-14659.
67. Alam, M. K.; Rahman, M. M.; Abbas, M.; Torati, S. R.; Asiri, A. M.; Kim, D.; Kim, C. G. Ultra-sensitive 2-nitrophenol detection based on reduced graphene oxide/ZnO nanocomposites. *J. Electroanal. Chem*. 2017, 788, 66-73.
68. Khan, A. A. P.; Khan, A.; Rahman, M. M.; Asiri, A. M.; Oves, M. Sensor development of 1,2 dichlorobenzene based on polypyrole/Cu-doped ZnO (PPY/CZO) nanocomposite embedded silver electrode and their antimicrobial studies. *Int. J. Biol. Macromol*. 2017, 98, 256-267.
69. Hussain, M. M.; Rahman, M. M.; Arshad, M. N.; Asiri, A. M. Hg^{2+} sensor development based on (*E*)-N′-nitrobenzylidine-benzenesulfonohydrazide (NBBSH) derivatives fabricated on a glassy carbon electrode with a nafion matrix. *ACS Omega*. 2017, 2, 420-431.

70. Rahman, M. M.; Balkhoyor, H. B.; Asiri, A. M. Phenolic sensor development based on chromium oxide-decorated carbon nanotubes for environmental safety. *J. Environ. Manag*. 2017, 188, 228-237.
71. Rahman, M. M.; Ahmed, J.; Asiri, A. M. Development of creatine sensor based on antimony-doped tin oxide (ATO) nanoparticles. *Sens. Actuators B*. 2017, 242, 167–175.
72. Akhter, H.; Murshed, J.; Rashed, M. A.; Oshima, Y.; Nagao, Y.; Rahman, M. M.; Asiri, A. M.; Hasnat, M. A.; Uddin, M. N.; Siddiquey, I. A. Fabrication of hydrazine sensor based on silica-coated Fe_2O_3 magnetic nanoparticles prepared by a rapid microwave irradiation method. *J. Alloys Compd*. 2017, 698, 921-929.
73. Arshad, M. N.; Sheikh, T. A.; Rahman, M. M.; Asiri, A. M.; Marwany, H. M.; Awual, M. R. Fabrication of cadmium ionic sensor based on (*E*)-4-methyl-N'-(1-(pyridin-2-yl)ethylidene) benzenesulfonohydrazide (MPEBSH) by electrochemical approach. *J. Organometal. Chem*. 2017, 827, 49-55.
74. Awual, M. R.; Hasan, M. M.; Eldesoky, G. E.; Khaleque, M. A.; Rahman, M. M.; Naushad, M. Facile mercury detection and removal from aqueous media involving ligand impregnated conjugate nanomaterials. *Chem. Eng. J*. 2016, 290, 243–251.
75. Khan, A.; Khan, A. A. P.; Rahman, M. M.; Asiri, A. M. High performance polyaniline/vanadyl phosphate (PANI-$VOPO_4$) nano composite sheets prepared by exfoliation/intercalation method for sensing applications. *Eur. Polymer J*. 2016, 75, 388-398.
76. Khan, A. A. P.; Khan, A.; Rahman, M. M.; Asiri, A. M.; Oves, M. Lead sensors development and antimicrobial activities based on graphene oxide/carbon nanotube/poly(O-toluidine) nanocomposite. *Int. J. Biol. Macromol*. 2016, 89, 198-205.
77. Akter, N.; Hossain, M. A.; Hassan, M. J.; Amin, M. K.; Elias, M.; Rahman, M. M.; Asiri, A. M.; Siddiquey, I. A.; Hasnat, M. A. Amine modified tannin gel for adsorptive removal of brilliant green dye. *J. Environ. Chem. Eng*. 2016, 4, 1231–1241.
78. Karim, M. R.; Takehira, H.; Rahman, M. M.; Asiri,A. M.; Amin, M. K.; Ohtani, R.; Hayami, S. Magnetic and liquid crystalline property of long-alkyl chain appended iron(II) imidazole complexes. *J. Organometal. Chem*. 2016, 808, 42-47.
79. Rahman, M. M.; Alamry, K. A.; Saleh, T. S.; Asiri, A. M. Sensitive and selective Cu^{2+} sensor based on 4-(3-(thiophen-2-yl)-9H-carbazol-9-yl)benzaldehyde (TPCBZ) conjugated copper-complex. *J. Organometal. Chem*. 2016, 817, 43-49.
80. Rahman, M. M.; Marwani, H. M.; Alshehri, A. A.; Albar, H. A.; Bisquert, J.; Asiri, A. M. Room temperature stable $ClPrNTf_2$ ionic liquid utilizing for chemical sensor development. *J. Organometal. Chem*. 2016, 811, 74–80.
81. Sheikh, T. A.; Arshad, M. N.; Rahman, M. M.; Asiri, A. M.; Alamry, K. A. Highly efficient Co^{2+} ions sensor based on *N,N'*-(ethane-1,2-diyl)bis(2,5-dimethoxyben-

zenesulfonamide) (EBDMBS) fabricated glassy carbon electrode. *J. Organometal. Chem.* 2016, 822, 53-61.
82. Khan, A. A. P.; Khan, A.; Rahman, M. M.; Asiri, A. M. Conventional surfactant-doped poly(o-anisidine)/GO nanocomposites for benzaldehyde chemical sensor development. *J. Sol-Gel Sci. Technol.* 2016, 77, 361-370.
83. Mamun, M. R. A.; Karim, M. R.; Rahman, M. M.; Asiri, A. M.; Torii, S. Methane enrichment of biogas by carbon dioxide fixation with calcium hydroxide and activated carbon. *J. Taiwan Inst. Chem. Engr.* 2016, 58, 476-481.
84. Rahman, M. M.; Ahmed, J.; Asiri, A. M.; Hasnat, M. A.; Siddiquey, I. A. Development of 4-methoxyphenol chemical sensor based on NiS_2-CNT nanocomposites. *J. Taiwan Inst. Chem. Eng.* 2016, 64, 157–165
85. Rahman, M. M.; Balkhoyor, H. B.; Asiri, A. M.; Sobahi, T. R. Development of selective chloroform sensor with transition metal oxide nanoparticle/multi-walled carbon nanotube nanocomposites by modified glassy carbon electrode. *J. Taiwan Inst. Chem. Eng.* 2016, 66, 336-346.
86. Rahman, M. M.; Asiri, A. M.; Youssef, T. E.; Marwani, H. M. Photocatalytic degradation of remazol brilliant orange 3R using wet-chemically prepared CdO-ZnO nanofibers for environmental remediation. *Mater. Express.* 2016, 6, 137-148.
87. Rahman, M. M.; Balkhoyor, H. B.; Asiri, A. M.; Marwani, H. M. A gold electrode modified with silver oxide nanoparticle decorated carbon nanotubes for electrochemical sensing of dissolved ammonia. *Microchim. Acta.* 2016, 183, 1677-1685.
88. Rahman, M. M.; Khan, A.; Marwani, H. M.; Asiri, A. M. Hydrazine sensor based on silver nanoparticle-decorated polyaniline tungstophosphate nanocomposite for use in environmental remediation. *Microchim. Acta*, 2016, 183, 1787-1796.
89. Rahman, M. M.; Hussain, M. M.; Asiri, A. M. A glutathione biosensor based on a glassy carbon electrode modified with CdO nanoparticles-decorated carbon nanotube in a nafion matrix. *Microchim. Acta*, 2016, 183, 3255–3263.
90. Hussain, M. M.; Rahman, M. M.; Asiri, A. M. Sensitive L-leucine sensor based on a glassy carbon electrode modified with SrO nanorods. *Microchim. Acta*, 2016, 183, 3265-3273.
91. Kim, T. Y.; Song, D.; Barea, E. M.; Lee, J. H.; Kim, Y. R.; Cho, W.; Lee, S.; Rahman, M. M.; Bisquert, J.; Kang, Y. S. Origin of high open-circuit voltage in solid state dye-sensitized solar cells employing polymer electrolyte. *Nano Energy.* 2016, 28, 455-461.
92. Hussain, M. M.; Rahman, M. M.; Asiri, A. M. Efficient 2-nitrophenol chemical sensor development based on Ce_2O_3 nanoparticles decorated CNT nanocomposites for environmental safety. *PLOS ONE*, 2016, 11, e0166265.
93. Hasnat, M. A.; Safwan, J. A.; Rashed, M. A.; Rahman, Z.; Rahman, M. M.; Nagao, Y.; Asiri, A. M. Inverse effects of supporting electrolytes on the electrocatalytic

nitrate reduction activities in a Pt|Nafion|Pt-Cu type reactor assembly. *RSC Adv*. 2016, 6, 11609–11617.

94. Rahman, M. M.; Balkhoyor, H. B.; Asiri, A. M. Ultrasensitive and selective hydrazine sensor development based on Sn/ZnO nanoparticles. *RSC Adv*. 2016, 6, 29342-29352.
95. Rahman, M. M.; Abu-Zied, M. M.; Hasan, M. M.; Asiri, A. M.; Hasnat, M. A. Fabrication of selective 4-aminophenol sensor based on H-ZSM-5 zeolites deposited silver electrodes. *RSC Adv*. 2016, 6, 48435–48444.
96. Balkhoyor, H. B.; Rahman, M. M.; Asiri, A. M. Effect of Ce-doping into ZnO nanostructures to enhance the phenolic sensor performances. *RSC Adv*. 2016, 6, 58236–58246.
97. Rahman, M. M.; Hussain, M. M.; Asiri, A. M. A novel approach towards the hydrazine sensor development by SrO·CNT nanocomposites. *RSC Adv*. 2016, *6*, 65338–65348.
98. Hussain, M. M.; Rahman, M. M.; Asiri, A. M.; Awual, M. R. Non-enzymatic simultaneous detection of L-glutamic acid and uric acid using mesoporous Co_3O_4 nanosheets. *RSC Adv*. 2016, 6, 80511-80521.
99. Rahman, M. M.; Ahmed, J.; Asiri, A. M.; Siddiquey, I. A.; Hasnat, M. A. Development of ultra-sensitive hydrazine sensor based on facile CoS_2-CNT nanocomposites. *RSC Adv*. 2016, 6, 90470-90479.
100. Marwani, H. M.; Alsafrani, A. E.; Asiri, A. M.; Rahman, M. M. Silica-gel particles loaded with an ionic liquid for separation of Zr(IV) prior to its determination by ICP-OES. *Sensors*. 2016, 16, 1001.
101. Khan, A. A. P.; Khan, A.; Asiri, A. M.; Azum, N.; Rub, M. A.; Rahman, M. M.; Khan, S. B. Mechanistic investigation of osmium(VIII) catalyzed oxidation of glutamic acid with sodium salt of N-chloro-4-methylbenzenesulfonamide in aqueous media: A practical approach. *Synth. React. Inorg. Metal Org. Nano Metal Chem.* 2016, 46, 10-18.
102. Rahman, M. M.; Hussein, M. A.; Alamry, K. A.; Al Shehry, F. M.; Asiri, A. M. Sensitive methanol sensor based on PMMA-G-CNTs nanocomposites deposited onto glassy carbon electrodes. *Talanta*. 2016, 150, 71–80.

Mohammad Musarraf Hussain

Present position

Associate Professor

Affiliation

Jagannath University, Dhaka, Bangladesh

Education

PhD in Chemistry

Research and Professional Experience

Material sciences, Natural Product Chemistry, and Medicinal Chemistry.

Short Biography

Dr. Mohammad Musarraf Hussain is working as an Associate Professor in the Department of Pharmacy at Jagannath University, Dhaka, Bangladesh. He is on teaching since August 10, 2006 to till date. He also completed MSc in Chemistry from University of Bergen, Norway in 2013. Before, he had completed B. Pharm. (Hon's) and M. Pharm. with first class from University of Dhaka, Bangladesh in 2005 and 2007, respectively. Dr. Hussain awarded PhD from King Abdulaziz Universiy, Saudi Arabia in 2018. His research areas are material science, natural product chemistry, and medicinal chemistry. Dr. Hussain has published few articles in ISI category journal and attended in the international conferences at home and abroad.

Professional Appointment

Associate Professor, Department of Pharmacy, Jagannath University.

Publication in Last Three Years (2016 -2018)

1. Hussain, M. M. A short review on phytoconstituents from the genera *Albizzia* and *Erythrina*. *Bangladesh Pharm. J.* 2018, 21, 160-172.
2. Asiri, A. M.; Hussain, M. M.; Arshad, M. N.; Rahman, M. M. Sensitive and selective heavy metal ion, Mn^{2+} sensor development based on the synthesized (*E*)-N′-chlorobenzylidene-benzenesulfonohydrazide (CBBSH) molecules modified with nafion matrix. *J. Indust. Eng. Chem.* 2018, 63, 312-321.
3. Rahman, M. M.; Alam, M. M.; Hussain, M. M.; Asiri, A. M.; Zayed, M. E. M. Hydrothermally prepared Ag_2O/CuO nanomaterial for an efficient chemical sensor development for environmental remediation. *Environ. Nanotechnol. Monitor. Manag.* 2018, 10, 1-9.
4. Hussain, M. M.; Arshad, M. N.; Asiri, A. M.; Rahman, M. M. Development of selective Co^{2+} ionic sensor based on variuos derivatives of benzenesulfonohydrazide (BSH) compounds: An electrochemical approach. *Chem. Engr. J.* 2018, 339, 133-143.
5. Asiri, A. M.; Hussain, M. M.; Arshad, M. N.; Rahman, M. M. A Ce^{2+} sensor based on napthalen-1-yl-methylene-benzenesulfonohydrazide (NMBSH) molecules: ecological sample analysis. *New J. Chem.* 2018, 42, 4465-4473.
6. Hussain, M. M.; Rahman, M. M.; Arshad, M. N.; Asiri, A. M. Fabrication of Ga^{3+} sensor probe based on methoxybenzylidene benzenesulfonohydrazide (MBBSH) by an electrochemical approach. *New J. Chem.* 2018, 42, 1169-1180.

7. Rahman, M. M.; Hussain, M. M.; Asiri, A.M. Bilirubin sensor based on CuO-CdO composites deposited in a nafion/glassy carbon electrode matrixes. *Progress Nat. Sci. Mater. Intl.* 2017, 27, 566-573.
8. Rahman, M. M.; Hussain, M. M.; Asiri, A. M. Fabrication of 3-methoxyphenol sensor based on Fe_3O_4 decorated carbon nanotube nanocomposites for environmental safety. *PLOS ONE,* 2017, 12, e0177817.
9. Hussain, M. M.; Rahman, M. M.; Arshad, M. N.; Asiri, A. M. Electrochemical detection of Ni^{2+} ions using synthesized (*E*)-*N′*-chlorobenzylidene-4-methylbenzene-sulfonohydrazide derivatives modified with a nafion matrix. *Chem. Select.* 2017, 2, 7455-7464.
10. Hussain, M. M.; Rahman, M. M.; Arshad, M. N.; Asiri, A. M. Trivalent Y^{3+} ionic sensor development based on (*E*)-methyl-*N′*-nitrobenzylidene-benzenesulphono-hydrazide (MNBBSH) derivatives modified with nafion matrix. *Sci. Rep.* 2017, 7, 5832.
11. Rahman, M. M.; Hussain, M. M.; Asiri, A. M. Ultrasensitive and label-free detection of creatine based on CdO nanoparticles: A real sample approach. *New J. Chem.* 2017, 41, 6667-6677.
12. Hussain, M. M.; Rahman, M. M.; Asiri, A. M. Ultrasensitive and selective 4-Aminophenol chemical sensor development based on nickel oxide nanoparticles decorated carbon nanotube nanocomposites for green environment. *J. Environ. Sci.* 2017, 53, 27-38.
13. Hussain, M. M.; Rahman, M. M.; Arshad, M. N.; Asiri, A.M. Hg^{2+} sensor development based on (*E*)-N′-nitrobenzylidinebenzenesulfonohydrazide (NBBSH) derivatives fabricated on a glassy carbon electrode with a nafion matrix. *ACS Omega.* 2017, 2, 420-431.
14. Baumann, M.; Hussain, M. M.; Henne, N.; Garrote, D. M.; Karlshøj, S.; Fossen, T.; Rosenkilde, M. M.; Våbenø, J.; Haug, B. E. Influence of chain length on the activity of tripeptidomimetic antagonists for CXC chemokine receptor 4 (CXCR4). *Bioorg. Med. Chem.* 2017, 25, 646-657.
15. Hussain, M. M.; Rahman, M. M.; Asiri, A. M. Efficient 2-nitrophenol chemical sensor development based on Ce_2O_3 nanoparticles decorated CNT nanocomposites for environmental safety. *PLOS ONE.* 2016, 11, e0166265.
16. Rahman, M. M.; Hussain, M. M.; Asiri, A. M. A glutathione biosensor based on a glassy carbon electrode modified with CdO nanoparticle-decorated carbon nanotube in a nafion matrix. *Microchim. Acta.* 2016, 183, 3255-3263.
17. Hussain, M. M.; Rahman, M. M.; Asiri, A. M. Sensitive L-Leucine sensor based on a glassy carbon electrode modifed with SrO nanorods. *Microchim. Acta.* 2016, 183, 3265-3273.

18. Hussain, M. M.; Rahman, M. M.; Asiri, A. M.; Awual, M. R. Non-enzymatic simultaneous detection of L-glutamic acid and uric acid using mesoporous Co_3O_4 nanosheets. *RSC Adv.* 2016, 6, 80511-80521.
19. Rahman, M. M.; Hussain, M. M.; Asiri, A. M. A novel approach towards the hydrazine sensor development by SrO·CNT nanocomposites. *RSC Adv.* 2016, 6, 65338-65348.
20. Hussain, M. M.; Tuhin, M. T. H.; Akter, F.; Rashid, M. A. Constituents of *Erythrina*- a potential source of secondary metabolites: A review. *Bangladesh Pharm. J.* 2016, 19, 237-253.
21. Hussain, M. M.; Tahia, F.; Rashid, M. A. Secondary metabolites from some species of *Albizzia*: A review. *Bangladesh Pharm. J.* 2016, 19, 1-8.

Abdullah M. Asiri

Present Position

Professor

Affiliation

King Abdulaziz University, Jeddah, Saudi Arabia

Education

PhD in Chemistry

Research and Professional Experience

His research interest covers color chemistry, synthesis of novel photochromic, thermochromic systems, synthesis of novel coloring matters and dyeing of textiles, materials chemistry, nanochemistry and nanotechnology, polymers and plastics. A major achievement of Professor Asiri is the discovery of tribochromic compounds, a class of compounds that change from slightly or colorless to deep colored when subjected to small pressure or when grind. This discovery was introduced to the scientific community as a new terminology published by IUPAC in 2000. Prof. Asiri is the principal supervisors of more than 20 MSc and six PhD thesis. He holds ten US patents, and author/coauthor of more than 1210 publications, 16 book chapters and 28 books. He is the Editor-in-Chief of King Abdulaziz University Journal of Science. He is also a member of the Editorial Board of the periodicals Pigments and Resin Technology, Organic Chemistry in Sight, Designed Monomers & Polymers, and Journal of Single Molecule Research. He is the Vice-President of Saudi Chemical Society (Western Province Branch).

Short Biography

Prof. Abdullah M. Asiri received BSc in Chemistry from King Abdulaziz University in 1990, and PhD from University of Wales, College of Cardiff (UK) in 1995. He was promoted to Professor in 2004. He is the head of the Chemistry Department at King

Abdulaziz University since October 2009, and the founder and the first director of the Center of Excellence for Advanced Materials Research (CEAMR). He is a Professor of Organic Photochemistry. He took a major role in the Advanced Materials Committee working for KACST in 2007. Prof. Asiri played a major role in advancing Chemistry education and research at KAU; he has obtained the Best Researchers award from KAU for the past five years. He also awarded the Young Scientist award from the Saudi Chemical Society in 2009; and the First Prize for Distinction in Science from the Saudi Chemical Society in 2012. He also received a recognition certificate from the American Chemical Society (Gulf region Chapter) for advancement of chemical science in the Saudi Arabia. Also, he received a Scopus certificate as the best publishing scientist in Saudi Arabia on the field of Chemistry in 2008. He was awarded the Shoman Prize for scientific research in 2018. He is the only Saudi researcher listed in the Highly Cited Researchers list of Clarivate Analytics in both Chemistry and Materials category in 2018.

Professional Appointment

Professor in Chemistry Department of King Abdulaziz University

Publications from the Last 3 Years

https://orcid.org/0000-0001-7905-3209

In: Glutathione
Editor: Pál Perjési
ISBN: 978-1-53614-740-7
© 2019 Nova Science Publishers, Inc.

Chapter 11

THE ROLE OF GLUTATHIONE IN SPERM RESISTANCE TO CRYOPRESERVATION

***Dario Santos*[1] *and Rita Payan-Carreira*[2,3,*]**
[1]CITAB & Department of Biology and Environment, Universidade de Trás-os-Montes e Alto Douro, Vila Real, Portugal
[2]CECAV, Universidade de Trás-os-Montes e Alto Douro, Vila Real, Portugal
[3]Department of Veterinary Medicine, School of Sciences and Technology, Évora University, Pole of Mitra, Évora, Portugal

ABSTRACT

The reduced form of glutathione (GSH), the most abundant intracellular antioxidant thiol in mammalian systems, participates in diverse reactions, including reacting with reactive oxygen (ROS) and nitrogen species (RNS), electrophiles, and other non-enzymatic antioxidant compounds, as well as protein deglutathionylation. GSH is a key-molecule in oxidative stress balance, as GSH effectively scavenges free radicals and other reactive oxygen species. Among its multiple roles, GSH is required for mitochondrial function and integrity.

Most of the cellular content of GSH (85–90%) is present in the cytosol, while the remainder can be found in several different organelles (such as the mitochondria, nuclear matrix, and peroxisomes). GSH transcription or activity is triggered in cells submitted to an oxidant or thermal stress, among other stimuli. However, due to their cellular characteristics, sperm are poorly endowed to react against oxidative stress. Both their scant cytosol content and transcriptional ability limits their production of GSH, rendering the sperm particularly sensitive to damage from the accumulation of reactive oxygen species in the medium. Moreover, the spermatozoon presents some weaknesses that make it an easy target for oxidative stress, such as the high levels of polyunsaturated fatty acids in the sperm plasmatic membrane, or the highly compacted mitochondrial sheet.

* Corresponding Author E-mail: rtpayan@gmail.com.

Notwithstanding, the seminal plasma contains high activities of several oxidative stress enzymes, which partially deflect spermatozoon sensitivity while bathed in this fluid. This sensitivity is a major factor to consider during sperm collection and cryopreservation.

Oxidative stress impacts sperm function by interfering with plasma membrane fluidity, inactivating membrane-enzymes, increasing its permeability for non-specific ions, affecting key intracellular enzymes, decreasing motility and damaging paternal DNA, leading to a loss of sperm fertility. In recent years, the incorporation of glutathione or other antioxidant molecules to the extender media has been assayed in multiple species to foster sperm survival for semen preservation, in particular for freezing-thawing procedures.

In this chapter, we will revise the part played by glutathione in male fertility and semen homeostasis, as well as addressing the role of GSH in sperm function and GSH-based strategies to improve sperm cryopreservation.

Keywords: glutathione, oxidative stress, sperm fertility, semen cryopreservation, sperm extender

INTRODUCTION

Glutathione, a tripeptide (Glu-Cys-Gly) molecule that participates in the oxidative stress pathway, may be found both in a reduced (GSH) and an oxidized (GSSG) form [1]. Even though the term glutathione is often used to refer to both molecules, this chapter focuses on the reduced form of glutathione, and the term is used in this context. GSH and GSSG act in concert with other redox-active compounds to regulate and maintain cellular redox status [2].

GSH is a tripeptide (γ-glutamyl-cysteinyl-glycine) with a molecular mass of 307. In mammalian cells, the cytosolic concentrations of GSH are in the range of 1-10 mM. GSH is also present in high concentrations in mitochondria and nuclei. In contrast, extracellular levels of GSH are much lower, and GSH levels in plasma are usually in the range of low micromolar concentrations. In mammalian cells, GSH synthesis involves two cytosolic enzymes, namely, γ-glutamylcysteine ligase (GCL) and glutathione synthetase (GSS) [3].

It was initially believed that GSH acts mainly as a reductant, but over the past two decades it has been demonstrated that GSH is one of the most important antioxidants, acting as a cofactor for glutathione peroxidase (GPx). GSH and GPx act as scavenging antioxidants in the male genital tract, as well as in the seminal plasma [4]. In mammalian cells and tissues, GSH participates in several reactions with reactive oxygen and nitrogen species (ROS and RNS). It also participates in antioxidant mechanism, as a cofactor of glutathione-dependent enzymes. Besides its role as co-factor, it is important to recognize that GSH itself plays a role. GSH can directly react with ROS/RNS, leading to detoxification of these reactive species and additionally reacts with reactive aldehydes derived from xenobiotic biotransformation as well as lipid peroxidation derivatives, forming less reactive adducts [5, 6].

Glutathione is the most abundant low-molecular-weight thiol in animal cells ranging from 0.5 to 10 mM [4]. Most cellular GSH (85–90%) is present in the cytosol, with the remainder found in various organelles (including mitochondria, nuclear matrix, and peroxisomes). Except for the bile, which may contain up to 10 mM GSH, extracellular concentrations of GSH are relatively low (e.g., 2–20 μM in plasma). Because of its cysteine residue, GSH is readily oxidized nonenzymatically to glutathione disulfide (GSSG) by redox active electrophilic substances (e.g., free radicals and reactive oxygen/nitrogen species). GSSG efflux from cells contributes to a net loss of intracellular GSH. Cellular GSH concentrations are markedly reduced in response to protein malnutrition, oxidative stress, and many pathological conditions. Like other sulfur-containing amino acid (AA) metabolites, GSH plus GSSG represent total GSH in cells, a significant amount of which (up to 15%), may be bound to proteins. The ratio of GSH to GSSG, which is often used as an indicator of the cellular redox state, is >10 under normal physiological conditions. GSH/GSSG is the major redox couple that determines antioxidant capacity of the cells, but its value can be affected by other redox couples, including $NADPH/NADP^+$ and $thioredoxin_{red}/thioredoxin_{ox}$ [7].

GSH has been detected in the male reproductive tissues and fluids (e.g., the testes, epididymis, and the seminal plasma) of different mammal species [8]. It has also been demonstrated that depletion of glutathione plays an important role in male infertility [9].

ROS and Glutathione in Male Fertility

The Physiological Role of ROS in the Male Reproductive System

Since oxygen is essential for animal life, all aspects of reproduction involve reactive species (RS) and are dependent on an equilibrium between RS and the antioxidant defense/repair systems [10]. Oxidative stress arises when an oxidant-generating system overpowers the defense system responsible for its neutralization and removal. In consequence, reactive oxygen species (ROS) accumulate locally. ROS accumulation may result from physiological or pathological mechanisms.

However, ROS possess bifunctional roles in several biological processes, such as cell proliferation, apoptosis or migration [11]. ROS can be both beneficial and harmful, according to their nature, concentration, location and length of exposure [12]. Transient increases in ROS and glutathione depletion can be associated with important physiologic actions requiring hypoxic states and, in the case of the reproductive system, they contribute to the maintenance of normal fertility [13, 14]. However, excessive production of ROS in the microenvironment surrounding the germ cells causes sperm dysfunction, which encompasses lipid peroxidation, loss of motility, and sperm DNA damage [15]. Thus, it can be said that maintenance of a high redox potential is a requirement for a healthily operating reproductive system [12]. In male reproduction, many important mechanisms of sperm physiology have been associated with ROS (for review, [16]), and therefore could be traced to glutathione actions.

Spermatogenesis is a replicative process that must account for both the self-renewal of spermatogonial stem cells and differentiation in the adult testis guaranteeing the constant production of spermatozoa (SPZ) with potential fertility [12]. The high rates of cell division inherent to spermatogenesis lead to a peak in mitochondrial oxygen consumption and therefore to reactive oxygen species (ROS) generation and GSH depletion [17]. It is well known that the microenvironment within the seminiferous tubules, during spermatogenesis, is crucial to sperm viability.

Several aspects of the spermatogenic process have been shown to be highly sensitive to oxidative stress disruptions in the spermatogenic microenvironment:

- Changes in chromatin condensation (e.g., replacement of histone by protamins, haploidy and the oxidative condensation of chromatin) and the packaging of the nucleus content shielding the chromatin from oxidative modification during the fertilization process [12, 18];
- Elimination of defective spermatogenic cells through apoptosis, which can be obtained througth the p53-dependent and -independent pathways [18];
- Changes in cell morphology (e.g., the formation of the sperm head) [18];
- Transcriptional control of regulators of cell junction dynamics [12].

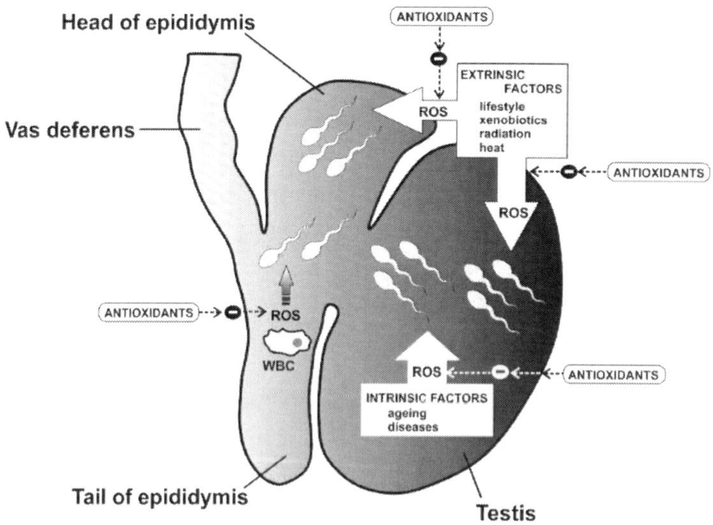

Figure 1. Multiple factors, both intrinsic (ageing, diseases) and extrinsic (e.g., related with lifestyle, hormonal imbalances, xenobiotics, or heat) generate reactive oxygen species (ROS) in the spermatozoa environment both during their formation in the testis and their maturation and storage in the epididymis and deferent ducts. The different sources of ROS can instigate defective sperm function and DNA damage. Antioxidants act at those sites to neutralize some of the ROS-mediated effects. WBC – white blood cells.

Oxidative stress disruption of spermatogenesis leads to defective sperm function (Figure 1), which has been related to sperm DNA damage and fragmentation [19-21], but also to induction of collateral damage to proteins and lipids in the sperm membranes [20-22]. But these disruptive effects are not restricted to the effect of oxidative stress on spermatogenesis (Figure 1); after leaving the testis, sperm maturation proceeds in the epididymis, where further packaging of the nuclear proteins occurs, *via* enzyme-mediated sulfoxidation [11]. In the epididymis, detoxification of lipid peroxidation products is crucial to protect sperm membranes from oxidative insults, in a process mediated by GSH [18]. In men, a 10% higher concentration of GSH has been reported in comparison to testicular levels [23]. Moreover, tight control of ROS is a major issue during storage of sperm in the epididymis and deferent duct, which has been associated with accumulation of GSH in the epididymal lumen [24]. Therefore, epididymal antioxidant status and degree of oxidative stress are also important determinants of male fertility.

Disruption of the antioxidant status of the testes has been found to underpin some pathological or stressful conditions, such as [17, 25]: (a) Cryptorchidism, which increases heat stress and impairs activity of scavenger oxidative stress enzymes; (b) Inflammatory processes and vascular diseases leading to hypoxia or ischemia (e.g., testicular torsion or varicocele) trigger increased ROS production, increased lipid peroxide formation and depletion of testicular glutathione, resulting in lipid peroxidation and sperm DNA damage; (c) Hyperthyroidism and Diabetes, which increase lipid peroxidation, decrease GSH levels and enhance H_2O_2 production; (d) Sex steroids imbalance, associated with endogenous or exogenous causes; (e) Systemic oxidative stress due to lifestyle and environmental causes (e.g., smoking, alcohol, xenobiotics (phthalate esters, bisphenol A, PCB/PCN, arsenic, lead, cadmium, formaldehyde, lindane, cisplatin, cyclophosphamide, acrylamide, monensin, aflatoxin, ozone)), as well as (f) Testicular or scrotal infection and (g) Excessive physical exercise, all converge to compromise integrity of the local antioxidant mechanisms due to depletion of the local antioxidant enzyme activity in response to the stressor stimuli [20]. In addition, electromagnetic radiation, including heat (infrared) and radio frequency radiation, affect ROS generation and provoke DNA damage [25, 26]. Furthermore, ageing tends to increase lipid peroxidation, which is associated with imperfect spermiation or incomplete cytoplasmic extrusion during terminal spermatogenesis and midpiece anomalies, which in turn may explain the age-related decline in male fertility reported in humans [24]. Age-related changes in the GSH content of seminal plasma has been reported in Arab stallions, suggesting that ageing effects in the antioxidant ability of the seminal fluid might occur across mammalian species [27].

Decreased glutathione content in SPZ and seminal plasma has been reported in men with infertility [4, 23]. In men with infection or varicocele, GSH content in the seminal fluid is altered, contributing to the oxidative damage of sperm [4]. The total GSH content in the semen of oligozoo-, asthenozoo- and oligoasthenozoo-spermic men is lower compared with normospermic men [28, 29].

The Physiological Role of ROS in Spermatozoa Functions

After being released from the germinal epithelium, isolated spermatozoa (SPZ) are highly sensitive to oxidative stress. The main reason for their vulnerability is that SPZ are divested of most of their initial cytoplasm in the final steps of spermatid differentiation. In majority of the somatic cells, most GSH cellular content (85–90%) is present in the cytosol, while the remainder can be found in several different organelles (such as the mitochondria, nuclear matrix, and peroxisomes) [30, 31]. Conversely, in spermatozoa, GSH and antioxidant enzymes are restricted to the sperm midpiece, in the vicinity of the mitochondria [32, 33]. Even though SPZ do not entirely lack antioxidant defenses (they do contain detectable SOD activity and an active glutathione system), the limited content in antioxidant enzymes and the limited transcriptional ability fragilizes their defenses, which are easily overwhelmed, with the resultant generation of oxidative stress [26, 32]. In men, differences in the GSH content between SPZ in the caput epididymis and ejaculated semen is about 75% [23].

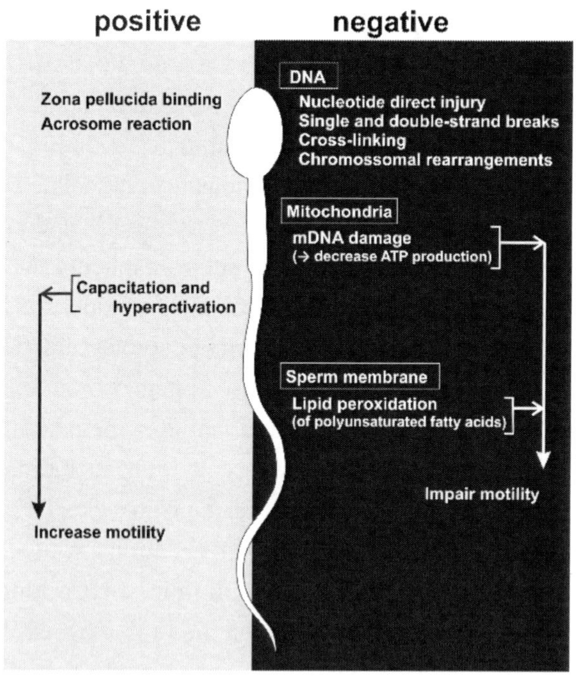

Figure 2. The effect of ROS upon sperm responds to a dual, balanced relationship (Yin/Yang effect): When physiological levels are balanced, ROS are crucial for sperm capacitation and the acquisition of progressive motility. When it is imbalanced, ROS induced damage leads to defective sperm functioning, due to increased lipid peroxidation of membranes, and DNA damage (DNA fragmentation and decreased motility).

Spermatozoa are particularly susceptible to oxidative stress because they are rich in unsaturated fatty acids and DNA, the major substrates for free radical attacks [26, 34]. As they have a limited capacity for DNA repair and a limited capacity for glutathione replenishment, SPZ depend on the surrounding environment to protect them from oxidative attack [32]. For this reason, epididymal and seminal plasma are among the richest media in antioxidant scavenger enzymes, including unique forms of glutathione peroxidase (GPx5) and extracellular SOD [32]. While in the male genital ducts, in physiological conditions, SPZ are protected against the attack of free radicals.

A dual relationship exists between ROS and SPZ functions (Figure 2). SPZ are spontaneously active generators of ROS [26], whose accumulation may harm the cell, but success of some SPZ processes requires production of controlled amounts of ROS, which are crucial to the processes of capacitation, acrosome reaction, and sperm binding to the oocyte [34, 35].

It is now commonly believed that mammalian sperm capacitation is an oxidative event [33, 36]. Capacitation is a crucial event in sperm fertilizing competency, which is triggered after deposition of SPZ in the female genital tract, particularly in the uterus. Entering a foreign environment, with a different pH and different concentrations in Na^+, Cl^- and HCO_3^-, the sperm initiates capacitation-signaling cascades [37]. Sperm capacitation involves loss of cholesterol from the SPZ plasma membrane and a consequent increase in membrane fluidity; activation of prostanoid pathways; changes in ion permeability, modification of the sperm membrane potential, hyperpolarization of the SPZ plasma membrane; and an increase in protein kinase A activity and protein tyrosine phosphorylation (for review, [37]), which is crucial for acquisition of hyperactivated motility. An increase in ROS generation triggers SPZ capacitation and leads to tyrosine phosphorylation [16]. While redox signaling is crucial for the capacitation process, excessive ROS production and oxidative stress can disrupt the event and compromise the fertilizing ability of SPZ, which can explain some infertility. Among other antioxidant partners, sufficient concentrations of GSH are necessary to guarantee redox signaling in the spermatozoon [34, 36].

Capacitation prepares the sperm for the acrosome reaction [38, 39]. Capacitated SPZ respond to a burst of extracellular superoxide anion, triggering the acrosome reaction [16]. A protein kinase-dependent transient increase in intracellular Ca^{2+} at the beginning of the acrosome reaction elicits a cascade of events that terminate with exocytosis of the acrosome content [37], and further stimulates production of superoxide anion [38]. In parallel, protein kinase also induces activation of phospholipase A_2, which extracts arachidonic acid from the membrane phospholipids [40], increasing fluidity of the SPZ membrane, in preparation for the interaction with the oocyte membrane, at fertilization [41]. Meanwhile, in response to an increase in Ca^{2+} and pH, the actin filaments located between the acrosomal and the cell membranes are broken [41], resulting loss of the actin cytoskeleton, allowing the fusion of acrosomal and cytoplasmic membranes, and exocytosis of the acrosome content [40]. Moreover, controlled lipid peroxidation of the SPZ membrane favors lipid regionalization within the SPZ membrane and exposure of particular proteins [41], which increase SPZ affinity to specific receptors in the oocyte membrane [38]. This is a particularly significant phenomenon for SPZ binding to the zona pellucida and fusion of the sperm/egg membranes.

ROS vs. Glutathione in Sperm Cryopreservation

Effects of GSH Depletion and Oxidative Stress on Semen Quality

After ejaculation, SPZ protection against oxidative stress relies mainly on the antioxidant buffering capacity of the seminal plasma, which is provided by enzymatic antioxidants and a variety of non-enzymatic antioxidants including GSH [4, 17], taurine, and hypotaurine [42]. It is commonly recognized that excessive H_2O_2 generation is a key factor in loss of sperm function [15, 20].

It is generally accepted that GSH content in SPZ varies between species [43] (Table 1). Moreover, some individual differences in GSH content in fresh semen have been reported in pigs [44, 45], which could account for the differences reported in males allowing them to be divided into good or bad freezers. Individual variations have also been reported in normospermic men [23].

Table 1. GSH levels in SPZ and seminal fluid in fertile males of various species

Male species	Concentration of GSH in fresh SPZ	Concentration of GSH in semen
Human	4.47 ± 0.46 nmol/10^8 cells [46] 4.0 pmol/10^6 cells [28] 6.8 nmol/10^8 cells [23] 1.29±0.28 nmol/mL [4]	
Rabbit	0.1 nmol/10^9 SPZ [47]	
Mouse	9 nmol/10^9 SPZ [47]	
Ram	4.5 ± 1.4 nmol/10^9 SPZ [47]	
Buck [epididymal SPZ]	13.00 ± 4.9 nmol/10^9 SPZ [47]	
Dog	5.3 ± 1.1 nmol/10^9 SPZ [47]	<0.2μM [47] 5.86±0.87 nM [48]
Bull	0.01 ± 0.002 mg/mg protein (in *Simmental-Fleckvieh breed*) [49] 566 ± 72 pmoles/mg protein (in *Holstein*) [50] 2.93 nmol /10^8 SPZ [8]	0.10 ± 0.02 mg/mg protein (in *Simmental-Fleckvieh breed*) [49] 17 ± 7 pmoles/mg protein(in *Holstein*) [50] 13–19 μM [8]
Stallion		23.33 ± 6.61 mg/mL [27]
Boar	3.84 ± 0.21 nM/10^8 sperm [44]	
Buffalo-bull		08.32 ± 0.53 μM [51]

Whenever ROS production exceeds the limited antioxidant defenses of SPZ, a state of oxidative stress prevails [33]. Oxidative stress attacks on sperm results in peroxidative damage to the sperm membranes [16, 40], loss of motility [25, 38, 40], DNA damage [25], and SPZ apoptosis [25]. Damage to the cell membranes may further contribute to leakage from the cell of antioxidant molecules, aggravating the inability to react to oxidative stress [43].

DNA damage occurs both during spermatogenesis, due to spermiogenesis errors, causing defective protamination and compaction of the sperm chromatin, which increases chromatin susceptibility to oxidative attack; and by a direct oxidative attack on the DNA after spermiation, leading to DNA fragmentation [19, 21, 32]. Since SPZ possess a limited capacity for DNA repair, once initiated, oxidative attacks on spermatozoa tend to be self-perpetuating and difficult to reverse [25]. The established critical threshold in humans for the number of SPZ with DNA fragmentation is 25% [52].

Consequently, infertility frequently results from the effect of an oxidative stress state on SPZ. It develops from the sperm´s inability to fertilize [53] and due to deleterious side-effects on early embryo development after fertilization and subsequent fetal development [54]. In humans, ROS contribution to male infertility is estimated to be between 30 and 80% [52].

Sperm dysfunction is a major cause of male infertility, as mentioned above, but its effects spread beyond those seen immediately. In humans, SPZ dysfunction has also been associated with pregnancy outcomes (paternal-mediated risk of miscarriage) and the health of offsprings, increasing the incidence of genetic disorders and childhood cancers [55].

The Effects of Oxidative Stress upon Semen Cryopreservation

It has long been recognized that cryopreservation damages spermatozoa by different mechanisms, specifically by formation of intracellular ice and osmotic stress [21, 56], but the pathways involved in sperm cryoinjury are diverse and far more complex. In the past three decades, attention has been paid to the effects of molecular stressors, including ROS and oxidative scavenging enzymes, and to the effects of extenders such as carriers that offer protective solutions for sperm, so as to guarantee adequate fertility following long term sperm preservation.

In view of the foregoing, preservation of SPZ fertilizing ability during storage depends on maintaining a controlled level of oxidative stress, to avoid damaging SPZ function. In general, use of fresh semen for artificial insemination (AI) is less harmful to sperm function than freezing/thawing because thermal treatments induce cell injuries that may be associated with formation of ROS [56, 57]. In general, mammalian SPZ are poorly tolerant to stressors associated with chilling or cryopreservation [58]. Compared to fresh semen, a larger number of sperm cells (8-fold higher in bovines) are needed in cryopreserved doses to achieve equivalent fertilization rates *in vivo* [59]. In other species, such as dogs and horses, the use of frozen semen has been postponed for many years because of the huge loss in sperm fertilizing ability associated with thermal shock [58, 60].

In most domestic farm species as well as in humans, it has been demonstrated that *in vitro* sperm storage compromises SPZ function to a degree that varies with the species and the conservation period [short-term (fresh or refrigerated semen) *vs.* long-term (cryopreservation)] [58].

In general, total antioxidant capacity (TAC) of seminal plasma, as well as peroxidation of membrane lipids of SPZ are affected by the freeze-thawing procedure [21]. In the process, seminal plasma is either removed or highly diluted [61] and the cellular antioxidant systems are depleted during freezing and thawing [43]. In pigs, Kim et al. [56] described decrease of intracellular $\bullet O_2$ and H_2O_2 in viable sperm submitted to cooling and storage at 15 and 5ºC; while in the frozen-thaw SPZ increase in the intracellular H_2O_2 (similarly to that detected in SPZ cooled at 5ºC) but not in $\bullet O_2$ was detected. The authors concluded that storage of liquid boar semen at 15ºC may induce destabilization of sperm membranes, whereas freezing-thawing would reduce sperm functions due to increased membrane damage and H_2O_2 accumulation. Such changes were shown to reflect loss of motility and decrease of viable acrosome-intact sperms in the frozen-thawed semen; no differences between fresh and cooled semens, however, were found [56].

The above-described species differences in resistance to oxidative stress might be related to differences in the relative amount of intracellular scavenging enzymes in the sperms. In boar semen, the content of intracellular superoxide dismutase (which rapidly dismutates $\bullet O_2$ into H_2O_2) is high, unlike that of catalase (which converts H_2O_2 into water and oxygen) [62]. These differences may account for collapse of intracellular antioxidant defenses during initial cooling of the freezing procedure; consequently, antioxidants are no longer available to protect the cells against cold injury during freezing [56].

In contrast to the boar semen, stallion SPZ contain both superoxide dismutase, catalase and glutathione peroxidase, while canine SPZ possess mainly SOD with only minor levels of glutathione peroxidase and reductase [48]. According to the work of Li [47], 80% of the total GSH found in the seminal fluid resides in the SPZ, which is the case in rams but not in humans, where 80% of the total GSH content is in the seminal plasma. Therefore, different contents within the enzymatic protective system will explain the differences found among species with regard to the effects of cryopreservation upon sperm fertility.

It has also been demonstrated in several mammalian species that freezing/thawing procedures boost ROS formation, such as in the bull [50, 63], stallion [64, 65], ram [66], boar [44], the stud dog [67] as well as in humans [46]. In consequence, sperm motility and viability decrease during storage [33], also compromising sperm survival in the female genital tract. Depletion of GSH and increased ROS leads to a notable reduction in progressive motility, due to compromised mitochondrial activity [21]. In humans, Gadea et al. [46] reported a 68% decrease of sperm progressive motility in frozen semen, in comparison to fresh semen; whereas in bulls, a decrease of about 50% has been reported [43].

Nevertheless, the impact of freezing/thawing procedures may vary depending on the species and composition of the extended media. It is also possible that the main effect of ROS on sperm functions varies with the species, arising from different SPZ susceptibilities. For instance, Guthrie and Welch [33] argue that boar sperm is less susceptible to lipid peroxidation during hypothermic storage in the absence of ROS generators than sperm of other species, such as the turkey, bull or ram. Similarly, proportion of DNA damage associated with freezing is lower in buffalo sperm in comparison to that of humans [33].

Moreover, some studies have reported changes in GSH content through the freezing/thawing procedures. Freezing leads to a 5-fold depletion of GSH in bull sperm [43, 50], thus bolstering interest in supplementing the extenders with antioxidant molecules to increase sperm resilience to cryopreservation. Similarly, a decrease in SPZ glutathione content upon cryopreservation has been reported in boars [44] and humans [46, 68]. The loss of GSH during freezing/thawing procedures represents a total depletion of up to 80% in bulls [50], 63% in humans [68] and 32% in boars [44], although chemical composition of the extender may affect the percentage of loss [43].

GSH Supplementation to Semen Extenders

Reinforcement of the protective effect of cooling or freezing extenders upon the oxidative damage following freezing/thawing procedures would be beneficial to mitigate ROS aggression in long-term preserved sperm. For some species, such as sheep or pigs, the damage to SPZ function induced by cryopreservation with traditional extenders has delayed development of freezing/thawing sperm preservation. Together with species-specific anatomical particularities, the liquid storage of semen remains the most used procedure in artificial insemination [69].

Supplementation of the extender with exogenous antioxidants has been tested and shown to have interesting positive results in sperm cryopreservation. Among the antioxidants tested, GSH has presented interesting results in diverse species. Supplementation of the extender with GSH has been shown to improve sperm quality in several species, in both cooled or frozen/thawed semen, as summarized in Table 2.

Increasing evidence has shown that addition of GSH to semen extenders increases quality of the SPZ. This effect is particularly significant in frozen semen, with stabilization of the nuclear and mitochondrial proteins, and improvement in the membrane stability, which decreases the percentage of capacitated APZ and increases the final proportion of SPZ with intact acrosome. Furthermore, some studies have also evidenced improvement of *in vivo* fertility rates [21].

Table 2. Evidence of the reported positive effects of GSH supplementation to extenders in different species

Species	GSH supplementation	Reported benefits	References
Men	GSH (1 mM) on a glycerol-based cryoprotectant (Sperm Freezing Medium); cryopreservation in pellets *Frozen semen*	Increased total motility and sperm viability Modified the SPZ motility pattern (assessed by casa) Reduced SPZ level of ROS No effects on sperm viability	[46]
Dog	GSH (5 mM) on a Tris-egg yolk-citrate extender-glycerol extender (20% egg yolk, Tris, citric acid, lactose, raffinose, antibiotics and 6.5% glycerol). *Frozen semen*	Improved motility Extended longevity Increased acrosome integrity Decreased lipid peroxidation after thawing Better fertility in vivo	[57]
Bull	GSH (5 mM) in a sperm-TALP extender *Frozen semen*	Higher number of non-capacitated viable Spermatozoa A reduction in ROS generation Lower DNA fragmentation Higher oocyte penetration rate *in vitro* Higher *in vitro* embryo production	[70]
Ram	GSH (100 and 200 mM) in a Tris-egg yolk extender *Cooled semen* *(Liquid storage at 5°C)*	Improved motility No differences in the functionality of the SPZ membranes (HOST test)	[69]
	GSH (2 mM) in a Tris-egg yolk extender *Cooled semen (Liquid storage at 5°C)*	Higher post-thaw motility and sperm survival Decreased free radical concentrations in the seminal doses Increased post-thaw activity of some antioxidant enzymes (SOD and GPx)	[71]
	GSH (2 mM) in a Tris-egg yolk extender *Frozen semen*	Improved motility and mitochondrial integrity Increased percentage of acrosome intact SPZ	[72]
Buck	GSH (1mM) in a soybean lecithin-based extender	Improved total motility Improved plasma membrane integrity Improved SPZ viability	[73]
	GSH (5 mM) in a Tris-egg yolk-citrate-glicerol extender *Frozen semen*	Improved post-thawing sperm motility and viability Improved DNA and acrosomal integrity Improved tail parameters	[74]
Boar	GSH (1 mM) in a mixed lactose egg yolk, Equex-Paste and glycerol extender *Frozen semen*	Improved the proportion of intact acrosome and viable SPZ, plasma membrane integrity and the non-capacitated SPZ Improved sperm motility (total and progressive)	[44, 75]
	GSH (2 mM) in Androstar Plus® *Frozen semen*	Decreased DNA fragmentation levels Improved sperm viability, sperm motility and acrosome integrity Decreased the post-thawn ROS concentrations Improved in vivo fertility rates (farrowing rate and prolificity)	[76]

[18] Tsunoda, S.; Kimura, N.; Fujii, J. Oxidative stress and redox regulation of gametogenesis, fertilization, and embryonic development. *Reprod. Med. Biol.* 2014, 13, 71-79.

[19] Gandini, L.; Lombardo, F.; Paoli, D.; Caponecchia, L.; Familiari, G.; Verlengia, C.; Dondero, F.; Lenzi, A. Study of apoptotic DNA fragmentation in human spermatozoa. *Hum. Reprod.* 2000, 15, 830-839.

[20] Aitken, R. J.; Gibb, Z.; Baker, M. A.; Drevet, J.; Gharagozloo, P. Causes and consequences of oxidative stress in spermatozoa. *Reprod. Fertil. Dev.* 2016, 28, 1-10.

[21] Yeste, M. Sperm cryopreservation update: Cryodamage, markers, and factors affecting the sperm freezability in pigs. *Theriogenology.* 2016, 85, 47-64.

[22] Aitken, R. J.; Smith, T. B.; Jobling, M. S.; Baker, M. A.; De Iuliis, G. N. Oxidative stress and male reproductive health. *Asian J. Androl.* 2014, 16, 31-38.

[23] Ochsendorf, F. R.; Buhl, R.; Bästlein, A.; Beschmann, H. Glutathione in spermatozoa and seminal plasma of infertile men. *Hum. Reprod.* 1998, 13, 353-359.

[24] Sheweita, S. A.; Tilmisany, A. M.; Al-Sawaf, H. Mechanisms of male infertility: role of antioxidants. *Curr. Drug. Metab.* 2005, 6, 495-501.

[25] Aitken, R. Reactive oxygen species as mediators of sperm capacitation and pathological damage. *Mol. Reprod. Dev.* 2017, 84, 1039-1052.

[26] Aitken, R. J.; Baker, M. A.; De Iuliis, G. N.; Nixon, B. New insights into sperm physiology and pathology. *Handb. Exp. Pharmacol.* 2010, 99-115.

[27] El Sisy, G. A.; Abo El-Maaty, A. M.; Rawash, Z. M. Comparative blood and seminal plasma oxidant/antioxidant status of Arab stallions with different ages and their relation to semen quality. *Asian Pac. J. Reprod.* 2016, 5, 428-433.

[28] Fafula, R.; Onufrovych, O.; Iefremova, U.; Melnyk, O.; Nakonechnyi, I.; Vorobets, D.; Vorobets, Z. Glutathione content in sperm cells of infertile men. *Regul. Mech. Biosyst.* 2017, 8, 157-161.

[29] Atig, F.; Raffa, M.; Habib, B. A.; Kerkeni, A.; Saad, A.; Ajina, M. Impact of seminal trace element and glutathione levels on semen quality of Tunisian infertile men. *BMC Urol.* 2012, 12, 6.

[30] Smith, C. V.; Jones, D. P.; Guenthner, T. M.; Lash, L. H.; Lauterburg, B. H. Compartmentation of glutathione: Implications for the study of toxicity and disease. *Toxicol. Appl. Pharmacol.* 1996, 140, 1-12.

[31] Wu, G.; Fang, Y.-Z.; Yang, S.; Lupton, J. R.; Turner, N. D. Glutathione metabolism and its implications for health. *J. Nutr.* 2004, 134, 489-492.

[32] Aitken, R. J.; Curry, B. J. Redox regulation of human sperm function: from the physiological control of sperm capacitation to the etiology of infertility and DNA damage in the germ line. *Antiox. Redox Signal.* 2011, 14, 367-381.

[33] Guthrie, H. D.; Welch, G. R. Effects of reactive oxygen species on sperm function. *Theriogenology.* 2012, 78, 1700-1708.

Species	GSH supplementation	Reported benefits	References
	GSH (5 mM) in Duragen® extender *Frozen semen*	Improved overall sperm cryotolerance, sperm viability; the number of acrosome-intact SPZ, sperm motility (total and progressive), and fragmented DNA *Distinct effects of GSH supplementation in good and bad freezer samples*	[45]
	GSH (5 mM) in Modena extender *Refrigerated semen (Liquid storage at 17ºC)*	Improved sperm motility Increased effective survival period Improved plasma membrane and acrosome integrity	[77]
Stallion	GSH (2.5 mM) in the Botucrio freezing extender *Frozen semen*	Higher total motility, progressive motility, viability and membrane integrity [HOST test]	[78]

The positive effects obtained from adding GSH to semen extenders are dose-dependent. If the dose of GSH added is too low, it will not be sufficient to significantly improve sperm functions, while GSH supplemented to the semen extender in high doses will also negatively affect SPZ fertility, compared to controls. Such effects have been reported for stallions [78], rams [69], dogs [57], bulls [77], and boars [75].

When adding high concentrations of GSH to the extender, studies have shown a compromise in sperm motility (both total and progressive) [57], poorer mitochondrial and membrane integrity [72], as well as an increased number of capacitated SPZ [57]. It is possible that some of the actions reported to be negatively affected by excessive GSH supplementation of the semen extender may respect protein stability, while others relate to changes in lipid and protein layers of sperm plasma membranes.

GSH is a disulfide bond reducer and seems to assure reduction of disulfide bonds in sperm nuclei initiating chromatin decondensation, before male pronucleus formation [79, 80] after fertilization. Sperm nuclear decondensation depends on the presence of GSH and requires reduction of disulfide to thiol in protamines, followed by replacement of protamines with histones [81, 82]. GSH induces *in vitro* sperm decondensation in humans [83], bovines [84], hamsters [85] and porcines [86]. Precocious decondensation of sperm chromatin will result in DNA fragmentation, while changes in protein stability will further cause instability of the sperm tail structure [87], leading to flagellar angulation and impaired SPZ motility [88]. Moreover, sperm-egg interaction is mediated by protein disulfide isomerase inducing a thiol (S-H)/disulfide (S-S) exchange and producing conformational changes in fusion-activate proteins [89]. Inability to fuse the acrosomal and the outer cell membranes, or to fuse the SPZ and the oocyte membranes will impair fertilization. S-S rich polypeptides have also been detected in the mitochondrial membranes of mammalian sperm. Their reduction triggers detachment of the mitochondrial sheath [87], which might occur in SPZ preserved in extender with high levels of GSH. *In vitro* experiments have shown that treatment of SPZ with a reducing agent causes sperm dysfunction through disassembly of the connecting piece and decondensation of the sperm nucleus [87]. The latter was also obtained *in vitro*, by incubation of human sperm cells with heparin [90], a known capacitation inducing agent [91-93].

Furthermore, changes in the lipid and protein layers of sperm plasma membranes follow a dramatic imbalance in cellular physiology [40]. As mentioned earlier, GSH is a potent antioxidant protecting sperm against the effects of oxidative stress-induced damage due to increased ROS production and subsequent plasmalemma lipid peroxidation, intracellular enzyme leakage, and eventual chromatin damage [20]. However, high concentrations of antioxidants may destroy functional integrity of the sperm membranes [94], and damage unsaturated bonds of PUFA, triggering a lipid peroxidation chain reaction that reduces sperm motility [95].

Conclusion

GSH plays a major role in the ROS balance that supports spermatozoa (SPZ) from their development in the spermatogenic process until they are ejaculated. Due to the dual effect that ROS exert on cell survival, maintenance of adequate levels of ROS is crucial to ensure SPZ fertilizing potential and avoid sperm dysfunction after ejaculation, particularly in semen preservation programs.

The increasing number of available studies over the past two decades involving GSH supplementation to extenders demonstrate a role for GSH in protecting sperm against oxidative injury. The positive influences on SPZ function are, however, depended on adding precise doses of GSH; as insufficient levels fail to be fully protective, whereas excessive concentrations have been demonstrated to be deleterious for SPZ. Further research to establish species-specific thresholds for GSH supplementation for short- or long-term sperm preservation are thus foreseen, together with studies addressing the molecular particularities of the protective/damaging role of GSH on SPZ.

References

[1] Sies, H. Glutathione and its role in cellular functions. *Free Rad. Biol. Med.* 1999, 27, 916-921.

[2] Flohé, L. The fairytale of the GSSG/GSH redox potential. *Biochim. Biophys. Acta.* 2013, 1830, 3139-3142.

[3] Lu, S. C. Glutathione synthesis. *Biochim. Biophys. Acta*. 2013, 1830, 3143-3153.

[4] Micheli, L.; Cerretani, D.; Collodel, G.; Menchiari, A.; Moltoni, L.; Fiaschi, A. I.; Moretti, E. Evaluation of enzymatic and non-enzymatic antioxidants in seminal

plasma of men with genitourinary infections, varicocele and idiopathic infer *Andrology.* 2016, 4, 456-464.

[5] Anders, M. W. Chemical toxicology of reactive intermediates formed by glutathione-dependent bioactivation of halogen-containing compounds. *Chem. R Toxicol.* 2008, 21, 145-159.

[6] Forman, H. J.; Zhang, H.; Rinna, A. Glutathione: overview of its protective role, measurement, and biosynthesis. *Mol. Aspects. Med.* 2009, 30, 1-12.

[7] Jones, D. P.; Go, Y. M.; Anderson, C. L.; Ziegler, T. R.; Kinkade, J. M.; Kirlin, W. G. Cysteine/cystine couple is a newly recognized node in the circuitry for biologic redox signaling and control. *FASEB J.* 2004, 18, 1246-1248.

[8] Agrawal, Y. P.; Vanha-Perttula, T. Glutathione, L-glutamic acid and gamma-glutamyl transpeptidase in the bull reproductive tissues. *Int. J. Androl.* 1988, 11, 123-131.

[9] Irvine, D. S. Glutathione as a treatment for male infertility. *Rev. Reprod.* 1996, 1, 6-12.

[10] Halliwell, B.; Gutteridge, J. M. C. Reactive species can pose special problems needing special solutions: some examples. In *Free Radicals in Biology and Medicine*, 5th ed. Halliwell, B.; Gutteridge, J. M. C., Eds.; Oxford University Press: New York, NY, 2015; pp 354-410.

[11] Fujii, J.; Tsunoda, S.; Kimura, N. Antithetical roles of reactive oxygen species in mammalian reproduction. In *Systems Biology of Free Radicals and Antioxidants*. Laher, I., Ed.; Springer: Berlin, Heidelberg, 2014; pp 2705-2721.

[12] Guerriero, G.; Trocchia, S.; Abdel-Gawad, F. K.; Ciarcia, G, Roles of reactive oxygen species in the spermatogenesis regulation. *Front. Endocrinol.* (Lausanne) 2014, 5, 56.

[13] Santos, D.; Payan-Carreira, R. The role of superoxide dismutase in endometrial functional integrity. In *Superoxide Dismutase: Structure, Synthesis and Applications*. Magliozzi, S., Ed.; Nova Science Publishers: Hauppauge, NY, 2018; pp 1-44.

[14] Maiorino, M.; Ursini, F. Oxidative stress, spermatogenesis and fertility. *Biol. Chem.* 2002, 383, 591-597.

[15] Zini, A.; Libman, J. Oxidative stress and male infertility. In *Systems Biology of Free Radicals and Antioxidants*. Laher, I., Ed.; Springer: Berlin, Heidelberg, 2014; pp 2815-2833.

[16] Tafuri, S.; Ciani, F.; Iorio, E. L.; Esposito, L.; Cocchia, N. Reactive oxygen species (ROS) and male fertility. In *New discoveries in Embriology.* Wu, B., Ed.; InTech Open: Rijeka, Croatia, 2015; pp 19-40.

[17] Aitken, R. J.; Roman, S. D. Antioxidant systems and oxidative stress in the testes. *Oxid. Med. Cell. Longev.* 2008, 1, 15-24.

[34] Garrido, N.; Meseguer, M.; Alvarez, J.; Simón, C.; Pellicer, A.; Remohí, J. Relationship among standard semen parameters, glutathione peroxidase/glutathione reductase activity, and mRNA expression and reduced glutathione content in ejaculated spermatozoa from fertile and infertile men. *Fertil. Steril.* 2004, 82 Suppl 3, 1059-1066.

[35] Sansone, A.; Lombardo, F.; Lenzi, A. Free radicals and reproductive health. In *Systems Biology of Free Radicals and Antioxidants.* Laher, I., Ed.; Springer: Berlin, Heidelberg, 2014; pp 2723-2742.

[36] O'Flaherty, C. Redox regulation of mammalian sperm capacitation. *Asian J. Androl.* 2015, 17, 583-590.

[37] Stival, C.; Puga Molina, Ldel. C.; Paudel, B.; Buffone, M. G.; Visconti, P. E.; Krapf, D. Sperm capacitation and acrosome reaction in mammalian sperm. *Adv. Anat. Embryol. Cell. Biol.* 2016, 220, 93-106.

[38] de Lamirande, E.; Leclerc, P.; Gagnon, C. Capacitation as a regulatory event that primes spermatozoa for the acrosome reaction and fertilization. *Mol. Hum. Reprod.* 1997, 3, 175-194.

[39] Buffone, M. G.; Ijiri, T. W.; Cao, W.; Merdiushev, T.; Aghajanian, H. K.; Gerton, G. L. Heads or tails? Structural events and molecular mechanisms that promote mammalian sperm acrosomal exocytosis and motility. *Mol. Reprod. Dev.* 2012, 79, 4-18.

[40] Sanocka, D.; Kurpisz, M. Reactive oxygen species and sperm cells. *Reprod. Biol. Endocrinol.* 2004, 2, 12.

[41] du Plessis, S. S.; Harlev, A.; Mohamed, M. I.; Habib, E.; Kothandaraman, N.; Cakar, Z. Physiological roles of reactive oxygen species (ROS) in the reproductive system. In *Oxidative Stress in Human Reproduction: Shedding Light on a Complicated Phenomenon*. Agarwal, A.; Sharma, R.; Gupta, S.; Harlev, A.; Ahmad, G.; du Plessis, S. S.; Esteves, S. C.; Wang, S. M.; Durairajanayagam, D., Eds.; Springer International Publishing: Cham, 2017; pp 47-64.

[42] Eskiocak, S.; Gozen, A. S.; Yapar, S. B.; Tavas, F.; Kilic, A. S.; Eskiocak, M. Glutathione and free sulphydryl content of seminal plasma in healthy medical students during and after exam stress. *Hum. Reprod.* 2005, 20, 2595-2600.

[43] Stradaioli, G.; Noro, T.; Sylla, L.; Monaci, M. Decrease in glutathione (GSH) content in bovine sperm after cryopreservation: comparison between two extenders. *Theriogenology.* 2007, 67, 1249-1255.

[44] Gadea, J.; Sellés, E.; Marco, M. A.; Coy, P.; Matás, C.; Romar, R.; Ruiz, S. Decrease in glutathione content in boar sperm after cryopreservation. Effect of the addition of reduced glutathione to the freezing and thawing extenders. *Theriogenology.* 2004, 62, 690-701.

[45] Yeste, M.; Estrada, E.; Pinart, E.; Bonet, S.; Miró, J.; Rodríguez-Gil, J. E. The improving effect of reduced glutathione on boar sperm cryotolerance is related with the intrinsic ejaculate freezability. *Cryobiology.* 2014, 68, 251-261.

[46] Gadea, J.; Molla, M.; Selles, E.; Marco, M. A.; Garcia-Vazquez, F. A.; Gardon, J. C. Reduced glutathione content in human sperm is decreased after cryopreservation: Effect of the addition of reduced glutathione to the freezing and thawing extenders. *Cryobiology.* 2011, 62, 40-46.

[47] Li, T. K. The glutathione and thiol content of mammalian spermatozoa and seminal plasma. *Biol. Reprod.* 1975, 12, 641-646.

[48] Strzezek, R.; Koziorowska-Gilun, M.; Kowalówka, M.; Strzezek, J. Characteristics of antioxidant system in dog semen. *Pol. J. Vet. Sci.* 2009, 12, 55-60.

[49] Tvrdá, E.; Lukáč, N.; Schneidgenová, M.; Lukáčová, J.; Szabó, C.; Goc, Z.; Greń, A.; Massányi, P. Impact of seminal chemical elements on the oxidative balance in bovine seminal plasma and spermatozoa. *J. Vet. Med.* 2013, 2013, 125096.

[50] Bilodeau, J. F.; Chatterjee, S.; Sirard, M. A.; Gagnon, C. Levels of antioxidant defenses are decreased in bovine spermatozoa after a cycle of freezing and thawing. *Mol. Reprod. Dev.* 2000, 55, 282-288.

[51] Waheed, M. M.; Gouda, E. M.; Khalifa, T. A. Impact of seminal plasma superoxide dismutase and glutathione peroxidase on cryopreserved buffalo spermatozoa. *Anim. Reprod. Sci.* 2013, 142, 126-130.

[52] Menezo, Y. J.; Silvestris, E.; Dale, B.; Elder, K. Oxidative stress and alterations in DNA methylation: two sides of the same coin in reproduction. *Reprod. Biomed. Online* 2016, 33, 668-683.

[53] Aitken, R. J.; Baker, M. A. Oxidative stress, sperm survival and fertility control. *Mol. Cell. Endocrinol.* 2006, 250, 66-69.

[54] Lane, M.; McPherson, N. O.; Fullston, T.; Spillane, M.; Sandeman, L.; Kang, W. X.; Zander-Fox, D. L. Oxidative stress in mouse sperm impairs embryo development, fetal growth and alters adiposity and glucose regulation in female offspring. *PLoS One* 2014, 9, e100832.

[55] Bisht, S.; Faiq, M.; Tolahunase, M.; Dada, R. Oxidative stress and male infertility. *Nat. Rev. Urol.* 2017, 14, 470-485.

[56] Kim, S.; Lee, Y. J.; Kim, Y. J. Changes in sperm membrane and ROS following cryopreservation of liquid boar semen stored at 15 °C. *Anim. Reprod. Sci.* 2011, 124, 118-124.

[57] Ogata, K.; Sasaki, A.; Kato, Y.; Takeda, A.; Wakabayashi, M.; Sarentonglaga, B.; Yamaguchi, M.; Hara, A.; Fukumori, R.; Nagao, Y. Glutathione supplementation to semen extender improves the quality of frozen-thawed canine spermatozoa for transcervical insemination. *J. Reprod. Dev.* 2015, 61, 116-122.

[58] Gibb, Z.; Aitken, R. J. The Impact of sperm metabolism during in vitro storage: The stallion as a model. *Biomed. Res. Int.* 2016, 2016, 9380609.

[59] Bailey, J. L.; Bilodeau, J. F.; Cormier, N. Semen cryopreservation in domestic animals: a damaging and capacitating phenomenon. *J. Androl.* 2000, 21, 1-7.

[60] Borges, P.; Mior, F.; Fontbonne, A.; Payan-Carreira, R. Subcellular distribution of heat shock protein 70 (hsp70) in frozen and chilled semen is related with dog sperm quality. In *Spermatozoa: Biology, Motility and Function and Chromosomal Abnormalities.* Ericson, B. T., Ed.; Nova Science Publishers: Hauppauge, NY, 2014; pp 75-98.

[61] Soini, K.; Diaz, C.; Gandini, G.; de Haas, Y.; Lilja, T.; Martin-Collado, D.; Pizzi, F.; Hiemstra, S. J.; Consortium, E. Developing a typology for local cattle breed farmers in Europe. *J. Anim. Breed. Genet.* 2012, 129, 436-447.

[62] Wales, R. G.; White, I. G.; Lamond, D. R. The spermicidal activity of hydrogen peroxide in vitro and in vivo. *J. Endocrinol.* 1959, 18, 236-244.

[63] Brouwers, J. F.; Gadella, B. M. In situ detection and localization of lipid peroxidation in individual bovine sperm cells. *Free Rad. Biol. Med.* 2003, 35, 1382-1391.

[64] Baumber, J.; Ball, B. A.; Gravance, C. G.; Medina, V.; Davies-Morel, M. C. The effect of reactive oxygen species on equine sperm motility, viability, acrosomal integrity, mitochondrial membrane potential, and membrane lipid peroxidation. *J. Androl.* 2000, 21, 895-902.

[65] Ball, B. A.; Vo, A. T.; Baumber, J. Generation of reactive oxygen species by equine spermatozoa. *Am. J. Vet. Res.* 2001, 62, 508-515.

[66] Christova, Y.; James, P. S.; Jones, R. Lipid diffusion in sperm plasma membranes exposed to peroxidative injury from oxygen free radicals. *Mol. Reprod. Dev.* 2004, 68, 365-372.

[67] Kim, S. H.; Yu, D. H.; Kim, Y. J. Effects of cryopreservation on phosphatidylserine translocation, intracellular hydrogen peroxide, and DNA integrity in canine sperm. *Theriogenology.* 2010, 73, 282-292.

[68] Mollá, M.; Selles, E.; Marco, M. A.; Remohi, J.; Ballesteros, A.; Gadea, J. Freezing procedure produces a reduction in the human spermatozoa glutathione content. *J. Androl.* 2004, 25(Suppl.), 45.

[69] Câmara, D. R.; Silva, S. V.; Almeida, F. C.; Nunes, J. F.; Guerra, M. M. Effects of antioxidants and duration of pre-freezing equilibration on frozen-thawed ram semen. *Theriogenology.* 2011, 76, 342-350.

[70] Gadea, J.; Gumbao, D.; Cánovas, S.; García-Vázquez, F. A.; Grullón, L. A.; Gardón, J. C. Supplementation of the dilution medium after thawing with reduced glutathione improves function and the in vitro fertilizing ability of frozen-thawed bull spermatozoa. *Int. J. Androl.* 2008, 31, 40-49.

[71] Zeitoun, M. M.; Al-Damegh, M. A. Effect of nonenzymatic antioxidants on sperm motility and survival relative to free radicals and antioxidant enzymes of chilled-stored ram semen. *Open J. Animal Sci.* 2014, 05, 50.

[72] Silva, S. V.; Soares, A. T.; Batista, A. M.; Almeida, F. C.; Nunes, J. F.; Peixoto, C. A.; Guerra, M. M. In vitro and in vivo evaluation of ram sperm frozen in tris egg-yolk and supplemented with superoxide dismutase and reduced glutathione. *Reprod. Domest. Anim.* 2011, 46, 874-881.

[73] Noei Razliqi, R.; Zhandi, M.; Shakeri, M.; Towhidi, A.; Sharafi, M.; Emamverdi, M.; Khodaei Motlagh, M. Protective role of glutathione in buck semen cryopreservation. *Iran. J. Vet. Res.* 2015, 16, 298-300.

[74] Rawash, Z.; Ibrahim, E.; El-Raey, M. Effects of reduced glutathione on Boer goat semen freezability. *Asia. Pac. J. Reproduct.* 2018, 7, 33-38.

[75] Gadea, J.; García-Vazquez, F.; Matás, C.; Gardón, J. C.; Cánovas, S.; Gumbao, D. Cooling and freezing of boar spermatozoa: supplementation of the freezing media with reduced glutathione preserves sperm function. *J. Androl.* 2005, 26, 396-404.

[76] Estrada, E.; Rodríguez-Gil, J. E.; Rocha, L. G.; Balasch, S.; Bonet, S.; Yeste, M. Supplementing cryopreservation media with reduced glutathione increases fertility and prolificacy of sows inseminated with frozen-thawed boar semen. *Andrology.* 2014, 2, 88-99.

[77] Zhang, X. G.; Liu, Q.; Wang, L. Q.; Yang, G. S.; Hu, J. H. Effects of glutathione on sperm quality during liquid storage in boars. *Anim. Sci. J.* 2016, 87, 1195-1201.

[78] de Oliveira, R. A.; Wolf, C. A.; de Oliveira Viu, M. A.; Gambarini, M. L. Addition of glutathione to an extender for frozen equine semen. *J. Equine Vet. Sci.* 2013, 33, 1148-1152.

[79] Funahashi, H.; Cantley, T. C.; Stumpf, T. T.; Terlouw, S. L.; Day, B. N. Use of low-salt culture medium for in vitro maturation of porcine oocytes is associated with elevated oocyte glutathione levels and enhanced male pronuclear formation after in vitro fertilization. *Biol. Reprod.* 1994, 51, 633-639.

[80] Sutovsky, P.; Schatten, G. Depletion of glutathione during bovine oocyte maturation reversibly blocks the decondensation of the male pronucleus and pronuclear apposition during fertilization. *Biol. Reprod.* 1997, 56, 1503-1512.

[81] Perreault, S. D.; Barbee, R. R.; Slott, V. L. Importance of glutathione in the acquisition and maintenance of sperm nuclear decondensing activity in maturing hamster oocytes. *Dev. Biol.* 1988, 125, 181-186.

[82] Zirkin, B. R.; Naish, S. J.; Perreault, S. D. 5-Formation and function of the male pronucleus during mammalian fertilization. In *The Molecular Biology of Fertilization.* Schatten, H., Ed.; Academic Press: 1989; pp 91-114.

[83] Reyes, R.; Rosado, A.; Hernández, O.; Delgado, N. M. Heparin and glutathione: physiological decondensing agents of human sperm nuclei. *Gamete Res.* 1989, 23, 39-47.

[84] Delgado, N. M.; Flores-Alonso, J. C.; Rodríguez-Hernández, H. M.; Merchant-Larios, H.; Reyes, R. Heparin and glutathione II: correlation between decondensation of bull sperm cells and its nucleons. *Arch. Androl.* 2001, 47, 47-58.

[85] Vendrell, F. J.; Rubio, C.; Tarín, J. J. The heparin-glutathione test: an alternative to the hypo-osmotic swelling test to select viable sperm for intracytoplasmic sperm injection. *Fertil. Steril.* 1998, 70, 1156-1161.
[86] Cheng, W. M.; An, L.; Wu, Z. H.; Zhu, Y. B.; Liu, J. H.; Gao, H. M.; Li, X. H.; Zheng, S. J.; Chen, D. B.; Tian, J. H. Effects of disulfide bond reducing agents on sperm chromatin structural integrity and developmental competence of in vitro matured oocytes after intracytoplasmic sperm injection in pigs. *Reproduction.* 2009, 137, 633-643.
[87] Sutovsky, P.; Tengowski, M. W.; Navara, C. S.; Zoran, S. S.; Schatten, G. Mitochondrial sheath movement and detachment in mammalian, but not nonmammalian, sperm induced by disulfide bond reduction. *Mol. Reprod. Dev.* 1997, 47, 79-86.
[88] Ijiri, T. W.; Vadnais, M. L.; Huang, A. P.; Lin, A. M.; Levin, L. R.; Buck, J.; Gerton, G. L. Thiol changes during epididymal maturation: a link to flagellar angulation in mouse spermatozoa? *Andrology.* 2014, 2, 65-75.
[89] Ellerman, D. A.; Myles, D. G.; Primakoff, P. A role for sperm surface protein disulfide isomerase activity in gamete fusion: Evidence for the participation of ERp57. *Dev. Cell.* 2006, 10, 831-837.
[90] Julianelli, V.; Farrando, B.; Alvarez Sedó, C.; Calvo, L.; Romanato, M.; Calvo, J. C. Heparin enhances protamine disulfide bond reduction during in vitro decondensation of human spermatozoa. *Hum. Reprod.* 2012, 27, 1930-1938.
[91] Parrish, J. J.; Susko-Parrish, J.; Winer, M. A.; First, N. L. Capacitation of bovine sperm by heparin. *Biol. Reprod.* 1988, 38, 1171-1180.
[92] Valencia, A.; Wens, M. A.; Merchant, H.; Reyes, R.; Delgado, N. M. Capacitation of human spermatozoa by heparin. *Arch. Androl.* 1984, 12 Suppl, 109-113.
[93] Varner, D. D.; Bowen, J. A.; Johnson, L. Effect of heparin on capacitation/acrosome reaction of equine sperm. *Arch. Androl.* 1993, 31, 199-207.
[94] Perumal, P.; Vupru, K.; Rajkhowa, C., Effect of Addition of Taurine on the Liquid Storage (5°C) of Mithun (Bos frontalis) *Semen. Vet. Med. Int.* 2013, 2013, 165348.
[95] Palani, A. F. Effect of serum antioxidant levels on sperm function in infertile male. *Middle East Fertil. Soc. J.* 2018, 23, 19-22.

Biographical Sketches

Dario Joaquim Simões Loureiro dos Santos

Present Position

Assistant Professor

Affiliation

Department of Biology and Environment, University of Trás-os-Montes e Alto Douro (UTAD), Vila Real, Portugal

Education

Biology (licenciate) - University of Coimbra, Coimbra - Portugal

Cellular and Molecular Biology (PhD) University of Coimbra, Coimbra - Portugal

Research and Professional Experience

Cellular Biology and Biochemistry, Cell Physiology, Phytochemicals, Oxidative Stress and antioxidants.

Rita Payan Carreira

Present Position

Full Professor

Affiliation

University of Évora, Department of Veterinary Medicine, Évora, Portugal

Education

DVM

Research and Professional Experience

Rita is a researcher in the Animal and Veterinary Centre (CECAV-UTAD) of University of Évora, Portugal. Her main interests focus on the animal fertility and physiopathology, mainly the molecular interplay in embryo-maternal crosstalk.

Short Biography

Rita Payan Carreira earned her veterinary degree from the Faculty of Veterinary Medicine in Lisbon, Portugal, in 1985. For some years, she parted her activity as a veterinary practitioner and teacher. She worked at the University of Trás-os-Montes and Alto Douro, for more than 30 years. She is now full Professor at the University of Évora, at the department of Veterinary Medicine. Rita has participated in several joint academic activities with foreign universities, and she entered several international networks in veterinary theriogenology. Often, she supervises students preparing their master or doctoral thesis. She is also a frequent referee for diverse journals in the area.

Publications from the Last 3 Years

In the past two years, Rita is involved in three research projects talking animal reproduction and productivity. In the last three years, she has contributed to 14 book chapters, 26 full papers and multiple Congress communications. For detailed information on her scientific publications and projects, please address to http://orcid.org/0000-0001-5225-4510.

In: Glutathione
Editor: Pál Perjési

ISBN: 978-1-53614-740-7
© 2019 Nova Science Publishers, Inc.

Chapter 12

THE CHEMISTRY OF GST-CATALYZED REACTIONS

Pál Perjési[1,*]***, Caridad Noda Perez***[1,2]***, Aline Bernardes***[1,2]***, Giulio D. d'Oliveira***[1,2] ***and Fatemeh Kenari***[1]

[1]Institute of Pharmaceutical Chemistry, University of Pécs, Pécs, Hungary
[2]Instituto de Química, Universidade Federal de Goiás, Goiânia, Goiás, Brasil

ABSTRACT

Non-endogenous substances (xenobiotics) that enter living organisms can undergo chemical transformations (biotransformation). As a result, new chemical entities are formed in different organs, tissues and cell compartments of the bodies. The reactions are principally divided into Phase I and Phase II transformations. In Phase I reactions, the parent compounds are structurally modified mainly by oxidation, reduction and hydrolysis. Phase II reactions include conjugation of the parent compounds or their transformed derivatives with glucuronic acid, sulfate, glycine or glutathione, among others. This chapter summarizes the catalytic functions of the soluble glutathione transferase (GST) enzymes and gives examples for the mechanism and stereochemistry of the GST-catalyzed transformations of the most frequently occurring types of endogenous and exogenous substrates (alkyl-aryl-halides, epoxides, α,β-unsaturated carbonyls and quinones). The selected examples play important roles either in determination of GST activities or biological action of the substrates.

Keywords: glutathione, glutathione transferase (GST), phase II biotransformation, glutathione conjugation, acrolein (ACR), ethacrynic acid (ETK), acetaminophen (paracetamol), 4-hydroxynonenal (HNE), prostaglandins, chalcones

* Corresponding Author E-mail: pal.perjesi@gytk.pte.hu.

INTRODUCTION

Non-endogenous substances (xenobiotics) that enter living organisms can undergo chemical transformations (biotransformation). As a result, new chemical entities are formed in different organs, tissues and cell compartments of the bodies. The history of such research is dated back to 1841, when Alexander Ure reported on conversion of benzoic acid to hippuric acid in the body [1]. The chemistry of the transformation was later confirmed by Wilheim Keller [2]. Since then several different biotransformation pathways of xenobiotics have been identified. They are principally divided into Phase I and Phase II transformations, after Richard Tecwyn Williams [3, 4]. In Phase I reactions, the parent compounds are structurally modified mainly by oxidation, reduction and hydrolysis. The most common functional groups exposed or introduced in the Phase I reactions are the hydroxyl (-OH), amino (-NH-), and the carboxyl (-COOH) groups. Phase II reactions include conjugation of the parent compounds or their transformed derivatives with glucuronic acid, sulfate, glycine or glutathione, among others. Acetylation and methylation are also examples of Phase II transformations [5].

Formation of N-acetylcysteine conjugates of halogen-substituted benzene derivatives was discovered in 1879 [6-8]. However, the complete metabolic sequence leading to formation of N-acetylcysteine conjugates (mercapturic acids) remained unknown for over 80 years. The first publication demonstrating the link between the reaction of halogen substituted benzenes with glutathione and mercapturic acid formation of the studied aromatics was described by a team lead by Sibyl James and H. G. Bray [9-11]. In 1961, Booth et al. described the direct enzymatic formation of glutathione conjugates in the cytosol of liver and other organs [12]. This observation led to the work of Chasseaud and Boyland demonstrating the existence of distinct classes of glutathione transferases [13].

Glutathione transferases (GSTs) (EC 2.5.1.18) are represented by a superfamily of isoenzymes catalyzing conjugation of glutathione (GSH; γ-Glu-Cys-Gly) with a wide range of nonpolar compounds of exogenous and endogenous origin containing electrophilic atoms of carbon, sulfur, nitrogen or phosphorous, which assists in defense of the cell against possible toxic action of these compounds. In addition to conjugation reactions, a number of GST isoforms can catalyze GSH-dependent reduction of organic hydroperoxides and isomerization of various unsaturated compounds [13]. These enzymes also have some non-catalytic functions that relate to the sequestering of carcinogens, intracellular transport of a wide spectrum of hydrophobic ligands, and modulation of signal transduction pathways [14].

GSTs are found in all major eukaryotic taxa including fungi, plant and animals. In any given organism multiple GST isoforms are expressed. In mammals, GST is present in virtually all organs and tissues, but the highest enzyme activity is found in the liver. Mammalian GST superfamily includes three subfamilies of isoenzymes: the cytosolic GST, the mitochondrial GST, and the microsomal MAPEG [15].

The cytosolic GSTs are the most numerous and most extensively studied. Historically, these enzymes have been first identified in mammalian tissues [13]. Mammalian cytosolic GSTs are dimeric proteins, which -based on the amino acid sequence homology- are grouped into seven classes, designated Alpha (α), Mu (μ), Pi (π), Sigma (σ) Theta (θ), Zeta (ζ), and Omega (ω), and comprise 17 isoenzymes [15-17]. The mitochondrial (Kappa-class) GSTs, reported to be also present in peroxisomes, are soluble enzymes integrated into the mitochondrial matrix [18, 19]. The soluble GSTs appeared to be involved primarily in the metabolism of electrophilic foreign chemicals (xenobiotics) as well as in detoxication of endogenously derived electrophilic metabolites [20]. Microsomal GSTs are trimeric integral membrane proteins that are now called membrane-associated proteins in eicosanoid and glutathione metabolism (MAPEG) [21]. Likewise, the cytosolic and mitochondrial (soluble) GST isoenzymes, the microsomal isoenzymes can also catalyze conjugation of GSH with electrophilic compounds, but they principally participate in isomerization of unsaturated compounds and biosynthesis of leukotrienes and prostaglandins [20].

The soluble GSTs exist as dimeric proteins, with an approximate molecular weight of 25 000 dalton for each. Each of the subunits possesses a GSH binding site (G-site) and an adjacent relatively hydrophobic binding site (H-site) for the substrate. In addition, in various isoenzymes (particularly in the Alfa-, Mu- and the Pi-class) a non-substrate binding site has been recognized. This has been described as a transport site for large hydrophobic molecules. Among the non-catalytic ligands of GSTs are steroids, bile acids, hem, bilirubin, and a variety of organic dyes, drugs and other xenobiotics [22].

Usually the dimers are made from identical chains but heterodimers made of two different chains from the same class are also found [23]. The standardized GST nomenclature identifies the species with a lower-case initial (*h*GSTP1-2, for human), the class by a capital following GST (hGST*P*1-2, for Pi), followed by an Arabic numeral representing the subfamily of both subunits (hGSTP*1-2*, for a heterodimer in the Pi-class subfamily 1 and 2) [24].

This chapter summarizes the catalytic functions of the soluble GST enzymes and gives examples for the mechanism and stereochemistry of the GST-catalyzed transformations of the most frequently occurring types of substrates (alkyl-aryl-halides, epoxides, α,β-unsaturated carbonyls and quinones). The selected examples play an important role either in determination of GST activities, or biological action of the substrates.

FUNCTIONS OF GLUTATHIONE TRANSFERASES

Glutathione transferases possess three main functions in the living organisms: (a) catalytic activities, (b) binding of non-substrate ligands, and (c) involvement in protein-

protein interactions [25]. In this chapter only the catalytic functions of the GST enzymes are discussed.

One of the main catalytic functions of soluble GSTs is *participation in the cellular antioxidant system* [26]. They demonstrate activity toward a number of by-products of oxidative stress, including lipid- and DNA-hydroperoxides as well as the secondary peroxidation products such as acrolein (propenal) and 4-hydroxynonenal (HNE) [17, 22, 27]. The reactive peroxidation products (e.g., hydroperoxides of unsaturated fatty acids) are converted to less toxic alcohols *via* their GST-catalyzed reactions with reduced glutathione (GSH). The reduction is a two-step reaction. The first step (**1**) is an enzymic reduction of the peroxide to the corresponding alcohol, with the concomitant formation of the sulfenic acid derivative of glutathione (GSOH). The second step (**2**) is a spontaneous process in which the formed GSOH reacts with a molecule of GSH to form oxidized glutathione (GSSG) and water [28].

$$\text{R-OOH} + \text{GSH} = \text{R-OH} + \text{GSOH} \quad (1)$$

$$\text{GSOH} + \text{GSH} = \text{GSSG} + H_2O \quad (2)$$

The reactive α,β-unsaturated aldehydes react with GSH in a Michael-type addition reaction to result in formation of the respective GSH-conjugates [15]. Furthermore, oxidation of catecholamines results in formation of quinoidal compounds that are substrates of GST isoforms [29].

Another well-known function of GSTs is *isomerization of biologically important endogen molecules*. The enzymes can catalyze *cis-trans* isomerization reactions or *movement of a double bond* within a polycyclic molecule [22, 30].

Most notable in this regard is the GSH-dependent conversion of Δ^5-3-ketosteroids (e.g., androst-5-en-3,17-dione) to Δ^4-3-ketosteroids (androst-4-ene-3,17-dione) catalyzed by hGSTA3-3, which is involved in biosynthesis of steroid hormones such as testosterone and progesterone. Figure 1 shows the mechanism of the reaction, in which proton (H_β) transfer from carbon-4 of the steroid to the enzyme(Tyr-9)-bound, deprotonated GSH (GS^-) represents the initial step. It is followed by relocation of the negative charge on carbon-6 and protonation of the rearranged anion by the GSH-bound Tyr-9 of the enzyme. These latter two steps might either occur concerted or sequential [30-33].

A further example of the GST-catalyzed physiological isomerization reactions is the conversion of 13-*cis*-retinoic acid to all-*trans*-retinoic acid, a reaction that increases the affinity of the retinoid to its receptor [34]. Another distinct, physiologically important isomerization is the conversion of maleylacetoacetate to fumarylacetoacetate catalyzed by GSTZ1-1, which is part of catabolism of phenylalanine and tyrosine [35]. The reaction mechanism is proposed to involve transient addition of GSH to the substrate followed by rotation around the formed single bond and elimination of GSH (Figure 2) [36].

GST-catalyzed isomerization is also involved in biosynthesis of prostaglandins and leukotrienes. The first step of the biosynthesis is the prostaglandin endoperoxide synthase (PGHS)-catalyzed conversion of arachidonic acid to the common prostaglandin precursor prostaglandin H_2 (PGH_2). Prostaglandin endoperoxide synthase has two isoforms, PGHS-1 and PGHS-2. The enzymes have two catalytic activities: cyclooxygenase and hydroperoxidase. The two isoforms are frequently referred to as COX-1 and COX-2 [37]. PGH_2 can be converted to a number of products, among them to PGE_2 or to PGD_2 by intramolecular redox reactions (Figure 3).

Conversion of PGH_2 to PGD_2 is catalyzed by a Sigma class GST (GSTS1-1), which is a versatile repertoire of species-specific conjugase and reductase activities [38]. The H-site of GSTS1-1 is extended, resulting in a unique deep cleft where the prostaglandin was suggested to bind with its peroxide group pointing at the thiolate of GSH [39]. Several mechanisms of the reaction are taking into considerations [28]. The reaction of PGH_2 yielding PGE_2 is catalyzed by the MPEG enzyme PGE synthase (PTGES1) [40] or the soluble GSTM-2-2 and GSTM3-3 [41].

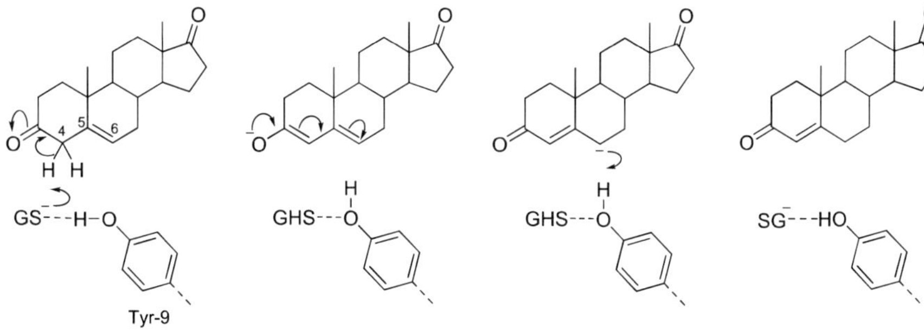

Figure 1. Mechanism of hGSTA3-3-catalyzed isomerization of Δ^5-3-ketosteroids. Based on reference [31].

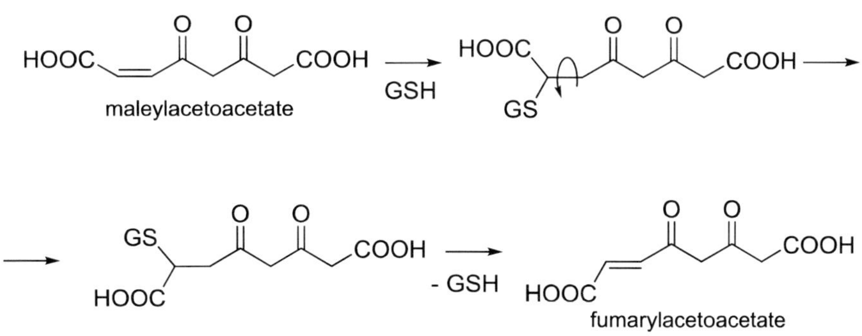

Figure 2. Proposed mechanism of GSH-dependent isomerization of maleylacetoacetate to fumarylacetoacetate. Based on reference [36].

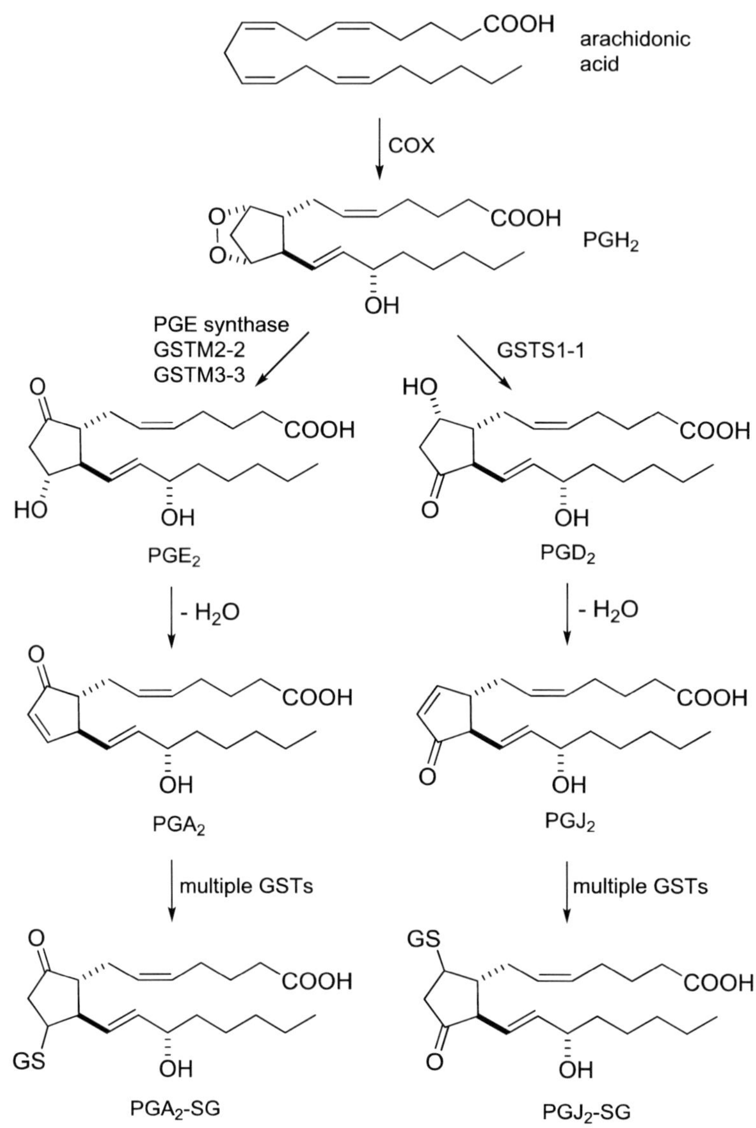

Figure 3. Conversion of PGH_2 to PGE_2, PGD_2, PGA_2 and PGJ_2 by GST enzymes.

The third type of catalytic activities of glutathione transferases is *conjugation of reduced glutathione (GSH) with endogenous or exogenous electrophilic species* involving the reactive thiol group of its cysteinyl moiety. Four groups of proteins are able to catalyze this characteristic reaction of glutathione transferases: (a) canonical soluble (cytosolic) GSTs, (b) distantly related soluble (mitochondrial) Kappa-class GSTs, (c) hydrophobic (microsomal) MAPEG family GSTs and (d) bacterial fosfomycin resistance proteins [42-46]. The sequence similarities of the six mammalian MAPEG classes are quite low [47]. Nevertheless, all MAPEG show a similar trimeric structure with four transmembrane helices for each subunit [28]. Mammalian MAPEG can be roughly subdivided into the MGST1/PTGES1 group and the MGST2/MGST3/LTC4/FLAP group [48, 49]. Among all families of GSH-dependent enzymes, GST and MAPEG are the most versatile catalysts converting a wide range of sulfur-, oxygen-, or carbon-containing electrophilic substances.

The catalytic events of GST's can be divided into two processes: (1) binding and activation of GSH, and (2) binding the electrophilic reaction partner. The first process is common in all soluble GSTs, while the second depends on the chemical nature and structure of the electrophile. It is worth mentioning that the former name "glutathione S-transferase" is misleading because the glutathionyl moiety (not a sulfur atom) is transferred [50].

BINDING AND ACTIVATION OF GSH

Cytosolic GSTs have been most extensively studied in human, rat and mouse tissues where they are the most abundant. The Pi-class enzyme is the most widely distributed isoenzyme and usually the most abundant [51]. The cytosolic enzymes have two active sites per dimer which behave independently of one another [52]. Each active site consists of at least two ligand binding regions: the GSH binding site, which is very specific for GSH, and the substrate binding site, which is less specific. The consequence of this latter characteristics is the ability of GSTs to react with a wide variety of agents.

The detailed understanding of the enzymatic mechanism needs a knowledge of the three-dimensional structure of each GST isoenzyme. There have been several comprehensive reviews on three-dimensional structure investigations of which results are not discussed here in details. Interested readers are directed towards these reviews which cover the structural, biochemical and genetic aspects of the field [53-55].

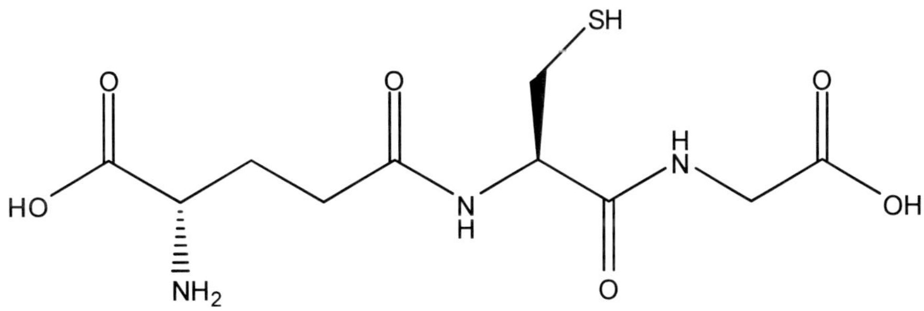

Figure 4. Structure of reduced glutathione (GSH).

Glutathione (GSH) has three acidic (thiol, glycinyl carboxyl, glutamyl carboxyl) and one basic (amino) functional groups (Figure 4). The macroscopic protonation constants (K_1-K_4) were determined by several authors. Each publication reported the pK_a of the sulfhydryl group of GSH in aqueous solution to be around pK_a 9.2 [56-59]. Accordingly, it is only a rather small percentage of the GSH molecules that exists in S-deprotonated form at physiological pH. It has been proved, however, that the pK_a of GSH bound to most GSTs is less than 7 [60]. Therefore, the enzyme-bound GSH is largely deprotonated at physiological pH – thus being activated for reaction with electrophilic substrates.

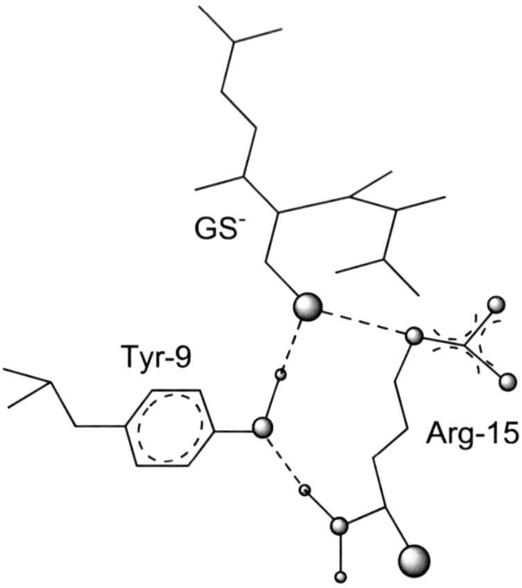

Figure 5. Schematic drawing of residues that interact with glutathione in the GSH binding site of Alfa-class GST. Based on reference [66].

The enzyme-bound thiolate of glutathione (GS^-) is stabilized by a hydrogen bond between the proton of a hydroxyl group in the protein and the sulfur atom of the thiolate [61, 62]. In most GSTs the hydrogen-bond donor is the hydroxyl group of a tyrosine residue close to the N-terminal of the protein: Tyr-7 in the Pi-class GST [63], Tyr-6 in the Mu-class GST [64] and Tyr-9 in the Alfa-class GST [65]. In Alfa-class GSTs, a secondary interaction of the ε-nitrogen atom of Arg-15 with the sulfur of GSH additionally helps to stabilize the thiolate [66]. Figure 5 shows a schematic drawing of residues that interact with glutathione in the GSH binding site of Alfa-class GST.

In contrast to binding and activation of GSH (at the G-site), which are not identical but show high degree of similarity [63-66], a variety of chemical and structurally diverse electrophilic substrates can bind to the electrophile binding site (H-site) of the GST enzymes. The H-site cavities of the Mu- and Alfa-class GSTs are generally hydrophobic, while those of the Pi-class enzymes are lined with both hydrophobic and hydrophilic surfaces [67].

Conjugation of GSH with Endogenous Electrophiles

GST-catalyzed reactions involve reactions of electrophilic oxidized products of endogenous substances (e.g., lipids, nucleic acids, neurotransmitters) with GSH. These reactive electrophiles can chemically modify the primary structure of cellular macromolecules (e.g., lipids, nucleic acid), being potentially toxic metabolites. One of the most investigated toxic lipid peroxidation products is 4-hydroxynonenal (HNE) (Figure 6). It is a reactive α,β-unsaturated aldehyde formed during oxidative degradation of unsaturated fatty acids [68, 69, 70]. Reactive α,β-unsaturated carbonyl compounds are suspected to play role in mutagenesis and carcinogenesis [71]. HNE can modulate a number of signaling processes mainly through forming covalent adducts with nucleophilic functional groups in proteins, nucleic acids, and membrane lipids. These properties have been extensively summarized in some excellent reviews [72-78].

There are three major detoxification pathways to convert HNE to less reactive chemical species and control their steady-state intracellular concentrations. Firstly, the main catabolic reaction is the formation of adducts with glutathione (GSH), which occurs spontaneously or can be catalyzed by glutathione-S-transferases (GSTs). Secondly, HNE can be reduced to 1,4-dihydroxy-2-nonene (DHN) by aldo-keto reductases (AKRs) or alcohol dehydrogenases (ADH). Thirdly, HNE can be oxidized to 4-hydroxy-2-nonenoic acid (HNA) by aldehyde dehydrogenase (ALDH) [79].

trans-4-Hydroxynon-2-enal (HNE), as an α,β-carbonyl compound, possesses an electrophilic beta-carbon atom due to the polar carbonyl group conjugated with the carbon-carbon double bond (Figure 6). Considering this type of electron polarizability, HNE is considered to be a soft electrophile that preferentially forms 1,4-Michael type adduct with soft nucleophiles. Cysteine sulfhydryl groups are the primary soft nucleophilic targets of HNE while lysine and histidine residues are harder biological nucleophiles [79]. Figure 6. shows the reaction of HNE with GSH.

HNE has a chirality at carbon-4. Although HNE is formed in tissues as racemate, enantiospecific HNE effects have not been widely investigated. Wakita et al. [80] investigated the stereochemistry of the cysteine adduct of HNE. Because HNE generated in lipid peroxidation is a racemic mixture of 4*R*- and 4*S*-enantiomers, the HNE Michael adducts, possessing three chiral centers at C-2, C-3, and C-5 in the tetrahydrofuran moiety (Figure 6), are composed of at least eight isomers. It was found that the open-chain adduct forms an equilibrium mixture of the anomeric cyclic hemiacetal diastereomers [80].

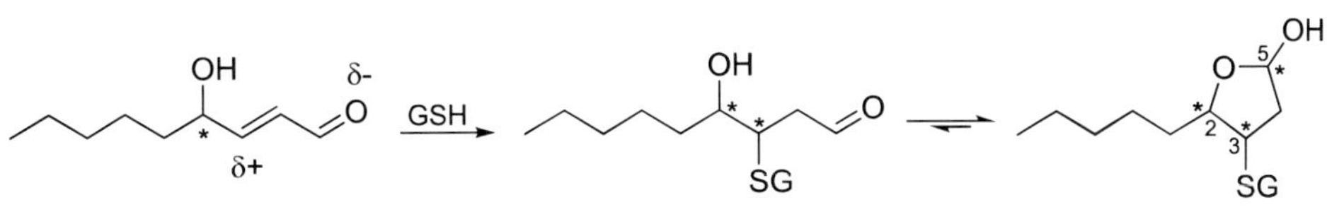

Figure 6. GST-catalyzed conjugate addition of GSH onto 4-HNE followed by cyclic hemiacetal formation.

The reaction is catalyzed by specialized isoforms of GSTs exemplified by the murine mGSTA4-4 [81, 82] and the human hGSTA4-4 [83]. These enzymes are catalyzing the conjugation of HNE with very high catalytic efficiency, exceeding 106 $M^{-1}s^{-1}$, which is among the highest catalytic efficiencies of any GST with any substrate. The possible reasons of the significantly high catalytic efficiencies are discussed in detail [22].

Another type of endogenous oxidized metabolites with electrophilic character are the 1,2- and 1,4-benzoquinone derivatives, which can be formed by oxidation of *o*- and *p*-dihydroxybenzenes and their substituted derivatives. Quinones represent a class of biotransformation products that can create a variety of hazardous effects *in vivo*, including acute cytotoxicity, immunotoxicity, and carcinogenesis. The mechanisms by which quinones cause these effects can be quite complex [84]. Quinones are Michael acceptors, and cellular damage can occur through alkylation by quinones of crucial cellular proteins and/or DNA. Alternatively, quinones are highly redox active molecules that can redox cycle with their semiquinone radicals, leading to formation of reactive oxygen species (ROS) [84, 85].

One of the well-known examples of formation of endogenous quinones is dopamine-*o*-quinone, which is suspected to be involved in development of oxidative stress-initiated neurodegenerative diseases, such as Parkinson-disease [86].

Dopamine oxidation seems to be a complex pathway in which dopamine *o*-quinone, aminochrome and 5,6-indolequinone are formed. Dopamine oxidation to dopamine *o*-quinone, aminochrome and 5,6-indolequinone seems to play an important role in the neurodegenerative processes of Parkinson's disease, as aminochrome induces: (i) mitochondria dysfunction, (ii) formation and stabilization of neurotoxic protofibrils of alpha synuclein, (iii) protein degradation dysfunction of both proteasomal and lysosomal systems and (iv) oxidative stress [87].

Dopamine-*o*-quinone is the precursor of formation of aminochrome and 5,6-indolequinone [83, 87]. Quinones, similar to α,β-unsaturated aldehydes and ketones, are soft nucleophiles, having preference to react with the soft nucleophilic thiol compounds. One of the possible protective mechanisms of preventing dopamine-induced neurotoxicity is the reaction of the unstable (reactive) dopamine-*o*-quinone with GSH. Dopamine-*o*-quinone can react with GSH catalyzed by GSTM2-2 (Figure 7) [87]. Conjugation reactions of GSH with quinones are additional examples of Michael additions. In these reactions glutathionyl hydroquinone derivatives are formed, which may be regarded as reduced quinones. GSH conjugation of *o*-quinones prevents redox cycling as well and can be considered as an important cytoprotective mechanism [88, 89].

Apart from conjugation reactions with electrophilic metabolites of xenobiotics to protect the cells, there are some GSH-conjugates that are formed in different tissues and play important role in biosynthesis of regulatory biomolecules or have such functions themselves. In this respect, the role of GSTs in metabolism of eicosanoids is remarkable [89].

As shown in Figure 3, the common precursor PGH_2 can be isomerized to PGE_2 and PGD_2. The isomerizations are GST-catalyzed processes which are catalyzed by different soluble GST isoforms. PGE_2 and PGD_2 can be transformed into the dehydration product prostaglandin A_2 (PGA_2) and prostaglandin J_2 (PGJ_2), respectively. Both PGA_2 and PGJ_2 - being α,β-unsaturated ketones - form GSH-conjugate, which reaction plays important roles in their biological activities [90, 91].

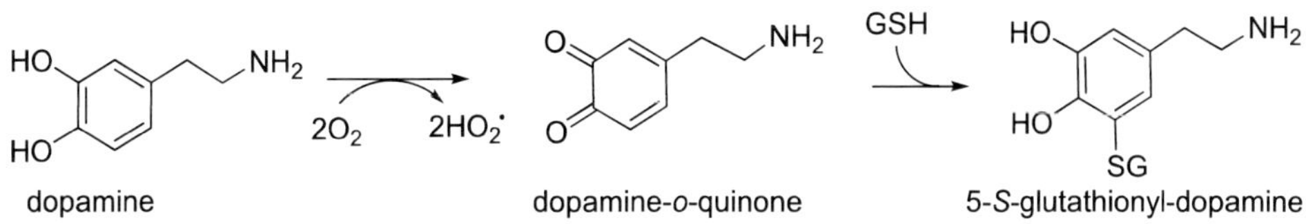

Figure 7. GST-catalyzed reaction of dopamine-*o*-quinone with GSH.

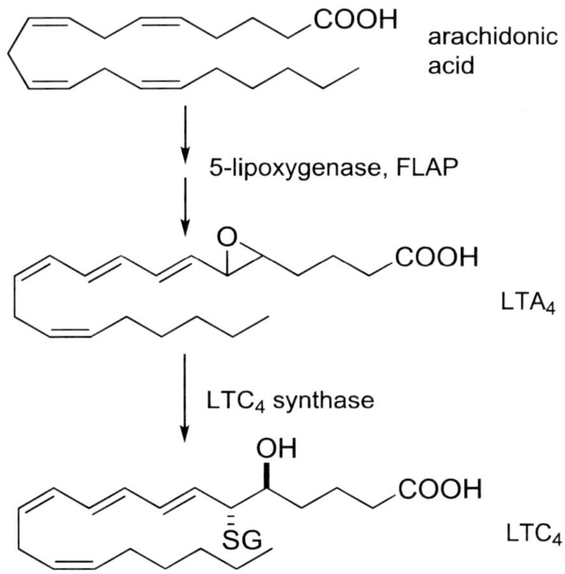

Figure 8. Formation of LTA_4 from arachidonic acid and conversion of LTA_4 to LTC4.

GST enzymes are also involved in biosynthesis of leukotrienes. Leukotrienes are also derived from arachidonic acid, abundant in biological membranes (Figure 8). The released arachidonic acid can be converted to the unstable epoxide leukotriene A_4 (LTA_4) by 5-lipoxygenase and the 5-lipoxygenase-activating protein (FLAP). One of the possible transformations of LTA_4 is stereoselective conjugation with GSH catalyzed by the membrane-bound enzyme LTC_4 synthase (Figure 8). LTC_4 synthase differs from conventional glutathione *S*-transferases (GST) by its selectivity for LTA_4 and its analogs, microsomal localization, and failure to conjugate xenobiotics [92, 93].

Hepoxilins are biologically active epoxy alcohols formed from arachidonic acid by 12-lipoxygenase-catalyzed hydroperoxidation and subsequent molecular rearrangement (Figure 9). The rearrangement is catalyzed by the ferric ion protoporphyrin subunit, which is present in hematin, hemoglobin and other hemeproteins. There are two positional isomeric hepoxilins, Hepoxilin A_3 (8-hydroxy-11,12-epoxyeicosa-(5*Z*,9*E*,14*Z*)-trienoic acid) and Hepoxilin B_3 (hydroxyl at C10) [94, 95]. The epoxide function of hepoxilins is subject to reaction with epoxide hydrolase or GST. It is worth mentioning that the GSH-conjugation occurs at C-11 rather than the allylic C-9 position [96].

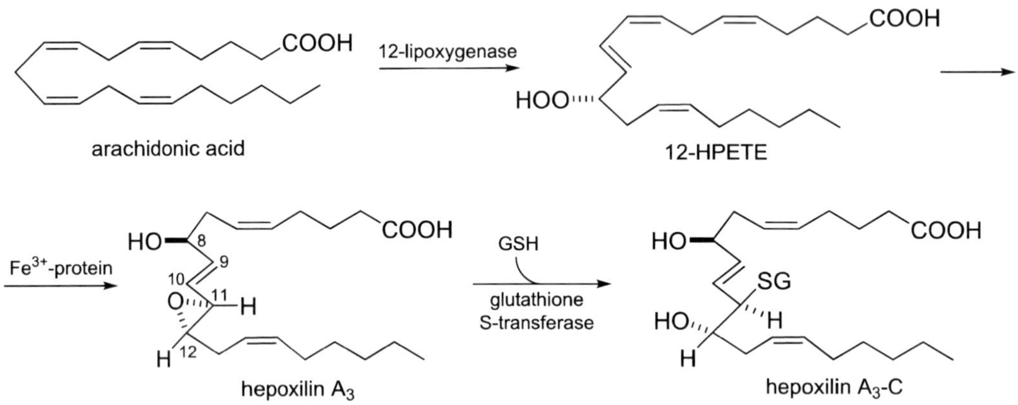

Figure 9. Formation of Hepoxilin A3 from arachidonic acid and conjugation of Hepoxilin A3 with GSH to form Hepoxilin A3-C.

CONJUGATION OF GSH WITH EXOGENEOUS ELECTROPHILES

Reaction with GSH of electrophilic species of exogenous origin (xenobiotics) is referred to as an example of Phase II biotransformations. In these reactions the electrophilic substrates (typically metabolites of the parent compound) react with GSH in a spontaneous or GST-catalyzed reaction to form GSH-conjugates in nucleophilic substitution or addition reactions. The products undergo further metabolic transformations to form either N-acetylcysteine (mercapturate) or thioalcohol (SH) derivatives [97]. It is the mercapturic acid (mercapturate) derivative that is typically the final metabolite, which is eliminated from the body in the urine. In general, glutathione conjugation is regarded as a detoxication reaction. However, depending on the chemical nature of the substrate, the pathways can generate reactive intermediates [98, 99].

From mechanistic aspects, conjugation reactions of GSH with xenobiotics can be classified into two major categories: (a) substitution, and (b) addition reactions. The most important structures involved in nucleophilic substitution reactions are (a) aromatic electrophiles, (b) aliphatic electrophiles, and (c) oxidative electrophiles. Those involved in addition reactions are (a) epoxides, (b) alkenes, (c) quinones and (d) organic isothiocyanates [97].

Nucleophilic Substitution Reactions

Aromatic Substitution

Nucleophilic substitution is the reaction of an electron pair donor (the nucleophile, :Nu) with an electron pair acceptor (the electrophile). There are four principal mechanisms of nucleophilic aromatic substitution reactions: (a) S_NAr mechanism; (b) S_N1 mechanism, (c) benzyne mechanism, and (d) $S_{RN}1$ mechanism [100, 101]. In the case of GST-catalyzed reactions, only the S_NAr mechanism can operate. In these reactions, an electron withdrawing group on the aromatic nucleus makes the system susceptible to nucleophilic attack. The mechanism is similar to that of the electrophilic aromatic substitution, except that an anion rather than a cation intermediate is involved. The nucleophile adds to the aromatic ring to afford a delocalized anion from which the leaving group is eliminated.

Physiologically relevant substrates that can undergo S_NAr reaction are rather rare, however, the reaction plays an important role in mechanistic studies of GSTs. A well-investigated example is the reaction between 1-chloro-2,4-dinitrobenzene (CDNB) and GSH (Figure 10).

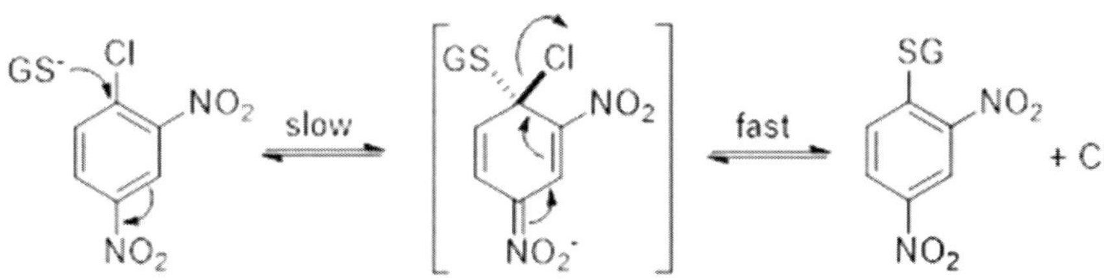

3 4 5

Figure 10. Aromatic substitution (S_NAr) reaction of 1-chloro-2,4-dinitrobenzene (CDNB) and GSH.

CDNB (**3**) reacts readily with deprotonated GSH (GS^-) even in a spontaneous, non-catalyzed reaction [102]. Furthermore, the size of CDNB is small compared with the H-site of most GSTs. For these reasons, CDNB is conjugated by most GSTs and is used in the laboratory as a universal substrate for assaying GSTs [103]. In the first step, reaction of CDNB with GSH results in formation of a Meisenheimer-complex type carbanion intermediate (**4**). In the second step, the leaving group (chloride ion) gets detached leading to regeneration of the aromatic ring (**5**) [22].

Reactions with sp^3-Hybridized Electrophilic Centers

Reactions with non-aromatic substrates with sp^3 hybridized electrophilic centers include, for example, haloalkanes, hydroperoxides and disulfides [28]. The reactions can, in principle, proceed *via* S_N2 mechanism with a carbon, oxygen or sulfur atom as the electrophilic center. The sp^3-hybridized electrophile must have a leaving group (:X) in order for the reaction to take place. In the case of the S_N2 reactions, the reaction proceeds through a transition state in which the negative charge of the sulfur atom is relocated to the leaving group. For example, when a tetrahedral carbon of an alkyl halide (X = halogen) is attacked, it is forced to adopt a bipyramidal configuration with the approaching sulfur in one apical position and the leaving group in the other (Figure 11). In the case of chiral carbon center, inversion of the configuration (Walden inversion) occurs [104].

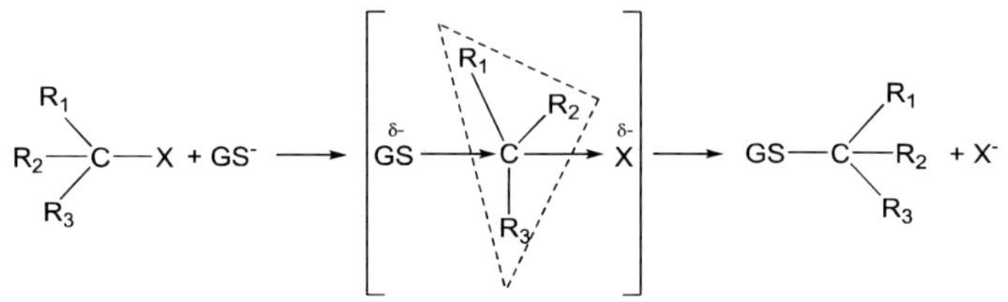

Figure 11. Walden inversion of the sp^3-hyridized carbon atom of an alkyl halide. Based on reference [104].

Haloalkanes and haloalkenes are important industrial chemicals and environmental contaminants. Many of them are toxic and suspected carcinogen. The toxicity of these chemicals is associated with their bioactivation to reactive intermediates by the cytochrome P450s or the GSTs [105]. GSTs, particularly the Theta-class GSTT1-1, catalyze the biotransformation of a range of haloalkanes. Dichloromethane (**6**) (an important industrial solvent), for example, undergoes GST-dependent biotransformation to formaldehyde [106]. The reaction mechanism involves the GSTT1-catalyzed displacement of chloride from dichloromethane to give S-(chloromethyl)glutathione (**7**). Hydrolysis of the formed conjugate gives S-(hydroxymethyl)glutathione (**8**), which is the hemithioacetal of formaldehyde and glutathione, which is in equilibrium with formaldehyde (**9**) and GSH (Figure 12).

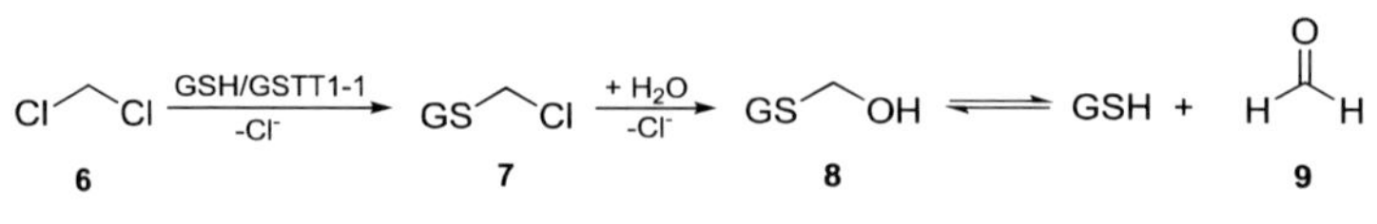

Figure 12. Glutathione transferase-dependent biotransformation of dichloromethane.

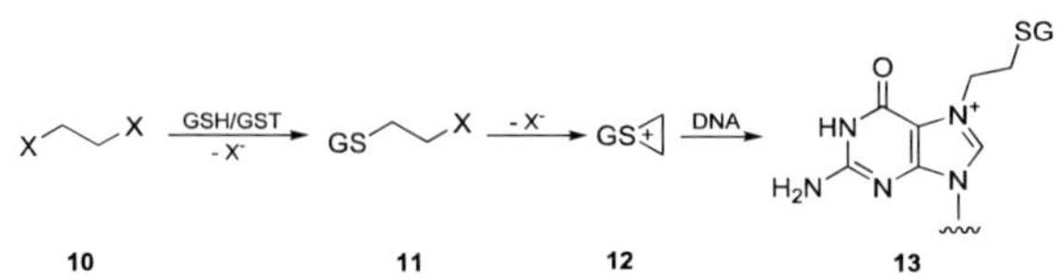

Figure 13. Glutathione-dependent biotransformation of 1,2-dihaloethanes. Reaction of the episulfonium ion metabolite with guanine.

1,2-Dihaloalkanes (**10**) are also an important group of chemicals. Many of them have been proved to be mutagenic and some of them are rodent carcinogenic. The mutagenic and, perhaps, the carcinogenic effect of these chemicals are associated with their biotransformation to reactive intermediates. 1,2-Dihaloalkanes (**10**) undergo GST-catalyzed biotransformation to S-(2-haloethyl)glutathione (**11**), which undergoes intramolecular cyclization to form thiiranium (episulfonium) ions (**12**). The reactive thiiranium ions thus formed may react with nucleophilic sites in DNA to give the adduct S-[2-(N^7-guanyl)ethyl]glutathione (**13**) (Figure 13) [107].

Nucleophilic Addition Reactions

Reaction with epoxides

GST-catalyzed reaction of epoxide metabolites of polycyclic aromatic hydrocarbons [108] and aflatoxins [109] represents one of the main detoxification pathways of these reactive Phase I metabolites. Benzo[a]pyrene (BaP) is strongly carcinogenic in certain extrahepatic tissues of rodents such as skin, lung, forestomach or mammary gland, but only weakly carcinogenic in the liver. BaP has to be activated by CYP enzymes in order to become cytotoxic and genotoxic. There is strong evidence that diol-epoxides, primarily the (+)-*anti*-7,8-dihydroxy-9,10-oxy-7,8,9,10-tetrahydrobenzo[a]pyrene (BaPDE) (**14**) (Figure 14) is the ultimate carcinogenic metabolite [108]. Human Alfa-class hGSTP1-1, hGSTA1-1 and hGSTA2-2 are able to conjugate (+)-*anti*-BaPDE with GSH resulting in formation of adduct (**15**) *via* stereoselective *trans* addition (Figure 14). The glutathionyl moiety is selectively linked to the benzyl (C-10) carbon atom. The stereochemistry of the reaction is in accordance with the S_N2 mechanism [110]. The transformation plays a protective role against the carcinogenicity of BaP and other polycyclic hydrocarbons [111].

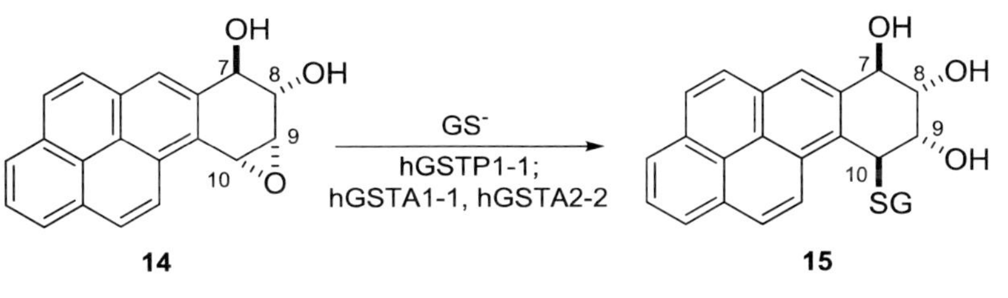

Figure 14. GSTA1-1-catalyzed epoxide ring opening and GSH addition on (+)-*anti*-BaPDE.

Reaction with N-Acetyl-p-Benzoquinone Imine (NAPQI)

The title compound, N-acetyl-p-benzoquinone imine (NAPQI), is a CYP450-catalyzed oxidation product of acetaminophen, the extensively used analgesics (Figure 15). The hepatotoxicity of the compound is a clinically important problem and an exemplary instance of how acquired factors (e.g., diet, drugs, diabetes, obesity) can enhance hepatotoxicity [112]. The first cases reporting the association between acetaminophen (paracetamol) overdose and acute liver injury and fatal hepatotoxicity were reported in the 1960s. Acute paracetamol overdose is now the most common cause of acute liver injury and acute liver failure in the UK, US, Canada, Australia, and Scandinavia [113]. Covalent binding of NAPQI to hepatic proteins is a widely accepted mechanism for the hepatotoxicity of this drug [112].

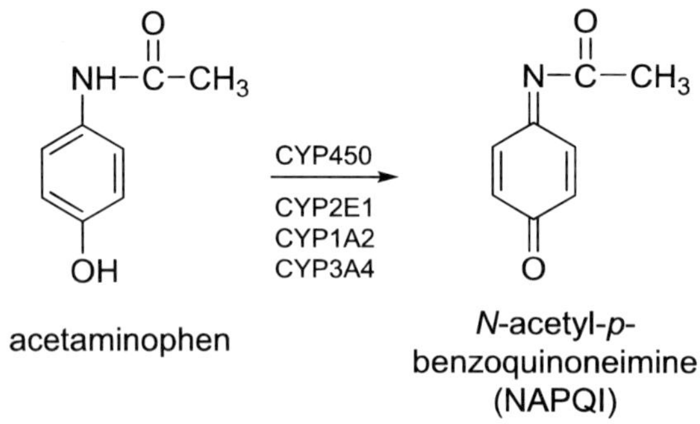

Figure.15. CYP450-dependent oxidation of acetaminophen (paracetamol).

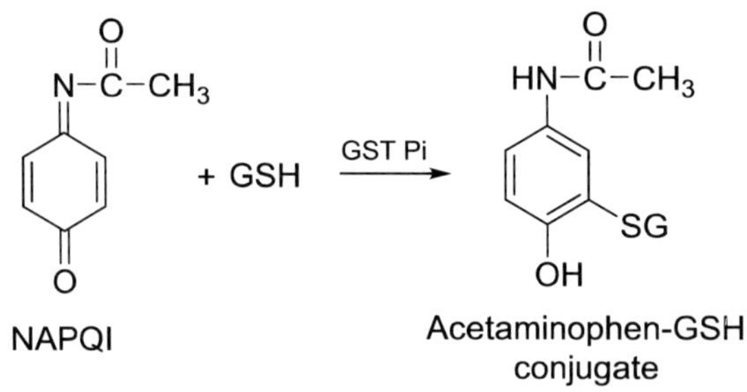

Figure 16. Conjugation of N-acetyl-*p*-benzoquinone-imine (NAPQI) with GSH.

Once formed within hepatocytes, NAPQI preferentially binds to the cysteine residue of glutathione in a reaction catalyzed by Pi-class glutathione S-transferases [114], resulting in the NAPQI-GSH conjugate. The reaction between GSH and NAPQI is a formal addition process in which the oxidized acetaminophen (NAPQI) is reduced to the aromatic GSH-substituted acetaminophen derivative (Figure 16). This conjugate is then metabolized stepwise by γ-glutamyl transpeptidase, dipeptidase, and N-acetylase, cleaving off the γ-glutamyl and glycine residues to ultimately form inert cysteine and mercapturate conjugates that are renally excreted [97].

Reaction with Acrolein

Acrolein (ACR; propenal) is a ubiquitous environmental pollutant, highly abundant in tobacco smoke, cooking, and automobile exhaust fumes. It is also generated endogenously by cellular metabolism and lipid peroxidation [115]. Moreover, ACR is also a metabolite of the anticancer drug cyclophosphamide, and is believed to be the main cause of its toxicity [116]. ACR is the most reactive electrophile among α, β-unsaturated aldehydes. Its cellular toxicity relies mainly on its ability to deplete glutathione and to form DNA and protein adducts [117]. Formation of ACR-protein covalent adducts is principally due to a Michael addition,, where the β-carbon of acrolein reacts with nucleophilic sites [117].

The main pathway for elimination of acrolein is conjugation with glutathione (GSH) in the liver, followed by enzymatic cleavage of the γ-glutamic acid and glycine residues, respectively, in the liver and in the kidney [115]. It has been pointed out by Esterbauer and others [118] that acrolein reacts 110-150 times faster with GSH than HNE. While GSH conjugation of acrolein can proceed without a catalyst, it is the same enzyme (GSTA4-4) that mediates formation of both the ACR-SG and the HNE-SG conjugate (Figure 17). Such reactivity was found to be a common metabolic pathway for several different propenal derivatives [119]. Overexpression of GSTA4-4 in mouse pancreatic islet endothelial cells resulted in an increase of cell viability. In addition, the GSTA4-4-transfected cells showed a significantly greater growth rate than the wild-type cells, indicating that acrolein is detoxified by GSTA4-4-mediated conjugation with GSH [120]. Acrolein is also a substrate for GSTA1-1, GSTM1-1, GSTP1-1, and GSTT1 [115].

Reaction with Ethacrynic Acid

Ethacrynic acid (EA) is a widely used diuretic drug, structurally characterized by presence of an α,β-unsaturated carbonyl moiety reactive towards sulphydryl groups (Figure 18). EA is known to be conjugated to GSH in spontaneous and GST-catalyzed reactions [121]. Both the parent compound (EA) and its GSH-conjugate (EA-SG) are potent inhibitors of GST enzymes. It was shown that EA-SG was an inhibitor of the GSTs due to its greater affinity for the enzymes than that of the parent molecule [122], whereas EA itself inhibits GSTP through reversible covalent interactions [123]. Because of these properties and because of the ability of EA to deplete cellular GSH, EA is potentially clinically useful as a modulator of drug resistance linked with the GSH/GST system [124].

The conjugation kinetics of glutathione (GSH) and ethacrynic acid (EA) were studied in rat liver perfusion studies. It was found that GSH conjugation of EA occurred spontaneously, but at a rate considerably less than that that of the enzyme-catalyzed GSH conjugation. GSH conjugation of EA accounted almost completely for the total removal of EA because unchanged EA was found minimally in the bile [125].

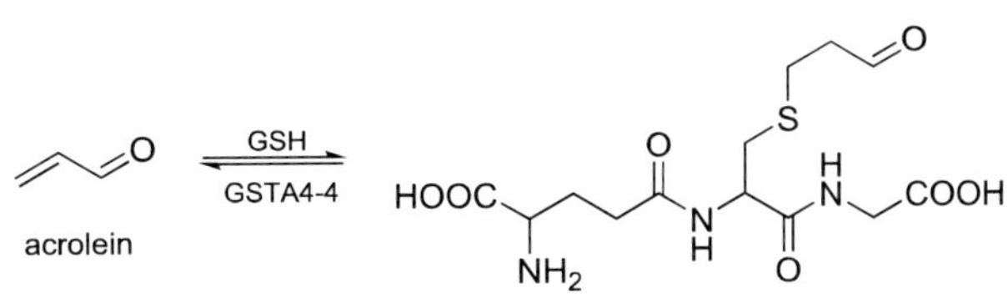

Figure 17. Reaction of acrolein with GSH.

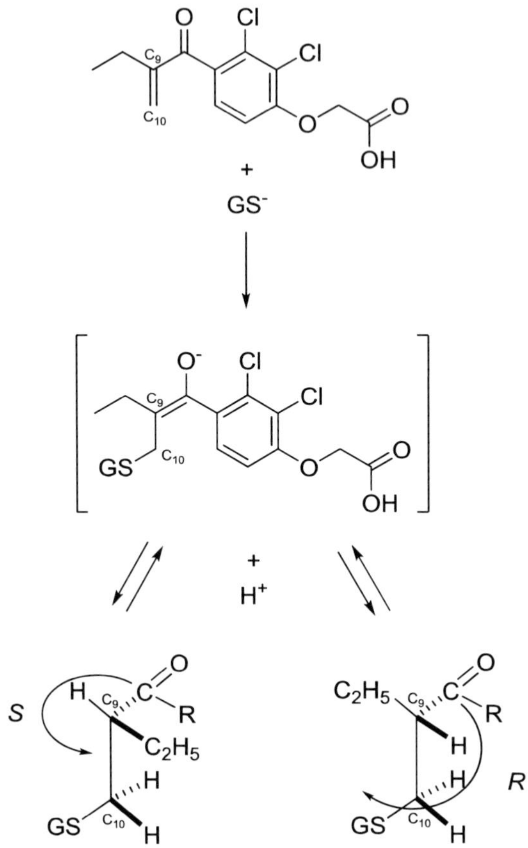

Figure 18. Reaction of ethacrynic acid with GSH.

The stereochemistry of GSH addition to EA catalyzed by GSTA1-1, GSTA1-2 and GSTP1-1 isoformes was studied by van Iersel et al. [126]. When GSH is added onto the C9-C10 double bond of EA, C-9 becomes an asymmetric carbon atom (Figure 18). Due to the chiral centers inherent in GSH, formation of the enantiomeric forms of the new chiral center results in formation of two diastereomeric adducts. The authors found that the addition reaction catalyzed by the GSTP1-1 isoform resulted in stereoselective formation of one of the adducts. The observed stereoselectivity was rationalized by enzyme-catalyzed protonation of the enolate intermedier [126].

Reaction with Chalcones and Chalcone Analogs

Hydroxy- and methoxy-substituted chalcones (**16**) are intermediary compounds of biosynthetic pathway of a very large and widespread group of plant constituents known collectively as flavonoids [127]. Among the naturally occurring chalcones and their synthetic analogs, several compounds displayed cytotoxic (cell growth inhibitor) activity towards cultured tumor cells. Chalcones are also effective *in vivo* as cell proliferating inhibitors, anti-tumor promoting, anti-inflammatory and chemopreventive agents [127-130].

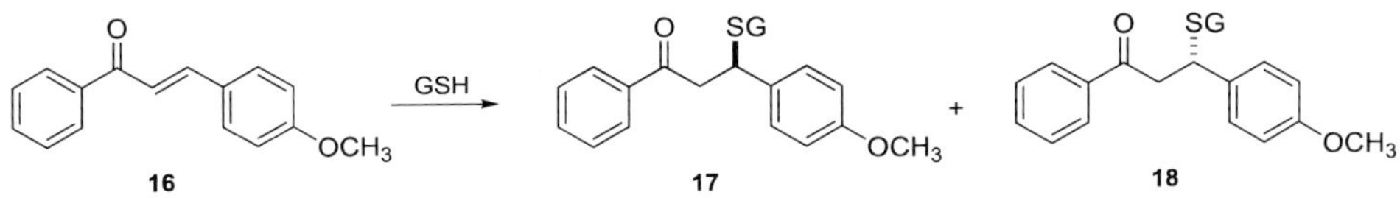

Figure 19. Reaction of 4-methoxychalcone (**16**) with GSH.

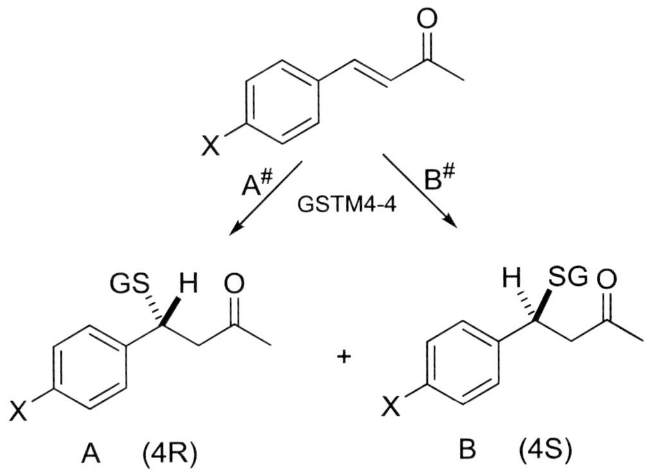

Figure 20. Stereochemistry of GSTM4-catalyzed addition of GSH onto (*E*)-(4'-X-phenyl)-3-butene-2-ones. Based on reference [135].

Nucleophilic addition of cellular thiols onto the polarized carbon-carbon double bond (Michael-reaction) is frequently associated with the biologic effects of chalcones and related α,β-unsaturated carbonyl compounds [127-130]. Such a reaction can alter intracellular redox status (redox signaling), which can modulate events such as DNA synthesis, enzyme activation, selective gene expression, and regulation of the cell cycle [131]. Several biological effects (e.g., NQO1 inducer [132], anti-inflammatory [133], GST P1-1 inhibitory [134]) of chalcones have been associated with their Michael-type reactivity towards reduced glutathione (Figure 19).

Earlier, formation of diastereomeric adducts was observed in the GSTM4-4-catalyzed reactions of GSH and the open-chain chalcone analog (*E*)-(4'-X-phenyl)-3-butene-2-ones (Figure 20). In the reactions the more polar adducts were formed in the case of each investigated open-chain chalcone analog. The observed (diastereo)selectivity (9:1) was presumed to be the result of selective enzyme-catalyzed protonation of the enolate intermediate of the more polar adducts with more extended dispersion of charge in the transition state [135].

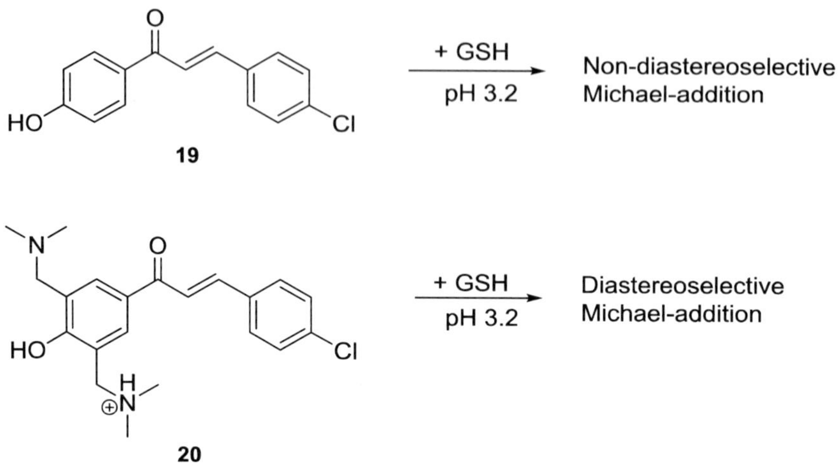

Figure 21. Reaction 4-hydroxychalcone (**19**) and its bis-Mannich derivative (**20**) with GSH. Based on reference [136].

In our previous work, stereochemistry of non-enzyme-catalyzed addition reaction of the hydroxychalcone (**19**) and its bis-Mannich derivative (**20**) was compared at different pH (Figure 21). In the case of (**20**) remarkable diastereoselectivity was observed when the reaction was conducted under acidic conditions. The observed stereoselectivity can be rationalized by ion-pair formation between the protonated Mannich-nitrogens and the deprotonated GSH(glutamate)-carboxylate [136]. Similar selectivity could not be observed when non-Mannich chalcone derivatives have been investigated. The results support importance of non-covalent interactions in developing of the transition states of these addition reactions.

Acknowledgments

The present contribution is dedicated to the 650th anniversary of the foundation of the University of Pécs, Hungary. The authors express their special thanks to Zsuzsanna Moravecz for her skilled assistance in preparation of the manuscript.

References

[1] Ure, A. On gouty concentrations; with a new method of treatment. *Med. Chir. Trans.*, 1841, 24, 30–35.

[2] Keller, W. Ueber verwandlung der benzoesaure in hippursaure. [On the transformation of benzoic acid into hippuric acid] *Ann.*, 1842, 43, 108-111.

[3] Williams, R. T. *Detoxication mechanisms. The metabolism of drugs and allied organic compounds.* Chapman and Hall: London, UK, 1947.

[4] Williams, R. T. *Detoxication mechanisms. The metabolism and detoxication of drugs, toxic substances, and other organic compounds*, 2nd ed. Chapman and Hall: London, UK 1959.

[5] Conti, A.; Bickel, M. H. History of drug metabolism: discovery of the major pathways in the 19th century. *Drug Metab. Rev.*, 1977, 6, 1-50.

[6] Baumann, E.; Preuss, C. Ueber bromophenylmercaptursaure. [About bromo phenyl mercapturic acid.] *Ber. Deut. Chem. Ges.*, 1879, 12, 806-810.

[7] Jaffe, M. Ueber die nach einfuhrung von brombenzol und chlorbenzol im organismus entstehenden schwefelhaltigen sauren. [About sulphurous acids arising in organism after bromobenzene and chlorobenzene are introduced.] *Ber. Deut. Chem. Ges.*, 1879, 12, 1092-1098.

[8] Baumann, E. Ueber cystin and cysteine. [About cystine and cysteine.] *Z. Physiol. Chem.*, 1884, 8, 299-305.

[9] Barnes, M. M.; James, S. P.; Wood, P. B. The formation of mercapturic acids. 1. Formation of mercapturic acid and the levels of glutathione tissues. *Biochem. J.*, 1959, 71, 680-690.

[10] Bray, H. G.; Franklin, T. J.; James, SP. The formation of mercapturic acids. 2. The possible role of glutathionase. *Biochem. J.*, 1959, 71, 690-696.

[11] Bray, H. G.; Franklin, T. J.; James, S. P. The formation of mercapturic acids. 3. N-Acetylation of S-substituted cysteines in the rabbit, rat and guinea pig. *Biochem. J.*, 1959, 73, 465-473.

[12] Booth, J.; Boyland, E.; Sims, P. An enzyme from rat liver catalysing conjugations with glutathione. *Biochem. J.*, 1961, 79, 516-524.

[13] Boyland, E.; Chasseaud, L. F. The role of glutathione and glutathione S-transferases in mercapturic acid biosynthesis. *Adv. Enzymol. Relat. Areas Mol. Biol.*, 1969, 32, 173-219.

[14] Listovsky, I.; Arias, I. M. Ligandin and glutathione S-transferases: Historical milestones. In *Toxicology of glutathione transferases*; Awasthi, Y. C., Ed.; CRC Press: Boca Raton, FL, 2007, pp. 1-9.

[15] Hayes, J. D.; Flanagan, J. U.; Jowsey, J. R. Glutathione transferases. *Annu. Rev. Pharmacol. Toxicol.*, 2005, 45, 51-58.

[16] Testa, B.; Kraemer, S. D. The biochemistry of drug metabolism. – An introduction Part 1. Principles and overview. *Chem. Biodivers.*, 2006, 3, 1053-1101.

[17] Wu, B.; Dong, D. Human cytosolic glutathione transferases: Structure, function, and drug discovery. *Trends Pharmacol. Sci.*, 2012, 33, 656-668.

[18] Gallagher, E. P.; Gardner, J. L.; Barber, D. S. Several glutathione S-transferase isozymes that protect against oxidative injury are expressed in human liver mitochondria. *Biochem. Pharmacol.*, 2006, 71, 1619-1628.

[19] Morel, F.; Aninat, C. The glutathione transferase kappa family. *Drug Metab. Rev.*, 2011, 43, 281-291.

[20] Zimniak, P.; Singh, S. P. Families of glutathione transferases. In *Toxicology of glutathione transferases*; Awasthi, Y. C., Ed.; CRC Press: Boca Raton, FL, 2007, pp. 11-26.

[21] Morgenstern, R.; Zhang, J.; Johansson, K. Microsomal glutathione transferase 1: Mechanism and functional roles. *Drug Metab. Rev.*, 2011, 43, 300-306.

[22] Zimniak, P. Substrates and reaction mechanisms of glutathione transferases. In *Toxicology of glutathione transferases*; Awasthi, Y. C., Ed.; CRC Press: Boca Raton, FL, 2007, pp. 71-101.

[23] Oakley, A. Glutathione transferases: a structural perspective. *Drug. Metab. Rev.*, 2011, 43, 138-151.

[24] Mannervik, B.; Board, P. G.; Hayes, J. D.; Listowsky, I.; Pearson, W. R. Nomenclature for mammalian soluble glutathione transferases. *Methods Enzymol.*, 2005, 401, 1-8.

[25] Sherratt, P. J.; Hayes, J. D. Glutathion S-transferases. In *Enzyme systems that metabolize drugs and other xenobiotics*; Ioannides, C.., Ed.; Wiley and Sons: Chichester, UK, 2002.

[26] Sharma, R.; Ansari, G. A. S.; Awasthi Y. C. Physiological substrates of glutathione S. transferases. In *Toxicology of glutathione transferases*; Awasthi, Y. C., Ed.; CRC Press: Boca Raton, FL, 2007, pp. 179-203.

[27] Hubatsch, L.; Ridderstrom, M.; Mannervik, B. Human glutathione S-transferase A4-4. An alfa-class enzyme with high catalytic efficiency in the conjugation of 4-hydroxynonenal and other genotoxic products of lipid peroxidation. *Biochem. J.*, 1998, 330, 175-179.

[28] Deponte, M. Glutathione catalysis and the reaction mechanism of glutathione-dependent enzymes. *Biochim. Biophys. Acta.*, 2013, 1830, 3217-3266.

[29] Segura-Aguilar, J.; Bae, S.; Widersten, M.; Welch, C. J.; Mannervik, B. Human class Mu glutathione transferases, in particular isoenzyme M2-2, catalyze detoxication of the dopamine metabolite aminochrome, *J. Biol. Chem.*, 1997, 272, 5727–5731.

[30] Johansson, A. S.; Mannervik, B. Human glutathione transferase A3-3, a highly efficient catalyst of double-bond isomerization in the biosynthetic pathway of steroid hormones. *J. Biol. Chem.*, 2001, 276, 33061-33065.

[31] Gu, Y.; Guo, J.; Pal, A.; Pan, S. S.; Zimniak, P.; Singh, S. V.; Ji, X. Crystal structure of human glutathione S-transferase A3-3 and mechanistic implications for its high steroid isomerase activity. *Biochem.*, 2004, 43, 15673-15679.

[32] Taes, K.; Olin, B.; Mannervik, B. Structural basis for featuring of steroid isomerase activity in alfa class glutathione transferases. *J. Mol. Biol.*, 2010, 397, 332-340.

[33] Calvaresi, M.; Stenta, M.; Garavelli, M.; Altoe, P.; Bottoni, A. Computational evidence for the catalytic mechanism of human glutathione S-transferase A3-3: A QM/MM investigation. *ACS Catal.*, 2012, 2, 280-286.

[34] Chen, H.; Juchau, M. R. Biotransformation of 13-*cis*- and 9-*cis*-retinoic acid to all-*trans*-retinoic acid in rat conceptal homogenates. *Drug Metabol. Dispos.*, 1998, 26, 222-228.

[35] Fernandez-Canon, J. M.; Penalva, M. A. Characterization of a fungal maleylacetoacetate isomerase gene and identification of its human homologue. *J. Biol. Chem.*, 1998, 273, 329-337.

[36] Seltzer, S.; Lin, M. Maleylacetone cis-trans isomerase. Mechanism of the interaction of coenzyme glutathione and substrate malylacetone in the presence and absence of enzyme. *J. Am. Chem. Soc.*, 1979, 101, 3091-3097.

[37] Williams, C. S.; DuBois, R. N. Prostaglandin endoperoxide synthase: why two isoforms? *Am. J. Physiol.*, 1996, 270, G393-400.

[38] Jowsey, J. R.; Thomson, A. M.; Flanagan, J. U.; Murdock, P. R.; Moore, G. B. T; Meyer, D. J.; Murphy, G. J.; Smith, S. A.; Hayes, J. D. Mammalian class Sigma glutathione S-transferases: Catalytic properties and tissue-specific expression of human and rat GSH-dependent prostaglandin D2 synthases. *Biochem. J.*, 2001, 359, 507–516.

[39] Kanaoka, Y.; Ago, H.; Inagaki, E.; Nanayama, T.; Miyano, M.; Kikuno, R.; Fujii, Y.; Eguchi, N.; Toth, H.; Urade, Y.; Hayaishi, O. Cloning and crystal structure of hematopoietic prostaglandin D synthase. *Cell.*, 1997, 90, 1085-1095.

[40] Jakobsson, P. J.; Thorén, S.; Morgenstern, R.; Samuelsson, B. Identification of human prostaglandin E synthase: a microsomal, glutathione-dependent, inducible enzyme, constituting a potential novel drug target. *Proc. Natl. Acad. Sci. USA.*, 1999, 96, 7220-7225.

[41] Beuckmann, C. T.; Fujimori, K.; Urade, Y.; Hayaishi, O. Identification of mu-class glutathione transferases M2-2 and M3-3 as cytosolic prostaglandin E synthases in the human brain. *Neurochem. Res.*, 2000, 25, 733-738.

[42] Armstrong, R. N. Structure, catalytic mechanism, and evolution of the glutathione transferases. *Chem. Res. Toxicol.*, 1997, 10, 2-18.

[43] Ladner, J. E.; Parsons, J. F.; Rife, C. L.; Gilliland, G. L.; Armstrong, R. N. Parallel evolutionary pathways for glutathione transferases: structure and mechanism of the mitochondrial class kappa enzyme rGSTK1-1. *Biochemistry.*, 2004, 43, 352-361.

[44] Jowsey, I. R; Thomson, R. E; Orton, T. C; Elcombe, C. R; Hayes, J. D. Biochemical and genetic characterization of a murine-class Kappa glutathione S-transferase. *Biochem. J.*, 2003, 373, 559-569.

[45] Morgenstern, R.; Guthenberg, C.; DePierre, J. W. Microsomal glutathione S-transferase. Purification, initial characterization and demonstration that it is not

identical to the cytosolic glutathione S-transferases A, B and C. *Eur. J. Biochem.*, 1982, 128, 243-248.

[46] Bernat, B. A.; Laughlin, L. T.; Armstrong, R. N. Fosfomycin resistance protein (FosA) is a manganese metalloglutathione transferase related to glyoxalase I and the extradiol dioxygenases. *Biochemistry*, 1997, 36, 3050-3055.

[47] Pearson, W. R. Phylogenies of glutathione transferase families. *Methods Enzymol.*, 2005, 401, 186-204.

[48] Bresell, A.; Weinander, R.; Lundqvist, G.; Raza, H.; Shimoji, M.; Sun, T. H.; Balk, L.; Wiklund, R.; Eriksson, J.; Jansson, C.; Persson, B.; Jakobsson, P. J.; Morgenstern, R. Bioinformatic and enzymatic characterization of the MAPEG superfamily. *FEBS J.*, 2005, 272, 1688-1703.

[49] Martinez Molina, D.; Eshaghi, S.; Nordlund, P. Catalysis within the lipid bilayer-structure and mechanism of the MAPEG family of integral membrane proteins. *Curr. Opin. Struct. Biol.*, 2008, 18, 442-449.

[50] Mannervik, B.; Board, P. G.; Hayes, J. D.; Listowsky, I.; Pearson, W. R. Nomenclature for mammalian soluble glutathione transferases. *Methods Enzymol.*, 2005, 401, 1-8.

[51] Suzuki, T.; Coggan, M.; Show, D. C.; Board, P. G. Electrophoretic and immunological analysis of human glutathione S-transferase isozymes. *Ann. Hum. Genet.*, 1987, 51, 95-106.

[52] Danielson, U. H.; Mannervik, B. Kinetic independence of the subunits of cytosolic glutathione transferase from the rat. *Biochem. J.*, 1985, 231, 263-267.

[53] (a) Hayes, J. D.; Pulford, D. J. The Glutathione S-Transferase supergene family: Regulation of GST and the contribution of the isoenzymes to cancer chemoprotection and drug resistance Part I. *Crit. Rev. Biochem. Mol. Biol.*, 1995, 30, 445-520. (b) Hayes, J. D.; Pulford, D. J. The Glutathione S-Transferase supergene family: Regulation of GST and the contribution of the isoenzymes to cancer chemoprotection and drug resistance Part II. *Crit. Rev. Biochem. Mol. Biol.*, 1995, 30, 521-600.

[54] Higgins, L. G.; Hayes, J. D. Mechanisms of induction of cytosolic and microsomal glutathione transferase (GST) genes by xenobiotics and pro-inflammatory agents. *Drug Metabol. Rev.*, 2011, 43, 92-137.

[55] Mohana, K.; Achary, A. Human cytosolic glutathione-S-transferases: quantitative analysis of expression, comparative analysis of structures and inhibition strategies of isozymes involved in drug resistance. *Drug Metabol. Rev.*, 2017, 49, 318-337.

[56] Pirie, N. W.; Pinhey, K. G. The titration curve of glutathione. *J. Biol. Chem.*, 1929, 84, 321-333.

[57] Li, N. C.; Gawron, O.; Bascuas, G. Stability of zinc complexes with glutathione and oxidized glutathione. *J. Am. Chem. Soc.*, 1954, 76, 225-229.

[58] Martin, R. B.; Edsall, J. T. Glutathione: ionization in basic solutions and molecular rearrangement in strongly acid solution. *Bull. Soc. Chim. Biol.*, 1958, 40, 1763-1771.

[59] Mirzahosseini, A.; Somlyay, M.; Noszál, B. The comprehensive acid–base characterization of glutathione. *Chem. Phys. Lett.*, 2015, 622, 50–56.

[60] Chen, W. J.; Graminski, G. F.; Armstrong, R. N. Dissection of the catalytic mechanism of isoenzyme 4-4 of glutathione S-transferase with alternative substrates. *Biochem.*, 1988, 27, 647-654.

[61] Parsons, J. F.; Armstrong, R. N. Proton configuration in the ground state and transition state of a glutathione transferase-catalyzed reaction inferred from the properties of tetradeca(3-fluorotyrosyl)glutathione transferase. *J. Am. Chem. Soc.*, 1996, 118, 2295-2296.

[62] Thorson, J. S.; Shin, I.; Chapman, E.; Stenberg, G.; Mannervik, B.; Schultz, P. G. Analysis of the role of the active site tyrosine in human glutathione transferase A1-1 by unnatural amino acid mutagenesis. *J. Am. Chem. Soc.*, 1998, 120, 451-452.

[63] Reinemer, P.; Dirr, H. W.; Ladenstein, R.; Huber, R.; Lo Bello, M.; Federici, G.; Parker, M. W. Three-dimensional structure of class π glutathione S-transferase from human placenta in complex with S-hexylglutathione at 2.8 Å resolution. *J. Mol. Biol.*, 1992, 227, 214-226.

[64] Ji, X.; Zhang, P.; Armstrong, R. N.; Gilliland, G. L. The three-dimensional structure of a glutathione S-transferase from the Mu gene class. Structural analysis of the binary complex of isoenzyme 3-3 and glutathione at 2.2-.ANG. resolution. *Biochem.*, 1992, 31, 10169-10184.

[65] Sinning, I.; Kleywegt, G. J.; Cowan, S. W.; Reinemer, P; Dirr, H. W.; Huber, R.; Gilliland, G. L.; Armstrong, R. N.; Ji, X.; Board, P. G.; Olin, B.; Mannervik, B.; Jones, T. A. Structure determination and refinement of human alpha class glutathione transferase A1-1, and a comparison with the Mu and Pi class enzymes. *J. Mol. Biol.*, 1993, 232, 192-212.

[66] Bjornestedt, R.; Stenberg, G.; Widersten, M.; Board, P. G.; Sinning, I.; Jones, T. A.; Mannervik, B. Functional significance of arginine 15 in the active site of human alpha glutathione transferase A1-1. *J. Mol. Biol.*, 1995, 247, 765-773.

[67] Ji, X.; Tordova, M.; O'Donnell, R.; Parsons, J. F.; Hayden, J. B.; Gilliland, G. L.; Zimniak, P. Structure and function of the xenobiotic substrate-binding site and location of a potential non-substrate-binding site in a class π glutathione S-transferase. *Biochem.*, 1997, 36, 9690-9702.

[68] Witz, G. Biological interactions of α,β-unsaturated aldehydes. *Free Rad. Biol. Med.*, 1989, 7, 333-349.

[69] Spickett, C. M. The lipid peroxidation product 4-hydroxy-2-nonenal: Advances in chemistry and analysis. *Redox. Biol.*, 2013, 1, 145-152.

[70] Sousa, B. C.; Pitt, A. R.; Spickett, C. M. Chemistry and analysis of HNE and other prominent carbonyl-containing lipid oxidation compounds. *Free Rad. Biol. Med.*, 2017, 111, 294-308.

[71] Eder, E.; Scheckenbach, S.; Deininger, C.; Huffman, C. The possible role of α,β-unsaturated carbonyl compounds in mutagenesis and carcinogenesis. *Toxicol. Letters.*, 1993, 67, 87-103.

[72] Shoeb, M.; Ansari, N. H.; Srivastava, S. K.; Ramana, K. V. 4-hydroxynonenal in the pathogenesis and progression of human diseases. *Curr. Med. Chem.*, 2014, 21, 230–237.

[73] West, J. D; Marnett, L. J. Endogenous reactive intermediates as modulators of cell signaling and cell death. *Chem. Res. Toxicol.*, 2006, 19, 173–194.

[74] Barrera, G.; Pizzimenti, S.; Ciamporcero, E. S.; Daga, M.; Ullio, C.; Arcaro, A.; Cetrangolo, G. P.; Ferretti, C.; Dianzani, C.; Lepore, A.; Gentile, F. Role of 4-hydroxynonenal-protein adducts in human diseases. *Antiox. Redox Signal.*, 2015, 22, 1681-1702.

[75] Roede, J. R.; Jones, D. P. Reactive species and mitochondrial dysfunction: mechanistic significance of 4-hydroxynonenal. *Environmental and Molecular Mutagenesis.*, 2010, 51, 380–390.

[76] Guéraud, F.; Atalay, M.; Bresgen, N.; Cipak, A.; Eckl, P. M.; Huc, L.; Jouanin, I.; Siems, W.; Uchida, K. Chemistry and biochemistry of lipid peroxidation products. *Free Rad. Res.*, 2010, 44, 1098–1124.

[77] Chen, Z. H.; Niki, E. 4-hydroxynonenal (4-HNE) has been widely accepted as an inducer of oxidative stress. Is this the whole truth about it or can 4-HNE also exert protective effects? *IUBMB Life.*, 2006, 58, 372–373.

[78] Aldini, G.; Carini, M.; Yeum, K. J.; Vistoli, G. Novel molecular approaches for improving enzymatic and nonenzymatic detoxification of 4-hydroxynonenal: toward the discovery of a novel class of bioactive compounds. *Free Rad. Biol. Med.*, 2014, 69, 145–156.

[79] Schaur, R. J.; Siems, W.; Bresgen, N.; Eckl, P. M. 4-Hydroxy-nonenal – A bioactive lipid peroxidation product. *Biomol.*, 2015, 5, 2247-2337.

[80] Wakita, C.; Maeshima, T.; Yamazaki, A.; Shibata, T.; Ito, S.; Akagawa, M.; Ojika, M.; Yodoi, J.; Uchida, K. Stereochemical configuration of 4-hydroxy-2-nonenal-cysteine adducts and their stereoselective formation in a redox-regulated protein. *Biol. Chem.*, 2009, 284, 28810-28822.

[81] Zimniak, P.; Eckles, M. A.; Saxena, M.; Awasthi, Y. C. A subgroup of class alpha glutathione S-transferases: cloning of cDNA for mouse lung glutathione S-transferase GST 5.7. *FEBS Lett.*, 1992, 313, 173-176.

[82] Zimniak, P.; Shinghal, S. S.; Srivastava, S. K.; Awashti, S.; Sharma, R.; Hayden, J. B.; Awashti, Y. C. Estimation of genomic complexity, heterologous expression, and

enzymatic characterization of mouse glutathione S-transferase mGSTA4-4 (GST 5.7). *J. Biol. Chem.*, 1994, 269, 992-1000.

[83] Hubatsch, I.; Ridderstrom, M.; Mannervik, B. Human glutathione transferase A4-4: an alpha class enzyme with high catalytic efficiency in the conjugation of 4-hydroxynonenal and other genotoxic products of lipid peroxidation. *Biochem. J.*, 1998, 330, 175-179.

[84] Bolton, J. L.; Trush, M. A.; Penning, T. M.; Dryhurst, G.; Monks, T. J. Role of quinones in toxicology. *Chem. Res. Toxicol.*, 2000, 13, 135-160.

[85] Bolton, J. L.; Dunlap, T. Formation and biological targets of quinones: cytotoxic versus cytoprotective effects. *Chem. Res. Toxicol.*, 2017, 30, 13–37.

[86] Bisaglia, M.; Soriano, M. E.; Arduini, I.; Mammi, S.; Bubacco, L. Molecular characterization of dopamine-derived quinones reactivity toward NADH and glutathione: implications for mitochondrial dysfunction in Parkinson disease. *Biochim. Biophys. Acta.*, 2010, 1802, 699-706.

[87] Segura-Aguilar, J.; Paris, I.; Muñoz, P.; Ferrari, E.; Zecca, L.; Zucca, F. A. Protective and toxic roles of dopamine in Parkinson's disease. *J. Neurochem.*, 2014, 129, 898-915.

[88] Dagnino-Subiabre, A.; Cassels, B. K.; Baez, S.; Johansson, A. S.; Mannervik, B.; Segura-Aguilar, J. Glutathione transferase M2-2 catalyzes conjugation of dopamine and dopa *o*-quinones. *Biochem. Biophys. Res. Commun.*, 2000, 274, 32–36.

[89] Blair, I. A. Endogeneous glutathione adducts. *Curr. Drug Metrabol.*, 2006, 7, 853-872.

[90] Bogaards, J. J. P.; Venekamp, J. C.; van Bladeren, P. J. Stereoselective conjugation of prostaglandin A2 and prostaglandin J2 with glutathione, catalysed by the human glutathione S-transferases A1-1, A2-2, M1a-1a, and P1-1. *Chem. Res. Toxicol.*, 1997, 10, 310-317.

[91] Bowers-Gentry, R. C.; Deems, R. A.; Harkewicz, R.; Dennis, E. A. Eicosanoid lipidomics. In *Functional lipidomics*; Feng, L.; Prestwich, G. D., Eds.; CRC Press: Boca Raton, FL, 2006, pp. 79-100.

[92] Nicholson, D. W.; Ali, A.; Klemba, M. W.; Munday, N. A.; Zamboni, R. J.; Ford-Hutchinson, A. W. Human leukotriene C_4 synthase expression in dimethyl sulfoxide-differentiated U937 cells. *J. Biol. Chem.* 1992, 267, 17849-17857.

[93] Welsch, D. J.; Creely, D. P.; Hauser, S. D.; Mathis, K. J.; Krivi, G. G.; Isakson, P. C. Molecular cloning and expression of human leukotriene C_4 synthase. *Proc. Natl. Acad. Sci. USA.*, 1994, 91, 9745-9749.

[94] Pace-Asciak, C. R. Hepoxilins: a review on their cellular actions. *Biochim. Biophys. Acta.*, 1994, 1215, 1-8.

[95] Pace-Asciak, C. R. Arachidonic acid epoxides: Demonstartion throught [18O]oxygen studies of an intramolecular transfer of the terminal hydroxyl group of

(12S)-hydroperoxyeicosa-5,8,10,14-tetraenoic acid to form hydroxyepoxides. *J. Biol. Chem.*, 1984, 259, 8332-8337.

[96] Pace-Asciak, C. R.; Laneuville, O.; Su, W. G.; Corey, E. J.; Gurevich, N.; Wu, P.; Carlen, P. L. A glutathione conjugate of hepoxilin A3: formation and action in the rat central nervous system. *Proc. Natl. Acad. Sci.*, 1990, 87, 3037-3041.

[97] Parkinson, A. Biotransformation of xenobiotics. In *Casarett and Doull's Toxicology. The basic science of poisons*, 6th. Ed.; Klaassen, C. D., (Ed.); Mc Graw-Hill: New York, NY, 2001, pp. 133-224.

[98] Davis, C. D.; Hanumegowda, U. M. The role of drug metabolism in toxicity. In *Drug metabolism handbook. Concepts and applications.* Nassar, A. F.; Hollenberg, P. F.; Scatina, J. (Eds); Wiley: Hoboken, NJ, 2009, pp. 561-628.

[99] Park, B. K.; Boobis, A.; Clarke, S.; Goldring, C. E. P.; Jones, D.; Kenna, J. G.; Lambert, C.; Laverty, H. G.; Naisbitt, D. J.; Nelson, S.; Nicoll-Griffith, D. A.; Obach, R. S.; Routledge, P.; Smith, D. A.; Tweedie, D. J.; Vermeulen, N.; Williams, D. P.; Wilson, I. D.; Baillie, T. A. Managing the challenge of chemically reactive metabolites in drug development *Nat. Rev. Drug Discov.*, 2011, 10, 292–306.

[100] Pietra, F. Mechanisms for nucleophilic and photonucleophilic aromatic substitution reactions. *Q. Rev. Chem. Soc.*, 1969, 23, 504-521.

[101] Makosza, M. Nucleophilic aromatic substitution, general corrected mechanism and versatile synthetic tool. *Org. Med. Chem. I. J.*, 2017, 2, 555-576.

[102] Abdelaal, Y. A. I.; Roe, M. Conjugation of chlorodinitrobenzene with reduced glutathione in the absence and presence of glutathione transferase from larvae of the southern armyworm, *Spodoptera eridania. Pest. Biochem. Physiol.*, 1990, 36, 248-258.

[103] Jakoby, W. B. The glutathione S-transferases: a group of multifunctional detoxification proteins. *Adv. Enzymol. Rel. Areas Mol. Biol.*, 1978, 46, 383-414.

[104] Ivarrson, Y.; Mannervick, B. Combinatorial protein chemistry in three dimensions: A paradigm fort the evolution of glutathione transferases with novel activities. In *Toxicology of glutathione transferases*; Awasthi, Y. C., Ed.; CRC Press: Boca Raton, FL, 2007, pp. 47-69.

[105] Anders, M. W. Glutathione-dependent bioactivation of haloalkanes and haloalkenes. *Drug. Metab. Rev.*, 2004, 36, 583-594.

[106] Hashmi, M.; Dechert, S.; Dekant, W.; Anders, M. W. Bioactivation of [13C] dichloromethane in mouse, rat, and human liver cytosol. 13C Nuclear magnetic resonance spectroscopic studies. *Chem. Res. Toxicol.*, 1994, 7, 291-296.

[107] Ozawa, N.; Guengerich, F. P. Evidence for formation of an S-[2-(N7-guanyl)ethyl]glutathione adduct in glutathione mediated binding of the carcinogen 1,2-dibromoethane to DNA. *Proc. Natl. Acad. Sci. USA.*, 1983, 80, 5266-5270.

[108] Hesse, S.; Krupski-Brennstuhl, G.; Cumpelik, O.; Mezger, M.; Wiebel, F. J. Glutathione depletion suppresses conjugation of benzo[a]pyrene metabolites and

(±)-trans-7,8-dihydroxy-7,8-dihydrobenzo[a]pyrene metabolites with glutathione but does not affect their binding to DNA in C3H/10T1/2 mouse fibroblasts. *Carcinogen.*, 1987, 8, 1051-1058.

[109] Ziglari, T.; Allameh, A. The significance of glutathione conjugation in aflatoxin metabolism. In *Aflatoxins. Recent advances and future prospects*; Razzaghi-Abayaneh, M. (Ed.); IntechOpen: London, UK, 2013, pp 267-286.

[110] Jernström, B.; Funk, M.; Frank, H.; Mannervik, B.; Seidel, A.. Glutathione S-transferase A1-1-catalysed conjugation of bay and fjord region diol epoxides of polycyclic aromatic hydrocarbons with glutathione. *Carcinogen.*, 1996, 17, 1491-1498.

[111] Sundberg, K.; Widersten, M.; Seidel, A.; Mannervik, B.; Jernström, B. Glutathione conjugation of bay- and fjord-region diol epoxides of polycyclic aromatic hydrocarbons by glutathione transferases M1-1 and P1-1. *Chem. Res. Toxicol.*, 1997, 10, 1221-1227.

[112] Kalsi, S. S.; Dargan, P. I.; Waring, W. S.; Wood, D. M. A review of the evidence concerning hepatic glutathione depletion and susceptibility to hepatotoxicity after paracetamol overdose. *Open Access Emerg. Med.*, 2011, 3, 87–96.

[113] Ramachandran, A.; Jaeschke, H. Mechanisms of acetaminophen hepatotoxicity and their translation to the human pathophysiology. *J Clin Transl Res.*, 2017 Feb, 3(Suppl 1), 157–169.

[114] Henderson, C. J.; Wolf, C. R.; Kitteringham, N.; Powell, H.; Otto, D.; Park, B. K. Increased resistance to acetaminophen heptotoxicity in mice lacking glutathione S-transferase Pi. *Proc. Natl. Acad. Sci. USA.*, 2000, 97, 12741-12745.

[115] Stevens, J. F.; Maier, C. S. Acrolein: Sources, metabolism, and biomolecular interactions relevant to human health and disease. *Mol. Nutr. Food Res.*, 2008, 52, 7–25.

[116] Kehrer, J. P.; Biswal, S. S. The molecular effects of acrolein. *Toxicol. Sci.*, 2000, 57, 6-15.

[117] Cai, J.; Bhatnagar, A.; Pierce, W. M. Jr. Protein modification by acrolein: formation and stability of cysteine adducts. *Chem. Res. Toxicol.*, 2009, 22, 708-716.

[118] Esterbauer, H.; Schaur, R. J.; Zollner, H. Chemistry and biochemistry of 4-hydroxynonenal, malonaldehyde and related aldehydes. *Free. Rad. Biol. Med.*, 1991, 11, 81–128.

[119] Berhane, K.; Widersten, M.; Engström, A.; Kozarich, J. W.; Mannervik, B. Detoxification of base propenals and other alpha, beta-unsaturated aldehyde products of radical reactions and lipid peroxidation by human glutathione transferase. *Proc. Natl. Acad. Sci. USA.*, 1994, 91, 1480-1484.

[120] Yang, Y.; Yang, Y.; Trent, M. B; He, N.; Lick, S. D.; Zimniak, P.; Awashti, Y. C.; Boor, P. J. Glutathione-S-transferase A4-4 modulates oxidative stress in

endothelium: possible role in human atherosclerosis. *Atherosclerosis.*, 2004, 173, 211–221.

[121] Ploemen, J. H. T. M.; Van Ommen, B.; Bogaards, J. J. P.; Van Bladeren, P. J. Ethacrynic acid and its glutathione conjugate as inhibitors of glutathione S-transferases. *Xenobiotica.*, 1993, 23, 913-923.

[122] Ploemen, J. H. T. M.; Van Ommen, B.; Van Bladeren, P. J. Inhibition of rat and human glutathione S-transferase isoenzymes by ethacrynic acid and its glutathione conjugate. *Biochem Pharmacol.*, 1990, 40, 1631–1635.

[123] Ploemen, J. H. T. M; Van Schanke, A.; Van Ommen, B.; Van Bladeren, P. J. Reversible conjugation of ethacrynic acid with glutathione and human glutathione S-transferase P1–1. *Cancer Res.*, 1994, 54, 915–919.

[124] O'Dwyer, P. J.; LaCreta, F.; Nash, S.; Tinsley, P. W.; Schilder, R.; Clapper, M. L.; Tew, K. D.; Panting, L.; Litwin, S.; Comis, R. L.; Ozols, R. F. Phase I study of thiotepa in combination with the glutathione transferase inhibitor ethacrynic acid. *Cancer Res.*, 1991, 51, 6059 – 6065.

[125] Tirona, R. G.; Pang, K. S. Bimolecular glutathione conjugation kinetics of ethacrynic acid in rat liver: *in vitro* and perfusion studies. *J. Pharmacol. Exp. Ther.*, 1999, 290, 1230-1241.

[126] van Iersel, M. L. P. S.; van Lipzig, M. M. H.; Rietjens, I. M. C. M.; Vervoort, J; van Bladeren, P. J. GSTP1-1 stereospecifically catalyzes glutathione conjugation of ethacrynic acid. *FEBS Letters.*, 1998, 441, 153-157.

[127] Rozmer, Z.; Perjési, P. Naturally occurring chalcones and their biological activities. *Phytochem. Rev.*, 2016, 15, 87-120.

[128] Go, M. L.; Wu, X.; Liu, X. L. Chalcones: an update on cytotoxic and chemoprotective properties. *Curr. Med. Chem.*, 2005, 12, 483-499.

[129] Nowakowska, Z. A review of anti-infective and anti-inflammatory chalcones. *Eur. J. Med. Chem.*, 2007, 42, 125-137.

[130] Rahman, M. A. Chalcone: A valuable insight into the recent advances and potential pharmacological activities. *Chem. Sci. J.*, 2011, 29, 1-16.

[131] Aquilano, K.; Baldelli, S.; Ciriolo, M. R. Glutathione: new roles in redox signaling for an old antioxidant. *Front. Pharmacol.*, 2014, 5, 196.

[132] Dinkova-Kostova, A. T.; Massiah, M. A.; Bozak, R. E.; Hicks, R. J.; Talalay, P. Potency of Michael reaction acceptors as inducers of enzymes that protect against carcinogenesis depends on their reactivity with sulfhydryl groups. *Proc. Natl. Acad. Sci.*, 2001, 98, 3404-3409.

[133] Jin, Y. L.; Jin, X. Y.; Jin, F.; Sohn, D. H.; Kim, H. S. Structure activity relationship studies of anti-inflammatory TMMC derivatives: 4-dimethylamino group on the B ring responsible for lowering the potency. *Arch. Pharm. Res.*, 2008, 31, 1145-1152.

[134] Wang, J.; Wang, S.; Song, D.; Zhao, D.; Sha, Y.; Jiang, Y.; Jing, Y.; Cheng, M. Chalcone derivatives inhibit glutathione S-transferase P1-1 activity: insights into the

interaction mode of alpha, beta-unsaturated carbonyl compounds. *Chem. Biol. Drug Des.*, 2009, 73, 511-514.

[135] Kubo, Y.; Armstrong, R. N. Observation of a substituent effect on the stereoselectivity of glutathione S-transferase toward para-substituted 4-phenyl-3-buten-2-ones., *Chem. Res. Toxicol.*, 1989, 2, 144-145.

[136] Bernardes, A.; D'Oliveira, G. D. C.; Silezin, A.; Kuzma, M.; Molnár, S.; Noda Pérez, C.; Perjési. P. Reagent-induced asymmetric induction in addition reaction of reduced glutathione onto bis-Mannich chalcones. *Arch. Pharm.*, 2018, 351, e1700386.

Biographical Sketches

Pál Perjési

Affiliation

Institute of Pharmaceutical Chemistry, University of Pécs, Pécs, Hungary

Education

1979: Pharmacy, MS, Albert Szent-Györgyi University, Szeged, Hungary;

1983: Pharmacy, University Doctor (DrUniv), Albert Szent-Györgyi University, Szeged, Hungary;

1994: Chemistry: Candidate of Sciences (CSc), Hungarian Academy of Sciences, Budapest, Hungary;

1996: Chemistry: PhD, Albert Szent-Györgyi University, Szeged, Hungary;

2000: Chemistry: Habilitation, University of Pécs, Medical School, Pécs, Hungary

Research and Professional Experience

1979-1981: Institute of Pharmaceutical Chemistry, Albert Szent-Györgyi University, Szeged, Hungary;

1981-2001: Institute of Medical Chemistry, University of Pécs, Medical School, Pécs, Hungary;

2002- Institute of Pharmaceutical Chemistry, University of Pécs, Faculty of Medicine, Pécs, Hungary;

2016- Institute of Pharmaceutical Chemistry, University of Pécs, Faculty of Pharmacy, Pécs, Hungary

Professional Appointments

2002: Head of department, Institute of Pharmaceutical Chemistry, University of Pécs, Pécs, Hungary;

2016: Dean, University of Pécs, Faculty of Pharmacy, Pécs, Hungary;

2018: Editor-in-Chief, Journal of Pharmaceutical and Biopharmaceutical Research, SyncSci Publishing, Singapore

Honors

1990-1991: Fogarty, International Research Fellowship, National Institutes of Health; Department of Medicinal Chemistry, University of Washington, Seattle, WA, USA;

2001: Postdoctoral Fellowship, Center for Drug Discovery, University of Florida, Gainesville, FL, USA;

2003: Postdoctoral Fellowship, Department of Medicinal Chemistry, University of Florida, Gainesville, FL, USA;

2012: Postdoctoral Fellowship, German Academic Exchange Service (DAAD); Kekule Institute, University of Bonn, Bonn, Germany

Publications from the Last 3 Years (selection):

1. Almasi, A.; Pinto, E. D. L. N.; Kovacs, N. P.; Fischer, T.; Markovics, Z.; Fischer, E.; Perjési, P. Changes in hepatic metabolic enzyme activities and biliary excretion of 4-nitrophenol in streptozotocin induced diabetic rats. *Braz. J. Pharm. Sci.* 2018, 54, e17347.
2. Bernardes, A.; D'Oliveira, G. D. C.; Silezin, A.; Kuzma, M.; Molnar, S.; Noda-Perez, C.; Perjési, P. Reagent-induced asymmetric induction in addition reaction of reduced glutathione onto bis-Mannich chalcones. *Archiv Pharm.* 2018, 351, e1700386.
3. Kuzma, M.; Kovács, N.; Sziva, L.; Maasz, G.; Avar, P.; Perjési, P. Oxidation of hydroxy- and dihydroxybenzoic acids under the Udenfriend's conditions. An HPLC study. *Open Med. Chem. J.* 2018, 12, 13-22.
4. Rozmer, Z.; Marton, E.; Perjési, P. (*E*)-2-Benzylidenecyclanones: part XIV. Study on interaction of some (*E*)-2-benzylidenebenzosuberone derivatives with calf thymus DNA by TLC and UV–Vis methods, a DNA cleavage study. *Med. Chem. Res.* 2017, 26, 2172-2179.
5. Rozmer Zs, Perjési P. Naturally occurring chalcones and their biological activities. *Phytochem. Rev.* 2016, 15, 87-120.

(For detailed information on list of publications, please address to https://m2.mtmt.hu/gui2/?type=authors&mode=browse&sel=10008303)

Caridad Noda Perez

Present Position

Professor

Affiliation

Chemical Institute, Federal University of Goiás, Goiânia, Goiás, Brazil

Education

Graduated in Chemical Engineering in 1983, Master's (MSc) degree (1997) and PhD (2001) in Chemical Engineering.

Research and Professional Experience

Synthesis of chalcones and their derivatives. Synthesis of Mannich bases derived from chalcones. Physic-chemical characterization of synthesized compounds. Evaluation of antitumor activity of chalcones and their derivatives. Study of genotoxic and antigenotoxic activities of chalcones and derivatives. Study of basic solid catalysts for the synthesis of chalcones and their derivatives.

Short Biography

Graduated in Chemical Engineering from Instituto Superior Politécnico José Antonio Echevarria (1983), in Havana, Cuba. Master's (MSc) degree (1997) and PhD (2001) in Chemical Engineering at COPPE by Federal University of Rio de Janeiro. Currently is professor of the Graduate Programs in Chemical Engineering and Chemistry and lecturer in the Chemical Engineering course at the Federal University of Goias. Has experience in kinetics and heterogeneous catalysis. Research in the area of basic solid catalysts for the condensation reactions of aldehydes and ketones. The main area of research is synthesis, characterization and antitumor evaluation of chalcones and their derivatives. Within this line of research she works on Michael addition reactions of glutathione and chalcones and their derivatives.

Publications from the Last 3 Years

1 Michelini, L. J.; Castro, M. R. C.; Custodio, J. M. F.; Naves, L. F. N.; Vaz, W. F.; Lobón, G. S.; Martins, F. T.; Perez, C. N.; Napolitano, H. B. A Novel Potential Anticancer Chalcone: Synthesis, Crystal Structure and Cytotoxic Assay. *J. Mol. Struct.* 2018, 1168, 309–315.

2 Lemes, S. R.; eSilva, C. R.; Véras, J. H.; Chen-Chen, L.; Lima, R. S.; Perez, C. N.; Sousa, M. A. M. de; Reis, P. R. de M.; Junior, N. J. da S. Presence of antigenotoxic and anticytotoxic effects of the chalcone 1*E*,4*E*-1-(4-chlorophenyl)-5-(2,6,6-trimethylcyclohexen-1-yl)penta-1,4-dien-3-one using in vitro and in vivo assays. *Drug Chem. Toxicol.* 2018. https://doi.org/10.1080/01480545.2018.1497046

3 Bernardes, A.; d'Oliveira, G. D. C.; Silezin, A.; Kuzma, M.; Molnár, S.; Pérez, C. N.; Perjési, P. Reagent-induced asymmetric induction in addition reaction of reduced glutathione onto bis-Mannich chalcones. *Arch. Pharm.* 2018, 351, 1700386.

4 Custodio, J. M. F.; Michelini, L. J.; Castro, M. R. C. de; Vaz, W. F.; Neves, B. J.; Cravo, P. V. L.; Barreto, F. S.; Filho, M. O. M.; Perez, C. N.; Napolitano, H. B. Structural insights into a novel anticancer sulfonamide chalcone. *New J. Chem.* 2018, 42, 3426–3434.

5 d'Oliveira, G.; Moura, A.; de Moraes, M.; Perez, C.; Lião, L. Synthesis, characterization and evaluation of in vitro antitumor activities of novel chalcone-quinolinone hybrid compounds. *J. Brazil. Chem. Soc.* 2018, 29, 2308-2325.
6 da Silva Lima, D. C.; do Vale, C. R.; Véras, J. H.; Bernardes, A.; Pérez, C. N.; Chen-Chen, L. Absence of genotoxic effects of the chalcone (*E*)-1-(2-hydroxyphenyl)-3-(4-methylphenyl)-prop-2-en-1-one) and its potential chemoprevention against DNA damage using in vitro and in vivo assays. *PloS one*. 2017, 12, e01712247.
7. Carneiro, C. C.; Véras, J. H.; Góes, B. R. L.; Pérez, C. N.; Chen-Chen, L.; Carneiro, C. C.; Véras, J. H.; Góes, B. R. L.; Pérez, C. N.; Chen-Chen, L. Mutagenicity and antimutagenicity of *Salacia Crassifolia* (Mart. Ex. Schult.) *G. Don*. evaluated by Ames test. *Brazil. J. Biol.* 2018, 78, 345–350.
8. Damasceno, C. T.; Jordao, G. R.; Noda-Perez, C. Simulação de uma unidade piloto de destilação de água para uso em laboratórios químicos empregando energia solar. *Rev. Proc. Quim.*, 2017, 22, 11-24.
9. de Castro, M. R.; Naves, R. F.; Bernardes, A.; da Silva, C. C.; Perez, C. N.; Moura, A. F.; Moraes, M. O.; Martins, F. T. Tandem chalcone-sulfonamide hybridization, cyclization and further Claisen–Schmidt condensation: Tuning molecular diversity through reaction time and order and catalyst. *Arab. J. Chem.* 2017. https://doi.org/10.1016/j.arabjc.2017.11.005
10. Winter, C.; Caetano, J. N.; Araújo, A. B. C.; Chaves, A. R.; Ostroski, I. C.; Vaz, B. G.; Pérez, C. N.; Alonso, C. G. Activated carbons for chalcone production: Claisen-Schmidt condensation reaction. *Chem. Engin. J.* 2016, 303, 604–610.
11. Lima, R. S.; Perez, C. N.; da Silva, C. C.; Santana, M. J.; Queiroz Júnior, L. H. K.; Barreto, S.; de Moraes, M. O.; Martins, F. T. Structure and cytotoxic activity of terpenoid-like chalcones. *Arab. J. Chem.* 2016. https://doi.org/10.1016/j.arabjc.2016.02.013.
12. Bernardes, A.; Pérez, C.; Mayer, M.; da Silva, C.; Martins, F.; Perjési, P. Study of reactions of two Mannich bases derived of 4'-hydroxychalcones with glutathione by RP-TLC, RP-HPLC and RP-HPLC-ESI-MS analysis. *J. Brazil. Chem. Soc.* 2017, 28, 1048-1062.

Aline Bernardes

Present Position

Assistant Professor

Affiliation

Federal Institute of Science and Technology of Mato Grosso, Cuiabá Bela Vista campus, Mato Grosso, Brazil

Education

MSc (Chemistry) - State University of Goiás (2010)
PhD (Chemistry) - Federal University of Goiás (2018)

Short Biography

Aline Bernardes was born in Goiânia, Brazil in 1988. She graduated in Industrial Chemistry from the State University of Goiás in 2010 and began the studies on chalcones in her monograph. She joined the research group of Prof. Dr. Caridad Noda Pérez in the MSc program in Molecular Sciences in 2010 and received her MSc degree in 2012 from the State University of Goiás. During her MSc studies, she joined the Department of Chemistry at Federal University of Goiás as an Assistant Professor for 18 months until she started her PhD studies in Chemistry from the Federal University of Goiás in 2013. During her PhD studies, she was a visiting researcher at the Institute of Pharmaceutical Chemistry at the University of Pécs, Hungary (2015-2016). Here she investigated bioactivities and glutathione-reactivity of chalcones and their Mannich derivatives under the supervision of Professor Pál Perjési. She received her PhD degree in 2018 and became a Professor at the Federal Institute of Science and Technology of Mato Grosso. Her research interests are synthesis and investigation of bioactivities of chalcones and chalcone analogs, reactivity of glutathione and analysis of water and effluents quality.

Professional appointments

Assistant Professor (Federal University of Goiás – 2012-2013)
Assistant Professor (Federal Institute of Science and Technology of Mato Grosso - 2017 – currently)

Publications from the Last 3 Years:

1. Custodio, J. M.; Vaz, W. F.; de Castro, M. R.; Bernardes, A.; Naves, R. F.; Moura, A. F.; Moraes, M. O.; da Silva, C. C.; Martins, F. T.; Perez, C. N.; Napolitano, H. B. Solvent-driven structural adaptation in a novel anticancer sulfonamide chalcone. *J. Mol. Struct.* 2019, 1175, 389-397.
2. Bernardes, A.; D'Oliveira, G. D.; Silezin, A.; Kuzma, M.; Molnár, S.; Noda Pérez, C.; Perjési, P. Reagent- induced asymmetric induction in addition reaction of reduced glutathione onto bis- Mannich chalcones. *Arch. Pharm.* 2018, 1700386.
3. de Castro, M. R.; Naves, R. F.; Bernardes, A.; da Silva, C. C.; Perez, C. N.; Moura, A. F.; Moraes, M. O.; Martins, F. T. Tandem chalcone-sulfonamide hybridization, cyclization and further Claisen–Schmidt condensation: Tuning molecular diversity through reaction time and order and catalyst. *Arab. J. Chem.* 2017, https://doi.org/10.1016/j.arabjc.2017.11.005
4. da Silva Lima, D. C.; do Vale, C. R.; Véras, J. H.; Bernardes, A.; Pérez, C. N.; Chen-Chen, L. Absence of genotoxic effects of the chalcone (*E*)-1-(2-hydroxyphenyl)-3-(4-

methylphenyl)-prop-2-en-1-one) and its potential chemopre-vention against DNA damage using in vitro and in vivo assays. *PloS one*. 2017, 12, e0171224.
5. Bernardes, A.; Pérez, C. N.; Mayer, M.; Silva, C. C. D.; Martins, F. T.; Perjési, P. Study of reactions of two Mannich bases derived of 4'-hydroxychalcones with glutathione by RP-TLC, RP-HPLC and RP-HPLC-ESI-MS analysis. *J. Brazil. Chem. Soc.* 2017, 28, 1048-1062.
6. e Silva, C. R.; Borges, F. F. V.; Bernardes, A.; Perez, C. N.; e Silva, D. D. M., Chen-Chen, L. Genotoxic, cytotoxic, antigenotoxic, and anticytotoxic effects of sulfonamide chalcone using the Ames test and the mouse bone marrow micronucleus test. *PloS one*. 2015, 10, e0137063.

Giulio Demetrius Creazzo d'Oliveira

Present Position

PhD Student

Affiliation

Federal University of Goiás, Goiânia, Goiás, Brazil

Education

Graduated in Industrial Chemistry (2011) and Master's (MSc) degree in Chemistry (2013).

Research and Professional Experience

Development and validation of analytical methods for application on local pharmaceutical industry. Physic-chemical characterization of synthesized compounds. Evaluation of reactivity of chalcones derivatives toward reaction with glutathione.

Short Biography

Graduated in Industrial Chemistry (Universidade Estadual de Goiás - 2011) and MSc degree in Chemistry (Universidade Estadual de Goiás - 2013), acting on development of method for quantification of vitamins on solid pharmaceutical form by Ultra High-Performance Liquid Chromatography. PhD in progress in Organic Chemistry (Universidade Federal de Goiás – since 2015), acting on synthesis and physico-chemical characterization of chalcone-quinolinone hybrid compounds and antitumor evaluation of those compound, as well evaluation of reactivity of some of them toward reaction with glutathione (University of Pécs – Hungary, as Invited Researcher). Laboratory analyst/researcher (Brainfarma S.A. – from 12/2010 to 03/2015; Halex Istar S.A. from 07/2018 to present), acting on development and validation of analytical methods for solid, semi-solid and liquid pharmaceutical forms with emphasis on development of stability indicating methods.

Publications from the Last 3 Years

1 Bernardes, A.; d'Oliveira, G. D. C.; Silezin, A.; Kuzma, M.; Molnár, S.; Pérez, C. N.; Perjési, P. Reagent-Induced Asymmetric Induction in Addition Reaction of Reduced Glutathione onto Bis-Mannich Chalcones. *Arch. Pharm.* 2018, 351, 1700386.

2 d'Oliveira, G.; Moura, A.; de Moraes, M.; Perez, C.; Lião, L. Synthesis, characterization and evaluation of in vitro antitumor activities of novel chalcone-quinolinone hybrid compounds. *J. Brazil. Chem. Soc.* 2018, 29, 2308-2325.

Fatemeh Kenari

Present Position

Assistant Professor, Institute of Pharmaceutical Chemistry, University of Pécs, Pécs, Hungary

Affiliation

Institute of Pharmaceutical Chemistry, Faculty of Pharmacy, University of Pécs, Pécs, Hungary

Education

PharmD (2018), University of Pécs, Pécs, Hungary

Research and Professional Experience

Glutathione conjugation, Chalcones, Stereochemistry of Michael additions

Short Biography

Through her PharmD studies, she has gained two years of research experience at the Pharmaceutical Chemistry Institute of University of Pecs working on the topic titled "Study on GSH-conjugation reaction of chalcone and cyclic chalcone analogs", In 2017, she has participated the Undergraduate Research Society's Conference of University of Pécs as well as the 7th BBBB International Pharmaceutical Conference (Balatonfüred, Hungary). Furthermore, she has spent two weeks at the University of Angers (Angers, France) participating in an undergraduate research summer course. In 2018 she participated the Otto Clauder Memorial Contest (Budapest, Hungary). In 2018, she was admitted to the PhD School of the University of Pecs and currently works as an assistant professor in the Institute of Pharmaceutical Chemistry, University of Pecs, Pécs, Hungary.

About the Editor

Pál Perjési, PhD

Affiliation: Institute of Pharmaceutical Chemistry
University of Pécs, Pécs, Hungary

Education: 1979: Pharmacy, MS,

Albert Szent-Györgyi University, Szeged, Hungary; 1983: Pharmacy, University Doctor (UDr),

Albert Szent-Györgyi University, Szeged, Hungary; 1994: Chemistry: Candidate of Sciences (CSc),

Hungarian Academy of Sciences, Budapest, Hungary;1996: Chemistry: PhD,

Albert Szent-Györgyi University, Szeged, Hungary; 2000: Chemistry: Habilitation, University of Pécs, Medical School, Pécs, Hungary

Research and Professional Experience:

1979-1981: Institute of Pharmaceutical Chemistry, Albert Szent-Györgyi University, Szeged, Hungary;

1981-2001: Institute of Medical Chemistry, University of Pécs, Medical School, Pécs, Hungary;

2002- Institute of Pharmaceutical Chemistry, University of Pécs, Faculty of Medicine, Pécs, Hungary;

2016- Institute of Pharmaceutical Chemistry, University of Pécs, Faculty of Pharmacy, Pécs, Hungary

Professional Appointments:

2002: Head of department, Institute of Pharmaceutical Chemistry, University of Pécs, Pécs, Hungary;

2016: Dean, University of Pécs, Faculty of Pharmacy, Pécs, Hungary;

2018: Editor-in-Chief, Journal of Pharmaceutical and Biopharmaceutical Research, SyncSci Publishing, Singapore

Honors:

1990-1991: Fogarty, International Research Fellowship, National Institutes of Health; Department of Medicinal Chemistry, University of Washington, Seattle, WA, USA;

2001: Postdoctoral Fellowship, Center for Drug Discovery, University of Florida, Gainesville, FL, USA;

2003: Postdoctoral Fellowship, Department of Medicinal Chemistry, University of Florida, Gainesville, FL, USA;

2012: Postdoctoral Fellowship, German Academic Exchange Service (DAAD); Kekule Institute, University of Bonn, Bonn, Germany

Publications from the Last 3 Years (selection):

Almasi, A.; Pinto, E. D. L. N.; Kovacs, N. P.; Fischer, T.; Markovics, Z.; Fischer, E.; Perjési, P. Changes in hepatic metabolic enzyme activities and biliary excretion of 4-nitrophenol in streptozotocin induced diabetic rats. Braz. J. Pharm. Sci. 2018, 54, e17347.

Bernardes, A.; D'Oliveira, G. D. C.; Silezin, A.; Kuzma, M.; Molnar, S.; Noda-Perez, C.; Perjési, P. Reagent-induced asymmetric induction in addition reaction of reduced glutathione onto bis-Mannich chalcones. Archiv Pharm. 2018, 351, e1700386.

Kuzma, M.; Kovács, N.; Sziva, L.; Maasz, G.; Avar, P.; Perjési, P. Oxidation of hydroxy- and dihydroxybenzoic acids under the Udenfriend's conditions. An HPLC study. Open Med. Chem. J. 2018, 12, 13-22.

Rozmer, Z.; Marton, E.; Perjési, P. (*E*)-2-Benzylidenecyclanones: part XIV. Study on interaction of some (*E*)-2-benzylidenebenzosuberone derivatives with calf thymus

DNA by TLC and UV–Vis methods, a DNA cleavage study. Med. Chem. Res. 2017, 26, 2172-2179.

Rozmer Zs, Perjési P. Naturally occurring chalcones and their biological activities. Phytochem. Rev. 2016, 15, 87-120.

(For detailed information on list of publications, please address to https://m2.mtmt.hu/gui2/?type=authors&mode=browse&sel=10008303)

INDEX

#

β

γ

A

B

C

D

E

F

G

H

I

J

K

L

M

N

O

P

Q

R

S

T

U

V

W

X

Y

Z

Glutathione: Biochemistry, Mechanisms of Action and Biotechnological Implications

Editors: Nikolaos Labrou and Emmanouil Flemetakis (Laboratory of Enzyme Technology, Department of Agr.

Series: Microbiology Research Advances

Book Description: This book discusses cutting-edge developments and deals with the information gained over the last few years on the physiological, biochemical and molecular roles of the GSH and its significance in mammalian, microbial and plant cells.

Hardcover ISBN: 978-1-62417-460-5
Retail Price: $270

Glutathione: Dietary Sources, Role in Cellular Functions and Therapeutic Effects

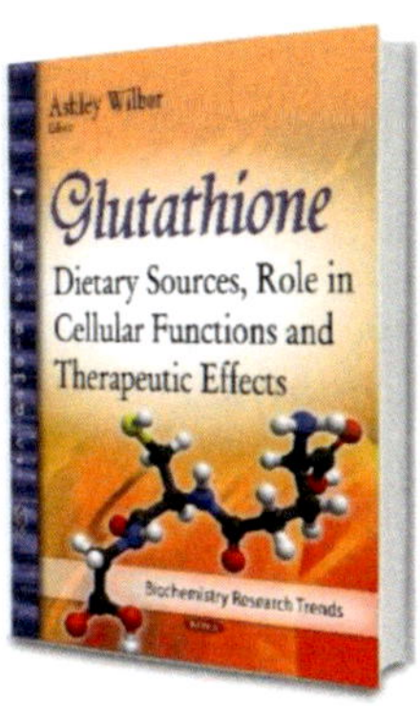

Editor: Ashley Wilber

Series: Biochemistry Research Trends

Book Description: This book discusses dietary sources of glutathione as well as the role it plays in cellular functions, and it's therapeutic effects.

Hardcover ISBN: 978-1-63463-372-7
Retail Price: $230

Biopharmaceutics and Drug Hypersensitivity

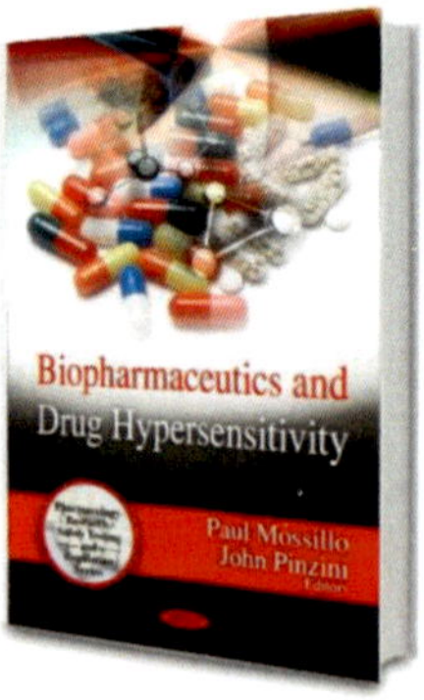

Editors: Paul Mossillo and John Pinzini

Series: Pharmacology - Research, Safety Testing and Regulation

Book Description: This book describes the pharmacopoeial quality of three EU registered and three Russian Federation registered aciclovir formulations. The data indicates that the pharmacopoeial quality of the antimicrobial and antiviral formulations from the Russian Federation drug market is similar to the European competitors.

Hardcover ISBN: 978-1-60741-830-6
Retail Price: $265

Topics in Mediators Pharmacology

Author: Jagdish N. Sharma (Kuwait University, Kuwait)

Series: Pharmacology - Research, Safety Testing and Regulation

Book Description: This book has been written with the aim of giving an overview of the scientific developments which have occurred in mediators pharmacology. Pharmacology can be defined as a discipline of paramedical sciences that teaches how the use of drugs can bring about changes from pathological to physiological states.

Hardcover ISBN: 978-1-61668-961-2
Retail Price: $95